TODAY'S TECHNICIAN ™

CLASSROOM MANUAL FOR
AUTOMATIC TRANSMISSIONS AND TRANSAXLES

FIFTH EDITION

TODAY'S TECHNICIAN ™

CLASSROOM MANUAL FOR
AUTOMATIC TRANSMISSIONS
AND TRANSAXLES

FIFTH EDITION

JACK ERJAVEC

DELMAR
CENGAGE Learning™

Australia • Canada • Mexico • Singapore • Spain • United Kingdom • United States

Today's Technician™: Automatic Transmissions and Transaxles, 5th Edition
Jack Erjavec

Vice President, Career and Professional Editorial: Dave Garza

Director of Learning Solutions: Sandy Clark

Executive Editor: David Boelio

Managing Editor: Larry Main

Senior Product Manager: Matthew Thouin

Editorial Assistant: Jillian Borden

Vice President, Career and Professional Marketing: Jennifer McAvey

Executive Marketing Manager: Deborah S. Yarnell

Marketing Manager: Katie Hall

Associate Marketing Manager: Mark Pierro

Production Director: Wendy Troeger

Production Manager: Mark Bernard

Content Project Manager: Cheri Plasse

Art Director: Benj Gleeksman

For product information and technology assistance, contact us at
Professional Group Cengage Learning Customer & Sales Support, 1-800-354-9706

For permission to use material from this text or product,
submit all requests online at **cengage.com/permissions.**
Further permissions questions can be e-mailed to
permissionrequest@cengage.com.

Library of Congress Control Number: 2010924965

ISBN-13: 978-1-4354-8103-9
ISBN-10: 1-4354-8103-8

Delmar
5 Maxwell Drive
Clifton Park, NY 12065-2919
USA

Cengage Learning is a leading provider of customized learning solutions with office locations around the globe, including Singapore, the United Kingdom, Australia, Mexico, Brazil and Japan. Locate your local office at:
international.cengage.com/region

Cengage Learning products are represented in Canada by Nelson Education, Ltd.

For your lifelong learning solutions, visit **delmar.cengage.com**

Visit our corporate website at **cengage.com.**

Notice to the Reader

Publisher does not warrant or guarantee any of the products described herein or perform any independent analysis in connection with any of the product information contained herein. Publisher does not assume, and expressly disclaims, any obligation to obtain and include information other than that provided to it by the manufacturer. The reader is expressly warned to consider and adopt all safety precautions that might be indicated by the activities described herein and to avoid all potential hazards. By following the instructions contained herein, the reader willingly assumes all risks in connection with such instructions. The publisher makes no representations or warranties of any kind, including but not limited to, the warranties of fitness for particular purpose or merchantability, nor are any such representations implied with respect to the material set forth herein, and the publisher takes no responsibility with respect to such material. The publisher shall not be liable for any special, consequential, or exemplary damages resulting, in whole or part, from the readers' use of, or reliance upon, this material.

Printed in the United States of America
3 4 5 6 XX 16 15 14 13

CONTENTS

CONTENTS

Thanks to the support the Today's Technician Series has received from those who teach automotive technology, Delmar Cengage Learning, the leader in automotive related textbooks, is able to live up to its promise to provide new editions of the series every few years. We have listened and responded to our critics and our fans and present this new updated and revised fifth edition. By revising this series on a regular basis, we can respond to changes in the industry, changes in technology, changes in the certification process, and to the ever-changing needs of those who teach automotive technology.

We also listened to instructors when they said something was missing or incomplete in the last edition. We responded to those and the results are included in this fifth edition.

The Today's Technician Series, by Delmar Cengage Learning, features textbooks that cover all mechanical and electrical systems of automobiles and light trucks. Principally the individual titles correspond to the certification areas for 2009 areas of ASE (National Institute for Automotive Service Excellence) certification.

Additional titles include remedial skills and theories common to all of the certification areas and advanced or specific subject areas that reflect the latest technological trends.

This new edition, like the last, was designed to give students a chance to develop the same skills and gain the same knowledge that today's successful technician has. This edition also reflects the changes in the guidelines established by the National Automotive Technicians Education Foundation (NATEF) in 2008.

The purpose of NATEF is to evaluate technician training programs against standards developed by the automotive industry and recommend qualifying programs for certification (accreditation) by ASE (National Institute for Automotive Service Excellence). Programs can earn ASE certification upon the recommendation of NATEF. NATEF's national standards reflect the skills that students must master. ASE certification through NATEF evaluation ensures that certified training programs meet or exceed industry-recognized, uniform standards of excellence.

The technician of today and for the future must know the underlying theory of all automotive systems and be able to service and maintain those systems. Dividing the material into two volumes, a Classroom Manual and a Shop Manual, provides the reader with the information needed to begin a successful career as an automotive technician without interrupting the learning process by mixing cognitive and performance learning objectives into one volume.

The design of Delmar's Today's Technician Series was based on features that are known to promote improved student learning. The design was further enhanced by a careful study of survey results, in which the respondents were asked to value particular features. Some of these features can be found in other textbooks, while others are unique to this series.

Each Classroom Manual contains the principles of operation for each system and subsystem. The Classroom Manual has discussions on design variations of key components used by the different vehicle manufacturers. It also looks into emerging technologies that will be standard or optional features in the near future. This volume is organized to build upon basic facts and theories. The primary objective of this volume is to allow the reader to gain an understanding of how each system and subsystem operates. This understanding is necessary to diagnose the complex automobiles of today and tomorrow. Although the basics contained in the Classroom Manual provide the knowledge needed for diagnostics, diagnostic procedures appear only in the Shop Manual. An understanding of the underlying theories is also a requirement for competence in the skill areas covered in the Shop Manual.

A coil-ring–bound Shop Manual covers the "how-to's." This volume includes step-by-step instructions for diagnostic and repair procedures. Photo Sequences are used to illustrate some of the common service procedures. Other common procedures are listed and are accompanied with fine line drawings and photos that allow the reader to visualize and conceptualize the finest details of the procedure. This volume also contains the reasons for performing the procedures, as well as when that particular service is appropriate.

The two volumes are designed to be used together and are arranged in corresponding chapters. Not only are the chapters in the volumes linked together, the contents of the chapters are also linked. This linking of content is evidenced by marginal callouts that refer the reader to the chapter and page that the same topic is addressed in the other volume. This feature is valuable to instructors. Without this feature, users of other two-volume textbooks must search the index or table of contents to locate supporting information in the other volume. This is not only cumbersome, but also creates additional work for an instructor when planning the presentation of material and when making reading assignments. It is also valuable to the students, with the page references they also know exactly where to look for supportive information.

Both volumes contain clear and thoughtfully selected illustrations. Many of which are original drawings or photos specially prepared for inclusion in this series. This means that the art is a vital part of each textbook and not merely inserted to increase the numbers of illustrations.

The page layout, used in the series, is designed to include information that would otherwise break up the flow of information presented to the reader. The main body of the text includes all of the "need-to-know" information and illustrations. In the wide side margins of each page are many of the special features of the series. Items that are truly "nice-to-know" information such as: simple examples of concepts just introduced in the text, explanations or definitions of terms that are not defined in the text, examples of common trade jargon used to describe a part or operation, and exceptions to the norm explained in the text. This type of information is placed in the margin, out of the normal flow of information. Many textbooks attempt to include this type of information and insert it in the main body of text; this tends to interrupt the thought process and cannot be pedagogically justified. By placing this information off to the side of the main text, the reader can select when to refer to it.

Jack Erjavec

HIGHLIGHTS OF THIS EDITION—CLASSROOM MANUAL

The text was updated throughout, to include the latest developments. Some of these new topics include the various transmission designs used (or planned to be used) in hybrid vehicles, the Lepelletier gear setup (which results in five-, six-, seven-, and eight-speed transmissions), power split units, and various constantly variable transmission designs. There is also more information on current electronic control systems, including the various protocols used for multiplexing.

The first chapter introduces the purpose of automatic transmissions and how they link to the rest of the vehicle. The chapter also describes the purpose and location of the subsystems, as well as the major components of the system and subsystems. The goal of this chapter is to establish a basic understanding for students to base their learning on. All systems and subsystems that are discussed in detail later in the text are introduced and their primary purpose described. The second chapter covers the underlying basic theories of operation as the topic of the text. This is valuable to the student and the instructor because it covers the theories that other textbooks assume the reader knows. All related basic physical, chemical, and thermodynamic theories are covered in this chapter.

The third chapter applies those theories to the operation of an automatic transmission. Great emphasis is placed on hydraulics. The fourth chapter goes deeply into the electronics involved in today's transmissions. This is a chapter that was greatly updated, since the manufacturers are constantly adding more sophisticated electronics to transmissions.

The chapters that follow cover the major components of an automatic transmission and transaxle, such as torque converters, pumps, hydraulic circuits, gears and shafts, and reaction and friction units. The last chapter takes a look at the commonly used transmissions and transaxles. This includes their mechanical, hydraulic, and electronic systems.

Current model transmissions are used as examples throughout the text. Many are discussed in detail. This includes five-, six-, seven-, and eight-speed and constantly variable transmissions. This new edition also has more information on nearly all automatic transmission-related topics. Finally, the art has been updated throughout the text to enhance comprehension and improve visual interest.

HIGHLIGHTS OF THIS EDITION—SHOP MANUAL

Along with the Classroom Manual, the Shop Manual was updated to match current trends. Service information related to the new topics covered in the Classroom Manual is included in this manual. In addition, several new photo sequences were added. The purpose of these detailed photos is to show students what to expect when they perform the same procedure. They also can provide a student with familiarity of a system or type of equipment they may not be able to perform at their school. Although the main purpose of the textbook is not to prepare someone to successfully pass an ASE exam, all of the information required to do so is included in the textbook.

Chapters 1 and 2 cover the need-to-know transmission-related information about tools, safety, and typical services procedures. The first chapter covers safety issues. To stress the importance of safe work habits, one full chapter is dedicated to safety. Included in this chapter are common shop hazards, safe shop practices, safety equipment, and the legislation concerning and the safe handling of hazardous materials and wastes. Chapter 2 covers the basics of things a transmission technician does to earn a living, including basic diagnostics. Also included in this chapter are those tools and procedures that are commonly used to diagnose and service automatic transmissions and transaxles.

Chapters 3 and 4 have been heavily revised and updated. This is due to the many new developments that have occurred in transmission controls.

The rest of the chapters have been thoroughly updated. Much of the updating focuses on the diagnosis and service to new systems, as well as those systems instructors have said they need more help in.

New photo sequences on reprogramming a TCM and overhauling a five- and six-speed transmission have been added. Currently accepted service procedures are used as examples throughout the text. These procedures also served as the basis for new job sheets that are included in the text. Finally, the art has been updated throughout the text to enhance comprehension and improve visual interest.

Features of this manual include:

COGNITIVE OBJECTIVES

These objectives define the contents of the chapter and define what the student should have learned upon completion of the chapter.

Each topic is divided into small units to promote easier understanding and learning.

CROSS-REFERENCES TO THE SHOP MANUAL

Reference to the appropriate page in the Shop Manual is given whenever necessary. Although the chapters of the two manuals are synchronized, material covered in other chapters of the Shop Manual may be fundamental to the topic discussed in the Classroom Manual.

MARGINAL NOTES

These notes add "nice-to-know" information to the discussion. They may include examples or exceptions, or may give the common trade jargon for a component.

AUTHOR'S NOTES

This feature includes simple explanations, stories, or examples of complex topics. These are included to help students understand difficult concepts.

Chapter 1

DRIVETRAIN BASICS

UPON COMPLETION AND REVIEW OF THIS CHAPTER, YOU SHOULD BE ABLE TO:

- Identify the major components of a vehicle's drivetrain.
- State the purpose of a transmission.
- Describe the major differences between a transmission and a transaxle.
- Describe the construction and operation of CVTs.
- Explain how a set of gears can increase torque.
- Define the term *gear ratio* and explain what happens with speed and torque when two gears are meshed.

- Describe the basic operation of a planetary gearset.
- State the purpose of a torque converter assembly.
- Describe the differences between a typical FWD and RWD car.
- State and understand the purpose of U and CV joints.
- State the purpose of a differential.
- Identify and describe the various gears used in modern drivetrains.
- Identify and describe the various bearings used in modern drivetrains.

INTRODUCTION

An automobile can be divided into four major systems: (1) the engine, which serves as the source for propulsion power; (2) the drivetrain, which transmits the engine's power to the car's wheels; (3) the chassis, which supports the engine and body and includes the brake, steering, and suspension systems; and (4) the car's body, interior, and accessories, which include the seats, heater and air conditioner, lights, windshield wipers, and other comfort and safety features.

The drivetrain has four main purposes: to connect and disconnect the engine's power to the wheels, to select different speed ratios, to provide a way to move the car in reverse, and to control the power to the drive wheels for safe turning of the vehicle. The main components of the drivetrain are the transmission, differential, and drive axles (Figure 1-1).

Today, most cars are **front-wheel-drive (FWD)**. Power flow through the drivetrain of a FWD vehicle passes through the clutch or torque converter, through the transmission, and then through a front differential, the driving axles, and onto the front wheels. The transmission and differential are housed in a single unit (Figure 1-2) called a **transaxle**. The gearsets in the transaxle provide the required gear ratios and direct the power flow into the differential. The differential gearing provides the final gear reduction and splits the power flow between the left and right drive axles. The drive axles extend from the sides of the transaxle. The outer ends of the axles are fitted to the hubs of the drive wheels. **Constant velocity (CV) joints** mounted on each end of the drive axles allow for changes in length and angle without affecting the power flow to the wheels.

The combination of the engine and the drivetrain is sometimes referred to as the vehicle's powertrain.

A few RWD vehicles have the engine mounted in the midsection of the vehicle and are called "mid-engined" and a few have the engine mounted at the rear of the vehicle and are called "rear-engined."

1

TERMS TO KNOW DEFINITIONS

Many of the new terms are pulled out into the margin and defined.

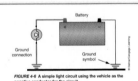

FIGURE 4-6 A simple light circuit using the vehicle as the negative conductor for the circuit.

FIGURE 4-7 A simple parallel circuit.

The legs of a parallel circuit can contain a series circuit. To determine the resistance of that leg, the resistance values are added together. The resistance values of each leg are used to calculate the total resistance of a parallel circuit. Total circuit current flows only through the common power and ground paths; therefore, a change in a branch's resistance will not only affect the current in the branch but will also affect total circuit current.

Electrical Problems

Normally, parallel circuits in an automobile have a circuit protection device placed in a common path from the positive side of the battery. This protection device is usually a fuse, fusible link, or a circuit breaker. They are designed to protect the wires and components from damage due to excessive current flow. When a great amount of current flows through the fuse or breaker, an element will burn out or open, causing the circuit to be opened and stopping current flow. This action prevents the high current from burning up the wires or components in the protected circuit.

Low resistance causes high current. A decrease in the amount of resistance is typically the result of a short. A *short* is best defined as an additional and unwanted path to ground. Most shorts, such as a bare wire contacting the frame of the car, create an extremely low resistance parallel branch. Low resistance and high current can also be caused by a slow-turning motor.

A short to ground can be present before the load in the circuit or internally within the load or component. A short can also connect two or more circuits together, causing additional parallel legs and uncontrolled operation of components. An example of a possible result from a wire-to-wire short would be the horn blowing each time the brake pedal was depressed (Figure 4-8). This could be caused by a wire-to-wire short between the horn and brake light circuits. Shorts are one of the three common types of electrical problems.

Another common electrical fault is the *open*. An open causes an incomplete circuit and can result from a broken or burned wire, loose connection, or a faulty component. If a circuit is open, there is no current flow and the component will not operate. If there is an open in one leg of a parallel circuit, the remaining part of that parallel circuit will operate normally.

Excessive resistance at a connector, internally in a component, or within a wire is also a common electrical problem. High, unwanted resistance will cause low current flow and the component will not be able to operate normally, if at all.

Shop Manual
Chapter 4, page 000

Motors attempting to move a heavier than normal load or an immovable object, such as a binding window, will rotate very slowly or not at all and will draw excessive amounts of current.

AUTHOR'S NOTE: High resistance problems always cause low circuit current and lower than normal voltage drops across the intended load. Opens always result in zero current flow and therefore there are no measurable voltage drops in the circuit. A short always increases current flow but the voltage drops will be about normal. Voltage drops will be normal because a short is an unwanted parallel branch in the circuit.

106

The transmission's input shaft is supported by bushings in the stator support inside the torque converter. There is no mechanical link between the output of the engine and the input of the transmission. The fluid connects the power from the engine to the transmission. The combined weight of the fluid, torque converter, and flexplate serves as the flywheel for the engine.

Torque Converter Construction

A typical torque converter consists of three elements sealed in a single housing: the impeller, the turbine, and the stator. The impeller is the drive member of the unit and its fins are attached directly to the converter cover. Therefore, the impeller is the input device for the converter and always rotates at engine speed.

The turbine is the converter's output member and is coupled to the transmission's input shaft (Figure 6-7). The turbine is driven by the fluid flow from the impeller and always turns at its own speed. The fins of the turbine face toward the fins of the impeller. The impeller and the turbine have internal fins, but the fins point toward each other.

The stator is the reaction member of the converter (Figure 6-8). This assembly is about one-half the diameter of the impeller or turbine and is positioned between the impeller and turbine. The stator is not mechanically connected to either the impeller or turbine; rather, it fits between the turbine outlet and the inlet of the impeller. All of the fluid returning from the turbine to the impeller must pass through the stator. The stator redirects the fluid leaving the turbine back to the impeller (Figure 6-9). By redirecting the fluid so that it is flowing in the same direction as engine rotation, it allows the impeller to rotate more efficiently, creating torque multiplication.

A BIT OF HISTORY

This feature gives the student a sense of the evolution of the automobile. This feature not only contains nice-to-know information, but also should spark some interest in the subject matter.

A BIT OF HISTORY

Variations on the basic three-element torque converter have been used. The 1948 Buick Dynaflow had two impellers, two stators, and one turbine. In 1953 the Twin-Turbine Dynaflow was released, with two turbines, one impeller, and one stator. In 1956, Buick introduced a multiple-turbine torque converter that had a variable pitch stator. By the late 1960s, the industry, including Buick, had returned to the basic three-element converter.

SUMMARIES

Each chapter concludes with a summary of key points from the chapter. These are designed to help the reader review the chapter contents.

SUMMARY

- The drivetrain has four primary purposes: to connect the engine's power to the drive wheels, to select different speed ratios, to provide a way to move the vehicle in reverse, and to control the power to the drive wheels for safe turning of the vehicle.
- The main components of the drivetrain are the transmission, differential, and drive axles.
- The rotating or turning effort of the engine's crankshaft is called engine torque.
- The amount of engine vacuum formed during the intake stroke is determined largely by the amount of load on the engine.
- Gears are used to apply torque to other rotating parts of the drivetrain and to multiply torque.
- Torque is calculated by multiplying the applied force by the distance from the center of the shaft to the point where the force is exerted.
- Gear ratios express the mathematical relationship, in size and number of teeth, of one gear to another.
- Transmissions offer various gear ratios through the meshing of various-sized gears.
- Reverse gear is accomplished by adding a gear to a two-gear set.
- Like manual transmissions, automatic transmissions provide various gear ratios, which match engine speed to the vehicle's speed. However, an automatic transmission is able to shift between gear ratios by itself and there is no need for a manually operated clutch to assist in the change of gears.
- A planetary gearset consists of a ring gear, a sun gear, and several planet gears, all mounted in the same plane.
- The ring gear has its teeth on its inner surface and the sun gear has its teeth on its outer surface. The planet gears are spaced evenly around the sun gear and mesh with both the ring and sun gears.
- By applying the engine's torque to one of the gears in a planetary gearset and preventing another member of the set from moving, torque multiplication, speed increase, or change of rotational direction is available on the third set of gears.
- Brake bands or multiple-friction disc packs attached to the individual gear carriers and shafts are hydraulically activated to direct engine power to any of the gears and to hold any of the gears from rotating. This allows gear ratio changes and the reversing of power flow while the engine is running.
- An oil pump in the transmission provides the hydraulic fluid needed to activate the various brake bands and clutch packs.
- The valve body controls the flow of the fluid throughout the transmission and acts on the vacuum and mechanical signals it receives about engine and vehicle speeds and loads.
- Most new automatic transmissions rely on data received from electronic sensors and use an electronic control unit to operate solenoids in the valve body to shift gears.
- In FWD cars, the transmission and drive axle is located in a single assembly called a transaxle. In RWD cars, the drive axle is connected to the transmission through a driveshaft.
- The driveshaft and its joints are called the driveline of the car.
- Universal joints allow the driveshaft to change angles in response to movements of the car's suspension and rear axle assembly.
- The rear axle housing encloses the entire rear-wheel driving axle assembly.
- The primary purpose of the differential is to allow a difference in driving wheel speed when the vehicle is rounding a corner or curve. The ring and pinion in the drive axle also multiplies the torque it receives from the transmission.
- On FWD cars, the differential is part of the transaxle assembly.

TERMS TO KNOW

All-wheel-drive (AWD)
Annulus
Automatic transmission fluid (ATF)
Ball bearing
Bevel gear
Bushing
Combustion
Constant velocity (CV) joint
Continuously variable transmission (CVT)
Differential
Direct drive
Driveshaft
Engine load
Engine torque
Flexplate
Flywheel
Four-wheel-drive (4WD)
Front-wheel-drive (FWD)
Gear ratio
Helical gear
Herringbone gear
Horsepower
Hybrid electric vehicle (HEV)
Hypoid gear
Limited-slip
Miscibility
Neutral
Newton-meters (Nm)
Overdrive
Overall gear ratio
Planetary carrier
Planetary gear
Planetary pinions
Pounds-foot (lbs.-ft.)
Pulley
Rear-wheel-drive (RWD)
Ring gear

TERMS TO KNOW LIST

A list of new terms appears next to the Summary.

REVIEW QUESTIONS

Short answer essay, fill-in-the-blank, and multiple-choice questions are found at the end of each chapter. These questions are designed to accurately assess the student's competence in the objectives stated at the beginning of the chapter.

REVIEW QUESTIONS

Short-Answer Essays

1. A computer relies on many different reference voltage sensors. These can be divided into types. Name the types and give a brief description of their operation.

2. Although computers receive different information from a variety of sensors, the decisions for shifting are actually based on more than the inputs. What are they based on?

3. What is the purpose of a protection device in an electrical circuit?

4. Some transmissions receive information through multiplexing. How does this work?

5. How does a solenoid in an electronically controlled transmission work?

6. Most late-model transmission control systems have adaptive learning. What does this mean?

7. What inputs are of prime importance to a computer in deciding when to shift gears?

8. What are the advantages of using electronic controls rather than relying on conventional hydraulic controls in a transmission?

9. What is so unique about a Honda transmission?

10. What is the major difference between the Honda IMA system and a typical ISAD system?

Fill-in the Blanks

1. Amperage in a series circuit is always _____.

2. The voltage required to push amperage through a resistance is called _____.

3. A VSS and an OSS are examples of _____ _____ sensors.

4. A circuit that has more than one path for current to flow through is a _____.

5. A computer is an electronic device that _____, _____, _____, and _____ information.

6. The input devices used in electronic control systems vary with each system; however, they can be grouped into distinct categories: _____, _____ and _____.

7. Common voltage reference sensors include _____ switches, _____, _____, and _____ sensors.

8. In an electronic control system, the typical output devices are _____ and _____ that cause something _____ or _____ to change.

9. Most electronically controlled transmissions rely on _____ _____ solenoids to control all forward gears.

10. Voltage generation devices are typically used to monitor _____ _____.

MULTIPLE CHOICE

1. Voltage generation devices are typically used to monitor rotational speeds. Which of the following is NOT a voltage-generating type sensor?
   A. Vehicle speed sensor   C. MAP
   B. OSS   D. ISS

2. While discussing transmission solenoids:
   Technician A says an EPC solenoid replaces the conventional TV cable setup to provide changes in fluid pressure in response to engine load.
   Technician B says a TCC solenoid can be modulated to smooth the engagement of the torque converter clutch.
   Who is correct?
   A. A only   C. Both A and B
   B. B only   D. Neither A nor B

145

To stress the importance of safe work habits, the Shop Manual also dedicates one full chapter to safety. Other important features of this manual include:

PERFORMANCE-BASED OBJECTIVES

These objectives define the contents of the chapter and define what the student should have learned upon completion of the chapter. These objectives also correspond to the list of required tasks for ASE certification.

Although this textbook is not designed simply to prepare someone for the certification exams, it is organized around the ASE task list. These tasks are defined generically when the procedure is commonly followed and specifically when the procedure is unique for specific vehicle models. Imported and domestic model automobiles and light trucks are included in the procedures.

BASIC TOOLS LIST

Each chapter begins with a list of the basic tools needed to perform the tasks included in the chapter.

TERMS TO KNOW DEFINITIONS

Many of the new terms are pulled out into the margin and defined.

MARGINAL NOTES

These notes add "nice-to-know" information to the discussion. They may include examples or exceptions, or may give the common trade jargon for a component.

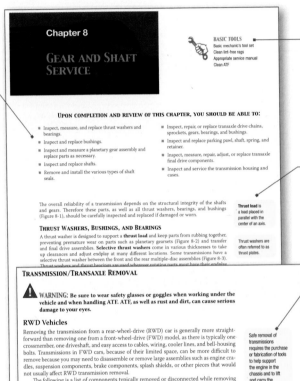

PHOTO SEQUENCES

Many procedures are illustrated in detailed photo sequences. These detailed photographs show the students what to expect when they perform particular procedures. They also can provide for the student a familiarity with a system or type of equipment that their school may not have.

SPECIAL TOOLS LIST

Whenever a special tool is required to complete a task, it is listed in the margin next to the procedure.

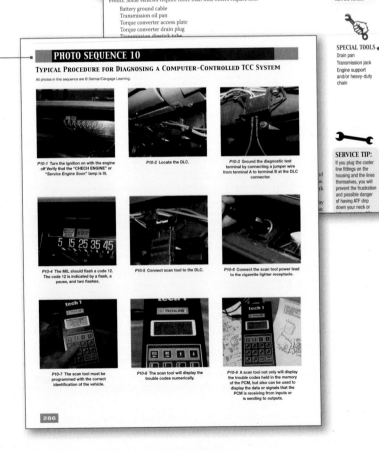

there is little air flow through the cooler, the radiator or external cooler must be removed and flushed or replaced.

GENERAL CONVERTER CONTROL DIAGNOSTICS

All testing of TCC controls should begin with a basic inspection of the engine and transmission. Too often, technicians skip this basic inspection and become frustrated during diagnostics because of conflicting test results. The basic inspection should include the following:

1. A road test to verify the complaint and further define the problem.
2. A careful inspection of the engine and transmission.
3. A check of the PCM for codes.
4. A check of the mechanical condition of the engine, the output of ignition system, and the operation of the fuel system.
5. An idle speed and ignition timing check. If the timing is nonadjustable, check the operation of the electronic spark control system.
6. A check of the entire intake system for vacuum leaks.

When inspecting wires, look for burnt spots, bare wires, and damaged or pinched wires. Make sure the wiring harness to the electronic control unit has a tight and clean connection. Also, check the source voltage at the battery before beginning any detailed tests. If the voltage is too low or too high, the system cannot function properly.

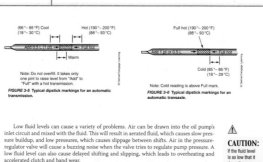

Note: Do not overfill. It takes only one pint to raise level from "Add" to "Full" with a hot transmission.

FIGURE 3-5 Typical dipstick markings for an automatic transmission.

Note: Cold reading is above Full mark.

FIGURE 3-6 Typical dipstick markings for an automatic transaxle.

Low fluid levels can cause a variety of problems. Air can be drawn into the oil pump's inlet circuit and mixed with the fluid. This will result in aerated fluid, which causes slow pressure buildup, and low pressures, which causes slippage between shifts. Air in the pressure-regulator valve will cause a buzzing noise when the valve tries to regulate pump pressure. A low fluid level can also cause delayed shifting and slipping, which leads to overheating and accelerated clutch and band wear.

Excessively high fluid levels can also cause **aeration**. As the planetary gears rotate in high fluid levels, air can be forced into the fluid. Aerated fluid can foam, overheat, and oxidize. All of these problems can interfere with normal valve, clutch, and servo operation. Foaming may be evident by fluid leakage from the transmission's vent.

> **CUSTOMER CARE:** Customers should be made aware that fluid level and condition should be checked at least every six months. Temperature fluctuations from summer to winter can cause a thermal breakdown of ATF. Even high-quality fluids can experience breakdown as a result of these frequent and extreme temperature changes. It is said that nearly 90 percent of all transmission failures are caused by fluid breakdown or oxidation.

The condition of the fluid should be checked while checking the fluid level. Examine the fluid carefully. The normal color of ATF is pink or red. If the fluid has a dark brownish or blackish color or a burned odor, the fluid has been overheated. A milky color indicates that water has mixed with the fluid, possibly from engine coolant leaking into the transmission's cooler in the radiator. If there is any question about the condition of the fluid, drain out a sample for closer inspection.

If there is evidence that there is water or moisture in the fluid, the transmission must be completely disassembled and the following parts should be cleaned or replaced:

- The torque converter
- All internal and external seals
- All transmission fluid filters
- All clutches and bands that have friction material
- All solenoids

After the transmission is reassembled, the transmission's fluid cooler(s) and its hoses and tubes should be flushed and cleaned.

After checking the ATF level and color, wipe the dipstick on absorbent white paper and look at the stain left by the fluid. Dark particles are normally band or clutch material, while

---

Classroom
Manual
Chapter 6, page 000

SERVICE TIP:
It is important to realize that some oil coolers cannot be flushed effectively. Check the service information for any notations regarding this.

SERVICE TIP:
Nearly all electronic converter and transmission controls have a self-test mode and are capable of displaying DTCs. However, the basic inspection is important because the computer will not display codes for low compression or other common engine problems.

SPECIAL TOOLS
Hydraulic
pressure gauge

⚠️
CAUTION:
If the fluid level is so low that it doesn't appear on the dipstick or is at the very bottom of the dipstick, the vehicle should not be driven until the level is brought up to normal. Driving with a very low level can cause internal transmission damage.

Aeration is the process of mixing air with a liquid.

Oxidation occurs when something is mixed with oxygen to produce an oxygen-containing compound. Oxidation is also the term given to the chemical breakdown of a substance or compound caused by its combination with oxygen.

CROSS-REFERENCES TO THE CLASSROOM MANUAL

Reference to the appropriate page in the Classroom Manual is given whenever necessary. Although the chapters of the two manuals are synchronized, material covered in other chapters of the Classroom Manual may be fundamental to the topic discussed in the Shop Manual.

SERVICE TIPS

Whenever a shortcut or special procedure is appropriate, it is described in the text. These tips are generally those things commonly done by experienced technicians.

CAUTIONS AND WARNINGS

Throughout the text, warnings are given to alert the reader to potentially hazardous materials or unsafe conditions. Cautions are given to advise the student of things that can go wrong if instructions are not followed or if a nonacceptable part or tool is used.

CUSTOMER CARE

This feature highlights those little things a technician can do or say to enhance customer relations.

JOB SHEET   **5**

Name _____ Date _____

FILLING OUT A WORK ORDER

Upon completion of this job sheet, you will be able to prepare a service work order based on customer input, vehicle information, and service history.

ASE Correlation

This job sheet is related to the ASE Automatic Transmission and Transaxle Test's Content Area *In-Vehicle Transmission and Transaxle Repair.*
Task: Complete work order to include customer information, vehicle identifying information, customer concern, related service history, cause, and correction.

Tools and Materials

An assigned vehicle or the vehicle of your choice
Service work order or computer-based shop management package
Parts and labor guide

Work Order Source:

Describe the system used to complete the work order. If a paper repair order is being used, describe the source.

_____
_____

Procedure   Task Completed

1. Prepare the shop management software for entering a new work order or obtain a blank paper work order.   ☐

2. Enter customer information, including name, address, and phone numbers onto the work order.   ☐

3. Locate and record the vehicle's VIN.   ☐

4. Enter the necessary vehicle information, including year, make, model, engine type and size, transmission type, license number, and odometer reading.   ☐

5. Does the VIN verify that the information about the vehicle is correct? _____

6. Normally, you would interview the customer to identify his or her concerns. However to complete this job sheet, assume the customer desires to have the fluid and filter changed in the transmission.   ☐

7. The history of service to the vehicle can often help diagnose problems as well as indicate possible premature part failure. Gathering this information from the customer can provide some of this information. For this job sheet assume the vehicle has not had a transmission problem and was not recently involved in a collision. Service history is further obtained by searching files based on customer name, VIN, and license number. Check the files for any related service work.   ☐

79

JOB SHEETS

Located at the end of each chapter, the Job Sheets provide a format for students to perform procedures covered in the chapter. A reference to the ASE Task addressed by the procedure is referenced on the Job Sheet.

TERMS TO KNOW LIST

A list of new terms appears after the case study.

ASE-STYLE REVIEW QUESTIONS

Each chapter contains ASE-style review questions that reflect the performance-based objectives listed at the beginning of the chapter. These questions can be used to review the chapter as well as to prepare for the ASE certification exam.

ASE PRACTICE EXAMINATION

A 50-question ASE practice exam, located in the appendix of the Shop Manual, is included to test students on the contents of the Shop Manual.

CASE STUDIES

Case Studies concentrate on the ability to properly diagnose the systems. Beginning with Chapter 3, each chapter ends with a case study in which a vehicle has a problem, and the logic used by a technician to solve the problem is explained.

ASE CHALLENGE QUESTIONS

Each technical chapter ends with five ASE challenge questions. These are not review questions, rather they test the student's ability to apply general knowledge to the contents of the chapter.

CASE STUDY

A customer with a late-model Ford pickup equipped with an E4OD transmission complained that every time he hit a bump, the transmission would downshift. Then, when driven on smooth surfaces, the transmission would begin to cycle on its own between third and fourth gears.

The technician originally thought the problem was caused by a bad fourth gear clutch that couldn't hold the planetary gearset in overdrive or by a problem in the lockup torque converter circuit. Beginning the diagnosis with a visual inspection of the transmission and the many sensors that provide information to the E4OD's control module, the technician found faulty signals from the manual lever position (MLP) sensor. These signals explained the erratic shifting, because the computer uses the sensor to determine what gear the transmission should be in and how much modified line pressure to supply.

The technician used the required special tool to realign the sensor with the gear selector, but still found that the resistance readings were out of specifications. The sensor was replaced and the problem of erratic shifting was solved.

TERMS TO KNOW

Aeration
Manifold absolute pressure (MAP) sensor
Oxidation
Throttle position sensor
Vehicle speed sensor
Variable force motor (VFM)

ASE-STYLE REVIEW QUESTIONS

1. While diagnosing noises apparently coming from a transaxle assembly:
   Technician A says a knocking sound at low speeds is probably caused by worn CV joints.
   Technician B says a clicking noise heard when the vehicle is turning is probably caused by a worn or damaged outboard CV joint.
   Who is correct?
   A. A only   C. Both A and B
   B. B only   D. Neither A nor B

2. *Technician A* says if the shift for all forward gears is delayed, a slipping front or forward clutch is normally indicated.
   Technician B says a slipping rear clutch is indicated when there is a delay or slip when the transmission shifts into any forward gear.
   Who is correct?
   A. A only   C. Both A and B
   B. B only   D. Neither A nor B

3. While discussing the results of an oil pressure test:
   Technician A says when the fluid pressures are high, internal leaks, a clogged filter, low oil pump output, or a faulty pressure regulator valve are indicated.
   Technician B says if the fluid pressure increased at the wrong time, an internal leak at the servo or clutch seal is indicated.

4. *Technician A* says low engine vacuum will cause a vacuum modulator to sense a load condition when it actually is not present, causing delayed and harsh shifts.
   Technician B says poor engine performance can cause delayed shifts through the action of the TV assembly.
   Who is correct?
   A. A only   C. Both A and B
   B. B only   D. Neither A nor B

5. *Technician A* says delayed shifting can be caused by worn planetary gearset members.
   Technician B says delayed shifts or slippage may be caused by leaking hydraulic circuits or sticking spool valves in the valve body.
   Who is correct?
   A. A only   C. Both A and B
   B. B only   D. Neither A nor B

6. While discussing proper band adjustment procedures:
   Technician A says that on some vehicles the bands can be adjusted externally with a torque wrench.
   Technician B says a calibrated pound-inch torque wrench is normally used to tighten the band-adjusting bolt to a specified torque.
   Who is correct?
   A. A only   C. Both A and B

ASE CHALLENGE QUESTIONS

1. While servicing a final drive unit:
   Technician A checks gear and bearing endplay whenever new bearings are installed in the unit.
   Technician B reuses the bearings, seals, and thrust washers if the bearing preload and endplay are fine, as is the condition of the bearings.
   Who is correct?
   A. A only   C. Both A and B
   B. B only   D. Neither A nor B

2. *Technician A* says a drive chain that is too loose should be shortened by removing a pair of links in the chain.
   Technician B says the drive sprockets should be replaced if the gear teeth are polished or show any other signs of wear.
   Who is correct?
   A. A only   C. Both A and B

5. *Technician A* says all hubs, drums, and shells should be carefully examined for wear and damage.
   Technician B says minor scoring or burrs on band application surfaces of a drum can be removed by lightly polishing the surface with a 200-grit crocus cloth.
   Who is correct?
   A. A only   C. Both A and B
   B. B only   D. Neither A nor B

APPENDIX A   ASE PRACTICE EXAMINATION

Final Exam Automatic Transmission/Transaxle A2

1. Which of the following is the *least* likely cause for a buzzing noise from a transmission?
   A. Improper fluid level or condition
   B. Defective oil pump
   C. Defective flexplate
   D. Damaged planetary gearset

2. A vehicle experiences engine flare in low gear only.
   Technician A says the torque converter lockup clutch is slipping.
   Technician B says the transmission oil pump is not providing the required pressure.
   Who is correct?
   A. A only   C. Both A and B
   B. B only   D. Neither A nor B

3. The results of a hydraulic pressure test are being discussed:
   Technician A says low idle pressure may be caused by a defective exhaust gas recirculation system.
   Technician B says low neutral and park pressures may indicate a fluid leakage past the clutch and servo seals.
   Who is correct?
   A. A only   C. Both A and B
   B. B only   D. Neither A nor B

4. The vehicle creeps in neutral.
   Technician A says a too high engine idle speed could be the cause.
   Technician B says a too tight clutch pack may be the problem.
   Who is correct?
   A. A only   C. Both A and B
   B. B only   D. Neither A nor B

5. *Technician A* says overtorqued valve body fasteners may cause a lack of engine braking in manual low.
   Technician B says a lack of engine braking in manual third may be caused by a faulty overrunning clutch.
   Who is correct?
   A. A only   C. Both A and B
   B. B only   D. Neither A nor B

6. The transmission's output shaft and its sealing components are being discussed:
   Technician A says the shaft and all of its sealing components must be replaced if nicks and scratches are found in the shaft's sealing area.
   Technician B says all of the shaft's seals and rings must be replaced during a rebuild.
   Who is correct?
   A. A only   C. Both A and B
   B. B only   D. Neither A nor B

7. The vehicle will only upshift to second at full throttle. This could be caused by any of the following EXCEPT:
   A. Clogged oil passages   C. Bad clutch pack
   B. Low fluid level   D. Open upshift switch

8. The vehicle will not move in any gear.
   Technician A says a misadjusted TV cable could be the cause.
   Technician B says leakage at the oil pump and/or valve body could cause this condition.
   Who is correct?
   A. A only   C. Both A and B
   B. B only   D. Neither A nor B

9. Sensors are being discussed:
   Technician A says most speed sensors are AC generators.
   Technician B says most speed sensors use a stationary magnet, rotor, and a voltage sensor.
   Who is correct?
   A. A only   C. Both A and B
   B. B only   D. Neither A nor B

10. *Technician A* says the PCM monitors the amount of voltage generated by the speed sensor to calculate the vehicle's speed.
    Technician B says the output of a speed sensor is pulsed as an on/off voltage signal when displayed on a DSO.
    Who is correct?
    A. A only   C. Both A and B
    B. B only   D. Neither A nor B

539

INSTRUCTOR RESOURCES

The Instructor Resources DVD is a robust ancillary that contains all preparation tools to meet any instructor's classroom needs. It includes chapter outlines in PowerPoint with images, video clips and animations that coincide with each chapter's content coverage, chapter tests in ExamView with hundreds of test questions, a searchable Image Library with all photos and illustrations from the text, theory-based Worksheets in Word that provide homework or in-class assignments, the Job Sheets from the Shop Manual in Word, a NATEF correlation chart, and an Instructor's Guide in electronic format.

WEBTUTOR ADVANTAGE

Newly available for this title and to the Today's Technician™ Series is the *WebTutor Advantage* for Blackboard and Angel online course management systems. The *WebTutor for Today's Technician: Automatic Transmissions & Transaxles, 5e* will include presentations in PowerPoint with video clips and animations, end-of-chapter review questions, pre-tests and post-tests, worksheets, discussion springboard topics, job sheets, and more. The *WebTutor* is designed to enhance the classroom and shop experience, engage students, and help them prepare for ASE certification exams.

REVIEWERS

The author and publisher would like to extend a special thanks to the individuals who reviewed this text and offered their invaluable feedback:

Timothy Belt
University of Northwestern Ohio
Lima, OH

Eric Harper
American River College
Sacramento, CA

Thomas L. Corban
University of Northwestern Ohio
Lima, OH

Michael Ronan
Alfred State College
Alfred, NY

Chapter 1

DRIVETRAIN BASICS

UPON COMPLETION AND REVIEW OF THIS CHAPTER, YOU SHOULD BE ABLE TO:

- Identify the major components of a vehicle's drivetrain.
- State the purpose of a transmission.
- Describe the major differences between a transmission and a transaxle.
- Describe the construction and operation of CVTs.
- Explain how a set of gears can increase torque.
- Define the term *gear ratio* and explain what happens with speed and torque when two gears are meshed.

- Describe the basic operation of a planetary gearset.
- State the purpose of a torque converter assembly.
- Describe the differences between a typical FWD and RWD car.
- State and understand the purpose of U and CV joints.
- State the purpose of a differential.
- Identify and describe the various gears used in modern drivetrains.
- Identify and describe the various bearings used in modern drivetrains.

INTRODUCTION

An automobile can be divided into four major systems: (1) the engine, which serves as the source for propulsion power; (2) the drivetrain, which transmits the engine's power to the car's wheels; (3) the chassis, which supports the engine and body and includes the brake, steering, and suspension systems; and (4) the car's body, interior, and accessories, which include the seats, heater and air conditioner, lights, windshield wipers, and other comfort and safety features.

The drivetrain has four main purposes: to connect and disconnect the engine's power to the wheels, to select different speed ratios, to provide a way to move the car in reverse, and to control the power to the drive wheels for safe turning of the vehicle. The main components of the drivetrain are the transmission, differential, and drive axles (Figure 1-1).

Today, most cars are **front-wheel-drive (FWD)**. Power flow through the drivetrain of a FWD vehicle passes through the clutch or torque converter, through the transmission, and then through a front differential, the driving axles, and onto the front wheels. The transmission and differential are housed in a single unit (Figure 1-2) called a **transaxle**. The gearsets in the transaxle provide the required gear ratios and direct the power flow into the differential. The differential gearing provides the final gear reduction and splits the power flow between the left and right drive axles. The drive axles extend from the sides of the transaxle. The outer ends of the axles are fitted to the hubs of the drive wheels. **Constant velocity (CV) joints** mounted on each end of the drive axles allow for changes in length and angle without affecting the power flow to the wheels.

The combination of the engine and the drivetrain is sometimes referred to as the vehicle's powertrain.

A few **RWD** vehicles have the engine mounted in the midsection of the vehicle and are called "mid-engined" and a few have the engine mounted at the rear of the vehicle and are called "rear-engined."

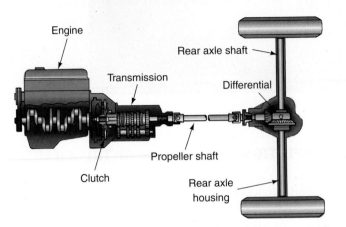

FIGURE 1-1 The drivetrain components for a typical rear-wheel-drive vehicle.

CV joints are **constant velocity joints.** They allow the angle of the axle shafts to change with no loss in rotational speed.

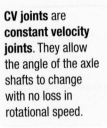

A BIT OF HISTORY

The French-built Panhard (1892) was the first vehicle to have its power generated by a front-mounted, liquid-fueled internal combustion engine and transmitted to the rear-driving wheels by a clutch, transmission, differential, and driveshaft.

A transfer case is an auxiliary transmission mounted behind the main transmission. It is used to divide power and transfer it to both the front and rear differential units.

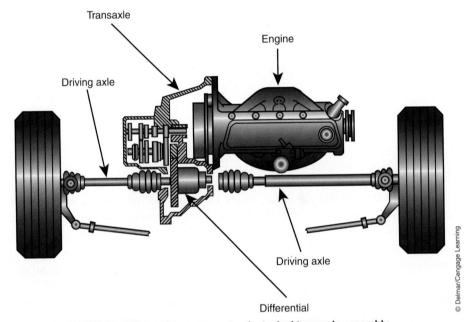

FIGURE 1-2 The main components of a typical transaxle assembly.

Some larger luxury and many performance cars are **rear-wheel-drive (RWD)**. Most pickup trucks, minivans, and SUVs are also RWD vehicles. Power flow in a RWD vehicle passes through the clutch or torque converter, the manual or automatic transmission, and the driveline (driveshaft assembly). Then it goes through the rear differential, the rear-driving axles, and onto the rear wheels.

Four-wheel-drive (4WD) or **all-wheel-drive (AWD)** vehicles combine features of both rear- and front-wheel-drive systems. Normally, when a pickup or full-size SUV has 4WD, the drive train is based on a RWD vehicle. When a smaller SUV or car has AWD or 4WD, the drive train is based on a FWD platform. In either case, there is a drive axle at both the front and the rear of the vehicle and an assembly that transfers the engine's power to both drive axles.

ENGINE

Although the engine is a major system by itself (Figure 1-3), its output should be considered a component of the drive train. The engine provides the power to drive the wheels of the vehicle. An engine develops a rotary motion or **torque** that, when multiplied by the transmission gears, will move the car under a variety of conditions. The engine produces power by burning a mixture of fuel and air in its combustion chambers. Combustion causes a high pressure in the cylinders, which forces the pistons downward. Connecting rods transfer the downward movement of the pistons to the crankshaft, which rotates by the force on the pistons.

All automobile engines, both gasoline and diesel, are classified as internal combustion engines because the combustion or burning that creates energy takes place inside the engine. **Combustion** is the burning of an air and fuel mixture. As a result of combustion, large amounts of pressure are generated in the engine. This pressure or energy is used to power the car. The engine must be built strong enough to hold the pressure and temperatures formed by combustion.

Diesel engines have been around a long time and are mostly found in big heavy-duty trucks. However, they are also used in some pickup trucks and will become more common in automobiles in the future. Diesel powered automobiles are quite common in Europe and other countries (Figure 1-4). Although the construction of a gasoline and diesel engine are similar, their operation is quite different.

A gasoline engine relies on a mixture of fuel and air that is ignited by a spark to produce power. A diesel engine also uses fuel and air, but does not need a spark to cause ignition. Diesel engines are often called compression ignition engines. This is because its incoming air is tightly compressed, which greatly raises its temperature. The fuel is then injected into the compressed air. The heat of the compressed air ignites the fuel and combustion takes place. The following sections cover the basic parts and the major systems of a gasoline engine.

Torque acts in a plane perpendicular to the axis.

A *stroke* is one complete up or down movement of the piston in its cylinder.

The low pressure formed on the intake stroke is commonly referred to as *engine vacuum*.

FIGURE 1-3 A typical late-model V-6 engine.

FIGURE 1-4 A European automotive diesel engine.

Most automotive engines are four-stroke cycle engines. The opening and closing of the intake and exhaust valves are timed to the movement of the piston. As a result, the engine passes through four different events or strokes during one combustion cycle. These four are the intake, compression, power, and exhaust strokes.

On the intake stroke, the piston moves downward, and a charge of air/fuel mixture is introduced into the cylinder. As the piston travels upward, the air/fuel mixture is compressed in preparation for burning. Just before the piston reaches the top of the cylinder, ignition occurs and combustion starts. The pressure of expanding gases forces the piston downward on its power stroke. When it reciprocates, or moves upward again, the piston is on the exhaust stroke. During the exhaust stroke, the piston pushes the burned gases out of the cylinder. As long as the engine is running, this cycle of events repeats itself, resulting in the production of engine torque.

The amount of **vacuum** formed on the intake stroke depends on the speed of the engine and the amount of air that is able to enter into the cylinders. It also depends on the cylinder's ability to seal when the piston is on its intake stroke. The throttle plates control both of these. Under normal conditions, the plates control engine speed by controlling the amount of air that enters into the cylinders. The amount of load on the engine determines how much the plates must be opened to maintain a particular engine speed. When there is a light load, such as while the vehicle is maintaining a cruising speed on a highway, the throttle plates need to be only slightly open to maintain the desired speed. Therefore, large amounts of vacuum are formed in the cylinders during the intake stroke. When the engine is under a heavy load, the throttle plates must be opened further to maintain the same speed. This allows more air to enter the cylinders and decreases the amount of vacuum formed during the intake stroke. Therefore, amount of engine vacuum formed during the intake stroke is primarily controlled by **engine load**. As engine load increases, vacuum decreases.

Vacuum can be best defined as any pressure lower than atmospheric pressure.

Engine load is the amount of resistance or weight the engine's crankshaft must overcome to move the vehicle or change its speed.

Torque Multiplication

The rotating or turning effort of the engine's crankshaft is called **engine torque**. Engine torque is measured in **pounds per foot (lbs.-ft.)** or in the metric measurement **Newton-meters (Nm)**. Most engines produce a maximum amount of torque while operating within a range of engine speeds and loads. When an engine reaches the maximum speed of that range, torque is no longer increased. This range of engine speeds is normally referred to as the engine's *torque curve* (Figure 1-5). Ideally, the engine should operate within its torque curve at all times.

Measurements of **horsepower** indicate the amount of work being performed and the rate at which it is being done. The term *power* actually means a force that is doing work over a period of time. The driveline can transmit power and multiply torque, but it cannot multiply power. When power flows through one gear to another, the torque is multiplied in proportion to the different gear sizes. Torque is multiplied, but the power remains the same, because the torque is multiplied at the expense of rotational speed.

As gears with different numbers of teeth mesh, each rotates at a different speed and torque. Torque is calculated by multiplying the force by the distance from the center of the shaft to the point where the force is exerted. For example, if you tighten a bolt with a wrench that is 1 foot long and apply a force of 10 pounds to the wrench, you are applying 10 lbs.-ft. of torque to the bolt. Likewise, if you apply a force of 20 pounds to the wrench, you are applying 20 lbs.-ft. of torque. You could also apply 20 lbs.-ft. of torque by applying only 10 pounds of force if the wrench were 2 feet long (Figure 1-6).

If a tooth on the driving gear is pushing against a tooth on the driven gear with a force of 25 pounds and the force is applied at a distance of 1 foot, which is the radius of the driving gear, a torque of 25 lbs.-ft. is applied to the driven gear. The 25 pounds of force from the teeth of the smaller (driving) gear is applied to the teeth of the larger (driven) gear. If that same force were applied at a distance of 2 feet from the center, the torque on the shaft at the center of the driven gear would be 50 lbs.-ft. The same force is acting at twice the distance from the shaft center (Figure 1-7).

A drivetrain consisting of a driving gear with 24 teeth and a radius of 1 inch and a driven gear with 48 teeth and a radius of 2 inches will have a torque multiplication factor of 2 and a speed reduction of $\frac{1}{2}$. Thus, it doubles the amount of torque applied to it at half the speed (Figure 1-8). The radii between the teeth of a gear act as levers; therefore, a gear that is twice the size of another has twice the lever arm length of the other.

Gear ratios express the mathematical relationship of one gear to another. Gear ratios can be varied by changing the diameter and number of teeth of the gears in mesh. A gear ratio

Pounds-foot (lbs.-ft.) is the expression of how much torque is present at a point. The correct expression for the amount of torque is pounds-foot, but some literature list torque as units of ft.-lb. One pound-foot is the torque obtained by a force of one pound applied to a wrench handle 12 inches long.

One **horsepower** is the equivalent of moving 33,000 pounds one foot in one minute.

The meshing of gears describes the fit of one tooth of one gear fitting between two teeth of another gear.

Gear ratios are normally expressed in terms of some number to 1 and use a colon (:) to show the numerical comparison, for example, 3.5:1, 1:1, 0.85:1.

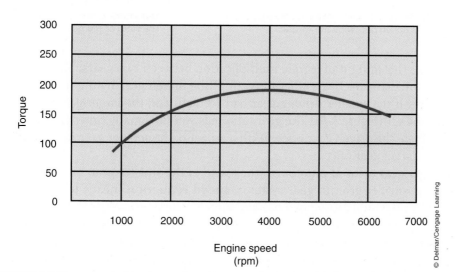

FIGURE 1-5 The amount of torque produced by an engine varies with the speed of the engine.

© Delmar/Cengage Learning

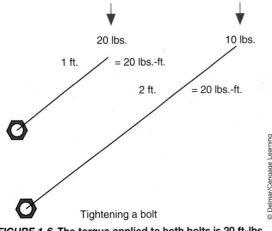

FIGURE 1-6 The torque applied to both bolts is 20 ft-lbs.

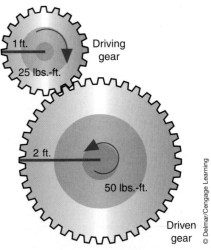

FIGURE 1-7 The driven gear will turn at half the speed but twice the torque because it is two times larger than the driving gear.

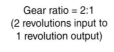

Gear ratio = 2:1
(2 revolutions input to
1 revolution output)

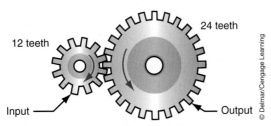

FIGURE 1-8 The smaller gear will turn the larger gear at half its speed but twice the torque.

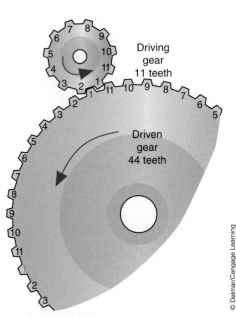

FIGURE 1-9 The driving gear must rotate four times to rotate the driven gear once. The ratio of the gearset is 4:1.

also expresses the amount of torque multiplication between two gears. The ratio is obtained by dividing the diameter or number of teeth of the driven gear by the diameter or teeth of the drive gear. If the smaller driving gear had 10 teeth and the larger gear had 40 teeth, the ratio would be 4:1 (Figure 1-9). The gear ratio tells you how many times the driving gear has to turn to rotate the driven gear once. With a 4:1 ratio, the smaller gear must turn four times to rotate the larger gear once.

The larger gear turns at one-fourth the speed of the smaller gear, but has four times the torque of the smaller gear. In gear systems, **speed reduction** means torque increases. For example, when a typical four-speed transmission is in first gear, there is a speed reduction of 12:1 from the engine to the drive wheels, which means that the crankshaft turns 12 times to turn the wheels once. The resulting torque is 12 times the engine's output; therefore, if the engine produces 100 pounds-feet of torque, a torque of 1200 pounds-feet is applied to the drive wheels (Figure 1-10).

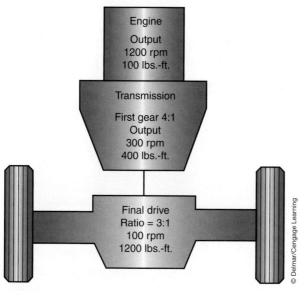

FIGURE 1-10 The torque multiplication and subsequent speed reduction of a typical vehicle's drivetrain.

TRANSMISSIONS

Transmissions contain several combinations of large and small gears. These provide for a variety of low and high gears. Lower (high-ratio) gears allow for lower vehicle speeds but more torque. Higher (low-ratio) gears provide less torque but higher vehicle speeds.

The transmission is mounted to the rear of the engine and is designed to allow the car to move forward and in reverse. There are two basic types of transmissions: automatic and manual. Automatic transmissions use a combination of a torque converter and a **planetary gear** system (Figure 1-11) to change gear ratios automatically. A manual transmission is an assembly of gears and shafts (Figure 1-12) that transmits power from the engine to the drive axle. Changes in gear ratios are controlled by the driver.

Transmissions are often called gearboxes.

Manual transmissions are commonly called standard shift or stick-shift transmissions.

Torque converters use fluid flow to connect and disconnect the engine's power to the transmission.

FIGURE 1-11 A six-speed automatic transmission.

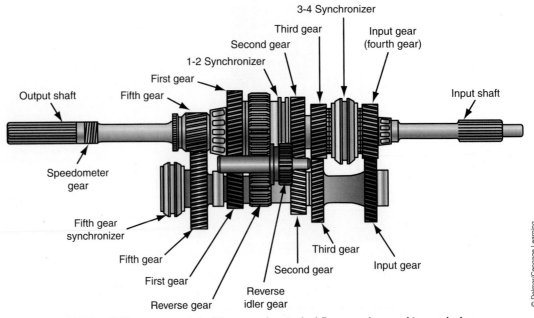

FIGURE 1-12 **The arrangement of the gears in a typical five-speed manual transmission.**

By moving the shift lever on a manual transmission and depressing the clutch pedal, various gear and speed ratios can be selected. The gears in a transmission are selected to give the driver a choice of both speed and torque. Like manual transmissions, automatic transmissions provide various gear ratios according to engine speed, power train load, vehicle speed, and other operating factors. However, with an automatic transmission, the driver has little to do because both upshifts and downshifts occur automatically and there is no need for a driver-operated clutch. The driver can manually select a lower forward gear, reverse, neutral, or park. Depending on the forward range selected, the transmission can also provide engine braking during deceleration. Also, an automatic transmission can remain engaged in a gear without stalling the engine while the vehicle is stopped.

Today's automatic transmissions have four to eight forward speeds. The most common units have five or six speeds. Different gear ratios are necessary because an engine develops relatively little power at low engine speeds. Without the aid of gears, the engine must be turning at a fairly high speed before it can deliver enough power to get the car moving. Through selection of the proper gear ratio, torque applied to the drive wheels can be multiplied. Increasing the number of forward speeds decreases fuel consumption and exhaust emissions.

Self-Shifting Manual Transmissions. Self-shifting manual transmissions are currently available on some passenger cars and are used in Formula One race cars. These transmissions work like typical manual transmissions except electronic or hydraulic actuators shift the gears and work the clutch (Figure 1-13). The driver shifts the gears using buttons or paddles on the steering wheel or a console-mounted shifter. Some units can also be operated in a fully automatic mode. It is important to realize that these are not automatic transmissions with manual controls!

These transmissions have computer-controlled actuators connected to the shift forks and a clutch actuator. The computer is programmed to shift the transmission automatically at the correct time, in the correct sequence, and to activate the clutch and allow for precise shifting when the driver selects the automatic mode.

The driver can also control gear changes by using the shifting mechanism. There is no gearshift linkage or cable; instead, a sensor at the shifter sends a signal to the computer. The computer, in turn, commands the actuators to engage or disengage the clutch and the gears with very fast response times. Engine torque is controlled during the shift by directly controlling the throttle or ignition/fuel injection system to provide smooth shifts.

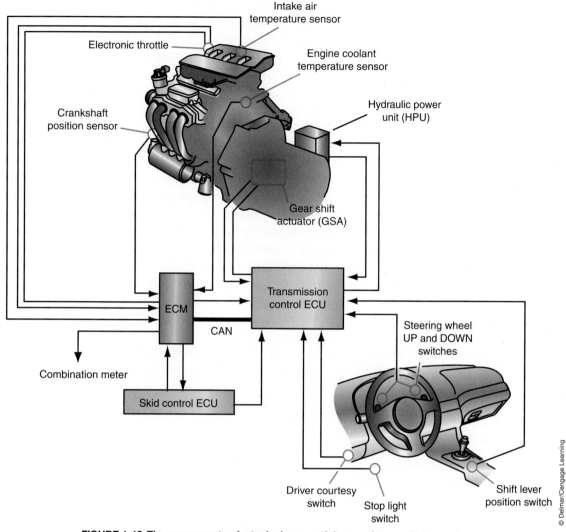

FIGURE 1-13 The components of a typical sequential manual transmission system.

Transmission Gears

A transmission has a **neutral** position that allows the engine to run without applying power to the drive wheels. Therefore, although there is input to the transmission when the vehicle is in neutral, there is no output from the transmission because the driving gears are not engaged to the output shaft.

In low or first gear, a small gear drives a large gear on another shaft. This reduces the speed of the larger gear but increases its turning force or torque and offers the proper gear ratio for starting movement or pulling heavy loads. First gear is primarily used to initiate movement. It has the lowest gear ratio of any gear in a transmission. It also allows for the most torque multiplication.

Wheel speed is decreased to about 1/12th of the speed of the engine. This combined with the torque increase, allows the vehicle to move at low speeds. It also provides a way to slow down the vehicle. By downshifting into a lower gear, the transmission slows the speed of the wheels and thereby helps stop the vehicle.

The ratio of second gear does not offer the same amount of torque multiplication as does first gear; however it does offer a substantial amount. Second gear is used when the need for torque multiplication is less than the need for vehicle speed and acceleration. Since the car is already in motion, less torque is needed to move the car.

First and second gears are the low gears in a typical transmission, whereas third, fourth, and fifth are the high gears.

Third gear allows for a further decrease in torque multiplication, while increasing vehicle speed and encouraging fuel economy. This gear may provide a **direct drive** (1:1) ratio, so that the amount of torque that enters the transmission is also the amount of torque that passes through and out of the transmission output shaft. This gear is used at cruising speeds and promotes fuel economy. While the car is in third gear, it lacks the performance characteristics of the lower gears.

The top gear or gears in most modern transmissions is an **overdrive** gear. Overdrive gears have ratios of less than 1:1. These ratios are achieved by using a small driving gear meshed with a smaller driven gear. Output speed is increased and torque is reduced. This allows the engine to run at a lower speed while the vehicle is traveling at highway speeds. This improves fuel economy, reduces exhaust emissions, and reduces operating noise while under light loads.

Through the use of an additional gear meshed with two other speed gears, the direction of the incoming torque is reversed and the transmission output shaft rotates in the opposite direction of the forward gears. Because reverse gear ratios are typically based on the drive and driven gears used for first gear, only low speeds can be obtained in reverse.

The transmission's gear ratios are further increased by the gear ratio of the ring and pinion gears in the drive axle assembly. Typical axle ratios are between 2.5 and 4.5:1. The final drive gear ratio is calculated by multiplying the transmission gear ratio by the final drive ratio. If a transmission is in first gear with a ratio of 3.63:1 and has a final drive ratio of 3.52:1, the **overall gear ratio** is 12.78:1. If third gear has a ratio of 1:1, using the same final drive ratio, the overall gear ratio is 3.52:1.

A manual transmission must be disconnected from the engine briefly each time the gears are shifted, by disengaging the clutch, but an automatic transmission does its gear shifting while it is engaged to the engine. This is accomplished through the use of constantly meshing planetary gears.

Planetary Gears

Planetary gearsets are commonly used in automatic transmissions, transfer cases, and many hybrid vehicle continuously variable transmissions (CVTs). A simple planetary gearset consists of three parts: a sun gear, a carrier with planetary pinions mounted to it, and an internally toothed ring gear or **annulus**. The **sun gear** is located in the center of the assembly (Figure 1-14). It meshes with the teeth of the planetary pinion gears. Planetary pinion gears are small gears fitted into a framework called the **planetary carrier**. The planetary

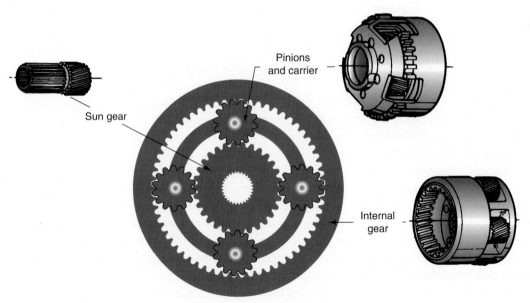

FIGURE 1-14 A planetary gearset.

© Delmar/Cengage Learning

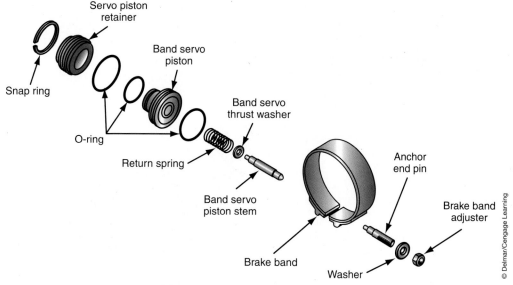

FIGURE 1-15 A typical band assembly.

© Delmar/Cengage Learning

carrier can be made of cast iron, aluminum, or steel plate and is designed with a shaft for each of the planetary pinion gears. (For simplicity, planetary pinion gears are called **planetary pinions**.)

The planetary pinions rotate on needle bearings positioned between the planetary carrier shaft and the planetary pinions. The carrier and pinions are considered one unit—the mid-size gear member. The planetary pinions surround the center axis of the sun gear and are surrounded by the annulus or ring gear, which is the largest part of the gearset. The ring gear acts like a band to hold the entire gearset together and provide great strength to the unit.

Many changes in speed and torque are possible with a planetary gearset. These depend on which parts are held stationary and which are driven. Any one of the three members can be used as the driving or input member. At the same time, another member might be kept from rotating and thus becomes the reaction, held, or stationary member. The third member then becomes the driven or output member. Depending on which member is the driver, which is held, and which is driven, either a torque increase (**underdrive**) or a speed increase (overdrive) is produced by the planetary gearset. Power transfer through a planetary gearset is only possible when one of the members is held at rest, or if two of the members are locked together.

Brake bands (Figure 1-15) or multiple-friction disc packs (Figure 1-16) attached to the individual gear carriers and shafts are hydraulically activated to direct power flow from the engine to any of the gears or to prevent one of the gears from rotating. One-way clutches are also used to prevent the rotation of a gear. These components allow for gear ratio changes and the reversing of power flow while the engine is running.

Planetary Gear Ratios

Calculating the gear ratios of a planetary gearset is much the same as calculating ratios of a spur gearset, except different formulas are used. These formulas are necessary because there are three sets of gears in mesh and some of the gears have internal teeth. However, the ratios are still based on the number of teeth on the drive and driven gears. In most cases, the carrier moves around a held gear. To make one complete rotation, the pinions of the carrier must mesh with all of the teeth of the held gear. Because the carrier links the drive and driven gears, the total number of teeth on the drive and driven gears must be used to calculate the gear ratio. To do this the number of teeth on the drive gear is added to the number of teeth on the

> A band is always used as a brake to hold or "brake" a drum. A multiple-friction disc pack can be used as a brake to stop or hold a component, or can be used as a drive clutch to engage or disengage two rotating components.

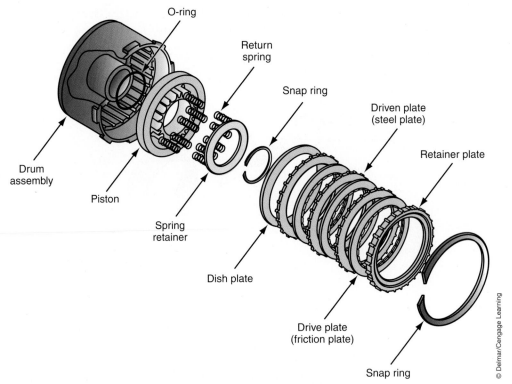

O-ring

Return
spring

Snap ring

Driven plate
(steel plate)

Retainer plate

Drum
assembly

Piston

Spring
retainer

Dish plate

Drive plate
(friction plate)

Snap ring

© Delmar/Cengage Learning

FIGURE 1-16 A typical multiple-disc clutch assembly.

driven gear. This sum is then divided by the number of teeth on the drive gear. Therefore, the formula for calculating the gear ratio of a planetary gearset is:

$$\text{gear ratio} = \frac{\text{drive gear} + \text{driven gear}}{\text{drive gear}}$$

For example, let's look at a gearset with a sun gear that has 25 teeth and a ring gear that has 75. In low or first gear, the sun gear is the drive gear. The ratio of this combination is calculated as follows:

$$\text{gear ratio} = \frac{\text{drive gear} + \text{driven gear}}{\text{drive gear}}$$

$$\text{gear ratio} = \frac{25 + 75}{25}$$

$$\text{gear ratio} = \frac{100}{25}; \text{ therefore, the gear ratio is 4:1.}$$

When the ring gear is the drive gear, such as in second gear operation, the ratio changes.

$$\text{gear ratio} = \frac{\text{drive gear} + \text{driven gear}}{\text{drive gear}}$$

$$\text{gear ratio} = \frac{25 + 75}{75}$$

$$\text{gear ratio} = \frac{100}{75}; \text{ therefore, the gear ratio is 1.33:1.}$$

When the gearset is in direct drive, two members of the gearset (typically the sun and ring gears) are locked together and rotate as a single member. The normal output gear is the carrier. No gear reduction takes place and the ratio is simply 1:1, also the direction of rotation is the same as the input.

When the transmission is in reverse, a simple drive to driven relationship exists. Using the sun and ring gears from the previous examples, the ratio of reverse gear would be 3:1 when the sun gear is the drive gear.

$$\text{gear ratio} = \frac{\text{driven}}{\text{drive}}$$

$$\frac{75}{25} = 3:1$$

Therefore, the gear ratio is 3:1 with output rotation in the opposite direction as the input.

When an overdrive gear is chosen, the sun gear is held, the carrier is the drive gear, and the ring gear is the driven. Since the carrier has no teeth and the pinion gears always act as idler gears, both are not related to the overall ratio of the planetary gearset. However, when the carrier is used as an input or output, it must be factored in when determining some gear ratios. To do this, use the sum of the sun and ring gear teeth. Therefore to calculate the ratio of this overdrive gear use this formula:

When the driven gear is the ring gear and the carrier is the drive gear

$$\text{gear ratio} = \frac{\text{driven gear}}{\text{drive gear (sun + ring)}}$$

$$\frac{75}{(75 + 25) = 100} = 0.75$$

Therefore, the gear ratio is 0.75:1

When both the ring and sun gears are the drive gears (this happens in overdrive), the ratio is determined by adding the number of teeth on the gears and using this formula:

When the driven gear is the ring gear

$$\text{gear ratio} = \frac{\text{driven gear}}{\text{drive gear}}$$

$$\text{gear ratio} = \frac{75}{100}; \text{ therefore, the gear ratio is 0.75:1.}$$

When the gearset is in direct drive, the members of the gearset are locked together and rotate as a single member. Because of this action, no gear reduction takes place and the ratio is simply 1:1.

When the transmission is in reverse, a simple drive to driven relationship exists. Using the sun and ring gears from the previous examples, the ratio of reverse gear would be 3:1 when the sun gear is the drive gear.

$$\frac{\text{driven gear}}{\text{drive gear}} = \text{gear ratio}$$

$$\frac{75}{25} = 3:1$$

Continuously Variable Transmission (CVT)

Pulleys can also be used to change speed and torque. Because they are typically connected by a drive belt, the direction of the driven pulley is the same as the direction of the drive pulley. However, the relationship of size has the same effect as the size of gears. When the drive pulley is the same diameter as the driven pulley, the two will rotate at the same speed and with the same torque. When the drive pulley is smaller than the driven pulley, the driven pulley will turn at a lower rotational speed but with greater torque. Likewise, when the drive pulley is larger than the driven pulley, the driven pulley will rotate faster but with

A **pulley** serves the same purpose as a gear except it is driven by a belt. Because pulleys do not mesh, the driven pulley rotates in the same direction as the drive pulley.

less torque. Pulleys are used with drive belts to operate some engine components such as generators, power steering pumps, and air conditioning compressors. Pulleys are also the basis for the operation of continuously variable ratio transmissions.

Continuously variable transmissions (CVTs) are found on some late-model cars and small SUVs. These transmissions automatically change torque and speed ranges without requiring a change in engine speed. The intent behind this transmission design is to keep the engine operating within a fixed speed range. This allows for improved fuel economy and decreased emission levels.

Basically, a CVT (Figure 1-17) is a transmission without fixed forward speeds. These transmissions are, however, fitted with a one-speed reverse gear. Many of these automatic-like transaxles do not have a torque converter; rather they use a manual transmission-type flywheel with a start clutch. Some CVTs offer the manual selection of a gear ratio. When a gear is selected by the driver, the pulleys in the transmission move to provide a fixed ratio. Normally, the size of the pulleys automatically responds to operating conditions.

One pulley is the driven member and the other is the driver. Each pulley has a moveable face and a fixed face. When the moveable face moves, the effective diameter of the pulley changes. The change in effective diameter changes the effective pulley (gear) ratio. A steel belt links the driven and drive pulleys.

To achieve a low pulley ratio, high hydraulic pressure works on the moveable face of the driven pulley to make it larger. In response to this high pressure, the pressure on the drive

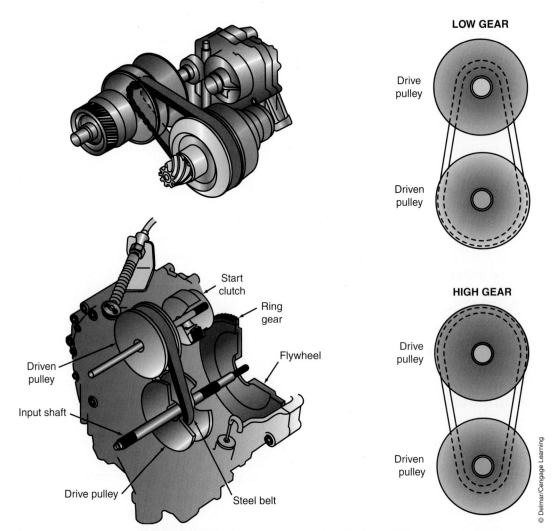

FIGURE 1-17 The basic construction of a belt-driven CVT.

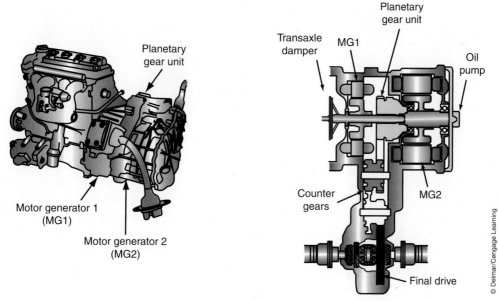

FIGURE 1-18 A hybrid CVT that uses the power output from two motor/generators and the engine to control the power applied to the drive wheels.

pulley is reduced. Since the belt links the two pulleys and proper belt tension is critical, the drive pulley reduces just enough to keep the proper tension on the belt. The increase of pressure at the driven pulley is proportional to the decrease of pressure at the drive pulley. The opposite is true for high pulley ratios. Low pressure causes the driven pulley to decrease in size, whereas high pressure increases the size of the drive pulley.

Different speed ratios are available any time the vehicle is moving. Since the size of the drive and driven pulleys can vary greatly, vehicle loads and speeds can be changed without changing the engine's speed. With this type transmission, attempts are made to keep the engine operating at its most efficient speed. This increases fuel economy and decreases emissions.

Some **hybrid electric vehicles (HEVs)** use a CVT that is not based on pulleys and belts. Rather electric motor/generators and the engine are directly connected to a planetary gearset (Figure 1-18). Variable drive ratios result from the amount of power applied to the different members of the planetary gearset.

Shift Control

The engine's power is transmitted to the transmission through a fluid coupler, called the **torque converter** (Figure 1-19). The torque converter drives an oil pump, which transmits fluid to a control-valve assembly. This valve assembly provides the hydraulic fluid needed to activate the various brake bands and multiple-friction disc packs. The **valve body** (Figure 1-20) controls the flow of the fluid throughout the transmission in response to the inputs it receives about engine and vehicle speeds and loads.

All new automatic transmissions rely on data received from electronic sensors and use an electronic control unit to operate solenoids in the valve body to shift gears. Older automatic transmissions rely on mechanical and vacuum signals, which determine when the transmission should shift.

Electronically controlled automatics have many advantages over the older designs, including more precise shifting of gears, greater fuel economy, and increased reliability. When shifting gears, older designs relied on the action of a cable-operated throttle valve or vacuum-controlled modulator to adjust fluid pressure in the transmission's lines to activate spring-loaded valves. This action took a small amount of time and delays in shifting and slippage resulted. The shifting of electronic transmissions is precisely controlled by a computer, which gathers information from many sensors, including those for throttle position,

A **valve body** is a complex assembly of valves and is mounted in the transmission's oil pan.

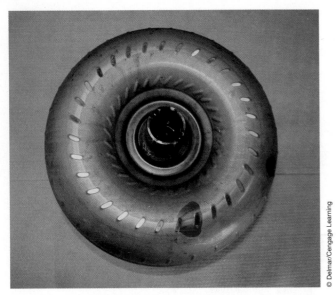

FIGURE 1-19 A torque converter.

FIGURE 1-20 A valve body for an automatic transmission.

temperature, engine load, and vehicle speed. The computer processes this information every few milliseconds and sends electrical signals to the shift solenoids, which control the shift valves of the valve body. Electrically controlled solenoids also match transmission line pressure to engine torque for better shift feel than mechanically controlled transmissions. The electronic controls also continuously monitor the automatic shifts and modify the hydraulic pressures to maintain a quality shift throughout the transmission's service life.

Torque Converter

An automatic transmission is connected to the engine by a fluid filled torque converter. The rotary motion of the engine's crankshaft is transferred from the **flexplate** (flywheel), through the torque converter, to the transmission (Figure 1-21). The flexplate is a thin disc that is designed to allow it to flex as the torque converter increases and decreases in size due to the

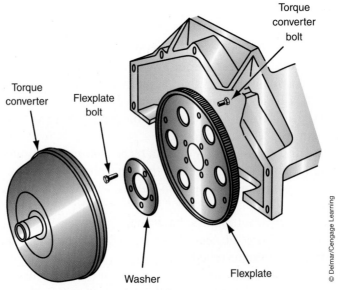

FIGURE 1-21 The torque converter is mounted on a flexplate which is attached to the engine's crankshaft.

changes in fluid heat and pressure inside the converter. The assembly of the torque converter mounted to the flexplate act to dampen engine crankshaft pulsations and add inertia to the crankshaft, just like the flywheel in a manual transmission equipped vehicles.

The rotary motion of the torque converter (T/C) is then delivered by the transmission to the **differential** and transferred by axle shafts to the tires, which push against the ground to move the car.

The torque converter (Figure 1-22) consists of an impeller, which is attached to the engine's crankshaft; a mating turbine, which is attached to the transmission's input shaft; and a torque-multiplying stator that is mounted between the turbine and the impeller.

The torque converter operates by hydraulic force generated by automatic transmission fluid. The torque converter changes or multiplies torque transmitted by the engine's crankshaft and directs it through the transmission. The torque converter also automatically engages and disengages engine power to the transmission in response to engine speed.

As the engine rotates, the impeller throws the transmission fluid at the blades of the turbine. The turbine spins in response to the force exerted by the moving fluid. Since the turbine is connected to the transmission's input shaft, engine output is transferred to the transmission.

While operating at normal idle speeds, the engine does not rotate fast enough to allow the impeller to throw fluid against the turbine with enough force to cause it to spin. This lack of hydraulic force enables the vehicle to stand still without stalling the engine when the gears

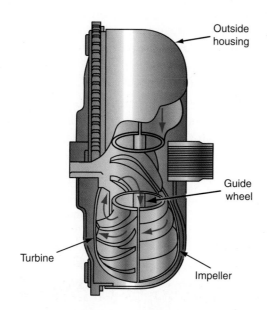

Outside housing

Guide wheel

Turbine

Impeller

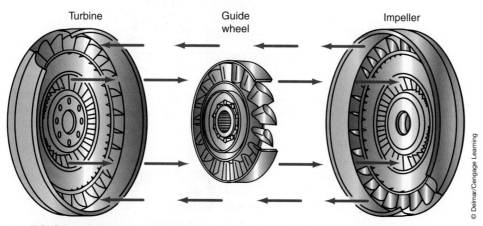

Turbine

Guide wheel

Impeller

FIGURE 1-22 The main components of a torque converter. Note: Some manufacturers refer to the impeller as the pump and the stator as a guide wheel.

are engaged. The hydraulic engaging and disengaging action of the impeller to the turbine performs a similar duty as the operation of the clutch in a manual transmission–equipped vehicle.

The clutch assembly of a vehicle equipped with a manual transmission is mounted to a **flywheel**, which is a large and heavy disc, attached to the rear end of the crankshaft. In addition to providing a friction surface and mounting for the clutch, the flywheel also dampens crankshaft vibrations, adds inertia to the rotation of the crankshaft, and serves as a large gear for the starter motor. Automatic transmissions do not require the use of a heavy flywheel; rather, the weight of the fluid-filled torque converter mounted to a lightweight flexplate is used for the same purposes.

DRIVELINE

Today's cars are designed to transfer the engine's power to either the front or rear wheels. In a FWD car, the transmission and driving axle are both located in one cast aluminum housing called a transaxle assembly (Figure 1-23). All of the driving components are located compactly at the front of the vehicle. One of the major advantages of front-wheel-drive is that the weight of the power train components is placed over the driving wheels, which provides for improved traction on slippery road surfaces.

RWD cars locate the powertrain, with the exception of the engine, beneath the body. The engine is mounted at the front of the chassis and the related powertrain components extend to the rear driving wheels. The transmission's internal parts are located within an aluminum or cast iron housing called the *transmission case assembly*. The driving axle is located at the rear of the vehicle, in a separate housing called the *rear axle assembly*. A driveshaft connects the output of the transmission to the rear axle.

Driveline for RWD Vehicles

The car's **driveshaft** (Figure 1-24) and its joints are often called the *driveline*. The driveline transmits torque from the transmission to the driving wheels. Rear-wheel-drive cars use a long driveshaft that connects the transmission to the rear axle. The engine and driveline of FWD cars are located between the front driving wheels.

A **flywheel** is a heavy circular component located on the rear of the crankshaft that keeps the crankshaft rotating during nonproductive strokes.

The **driveshaft** is an assembly of one or two universal joints connected to a shaft or tube; it is used to transmit power from the transmission to the differential. It is also called the *propeller shaft*.

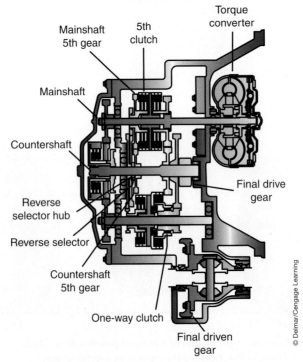

FIGURE 1-23 A cutaway of an automatic transaxle.

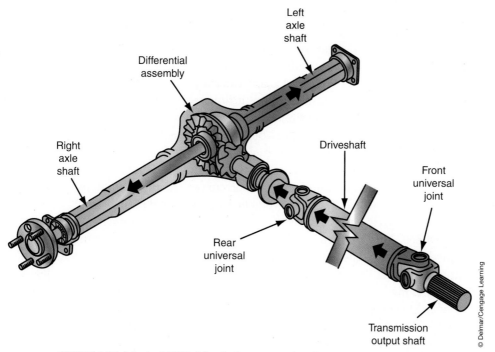

FIGURE 1-24 A typical RWD driveshaft connected to the rear axle assembly.

A driveshaft is a steel or aluminum tube normally consisting of two universal joints and a slip joint. The driveshaft transfers power from the transmission output shaft to the rear drive axle. A differential in the axle housing transmits the power to the rear wheels, which then move the car forward or backward.

Driveshafts differ in construction, lining, length, diameter, and type of slip joint. Typically, the driveshaft is connected at one end to the transmission and at the other end to the rear axle, which moves up and down with wheel and spring movement.

Driveshafts are typically made of hollow, thin-walled steel or aluminum tubing with the universal joint yokes welded at either end. **Universal joints (U-joints)** (Figure 1-25) allow the driveshaft to change angles in response to the movements of the rear axle assembly. As the angle of the driveshaft changes, its length must also change. The **slip yoke** normally fitted to the front universal joint allows the shaft to remain in place as its length requirements change.

Universal joints are most often called U-joints.

The **slip yoke** is a component with internal splines that slide on the transmission output shaft's external splines, allowing the driveline to adjust for variations in length as the rear axle assembly moves. Also called the slip joint.

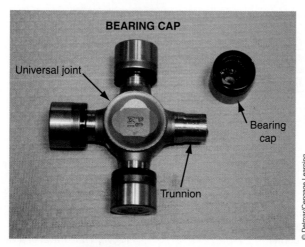

FIGURE 1-25 A universal joint.

The rear axle housing encloses the complete rear-wheel driving axle assembly. In addition to housing the parts, the axle housing also serves as a place to mount the vehicle's rear suspension and braking system. The rear axle assembly serves two other major functions: it changes the direction of the power flow 90 degrees and acts as the final gear reduction unit (Figure 1-26).

The rear axle consists of two sets of gears: the ring and pinion gearset and the differential gears. When torque leaves the transmission, it flows through the driveshaft to the ring and pinion gears, where it is further multiplied. By considering the engine's torque curve, the car's weight, and tire size, manufacturers are able to determine the best rear axle gear ratios for proper acceleration, hill-climbing ability, fuel economy, and noise level limits.

The primary purpose of the differential gearset is to allow a difference in rear driving wheel speed when the vehicle is rounding a corner or curve. The differential also transfers torque equally to both driving wheels when the vehicle is traveling in a straight line.

The torque on the ring gear is transmitted to the differential, where it is split and sent to the two driving wheels. When the car is traveling in a straight line, both driving wheels travel the same distance at the same speed. However, when the car is making a turn, the outer wheel must travel farther and faster than the inner wheel. When the car is steered into a 90-degree turn to the right and the inner wheel turns on a 30-foot radius, the inner wheel travels about 46 feet. The outer wheel, being nearly 5 feet from the inner wheel, travels nearly 58 feet (Figure 1-27).

Without some means for allowing the drive wheels to rotate at different speeds, the wheels would skid when the car was turning. This would result in little control during turns and in excessive tire wear. The differential eliminates these troubles by allowing the outer wheel to rotate faster as turns are made.

On RWD automobiles, the axle shafts or drive axles are located within the hollow horizontal tubes of the axle housing. The purpose of an axle shaft is to transmit the torque from the differential's side gears to the driving wheels. Axle shafts are heavy steel bars splined at the inner end to mesh with the axle side gear in the differential. The driving wheel is bolted to the wheel flange at the outer end of the axle shaft. The car's wheels rotate with the axles, which allow the car to move.

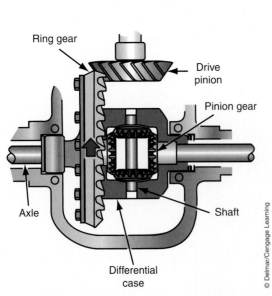

FIGURE 1-26 The gears in a differential assembly not only multiply the torque output but also transmit power "around the corner" to the drive wheels.

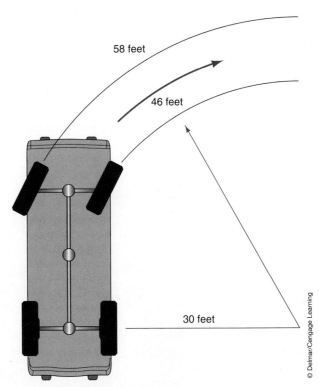

FIGURE 1-27 Travel of a vehicle's wheels when it is turning a corner.

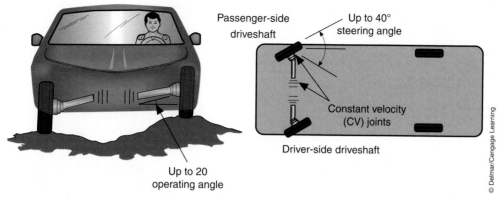

Passenger-side driveshaft

Up to 40° steering angle

Constant velocity (CV) joints

Driver-side driveshaft

Up to 20 operating angle

© Delmar/Cengage Learning

FIGURE 1-28 FWD driveshaft operating angles.

Driveline for FWD Vehicles

Front-wheel-drive vehicles do not need a driveshaft, as the engine and transaxle are located together above the drive wheels. Inside the transaxle is the differential, which has the same purpose as the differential in a rear drive axle. The drive axles of a FWD car extend from the differential through the sides of the transaxle to the drive wheels. Constant velocity (CV) joints are fitted to the axles to allow the axles to move with the car's suspension and steering systems (Figure 1-28).

Driveline for 4WD Vehicles

Four-wheel-drive vehicles can deliver power to all four wheels. There are many variations of 4WD found on today's vehicles. Four-wheel-drive vehicles designed for off-road use are normally RWD vehicles equipped with a **transfer case** (Figure 1-29), a front driveshaft, and front driving axles with a differential and axles. Same 4WD vehicles have three driveshafts. One short shaft connects the output from the transmission to the transfer case. The output from the transfer case is then sent to the front and rear drive axles through separate driveshafts. In many cases, the selection of RWD or 4WD is made by the driver.

Some high-performance vehicles are FWD models converted to 4WD. Normally, FWD cars are modified by adding a transfer case, a rear driveshaft, and a rear drive axle assembly with a differential. Although this is a typical modification, some cars are equipped with a center differential or **viscous clutch** in place of the transfer case. This differential allows the front and rear drive wheels to turn at different speeds and with different amounts of torque. In this type of setup, the engagement and disengagement of 4WD are not controlled by the driver.

Viscous clutches are not normally used in late-model vehicles. The resistance of the fluid is so great that it increases fuel consumption and emissions levels. Most of today's vehicles use an electromagnetic clutch in place of the viscous unit.

The transfer case is usually mounted to the side or rear of the transmission. When a driveshaft is not used to connect the transmission to the transfer case, a chain, or gear drive, within the transfer case, receives the engine's power from the transmission and transfers it to the driveshafts leading to the front and rear drive axles.

The transfer case itself is constructed like a transmission. It uses shift forks to select the operating mode, and splines, gears, shims, bearings, and other components found in transmissions. The housing is filled with lubricant that cuts friction on all moving parts. Seals hold the lubricant in the case and prevent leakage around shafts and yokes. Shims set up the proper clearance between the internal components and the case.

An electric switch or shift lever, located in the passenger compartment, controls the transfer case so that power is directed to the axles selected by the driver. Power can typically be directed to all four wheels, two wheels, or none of the wheels. On many vehicles, the driver can also select a low-speed range for extra torque while traveling in very adverse conditions.

A **transfer case** is a case or unit that has gears like a transmission. It is used to send power to only one set of wheels or both and may have the ability to lower the overall drive gear ratio when it is operating in the low range. A transfer case is normally mounted behind or next to the transmission.

A **viscous clutch** is basically a drum with some thick liquid in it that houses several closely fitted, thin steel discs. One set of discs is connected to the front wheels and the other set to the rear. The advantage of the viscous clutch is that it splits engine torque according to the needs of each axle.

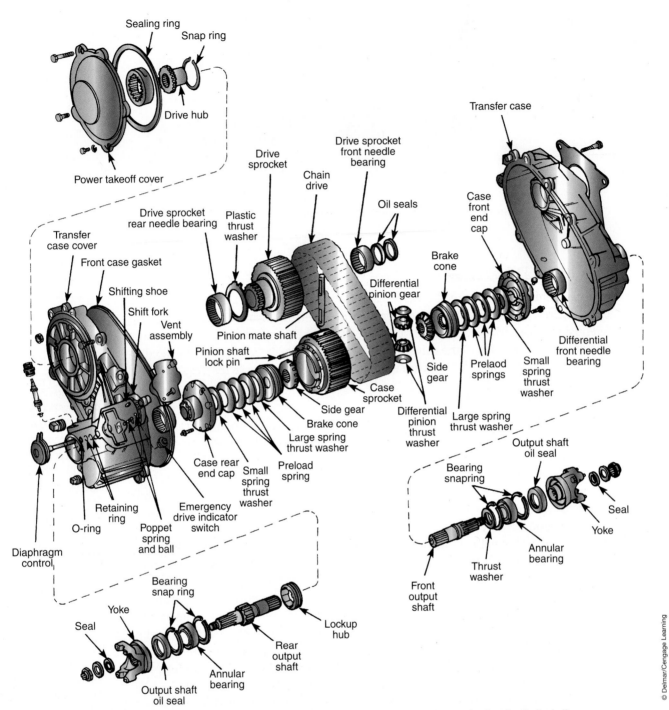

FIGURE 1-29 A four-wheel-drive transfer case with integral differential and cone brakes for limited slip.

The rear drive axle of a 4WD vehicle is identical to those used in two-wheel-drive vehicles. The front drive axle is also like a conventional rear axle, except that it is modified to allow the front wheels to steer. Further modifications are also necessary to adapt the axle to the vehicle's suspension system. The differential units housed in the axle assemblies are similar to those found in a RWD vehicle.

Differentials

On FWD cars, the differential unit is normally part of the transaxle assembly. On RWD cars, it is part of the rear axle assembly. A differential unit is located in a cast-iron casting, the differential case, and is attached to the center of the ring gear. Located inside the case are the differential

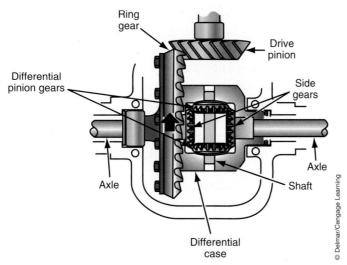

FIGURE 1-30 The components of a typical final drive unit.

Ring gear

Drive pinion

Differential pinion gears

Side gears

Axle

Axle

Shaft

Differential case

© Delmar/Cengage Learning

pinion shafts and gears and the axle side gears. The differential assembly revolves with the ring gear. Axle side gears are splined to the rear axle or front axle driveshafts (Figure 1-30).

When an automobile is moving straight ahead, both wheels are free to rotate. Engine power is applied to the pinion gear, which rotates the ring gear. Beveled pinion gears are carried around by the ring gear and rotate as one unit. Each axle receives the same power, so each wheel turns at the same speed.

When the car turns a sharp corner, only one wheel rotates freely. Torque still comes in on the pinion gear and rotates the ring gear, carrying the beveled pinions around with it. However, if one axle is held stationary and the beveled pinions are forced to rotate on their own axis and "walk" around their gear. The other side is forced to rotate because it is subjected to the turning force of the ring gear, which is transmitted through the pinions.

During one revolution of the ring gear, one gear makes two revolutions, one with the ring gear and another as the pinions "walk" around the other gear. As a result, when the drive wheels have unequal resistance applied to them, the wheel with the least resistance turns more revolutions. As one wheel turns faster, the other turns proportionally slower.

To prevent a loss of power on slippery surfaces, some differentials are fitted with clutches that lock the two axles together until the slippery spot is passed, at which point they are released. These differentials are referred as to **limited-slip** or **traction-lock** differentials. Rather than clutch discs, some units use cone clutches or gears to restrict the normal differential action and deliver torque to the nonslipping wheel. When the car is proceeding in a straight line, the differential gears are locked against rotation due to gear reaction. When the vehicle turns a corner or a curve, the differential pinion gears rotate around the differential pinion shaft. The differential pinion gears allow the inside axle shaft and driving wheels to slow down. On the opposite side, the pinion gears allow the outside wheels to accelerate. Both driving wheels resume equal speeds when the vehicle completes the corner or curve. This differential action improves vehicle handling and reduces driving wheel tire wear.

Types of Gears

An automobile relies on many different gears to transmit torque from one shaft to another. These shafts may operate in line, parallel to each other, or at an angle to each other. These different applications require a variety of gear designs, which vary primarily in the size and shape of the teeth.

In order for gears to mesh, they must have teeth of the same size and design. Meshed gears have at least one pair of teeth engaged at all times. Some gear designs allow for contact between more than one pair of teeth. Gears are normally classified by the type of teeth they have and by the surface on which the teeth are cut.

When manufacturers need additional strength in a planetary gearset, they will use straight-cut spur gears. For example, in GM's 4L80E transmission, a normal carrier will have helical gears but when the transmission is used in a motor home, it will have straight-cut gears.

Traction-lock and **limited-slip** are common names for differentials equipped with clutches that limit differential action to improve traction.

Automobiles use a variety of gear types to meet the demands of speed and torque. The most basic type of gear is the **spur gear**, which has its teeth parallel to and in alignment with the center of the gear. Early manual transmissions used straight-cut spur gears (Figure 1-31), which were easier to machine but were noisy and difficult to shift. Today these gears are used mainly for slow speeds to avoid excessive noise and vibration. They are commonly used in simple devices such as hand or powered winches.

Helical gears (Figure 1-32) are like spur gears except that their teeth have been twisted at an angle from the gear centerline. These gears get their name from being cut in a helix, which is a form of curve. This curve is more difficult to machine but is used because it reduces gear noise. Engagement of these gears begins at the tooth tip of one gear and rolls down the trailing edge of the teeth. This angular contact tends to cause side thrusts, which a bearing must absorb. However, helical spur gears are quieter in operation and have greater strength and durability than straight spur gears, simply because the contacting teeth are longer. Helical spur gears are widely used in transmissions today because they are quieter at high speeds and are durable.

Herringbone gears are actually double helical gears with teeth angles reversed on opposite sides. This causes the thrust produced by one side to be counterbalanced by the thrust produced by the other side. The two sets of teeth are often separated at the center by a narrow gap for better alignment and to prevent oil from being trapped at the apex. Herringbone gears are best suited for quiet, high-speed, low-thrust applications where heavy loads are applied. Large turbines and generators frequently use herringbone gears because of their durability.

Bevel gears are shaped like a cone with its top cut off. The teeth point inward toward the peak of the cone. These gears permit the power flow to "turn a corner." Spiral bevel gears (Figure 1-33) have their teeth cut obliquely on the angular faces of the gears. The most commonly used spiral beveled gearset is the ring and pinion gearset used in heavy truck differentials.

FIGURE 1-31 **Spur gears have teeth cut straight across the gear's edge and parallel to the shaft.**

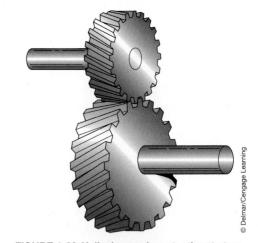

FIGURE 1-32 **Helical gears have teeth cut at an angle to the gear's axis of rotation.**

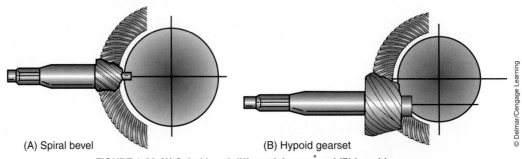

(A) Spiral bevel   (B) Hypoid gearset

FIGURE 1-33 **(A) Spiral bevel differential gears and (B) hypoid gears.**

The **hypoid gear** resembles the spiral bevel gear but the pinion drive gear is located below the center of the ring gear. Its teeth and general construction are the same as the spiral bevel gear. The most common use for hypoid gears is in modern differentials. Here, they allow for lower body styles by lowering the transmission driveshaft.

The worm gear is actually a screw capable of high speed reductions in a compact space. Its mating gear has teeth, which are curved at the tips to permit a greater contact area. Power is supplied to the worm gear, which drives the mating gear. Worm gears usually provide right-angle power flows.

Rack and pinion gears convert straight-line motion into rotary motion, and vice versa. Rack and pinion gears also change the angle of power flow with some degree of speed change. The teeth on the rack are cut straight across the shaft, while those on the pinion are cut like a spur gear. These gearsets can provide control of arbor presses and other devices where slow speed is involved. Rack and pinion gears also are commonly used in automotive steering boxes.

Internal gears have their teeth pointing inward and are commonly used in the planetary gear-set used in automatic transmissions and transfer cases. These are gearsets in which an outer ring gear has internal teeth that mate with teeth on smaller planetary gears with external teeth. These gears, in turn, mesh with a center or sun gear. Many changes in speed and torque are possible, depending on which parts are held stationary and which are driven. In a planetary gearset, one gear is normally the input, another is prevented from moving or held, and the third gear is the output gear. Planetary gears are widely used because each set is capable of more than one speed change. The gear load is spread over several gears, reducing stress and wear on any one gear.

BEARINGS

Gears are either securely attached to a shaft or designed to move freely on the shaft. The ease with which the gears rotate on the shaft or the shaft rotates with the gears partially determines the amount of power needed to rotate them. If they rotate with great difficulty due to high friction, much power is lost. High friction will also cause excessive wear to the gears and shaft. To reduce the friction, bearings are fitted to the shaft or gears.

The simplest type of bearing is a cylindrical hole formed in a piece of material, into which the shaft fits freely. The hole is usually lined with a brass or bronze lining, or **bushing**, which not only reduces the friction but also allows for easy replacement when wear occurs. Bushings usually have a tight fit in the hole in which they go.

Bushings are often referred to as plain bearings.

Ball or **roller bearings** are used wherever friction must be minimized (Figure 1-34). With these types of bearings, rolling friction replaces the sliding friction that occurs in plain bearings. Typically, two bearings are used to support a shaft instead of a single long bushing. Bearings have three purposes: they support a load, maintain alignment of a shaft, and reduce rotating friction.

Most bearings are capable of withstanding only loads that are perpendicular to the axis of the shaft. Such loads are called radial loads and bearings that carry them are called *radial* or *journal bearings*.

A *journal* is the area on a shaft that rides on the bearing.

To prevent the shaft from moving in the axial direction, shoulders or collars may be formed on it or secured to it. In most cases, bearings or washers are used to limit end thrusts. These bearings are called **thrust bearings** or washers.

Some thrust bearings look similar to a thick washer fitted with needle bearings on its flat surface. These are typically called *thrust needle bearings* or *thrust washers* and are commonly used in automatic transmissions (Figure 1-35). Automatic transmissions also use items that look like thrust needle bearings without the needle bearings. These are thrust washers and are used to control end clearance. These bearings are commonly referred to as *Torrington bearings*.

A *needle bearing* is a small roller bearing.

A single-row journal or radial ball bearing has an inner race made of a ring of case-hardened steel with a groove or track formed on its outer circumference for a number of

1. BALL BEARINGS:
Economical, widely used

Single row radial
for radial loads.

Single row angular contact
for radial and axial loads.

Axial thrust
for axial loads.

Double row
for heavier radial loads.

Self-aligning
for radial and axial loads,
large amounts of angular
misalignment.

2. ROLLER BEARINGS:
For shock, heavy load

Cylindrical
for relatively high speeds.

Needle
for low speeds,
intermittent loads.

Tapered
for heavy
axial (thrust) loading.

Spherical
for thrust loads and large
amounts of angular
misalignment.

Spherical thrust
to maintain alignment under
high thrust loads and
high speeds.

FIGURE 1-34 Many different types of roller and ball bearings are used in a vehicle's drive train.

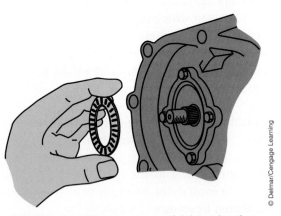

FIGURE 1-35 A Torrington-type axial thrust bearing.

hardened steel balls to run upon. The outer race is another ring, which has a track on its inner circumference. The balls fit between the two tracks and roll around in the tracks as either race turns. The balls are kept from rubbing against each other by some form of cage. These bearings can withstand radial loads and can also withstand a considerable amount of axial thrust. Therefore, they are often used as combined journal and thrust bearings.

A bearing designed to take only radial loads has only one of its races machined with a track for the balls. Other bearings are designed to take thrust loads in only one direction. If this type of bearing is installed incorrectly, the slightest amount of thrust will cause the bearing to come apart.

Another type of ball bearing uses two rows of balls. These are designed to withstand considerable amounts of radial and axial loads. Constructed as two single-row ball bearings joined together, these bearings are often used in rear axle assemblies.

Roller bearings are used wherever it is desirable to have a large bearing surface and low amounts of friction. Large bearing surfaces are needed in areas of extremely heavy loads. The rollers are usually fitted between a journal of a shaft and an outer race. As the shaft rotates, the rollers turn and rotate in the race. Tapered roller bearings (Figure 1-34) are commonly used in drive axle assemblies.

GENERAL MAINTENANCE

Most of the driveline needs little maintenance other than periodic oil changes. However, the fluid and filter of an automatic transmission should be replaced at a specific time and mileage intervals. Always follow the recommended service intervals of the manufacturer.

Excessive heat is the number one enemy of automatic transmissions. Transmission fluid transmits power from the engine and also cools and lubricates the internal parts of the transmission. If the fluid is too hot, it can no longer serve those functions and can lead to transmission failure. When the transmission is abused or malfunctions, the fluid may overheat, oxidize, and break down. This would cause damage to the internal seals, clutches, brakes, and other important parts of the transmission. The condition and level of the fluid should be checked on a frequent and regular basis.

Today's transmissions tend to run cooler because of the use of a converter clutch, which reduces the heat generated inside the T/C. Therefore the fluid in these transmissions will have a longer service life than the fluid used in transmissions without a T/C clutch.

AUTOMATIC TRANSMISSION FLUID

The fluid used in an automatic transmission's hydraulic system is called **automatic transmission fluid (ATF)**. ATF is a hydraulic oil designed specifically for automatic transmissions. Its primary purpose is to transmit pressure to activate the transmission's brakes and clutches. It also serves as a fluid connector between the engine and the transmission, removes heat from the transmission, and lubricates the transmission's moving parts.

Automatic transmission fluid can be a petroleum-based, partially synthetic, or totally synthetic oil. Nearly all domestic automobile manufacturers require a petroleum-based fluid in their automatic transmissions; however, some imported vehicles require the use of a partially synthetic fluid.

Automatic transmission fluid is a compound liquid that also includes special additives, which allow the lubricant to better meet the flow and friction requirements of an automatic transmission. ATF is normally dyed red, primarily to help distinguish it from engine oil when determining the source of fluid leaks.

Petroleum-based ATF typically has a clear red color and will darken when it is burnt or become milky when contaminated by water. Synthetic ATF is normally a darker red than petroleum-based fluid. Synthetic fluids tend to look and smell burnt after normal use; therefore the appearance and smell of these fluids is not a good indicator of the fluid's condition.

The various chemicals added to ATF ensure the durability and overall performance of the fluid. Zinc, phosphorous, and sulfur are commonly added to reduce friction. Detergent additives are added to ATF to help keep the transmission parts clean. Also added are dispersants, which keep contaminants suspended in the fluid so they can be trapped by the filter.

Automatic transmission fluid is often referred to as **ATF** and is at times simply called transmission oil.

All certified types and brands of ATF have been tested to ensure they meet the criteria set by the manufacturers. Some of these test standards apply to all types of ATF, such as oxidation resistance, corrosion and rust inhibition, flash and flame points, and resistance to foaming. Other standards are specific for a particular fluid rating or type.

Because some chemicals used in the composition of transmission fluid may adversely react with the fibers or synthetic materials used in the seals of the transmission, the compatibility of the fluids with specific transmissions is also tested. Incompatibility can result in external and internal transmission fluid leaks due to deterioration, swelling, or shrinking of the seals.

All brands of ATF are also tested for their **miscibility** or compatibility with other brands of ATF. Although the different brands of transmission fluids must meet the same set of standards, they may differ in their actual chemical composition and be incompatible with others. There should be no fluid separation, color change, or chemical breakdown when two different brands of ATF are mixed together. This level of compatibility is important to the service life of a transmission because it allows for the maintenance of fluid levels without the worry about switching to or mixing different brands of fluid.

Miscibility is the property of a fluid that allows it to mix and blend with other similar fluids.

Common Additives. The following is a summary of the common additives blended into the various types of ATF:

- *Antifoam agents:* These minimize foaming caused by the movement of the planetary gears and the fluid movement inside the torque converter.
- *Antiwear agents:* Zinc and other metals are blended into the fluid to minimize gear, bushing, and thrust washer wear.
- *Corrosion inhibitors:* These are added to prevent corrosion of the transmission's bushings and thrust washers, in addition to preventing fluid cooler line corrosion.
- *Dispersants:* These keep dirt suspended in the fluid, which helps prevent the buildup of sludge inside the transmission.
- *Friction modifiers:* Additives are blended into the base fluid to provide for an intentional amount of clutch and band slippage. This allows them to smoothly apply and release and prevents chatter.
- *Oxidation stabilizers:* To control oxidation of the fluid, additives are used to allow the ATF to absorb and dissipate heat. If the fluid is not designed to handle the high heat that is normally present in a transmission, the fluid will burn or oxidize. Oxidized fluid will cause severely damaged friction materials, clogged fluid filters, and sticky valves.
- *Seal swell controllers:* These additives control the swelling and hardening of the transmission's seals, while maintaining their normal pliability and tensile strength.
- *Viscosity index improvers:* These are blended into the fluid as an attempt to maintain the viscosity of fluid regardless of its temperature.

Recommended Applications

There are several ratings or types of ATF available; each is designed for a specific application. The different classifications of transmission fluid have resulted from the inclusion of new or different additives, which enhance the operation of the different transmission designs. Each automobile manufacturer specifies the proper type of ATF that should be used in their transmissions. Some manufacturers require a specially formulated fluid in their transmissions. Both the design of the transmission and the desired shift characteristics are considered when a specific ATF is chosen. Nearly, all of today's transmissions require Dextron/Mercon fluid.

Friction modifiers are additives that allow an oil to maintain its viscosity over a wide range of temperatures.

To reduce wear and friction inside a transmission, most commonly used transmission fluids are mixed with friction modifiers. Transmission fluids with these additives allow for the use of lower friction disc and brake application pressures, which in turn, provide for a very smooth-feeling shift. Transmission fluids without a friction modifier tend to have a firmer shift because higher friction disc and brake application pressures are required to avoid excessive slippage during gear changes.

If an ATF without friction modifiers is used in a transmission designed for friction-modified fluid, the service life of the transmission is not normally affected. However, firmer shifting will result and the driver may not welcome this change in shifting quality. Transmission durability is affected by using friction-modified fluid in a transmission designed for nonmodified fluids. This incorrect use of fluid will cause slippage, primarily when the vehicle is working under a load. Any amount of slippage can cause the clutches and brakes to wear prematurely. Also, because of the high heat generated by the slippage, the fluid may overheat and lose some of its lubrication and cooling qualities, which could cause the entire transmission to fail.

The formulation of an ATF must also be concerned with the viscosity of the fluid. Although the fluids are not selected according to viscosity numbers, proper flow characteristics of the fluid are important in operation of a transmission. If the viscosity is too low, the chances of internal and external leaks increases, parts can prematurely wear due to a lack of adequate lubrication, system pressure will be reduced, and overall control of the hydraulics will be less effective. If the viscosity is too high, internal friction will increase, resulting in an increase in the chance of building up sludge; hydraulic operation will be sluggish; and the transmission will require more engine power for operation.

The viscosity of a fluid is directly affected by temperature. Viscosity increases at low temperatures and decreases with higher temperatures. A transmission operates at many different temperatures. Since the fluid is used for lubricating as well as for shifting, it must be able to flow well at any temperature. ATF has a low viscosity but it viscous enough to prevent deterioration at higher temperatures. High-temperature performance is improved by many additives, such as friction modifiers. Low-temperature fluid flow of ATF is enhanced by mixing pour-point depressants into the base fluid. These additives are normally referred to as "viscosity index improvers."

The use of the correct ATF is critical to the operation and durability of automatic transmissions. We have already discussed the differences between friction-modified and non-friction-modified fluids. Certainly, this is one aspect of ATF that must be considered when putting fluid into a transmission. There are other considerations, too. Each type of ATF is specifically blended for a particular application. Each one has a unique mixture of additives that make it suitable for certain types of transmissions. Always fill a transmission with the fluid recommended by the transmission's manufacturer.

Through the years, as automatic transmissions have changed, so have their fluid requirements. In many cases, the development of new fluids has allowed automobile manufacturers to improve their transmission designs. In other cases, changes in the transmission have mandated the development of new fluid types.

Filtering

To trap dirt and metal particles from the circulating ATF, automatic transmissions have an oil filter (Figure 1-36), normally located inside the transmission case between the oil pump pickup and the bottom of the oil pan. If dirt, metal, and friction materials are allowed to circulate, they can cause valves to stick or cause premature transmission wear.

 WARNING: **Some automatic transmissions are equipped with an extra deep oil sump, which allows for improved cooling and increased capacity. If the transmissions is equipped with a deep pan, a special filter must be used that will reach into the bottom of the pan.**

Current automatic transmissions are fitted with one of three types of filter: a screen filter, paper filter, or felt filter. Screen filters use a fine wire mesh to trap the contaminants in the ATF. This type of filter is considered a surface filter because it traps the contaminants on its surface. As a screen filter traps dirt, metal, and other materials, fluid flow through the filter

Filter

© Delmar/Cengage Learning

FIGURE 1-36 Transmission fluid filters are attached to the transmission case by screws, bolts, retaining clips, and/or by the pickup tube.

Cellulose is a resin-based substance that serves as the basis for producing paper.

is reduced. The openings in the screen are relatively large so that only larger particles are trapped and small particles remain in the fluid. Although this does not remove all of the contaminants from the fluid, it does prevent quick clogging of the screen and helps to maintain normal fluid flow.

Another surface filter that is commonly used is the paper filter. This type of filter is more efficient than the screen type because it can trap smaller sized particles. Paper filters are typically made from a cellulose or Dacron fabric. Although this type of filter is quite efficient, it can quickly clog and cause a reduction in fluid flow through the transmission. Therefore, some older transmissions equipped with a paper filter have a bypass circuit, which allows contaminated fluid to circulate through the transmission if the filter becomes clogged and greatly restricts fluid flow.

The most commonly used filter in current model transmissions is the felt type. These are not surface filters; rather, they are considered depth filters because they trap contaminants within the filter and not just on its surface. Normally made from randomly spaced polyester materials, felt filters are able to trap both large and small particles and are less likely susceptible to clogging.

To protect vital transmission circuits and components, most transmissions are equipped with a secondary filter, located in a hydraulic passage, which helps keep dirt out of the pump, valves, and solenoids. Secondary filters are simply small screens fit into a passage or bore.

> **AUTHOR'S NOTE:** Customers should be made aware that transmission filters should be replaced at the mileage intervals recommended by the manufacturer. Failure to this will result in a loss of fluid flow and pressure, which can cause premature transmission failure.

SUMMARY

- The drivetrain has four primary purposes: to connect the engine's power to the drive wheels, to select different speed ratios, to provide a way to move the vehicle in reverse, and to control the power to the drive wheels for safe turning of the vehicle.
- The main components of the drivetrain are the transmission, differential, and drive axles.
- The rotating or turning effort of the engine's crankshaft is called engine torque.
- The amount of engine vacuum formed during the intake stroke is determined largely by the amount of load on the engine.
- Gears are used to apply torque to other rotating parts of the drivetrain and to multiply torque.
- Torque is calculated by multiplying the applied force by the distance from the center of the shaft to the point where the force is exerted.
- Gear ratios express the mathematical relationship, in size and number of teeth, of one gear to another.
- Transmissions offer various gear ratios through the meshing of various-sized gears.
- Reverse gear is accomplished by adding a gear to a two-gear set.
- Like manual transmissions, automatic transmissions provide various gear ratios, which match engine speed to the vehicle's speed. However, an automatic transmission is able to shift between gear ratios by itself and there is no need for a manually operated clutch to assist in the change of gears.
- A planetary gearset consists of a ring gear, a sun gear, and several planet gears, all mounted in the same plane.
- The ring gear has its teeth on its inner surface and the sun gear has its teeth on its outer surface. The planet gears are spaced evenly around the sun gear and mesh with both the ring and sun gears.
- By applying the engine's torque to one of the gears in a planetary gearset and preventing another member of the set from moving, torque multiplication, speed increase, or change of rotational direction is available on the third set of gears.
- Brake bands or multiple-friction disc packs attached to the individual gear carriers and shafts are hydraulically activated to direct engine power to any of the gears and to hold any of the gears from rotating. This allows gear ratio changes and the reversing of power flow while the engine is running.
- An oil pump in the transmission provides the hydraulic fluid needed to activate the various brake bands and clutch packs.
- The valve body controls the flow of the fluid throughout the transmission and acts on the vacuum and mechanical signals it receives about engine and vehicle speeds and loads.
- Most new automatic transmissions rely on data received from electronic sensors and use an electronic control unit to operate solenoids in the valve body to shift gears.
- In FWD cars, the transmission and drive axle is located in a single assembly called a transaxle. In RWD cars, the drive axle is connected to the transmission through a driveshaft.
- The driveshaft and its joints are called the driveline of the car.
- Universal joints allow the driveshaft to change angles in response to movements of the car's suspension and rear axle assembly.
- The rear axle housing encloses the entire rear-wheel driving axle assembly.
- The primary purpose of the differential is to allow a difference in driving wheel speed when the vehicle is rounding a corner or curve. The ring and pinion in the drive axle also multiplies the torque it receives from the transmission.
- On FWD cars, the differential is part of the transaxle assembly.

TERMS TO KNOW

All-wheel-drive (AWD)
Annulus
Automatic transmission fluid (ATF)
Ball bearing
Bevel gear
Bushing
Combustion
Constant velocity (CV) joint
Continuously variable transmission (CVT)
Differential
Direct drive
Driveshaft
Engine load
Engine torque
Flexplate
Flywheel
Four-wheel-drive (4WD)
Front-wheel-drive (FWD)
Gear ratio
Helical gear
Herringbone gear
Hypoid gear
Limited-slip
Miscibility
Neutral
Newton-meters (Nm)
Overdrive
Overall gear ratio
Planetary carrier
Planetary gear
Planetary pinions
Pounds-foot (lbs.-ft.)
Pulley
Rear-wheel-drive (RWD)
Ring gear
Roller bearing

TERMS TO KNOW
(continued)

Slip yoke

Speed reduction

Spur gear

Sun gear

Thrust bearing

Torque

Torque converter

Traction-lock

Transaxle

Transfer case

Underdrive

Universal joint (U-joint)

Vacuum

Valve body

Viscous clutch

SUMMARY

- The drive axles on FWD cars extend from the sides of the transaxle to the drive wheels. Constant velocity (CV) joints are fitted to the axles to allow the axles to move with the car's suspension.
- Four-wheel-drive vehicles typically use a transfer case, which relays engine torque to both a front and rear driving axle.
- An understanding of gears and bearings is the key to effective troubleshooting and repair of driveline components.
- Automatic transmission fluid (ATF) is hydraulic oil designed specifically for automatic transmissions. Its primary purpose is to transmit pressure to activate the transmission's bands and clutches. It also serves as a fluid coupling between the engine and the transmission, removes heat from the transmission, and lubricates the transmission's moving parts.
- Petroleum-based ATF typically has a clear red color and will darken when it is burnt or become milky when contaminated by water.
- To trap dirt and metal particles from the circulating ATF, automatic transmissions have an oil filter, normally located inside the transmission case between the oil pump pickup and the bottom of the oil pan.
- Current automatic transmissions are fitted with one of three types of filter: a screen filter, paper filter, or felt filter.

REVIEW QUESTIONS

Short-Answer Essays

1. What are the primary purposes of a vehicle's drivetrain?

2. Why does torque increase when a smaller gear drives a larger gear?

3. What mechanisms do most CVTs use to vary gear and speed ratios?

4. Why are transmissions equipped with many different forward gear ratios?

5. What is the primary difference between a transaxle and a transmission?

6. Why are U- and CV joints used in the driveline?

7. What does a differential unit do to the torque it receives?

8. What are the purposes of ATF?

9. What kind of gears are commonly used in today's automotive drivelines?

10. When are ball- or roller-type bearings used?

Fill in the Blanks

1. The main components of the drivetrain are the _____, _____, and _____ _____.

2. The rotating or turning effort of the engine's crankshaft is called _____ _____.

3. Gears are used to apply torque to other rotating parts of the drivetrain and to _____ torque.

4. Torque is calculated by multiplying the applied force by the _____ from the center of the _____ to the point where the force is exerted.

5. Current automatic transmissions are fitted with one of three types of filter: a _____ filter, _____ filter, or _____ filter.

6. Torque is measured in _____-_____ and _____-_____.

7. Gear ratios are determined by dividing the number of teeth on the _____ gear by the number of teeth on the _____ gear.

8. Reverse gear is accomplished by adding a _____ _____ to a two-gear set.

9. The torque converter assembly comprises _____, an _____, and a _____.

10. In FWD cars, the transmission and drive axle are located in a single assembly called a _____.

11. In RWD cars, the drive axle is connected to the transmission by a _____.

MULTIPLE CHOICE

1. Which of the following does NOT describe the purpose of a vehicle's drivetrain?
   A. It connects the engine's power to the drive wheels.
   B. It controls the power to the drive wheels for safe turning of the vehicle.
   C. It increases the power available to the drive wheels.
   D. It changes the direction of the power from the engine.

2. Which of the following can be used to multiply torque?
   A. pulleys
   B. gears
   C. levers
   D. all of the above

3. Which of the following is not a true statement about gear ratios?
   A. A high ratio is one that allows for great torque multiplication.
   B. A variety of gear ratios are needed to allow an engine to move a vehicle from a standing start to a cruising speed.
   C. Gear ratios express the mathematical relationship, according to the number of teeth, of one gear to another.
   D. Gear ratios express the size difference of two gears by stating the size ratio of the larger gear to the smaller gear.

4. The device used to transfer engine torque to the transmission is the:
   A. differential
   B. torque converter
   C. valve body
   D. flexplate

5. While discussing engine vacuum:
   Technician A says it always increases as engine load decreases.
   Technician B says it always increases with an increase of engine speed.
   Who is correct?
   A. A only
   B. B only
   C. Both A and B
   D. Neither A nor B

6. In an automatic transmission, the time to shift up or down is directly controlled by:
   A. engine load
   B. engine speed
   C. gear selector position
   D. operation of the torque converter

7. *Technician A* says all types of ATF become lighter when the friction modifiers are depleted.
   Technician B says petroleum-based ATF typically has a clear red color and will darken when it is burnt.
   Who is correct?
   A. A only
   B. B only
   C. Both A and B
   D. Neither A nor B

8. While discussing the purpose of a differential:
   Technician A says it allows for equal wheel speed while the vehicle is rounding a corner or curve.
   Technician B says the ring and pinion in the drive axle multiply the torque it receives from the transmission.
   Who is correct?
   A. A only
   B. B only
   C. Both A and B
   D. Neither A nor B

9. Which of the following parts is not within a FWD transaxle assembly?
   A. torque converter
   B. differential
   C. drive axles
   D. transmission

10. *Technician A* says 4WD vehicles typically use a transfer case to transfer engine torque to both a front- and a rear-driving axle.
    Technician B says 4WD vehicles normally have two transmissions, two driveshafts, and two differentials.
    Who is correct?
    A. A only
    B. B only
    C. Both A and B
    D. Neither A nor B

Chapter 2

DRIVETRAIN THEORY

UPON COMPLETION AND REVIEW OF THIS CHAPTER, YOU SHOULD BE ABLE TO:

- Describe how all matter exists.
- Explain what energy is and how energy is converted.
- Explain the forces that influence the design and operation of an automobile.
- Describe and apply Newton's laws of motion to an automobile.
- Define friction and describe how it can be minimized.

- Describe the various types of simple machines.
- Explain the difference between torque and horsepower.
- Explain Pascal's law and give examples of how it applies to an automobile.
- Explain how heat affects matter.
- Describe the origin and practical applications of electromagnetism.

INTRODUCTION

This chapter contains many of the things you have learned or will learn in other courses. It is not my intent to present this material in lieu of those other courses, but rather to emphasize those things that you will need to gain employment and be successful in an automotive career. Many of the facts presented in this chapter will be addressed again in this course in greater detail according to the topic covered. This chapter contains the theories that are the basis for the rest of the contents of this book. I highly recommend that you make sure you understand the contents of this chapter. Look at the end-of-chapter review questions and, if you have difficulty answering them, study the appropriate content in the chapter until you have a clear understanding and are able to answer the questions correctly.

MATTER

Matter is anything that occupies space. All matter exists as a gas, liquid, or solid. Gases and liquids are both considered *fluids* because they move or flow easily and easily respond to pressure. A gas has neither a shape nor volume of its own and tends to expand without limits. A liquid takes a shape and has volume. A solid is matter that does not flow.

AUTHOR'S NOTE: Knowing what sort of matter something is may not seem to matter, but it does! All matter reacts to changes in its environment and you need to know what will happen when certain things change.

Atoms and Molecules

All matter consists of countless tiny particles called **atoms**. An atom may be defined as the smallest particle of an element in which all the chemical characteristics of the element are present. Atoms are so small they cannot be seen with an electron microscope, which magnifies millions of times. A substance with only one type of atom is referred to as an **element**. Over one hundred elements are known to exist at present. Of the known elements, 92 occur naturally; the remaining elements have been manufactured in laboratories.

Small, positively charged particles called *protons* are located in the center, or nucleus, of each atom. In most atoms, neutrons are also located in the nucleus. Neutrons have no electrical charge, but they add weight to the atom. The positively charged protons tend to repel each other, and this repelling force could destroy the nucleus. The presence of the neutrons with the protons in the nucleus cancels the repelling action of the protons and keeps the nucleus together. Electrons are small, very light particles with a negative electrical charge. Electrons move in orbits around the nucleus of an atom.

A proton is approximately 1,840 times heavier than an electron, and this makes the electron much easier to move. Because the electrons are orbiting around the nucleus, centrifugal force tends to move the electrons away from the nucleus. However, the attraction between the positively charged protons and the negatively charged electrons holds the electrons in their orbits. The atoms of the different elements have different numbers of protons, electrons, and neutrons. Some of the lighter elements have the same number of protons and neutrons in the nucleus, but many of the heavier elements have more neutrons than protons.

The simplest atom is the hydrogen (H) atom, which has one proton in the nucleus and one electron orbiting around the nucleus (Figure 2-1). The nucleus of a copper (CU) atom contains 29 protons and 34 neutrons, while 29 electrons orbit in 4 different rings around the nucleus. Since 2, 8, and 18 electrons are the maximum number of electrons that can be on the first 3 electron rings next to the nucleus, the fourth ring must have 1 electron (Figure 2-2). The outer ring of an atom is called the *valence ring*, and the number of electrons on this ring determines the electrical characteristics of the element. Elements are listed on the atomic scale, or periodic chart, according to their number of protons and electrons. For example, hydrogen is number 1 on this scale, and copper is number 29.

Water is a compound that contains oxygen and hydrogen atoms. The chemical symbol for water is H_2O. This chemical symbol indicates that each molecule of water contains two atoms of hydrogen and one oxygen atom.

> The makeup of an atom is also an explanation of how electricity works.

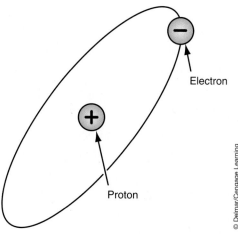

FIGURE 2-1 A hydrogen atom.

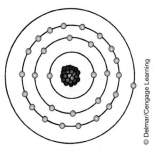

FIGURE 2-2 A copper atom.

© Delmar/Cengage Learning

States of Matter

The particles of a solid are held together in a rigid structure. When a solid dissolves into a liquid, its particles break away from this structure and mix evenly in the liquid, forming a **solution**. When they are heated, most liquids **evaporate**. This means that the atoms or molecules of which they are made break free from the body of the liquid to become gas particles. When all of the liquid in a solution has evaporated, the solid is left behind. The particles of the solid normally arrange in a structure called a *crystal*.

Absorption and Adsorption. Not all solids will dissolve in a liquid; rather the liquid will be either absorbed or adsorbed. The action of a sponge serves as the best example of absorption. When a dry sponge is put into water, the water is absorbed by the sponge. The sponge does not dissolve, the water merely penetrates into the sponge and the sponge becomes filled with water. There is no change to the atomic structure of the sponge, nor does the structure of the water change.

If you take a glass and put it into water, the glass does not absorb the water. The glass, however, still gets wet, as a thin layer of water adheres to the glass. This is adsorption. Materials that *absorb* fluids are **permeable** substances. **Impermeable** substances *adsorb* fluids. Some materials are impermeable to most fluids, while others are impermeable to just a few.

Since automatic transmissions rely on hydraulics, or liquids, the use of **impermeable** seals is critical to their operation.

ENERGY

Energy may be defined as the ability to do work. Since all matter consists of atoms and molecules that are in constant motion, all matter has energy. Energy is not matter, but it affects the behavior of matter. Everything that happens requires energy, and energy comes in many forms.

Each form of energy can change into other forms. However, the total amount of energy never changes; it can only be transferred from one form to another, not created or destroyed. This is known as the principle of conservation of energy.

Kinetic and Potential Energy

When energy is released to do work, it is called **kinetic energy**. This type of energy may also be referred to as *energy in motion*. Stored energy may be called **potential energy**.

There are many automobile components and systems that have potential energy and, at times, kinetic energy. The ignition system is a source for high electrical energy. The heart of the ignition system is the ignition coil, which has much potential energy when it has current flowing through it. When it is time to fire a spark plug, current flow is stopped and energy is released and becomes kinetic energy as it creates a spark across the gap of a spark plug.

Hybrid and electric vehicles capture the **kinetic energy** of a vehicle to recharge their batteries through a process called *regenerative braking*.

Energy Conversion

Energy conversion occurs when one form of energy is changed to another form. Since energy is not always in the desired form, it must be converted to a form you can use. Some of the most common automotive energy conversions are discussed here.

Chemical to Thermal Energy. Chemical energy in gasoline or diesel fuel is converted to thermal energy when the fuel burns in the engine cylinders.

Chemical to Electrical Energy. The chemical energy in a battery (Figure 2-3) is converted to electrical energy, enabling it to power many of the accessories on an automobile.

Electrical to Mechanical Energy. In the automobile, the battery supplies electrical energy to the starting motor, and this motor converts the electrical energy to mechanical energy to crank the engine.

Thermal to Mechanical Energy. The thermal energy that results from the burning of fuel in the engine is converted to mechanical energy (Figure 2-4), which is used to move the vehicle.

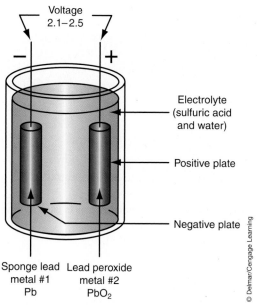

FIGURE 2-3 The chemical energy in a battery is converted to electrical energy.

Voltage 2.1–2.5

−   +

Electrolyte (sulfuric acid and water)

Positive plate

Negative plate

Sponge lead metal #1 Pb

Lead peroxide metal #2 PbO₂

© Delmar/Cengage Learning

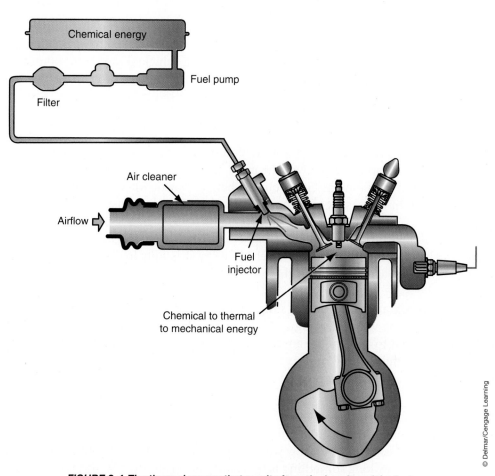

FIGURE 2-4 The thermal energy that results from the burning of the fuel in the engine is converted to mechanical energy.

Chemical energy

Fuel pump

Filter

Air cleaner

Airflow

Fuel injector

Chemical to thermal to mechanical energy

© Delmar/Cengage Learning

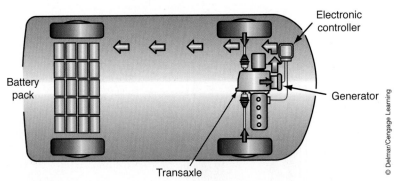

DECELERATION

Electronic controller

Battery pack

Generator

Transaxle

© Delmar/Cengage Learning

FIGURE 2-5 During deceleration, kinetic energy is used to drive a generator, which recharges the batteries.

A BIT OF HISTORY

Albert Einstein, in his theory of relativity, proposed an equation for energy that many have heard of but few understand. He stated that energy equals mass times the speed of light squared or $E = M \times C^2$.

Mechanical to Electrical Energy. The generator is driven by mechanical energy from the engine. The generator converts this energy to electrical energy, which powers the electrical accessories on the vehicle and recharges the battery.

Electrical to Radiant Energy. Radiant energy is light energy. In the automobile, electrical energy is converted to thermal energy, which heats up the inside of light bulbs so they illuminate and release radiant energy.

Kinetic to Thermal Energy. To stop a vehicle, the brake system must change the kinetic energy of the moving vehicle to kinetic and static thermal or heat energy. This is the result of friction, which will be discussed later in this chapter.

Kinetic to Mechanical to Electrical Energy. Hybrid vehicles have a system, called regenerative braking, that uses the energy of the moving vehicle (kinetic) to rotate a generator. The mechanical energy used to operate the generator is used to provide electrical energy to charge the batteries (Figure 2-5) or power the electric drive motor.

Mass and Weight

Mass is the amount of matter in an object. **Weight** is a force and is measured in pounds or kilograms. Gravitational force gives the mass its weight. As an example, a spacecraft can weigh five hundred tons (one million pounds) here on Earth where it is affected by Earth's gravitational pull. In outer space, beyond Earth's gravity and atmosphere, the spacecraft is nearly weightless (Figure 2-6).

Automobile specifications list the weight of a vehicle primarily in two ways: **Gross weight**, which is the total weight of the vehicle when it is fully loaded with passengers and cargo, **Curb weight**, which is the weight of the vehicle when it is not loaded with passengers or cargo.

To convert kilograms into pounds, simply multiply the weight in kilograms by 2.2046. For example, if something weighs 5 kilograms, it weighs about 11 pounds ($5 \times 2.2046 = 11.023$ pounds). If you want to express the answer in pounds and ounces, convert the 0.023 pounds into ounces. Since there are 16 ounces in a pound, multiply 16 by 0.023 ($16 \times 0.023 = 0.368$ ounces). Therefore, 5 kilograms is equal to 11 pounds, 0.368 ounces.

Size

The size of something is related to its mass. The size of an object defines how much space it occupies. Size dimensions are typically stated in terms of length, width, and height. Length is a measurement of how long something is from one end to another. Width is a measurement of how wide something is from one side to another. Obviously, height is a measurement of the

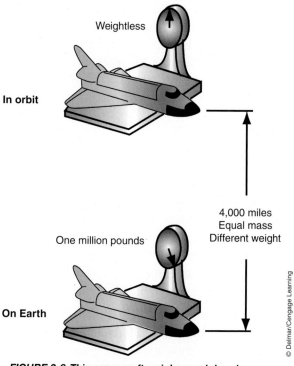

Weightless

In orbit

4,000 miles
Equal mass
Different weight

One million pounds

On Earth

© Delmar/Cengage Learning

FIGURE 2-6 This spacecraft weighs much less in
space than on earth although its mass is not changed.

distance from something's bottom to its top. All three of these dimensions are measured in inches, feet, yards, and miles in the imperial system and meters in the metric system.

To convert a meter into feet, multiply the number of meters by 3.281. If you want to convert the feet into inches, simply multiply the number of feet by 12. For example, to convert 0.01 mm into inches, begin by converting 0.01 mm into meters. Because 1 mm is equal to 0.001 meters, you need to multiply 0.01 by 0.001 ($0.001 \times 0.01 = 0.00001$). Then multiply 0.00001 meters by 3.281 ($0.00001 \times 3.281 = 0.00003281$ feet). Now convert feet into inches by multiplying by 12 ($0.00003281 \times 12 = 0.00039372$ inches).

An easier way to do this would be by using the conversion factor that equates 1 mm to 0.03937 inches. To use this conversion factor, multiply 0.01 mm by 0.03937 ($0.01 \times 0.03937 = 0.0003937$ inches).

Sometimes distance measurements are made with a rule that has fractional rather than decimal increments. Most automotive specifications are given decimally; therefore, fractions need to be converted into decimals. It is also easier to add and subtract dimensions if they are expressed in decimal form rather than in fractions. For example, assume you want to find the rolling circumference of a tire with a diameter of $20\frac{3}{8}$ inches. The distance around the tire is the *circumference*. It is equal to the diameter multiplied by a constant called pi (π). Pi is equal to approximately 3.14; therefore the circumference of the tire is equal to $20\frac{3}{8}$ inches multiplied by 3.14. This calculation is much easier if you convert the $20\frac{3}{8}$ inches into a whole number and a decimal. To convert $\frac{3}{8}$ to a decimal, divide 3 by 8 ($3 \div 8 = 0.375$). Therefore, the diameter of the tire is 20.375 inches. Now multiply the diameter by π ($20.375 \times 3.14 = 63.98$). The circumference of the tire is nearly 64 inches.

FORCE

A **force** is a push or pull, and can be large or small. Force can be applied to objects by direct contact or from a distance. Gravity and electromagnetism are examples of forces that are applied from a distance. Forces can be applied from any direction and with any intensity. For example, if a pulling force on an object is twice that of the pushing force, the object will be pulled at one-half of the pulling force. When two or more forces are applied to an object, the combined force is called the *resultant*. The resultant is the sum of the size and direction of the forces. For example, when a mass is suspended by two lengths of wire, each wire should carry half the weight of the mass. If you move the attachment of the wires so they are at an angle to the mass, each wire now carries more force. Each wire carries the force of the mass plus a force that pulls against the other wire.

Automotive Forces

When a vehicle is sitting still, gravity exerts a downward force on the vehicle. The ground exerts an equal and opposite upward force and supports the vehicle. When the engine is running and its power output transferred to the vehicle's drive wheels, the wheels exert a force against the ground in a horizontal direction. This force causes the vehicle to move but is opposed by the mass of the vehicle. To move the vehicle faster, the force supplied by the wheels must increase beyond the opposing forces. When the vehicle does move faster, it pushes against the air as it travels. This becomes a growing opposing force and the force at the drive wheels must overcome the force in order for the vehicle to increase speed. After the vehicle has achieved the desired speed, no additional force is required at the drive wheels.

Force—Balanced and Unbalanced. When the applied forces are balanced and there is no overall resultant force, the object is said to be in **equilibrium**. An object sitting on a solid flat surface is in equilibrium, because its weight is supported by the surface and there is no resultant force. If the surface is put on an angle, the object will tend to slide down the surface. If the surface is at a slight angle, the downward force will cause the object to slowly slide down the surface. If the surface is at a severe angle, the downward force will cause the object to quickly slide down the slope. In both cases, the surface is still supplying the force needed to support the object but the pull of gravity is greater and the resultant force causes the object to slide down the slope.

Turning Forces. Forces can cause rotation as well as straight-line motion. A force acting on an object that is free to rotate will have a turning effect, or turning force. This force is equal to the size of the force multiplied by the distance of the force from the turning point around which it acts.

Centrifugal/Centripetal Forces

When an object moves in a circle, its direction is continuously changing. All changes in direction require a force. The forces required to maintain circular motion are called *centripetal* and *centrifugal* force. The size of these forces depends on the size of the circle and the mass and speed of the object.

Centripetal force tends to pull the object toward the center of the circle; whereas centrifugal force tends to push the object away from the center. The centripetal force that keeps an object whirling around on the end of a string is caused by **tension** in the string. If the string breaks, there is no longer string tension and centripetal force and the object will fly off in a straight line because of the centrifugal force on it. Gravity is the centripetal force that keeps the planets orbiting around the sun. Without this centripetal force, Earth would move in a straight line through space.

Pressure

Pressure is a force applied against an object. It is measured in units of force per unit of surface area (pounds per square inch or kilograms per square centimeter). Mathematically, pressure is equal to the applied force divided by the area over which the force acts. Consider two 10-pound weights sitting on a table; one occupies an area of 1 square inch and the other an area of 4 square inches. The pressure exerted by the first weight would be 10 pounds per square inch (10 psi). The other weight, although it weighs the same, will exert only 2.5 psi (10 pounds per 4 square inches = 10 ÷ 4 = 2.5). This illustrates an important concept: a force acting over a large area will exert less **pressure** than the same force acting over a small area.

Since pressure is a force, all principles of force apply to pressure. If more than one pressure is applied to an object, the object will respond to the resultant force. Also, all matter (liquids, gases, and solids) will tend to move from an area of high pressure to a low-pressure area.

MOTION

If the forces on an object do not cancel each other out, they will change the motion of the object. The object's speed, direction of motion, or both will change. The greater the mass of an object, the greater the force needed to change its motion. This resistance to change in motion is called **inertia**. Inertia is the tendency of an object at rest to remain at rest, or the tendency of an object in motion to stay in motion. The inertia of an object at rest is called *static inertia*, whereas *dynamic inertia* refers to the inertia of an object in motion. Inertia exists in liquids, solids, and gases. When you push and move a parked vehicle, you overcome the static inertia of the vehicle. If you catch a ball in motion, you overcome the dynamic inertia of the ball.

When a force overcomes static inertia and moves an object, the object gains momentum. **Momentum** is the product of an object's weight and its speed. Momentum is a type of mechanical energy. An object loses momentum if another force overcomes the dynamic inertia of the moving object.

Rates

Speed is the distance an object travels in a set amount of time. It is calculated by dividing distance covered by time taken. We refer to the speed of a vehicle in terms of miles per hour (mph) or kilometers per hour (km/h). **Velocity** is the speed of an object in a particular direction. **Acceleration**, which only occurs when a force is applied, is the rate of increase in speed. Acceleration is calculated by dividing the change in speed by the time it took for that change. **Deceleration** is the reverse of acceleration, as it is the rate of a decrease in speed.

Ratios. Often automotive features are expressed as ratios. A ratio expresses the relationship between two things. If something is twice as large as another, there is a ratio of 2:1. Sometimes, ratios are used to compare the movement of an object. For example, if a one-inch movement by something causes something else to move two inches, there is a travel ratio of 1:2.

Newton's Laws of Motion

How forces change the motion of objects was first explained by Sir Isaac Newton. These explanations are known as Newton's Laws. Newton's first law of motion is called the law of inertia. It states that an object at rest tends to remain at rest and an object in motion tends to remain in motion, unless some force acts on it. When a car is parked on a level street, it remains stationary unless it is driven or pushed.

Newton's second law states when a force acts on an object, the motion of the object will change. This change in motion is equal to the size of the force divided by the mass of the object on which it acts. Trucks have a greater mass than cars. Since a large mass requires a larger force to produce a given acceleration, a truck needs a larger engine than a car.

Newton's third law says that for every action there is an equal and opposite reaction (Figure 2-7). A practical application of this law occurs when the wheel on a vehicle strikes a bump in the road surface. This action drives the wheel and suspension upward with a certain force, and a specific amount of energy is stored in the spring. After this action occurs, the spring forces the wheel and suspension downward with a force equal to the initial upward force caused by the bump.

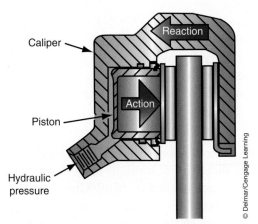

Caliper

Reaction

Piston

Action

Hydraulic pressure

© Delmar/Cengage Learning

FIGURE 2-7 The action of a disc brake caliper illustrates Newton's third law—for every action there is an equal but opposite reaction.

Friction

Friction is a force that slows or prevents motion of two moving objects or surfaces that touch. Friction may occur in solids, liquids, or gases. It is the joining or bonding of the atoms at each of the surfaces that causes the friction. When you attempt to pull an object across a surface, the object will not move until these bonds have been overcome. Smooth surfaces produce little friction; therefore, only a small amount of force is needed to break the bonds between the atoms. Rougher surfaces produce a larger friction force because stronger bonds are made between the two surfaces. To move an object over a rough surface, such as sandpaper, a great amount of force is required.

Lubrication. Friction can be reduced in two main ways: by lubrication or by the use of rollers. The presence of oil or another fluid between two surfaces keeps the surfaces apart. Because fluids (liquids and gases) flow, they allow movement between surfaces. The fluid keeps the surfaces apart, allowing them to move smoothly past one another (Figure 2-8).

Rollers. Rollers placed between two surfaces keep the surfaces apart. An object placed on rollers will move smoothly if pushed or pulled. Rollers actually use friction to grip the surfaces and produce rotation. Instead of sliding against one another, the surfaces produce turning forces, which cause each roller to spin. This leaves very little friction to oppose motion. *Bearings* are a type of roller used to reduce the friction between moving parts such as a wheel and its axle. As the wheel turns on the axle, the balls in the bearing roll around inside it, drastically reducing the friction between the wheel and axle.

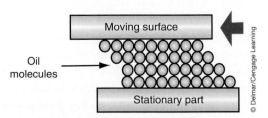

Moving surface

Oil molecules

Stationary part

© Delmar/Cengage Learning

FIGURE 2-8 Oil between two surfaces keeps the surfaces apart and reduces the friction between the surfaces as one or both move.

Work = Force x Distance

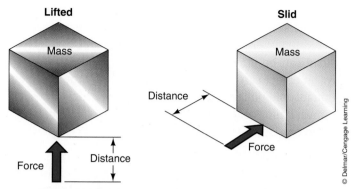

FIGURE 2-9 When work is performed; a mass is slid or lifted a certain distance.

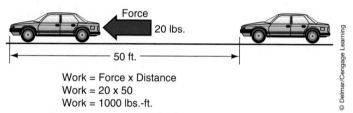

Work = Force x Distance
Work = 20 x 50
Work = 1000 lbs.-ft.

FIGURE 2-10 Work is measured in pounds per foot or as commonly expressed foot-pounds.

WORK

When a force moves a certain mass a specific distance, **work** is done. When work is accomplished, the mass may be lifted or slid on a surface against a resistance or opposing force (Figure 2-9). Work is equal to the applied force multiplied by the distance the object moved (force × distance = work) and is measured in pounds-foot (Figure 2-10), watts, or Newton-meters. For example, if a force moves a 3,000-pound car 50 feet, 150,000 pounds-foot of work was done.

During work, a force acts on an object to start, stop, or change the direction of the object. It is possible to apply a force to an object and not move the object. For example, you may push with all your strength on a car stuck in a ditch and not move the car. Under this condition, no work is done. Work is only accomplished when an object is started, stopped, or redirected by a force.

Simple Machines

A machine is any device that can be used to transmit a force and, in doing so, change the amount of force or its direction. The force applied to a machine is called the *effort*, while the force it overcomes is called the **load**. The effort is often smaller than the load, because a small effort can overcome a heavy load if the effort is moved a larger distance. The machine is then said to give a mechanical advantage. Although the effort will be smaller when using a machine, the amount of work done, or energy used, will be equal to or greater than that without the machine.

Inclined Plane. The force required to drag an object up a slope is less than that required to lift it vertically. However, the overall distance moved by the object is greater when pulled up the slope than if it were lifted vertically. A screw is like an inclined plane wrapped around a shaft. The force that turns the screw is converted to a larger one, which moves a shorter distance and drives the screw in.

Pulleys. A *pulley* is a wheel with a grooved rim in which a rope, belt, or chain runs to raise something by pulling on the other end of the rope, belt, or chain. A simple pulley changes the direction of a force but not its size. Also, the distance the force moves does not change. By using several pulleys connected together as a block and tackle, the size of the force can be changed, too, so that a heavy load can be lifted using a small force. With a double pulley, the applied force required to move an object can be reduced by one-half but the distance the force must be moved is doubled. A quadruple pulley can reduce the force by four times but the distance will be increased by four times. Pulleys of different sizes can change the required amount of applied force, as well as the speed or distance the pulley must travel to accomplish work (Figure 2-11).

Levers. A **lever** is a device made up of a bar turning about a fixed pivot point, called the *fulcrum*, that uses a force applied at one point to move a mass on the other end of the bar. Types of levers are divided into classes. In a class 1 lever, the fulcrum is between the effort and the load (Figure 2-12). The load is larger than the effort, but it moves through a smaller distance. A pair of pliers is an example of a class 1 lever.

In a class 2 lever, the load is between the fulcrum and effort. Here again, the load is greater than the effort and moves through a smaller distance (Figure 2-13). In a class 3 lever, the effort is between the fulcrum and the load. In this case, the load is less than the effort but it moves through a greater distance.

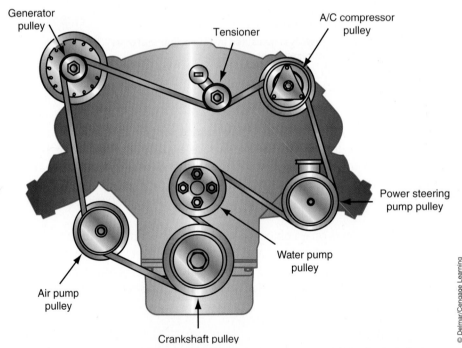

FIGURE 2-11 Pulleys of different sizes can change the amount of required applied force and the speed or distance the pulley needs to travel to accomplish work.

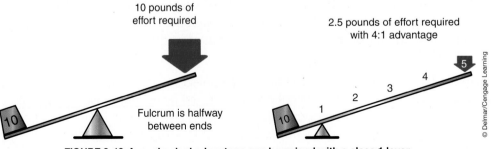

FIGURE 2-12 A mechanical advantage can be gained with a class 1 lever.

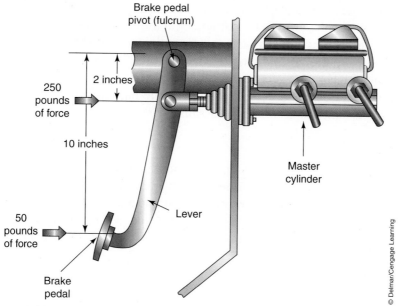

FIGURE 2-13 A brake pedal assembly is an example of a class 2 lever.

Gears. A gear is a toothed wheel that becomes a machine when it is meshed with another gear. The action of one gear is that of a rotating lever. It moves the other gear meshed with it. Based on the size of the gears in mesh, the amount of force applied from one gear to the other can be changed. Keep in mind that this does not change the amount of work performed by the gears because although the force changes, so does the distance of travel (Figure 2-14). The relationship of force and distance is inverse. Gear ratios express the mathematical relationship (diameter and number of teeth) of one gear to another.

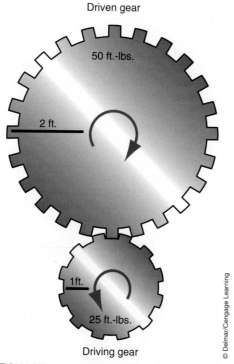

FIGURE 2-14 When a small gear drives a larger gear, the larger gear turns with more force but travels less, therefore the amount of work stays the same.

Wheels and Axles. The most obvious application of a wheel and axle is a vehicle's tires and wheels. These units revolve around an axle and limit the amount of area of a vehicle that contacts the road. Wheels function as rollers to reduce the amount of friction between a vehicle and the road. Basically, the larger the wheel, the less force is required to turn it. However, the wheel must move farther as it gets larger. An example of this is a steering wheel. A steering wheel that is twice the size of another will require one-half the force to turn it but will also require twice the distance to accomplish the same work.

Power

To convert pounds-foot to Newtons-meters, multiply the number of pounds-foot by 1.355.

Power is a measurement for the rate, or speed, at which work is done. The metric unit for power is the watt. A watt is equal to one Newton-meter per second. You can multiply the amount of torque in Newton-meters by the rotational speed to determine the power in watts. Power is a unit of speed combined with a unit of force. For example if you were pushing something with a force of 1 N and it moved at a speed of 1 meter per second, the power output would be 1 watt.

Horsepower

Horsepower is the rate at which torque is produced. James Watt is credited with being the first person to calculate horsepower. He measured the amount of work that a horse could do in a specific time. He found that a horse could move 330 pounds 100 feet in one minute (Figure 2-15). Put another way, he determined that one horse could do 33,000 pounds-foot of work in one minute. Thus, one horsepower is equal to 33,000 pounds-foot per minute, or 550 pounds-foot per second. Two horsepower could do this same amount of work in half a minute. If you push a 3,000-pound (1,360-kilogram) car for 11 feet (3.3 meters) in one-quarter of a minute, you produce four horsepower.

An engine that produces 300 pounds-feet of torque at 4,000 rpm produces 228 horsepower at 4,000 rpm. This is based on the formula that horsepower is equal to torque multiplied by engine speed and then divided by 5,252. Torque × engine speed ÷ 5,252 = horsepower. The constant, 5,252, is used to convert the rpm for torque and horsepower into revolutions per second.

> **AUTHOR'S NOTE:** Manufacturers are now rating their engines' outputs in watts. One horsepower is equal to approximately 746 watts. Therefore, an 228-hp engine is rated at 170,088 watts or about 170 kW.

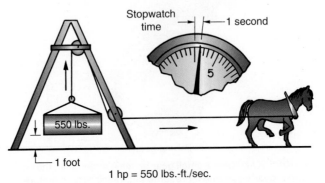

1 hp = 550 lbs.-ft./sec.

© Delmar/Cengage Learning

FIGURE 2-15 This is how James Watt defined one horsepower.

LIQUIDS

A fluid is something that does not have a definite shape; therefore liquids and gases are fluids. A characteristic of all fluids is that they will conform to the shape of their container. A major difference between a gas and a liquid is that a gas will always fill a sealed container whereas a liquid may not. A gas will also readily expand or compress according to the pressure exerted on it. Liquids are basically incompressible (Figure 2-16), which gives them the ability to transmit force (Figure 2-17). The pressure applied to a liquid in a sealed container is transmitted equally in all directions and to all areas of the system and acts with equal force on all areas. As a result, liquids can provide great increases in the force available to do work. They also always seek a common level. A liquid under pressure may also change into a gas in response to temperature changes.

Liquids exert pressure on immersed objects, resulting in an upward resultant force called *upthrust*. The upthrust is equal to the weight of the liquid displaced by the immersed object. If the upthrust on an object is greater than the weight of the object, then the object will float. Large ships float because they displace huge amounts of water, producing a large upthrust.

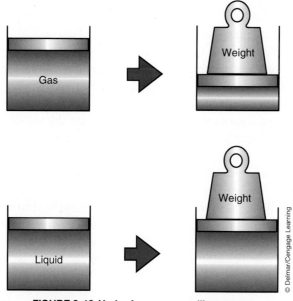

FIGURE 2-16 Under force, a gas will compress but a liquid will not.

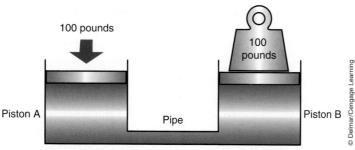

FIGURE 2-17 Liquids can be used to transfer force.

Laws of Hydraulics

Hydraulics is the study of liquids in motion. Liquids will predictably respond to pressures exerted on them. Their reaction to pressure is the basis of all hydraulic applications. This fact allows hydraulics to do work. A simple hydraulic system has liquid, a pump, lines to carry the liquid, control valves, and an output device. The liquid must be available from a continuous source, such as an oil pan or sump. An oil pump is used to move the liquid through the system. The lines that carry the liquid may be pipes, hoses, or a network of internal bores or passages in a single housing. Control valves are used to regulate hydraulic pressure and direct the flow of the liquid. The output device is the unit that uses the pressurized liquid to do work.

Over three hundred years ago a French scientist, Blaise Pascal, determined that if you had a liquid-filled container with only one opening and applied force to the liquid through that opening, the force would be evenly distributed throughout the liquid. This explains how pressurized liquid is used to operate and control systems, such as the brake system and automatic transmissions.

Pascal's law or principle states that the pressure of an incompressible fluid in a closed container is the same in all directions within that container. Further this law states that in a closed system, the pressure applied to a piston produces an equal pressure increase on another piston in the system. If the second piston has an area ten times that of the first, the force on the second piston is ten times greater, even though the applied pressure is the same as that applied to the first piston.

Pascal constructed the first known hydraulic device, which consisted of two sealed containers connected by a tube. The pistons inside the cylinders seal against the walls of each cylinder, preventing the liquid from leaking out of the cylinder and preventing air from entering into the cylinder. When the piston in the first cylinder has a force applied to it, the pressure moves everywhere within the system. The force is transmitted through the connecting tube to the second cylinder. The pressurized fluid in the second cylinder exerts force on the bottom of the second piston, moving it upward and lifting the load on the top of it. By using this device, Pascal found he could increase the force available to do work, just as could be done with levers or gears.

Pascal determined that force applied to liquid creates pressure, or the transmission of the force through the liquid. These experiments revealed two important aspects of a liquid when it is confined and put under pressure. First, the pressure applied to it is transmitted equally in all directions. Second, this pressure acts with equal force at every point in the container. If a liquid is confined and a force applied, pressure is produced. In order to pressurize a liquid, the liquid must be in a sealed container. Any leak in the container will decrease the pressure.

> **AUTHOR'S NOTE: Do your best to understand this law as it is the basis for understanding how an automatic transmission works.**

Mechanical Advantage with Hydraulics

Hydraulics are used to do work in the same way as a lever or gear does. All of these systems transmit energy. Since energy cannot be created or destroyed, these systems only redirect energy to perform work and do not create more energy. If a hydraulic pump provides 100 psi, there will be 100 pounds of pressure on every square inch of the system. If the system includes a piston with an area of 30 square inches, each square inch receives 100 pounds of pressure. This means there will be 3,000 pounds of force applied to that piston (Figure 2-18), but the output's travel will decrease proportionally. The use of the larger piston gives the system a **mechanical advantage** or increase in the force available to do work. The multiplication of force through a hydraulic system is directly proportional to the difference in the piston sizes throughout the system (Figure 2-19).

The brake pedal is a typical example of how a lever increases the force on a hydraulic system through its **mechanical advantage**.

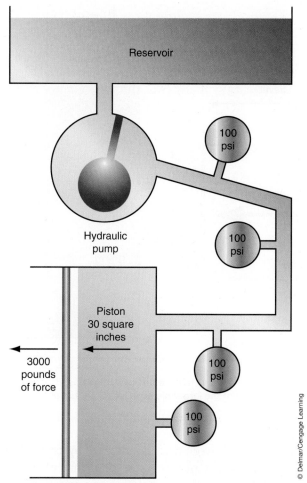

FIGURE 2-18 Pressure throughout this simple hydraulic system is the same, but the amount of force can be increased with an increase in the size of the output piston.

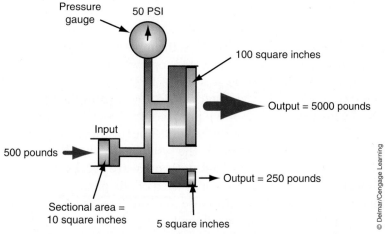

FIGURE 2-19 While keeping the pressure in the system constant, the output force will change with a change in the size of the output piston.

By changing the size of the pistons in a hydraulic system, force is multiplied, and as a result, low amounts of force can be used to move heavy objects. The mechanical advantage of a hydraulic system can be further increased by the use of levers to increase the force applied to a piston.

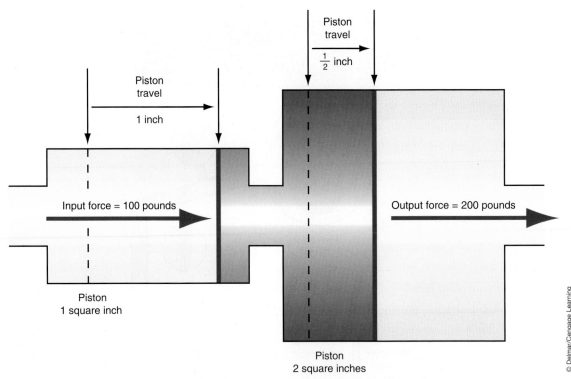

Piston
travel

$\frac{1}{2}$ inch

Piston
travel

1 inch

Input force = 100 pounds

Output force = 200 pounds

Piston
1 square inch

Piston
2 square inches

© Delmar/Cengage Learning

FIGURE 2-20 Although the force available to do work is increased by using a larger piston in one cylinder, the total movement of the larger piston is less than that of the smaller one.

Although the force available to do work is increased by using a larger piston in one cylinder, the total movement of the larger piston is less than that of the smaller one (Figure 2-20). A hydraulic system with two cylinders, one with a one-inch piston and the other with a two-inch one, will double the force at the second piston. However, the total movement of the larger piston will be half the distance of the smaller one.

The use of hydraulics to gain a mechanical advantage is similar to the use of levers or gears. Hydraulics are preferred when the size and shape of the system is of concern. In hydraulics, the force applied to one piston will transmit through the fluid and the opposite piston will have the same force on it. The distance between the two pistons in a hydraulic system does not affect the force in a static system. Therefore, the force applied to one piston can be transmitted without change to another piston located somewhere else.

A hydraulic system responds to the pressure or force applied to it. The mere presence of different-sized pistons does not always result in fluid power. Either the pressure applied to the pistons must be different or the size of the pistons must be different in order to cause fluid power. If an equal amount of pressure is exerted onto both pistons in a system and both pistons are the same size, neither piston will move. The system is balanced or is at equilibrium. The pressure inside the hydraulic system is called **static pressure** because there is no fluid motion.

When an unequal amount of pressure is exerted on the pistons, the piston with the least amount of pressure on it will move in response to the difference between the two pressures. Likewise, if the size of the two pistons is different and an equal amount of pressure is exerted on the pistons, the fluid will move. The pressure of the fluid while it is in motion is called **dynamic pressure**.

A common device used in transmissions is an orifice, which is a partial restriction to fluid flow. Orifices are typically plugs with a small hole bored in them. An orifice increases the dynamic pressure of the fluid while decreasing the actual volume of the fluid beyond its bore.

As fluid flows through a tube or hydraulic path, it has a steady velocity and pressure. When the fluid reaches the orifice, the fluid builds up behind the orifice's hole since only part of the volume can be pushed through the hole. The fluid before the orifice has a higher pressure than the fluid that passes through it. However, the speed of the fluid is greater after the orifice than it was before it. Once flow is established after the orifice, the fluid begins to expand and fill the passageway and the fluid's pressure and speed begins to decrease.

Orifices are commonly used to maintain smooth movement of hydraulic control valves and control shift timing by restricting fluid flow.

HEAT

Heat is a form of energy and is used in many ways. The main sources of heat are the sun, the Earth, chemical reactions, electricity, friction, and nuclear energy. Heat is the result of the kinetic energy that is present in all matter. Therefore, everything has heat. Cold objects have low kinetic energy because their atoms and molecules are moving very slowly, whereas hot objects have more kinetic energy because their atoms and molecules are moving fast.

Temperature is an indication of an object's intensity (not volume) of kinetic energy. Temperature is measured with a thermometer, which has either a (F) Fahrenheit or (C) Celsius (Centigrade) scale. At absolute zero ($-273°C$, also referred to as 0 Kelvin), particles of matter do not vibrate, but at all other temperatures, particles have motion. The temperature of an object is also a statement of how well the object will transfer heat or kinetic energy to or from another object. Heat and temperature are not the same thing. Heat is the volume commonly measured in British thermal units (BTUs) of kinetic energy something has. Heat is also measured in calories. One BTU is the amount of heat required to heat one pound of water by $1°F$ (Figure 2-21). One calorie is equal to the amount of heat needed to raise the temperature of one gram of water $1°C$.

Temperature is an indication of the intensity of kinetic energy something has. Energy from something hot will always move to an object that is colder, until both are at the same temperature. The greater the difference in temperature between the two objects, the faster the heat will flow from one to the other.

The Effects of Temperature Change

Any time the temperature of an object has changed, a transfer of heat has occurred. A transfer of heat may also cause the object to change size or its state of matter. The amount of heat required to raise the temperature of 1 gram of mass $1°C$ is called the *specific heat capacity*. Every substance has its own specific heat capacity and this factor is assigned to material based on its difference from water, which has a specific heat capacity of 1. For example, the temperature of one gram of water will increase by $10°C$ if 10 calories of heat are transferred to it.

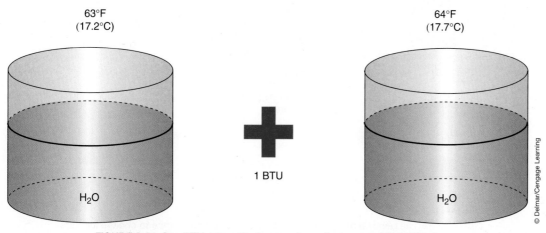

63°F
(17.2°C)

64°F
(17.7°C)

1 BTU

H$_2$O

H$_2$O

© Delmar/Cengage Learning

FIGURE 2-21 One BTU raises the temperature of one pound (0.4536 kg) of water (H$_2$O) by one degree Fahrenheit (0.56 °C).

But if 10 calories of heat were added to one gram of copper, the temperature would increase by 111°C. This is because copper has a specific heat capacity of only 0.09 as compared to the 1.0 specific heat capacity of water.

As heat moves in and out of a mass, the movement of atoms and molecules in that mass increases or slows down. With an increase in motion, the size of the mass tends to get bigger or expand. This is commonly called thermal **expansion**. Thermal **contraction** takes place when a mass has heat removed from it and the atoms and molecules slow down. All gases and most liquids and solids expand when heated, with gases expanding the most. Water is an exception; it expands when enough heat is removed to turn it into a solid—ice. Solids, because they are not fluid, expand and contract at a much lower rate. It is important to realize that all materials do not expand and contract at the same rate. For example, an aluminum component will expand at a faster rate than the same component made of iron. This explains why aluminum cylinder heads have different service requirements and procedures from iron cylinder heads.

Thermal expansion takes place every time fuel and air are burned in an engine's cylinders. The sudden temperature increase inside the cylinder causes a rapid expansion of the gases, which pushes the piston downward and causes engine rotation.

Typically, when heat is added to a mass, the temperature of the mass increases. This does not always happen, however. In some cases, the additional heat causes no increase in temperature but causes the mass to change its state (solid to liquid or liquid to gas). For example, if we take an ice cube and heat it to 32°F (0°C) and continue to apply heat to it, it will begin to melt (Figure 2-22). As heat is added to the ice cube, the temperature of the ice cube will not increase until it becomes a liquid. The heat added to the ice cube that did not raise its temperature but caused it to melt is called **latent heat** or the heat of fusion. Each gram of ice at 0°C requires 80 calories of heat to melt it to water at 0°C. As more heat is added to the 0°C water, the water's temperature will once again increase. This continues until the temperature of the water reaches 212°F (100°C). This is the boiling temperature of water. At this point, any additional heat applied to the water is latent heat causing the water to change its state to that of a gas. This added heat is called the heat of evaporation.

To change the water gas or steam back to liquid water, the same amount of heat required to change the liquid to a gas must be removed from the gas. At that point, the steam begins to

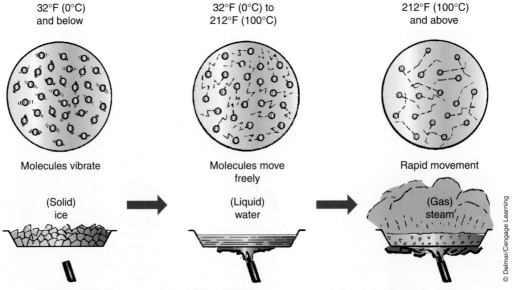

FIGURE 2-22 The removal or addition of heat may cause a substance to change its state.

condense into a liquid. As additional heat is removed, the temperature will drop until enough heat is removed to bring its temperature back down to freezing (melting in reverse) point. At that time latent heat must be removed from the liquid before the water turns to ice again.

CHEMICAL PROPERTIES

The properties of something describe or identify the characteristics of an object. Physical properties are characteristics that are readily observable, such as color, size, luster, and smell. Chemical properties are only observable during a chemical reaction and describe how one type of matter reacts with another to form a new and different substance. Chemical properties are quite different than physical properties. A chemical property of some metals is the ability to combine with oxygen to form rust (iron and oxygen) or tarnish (silver and sulfur). Another example is hydrogen's ability to combine with oxygen to form water.

A solution is a mixture of two or more substances. Most solutions are liquids, but solutions of gases and solids are possible. An example of a gas solution is the air we breathe; it is composed of mostly oxygen and nitrogen. Brass is a good example of a solid solution as it is composed of copper and zinc. The liquid in a solution is called the **solvent**, and the substance added is the solute. If both are liquids, the one present in the smaller amount is usually considered the solute. Solutions can vary widely in terms of how much of the dissolved substance is actually present. A heavily diluted (much water) acid solution has very little acid and may not be noticeably acidic.

Specific Gravity

Specific gravity is the heaviness or relative **density** of a substance as compared to water. If something is 3.5 times as heavy as an equal volume of water, its specific gravity is 3.5. Its density is 3.5 grams per cubic centimeter, or 3.5 kilograms per liter. Specific gravity checks of a battery's electrolyte are an indication of the battery's state of charge.

Density is a statement of how much mass there is in a particular volume. Water is denser than air; therefore there will be less air in a given container than water in that same container (Figure 2-23). The density of a material changes with temperature as well (Figure 2-24). This is the reason an engine runs more efficiently with cool intake air.

Substance	Density in g/cm³
Air	0.0013
Ice	0.92
Water	1.00
Aluminum	2.70
Steel	7.80
Gold	19.30

© Delmar/Cengage Learning

FIGURE 2-23 A look at the density of different substances as compared to water.

Temp °F	Temp °C	Approx. Change In Density
200°	93°	−21%
180°	82°	−16.8%
160°	71°	−12.6%
140°	60°	−8.4%
120°	49°	−4.2%
100°	38°	−
80°	27°	+4.2%
60°	16°	+8.4%
40°	4°	+12.6%
20°	−7°	+16.8%
0°	−18°	+21%

© Delmar/Cengage Learning

FIGURE 2-24 The effect that temperature has on the density of air at atmospheric pressure.

Chemical Reactions

Chemical changes, or chemical reactions, result in the production of another substance, such as wood turns to carbon after it has been completely burned. A chemical reaction is always accompanied by a change in energy. This means energy is given off or taken in during the reaction. Some reactions that release energy need some energy to get the reaction started. A reaction takes place when two or more molecules interact and the one of the following happens:

- A chemical change occurs.
- Single reactions occur as part of a large series of reactions.
- Ions, molecules, or pure atoms are formed.

Catalysts and Inhibitors

Reactions need a certain amount of energy to happen. A **catalyst** lowers the amount of energy needed to make a reaction happen. A catalyst is any substance that affects the speed of a chemical reaction without itself being consumed or changed. Catalysts tend to be highly specific, reacting with one substance or a small set of substances. In a car's catalytic converter, the platinum catalyst converts unburned hydrocarbons and nitrogen compounds into products that are harmless to the environment. Water, especially salt water, catalyzes oxidation and corrosion. An inhibitor is the opposite of a catalyst and stops or slows the rate of a reaction.

Acids/Bases

An ion is an atom or group of atoms with one or more positive or negative electric charges. Ions are formed when electrons are added to or removed from neutral molecules or other ions. Ions are what make something an **acid** or a **base**.

Acids are compounds which break into hydrogen (H^+) ions and another compound when placed in an aqueous (water) solution. They have a sour taste, are corrosive, react with some metals to produce hydrogen, react with carbonates to produce carbon dioxide, change the color of litmus from blue to red, and become less acidic when combined with alkalis. Most acids are slow reacting, especially if they are weak acids. Acids also react with bases to form salts.

Alkalis (bases) are compounds that release hydroxide ions (OH^-) and react with hydrogen ions to produce water, thus neutralizing each other. Most substances are neutral (not an acid or a base). Alkalis feel slippery, change the color of litmus from red to blue, and become less alkaline when they are combined with acids.

A hydroxide is any compound made up of one atom each of hydrogen and oxygen, bonded together and acting as the hydroxyl group or hydroxide anion (OH^-). An oxide is any chemical compound in which oxygen is combined with another element. Metal oxides typically react with water to form bases or with acids to form salts. Oxides of nonmetallic elements react with water to form acids or with bases to form salts.

A salt is a chemical compound formed when the hydrogen of an acid is replaced by a metal. Typically, an acid and a base react to form a salt and water.

pH. The **pH scale** is used to measure how acidic or basic a solution is. Its name comes from the fact that pH is the absolute value of the power of the hydrogen-ion concentration. The scale goes from 0 to 14. Distilled (pure) water is 7. Acids are found between 0 and 7 and bases are from 7 to 14. When the pH of a substance is low, the substance has many H^+ ions. When the pH is high, the substance has many OH^- ions. The pH value helps inform scientists, as well as technicians, of the nature, composition, or extent of reaction of substances.

The pH of something is typically checked with litmus paper. Litmus is a mixture of colored organic compounds obtained from several species of lichen. Lichen is a type of plant

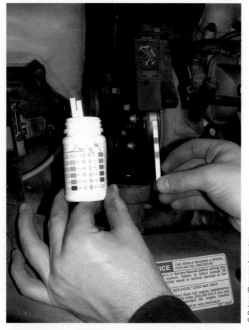

FIGURE 2-25 Litmus test strips can be used to check the condition of an engine's coolant.

© Delmar/Cengage Learning

that is actually a combination of a fungus and algae. Litmus test strips can be used to check the condition of the engine's coolant (Figure 2-25).

Reduction and Oxidation

Oxidation is a chemical reaction in which a substance combines with oxygen. Rapid oxidation produces heat fast enough to cause a flame. When fuel burns, it combines with oxygen to form other compounds. This chemical reaction is combustion, which produces heat and fire.

The addition of hydrogen atoms or electrons is **reduction**. Oxidation and reduction always occur simultaneously: one substance is oxidized by the other, which is reduced. During oxidation, a molecule provides electrons. During reduction, a molecule accepts electrons. Oxidation and reduction reactions are usually called **redox** reactions. Redox is any chemical reaction in which electrons are transferred. Batteries, also known as voltaic cells, produce an electrical current at a constant voltage through redox reactions.

An oxidizing agent is something that accepts electrons and oxidizes something else while being reduced in the process. A reducing agent is something that provides electrons and reduces something else while being oxidized.

Every atom or ion has an oxidation number. This value compares the number of protons and electrons in that atom. In many cases, the oxidation number reflects the actual charge on the atom, but there are many cases where it does not. The oxidation number is reduced during reduction, by adding electrons. The oxidation number is increased during oxidation, by removing electrons. All free, uncombined elements have an oxidation number of zero. Hydrogen, in all its compounds except hydrides, has an oxidation number of +1. Oxygen, in all its compounds except peroxides, has an oxidation number of −2.

Metallurgy

Metallurgy is the art and science of extracting metals from their ores and modifying them for a particular use. This includes the chemical, physical, atomic properties, and structures of metals and the way metals are combined to form alloys. An **alloy** is a mixture of two or more metals. Steel, is an alloy of iron plus carbon and other elements.

Metals have one or more of the following properties:

- Good heat and electric conduction.
- Malleability—can be hammered, pounded, or pressed into a shape without breaking.
- Ductility—can be stretched, drawn, or hammered without breaking.
- High light reflectivity—can make light bounce off its surface.
- The capacity to form positive ions in a solution and hydroxides rather than acids when their oxides meet water.

About three-quarters of the elements are metals. The most abundant metals are aluminum, iron, calcium, sodium, potassium, and magnesium.

Rust and Corrosion. The rusting of iron is an example of oxidation. Unlike fire, rusting occurs so slowly that little heat is produced. Iron combines with oxygen to form rust. The rate this occurs at depends on several factors: temperature, surface area (more iron exposed for oxygen to get at), catalysts (speed up a reaction but do not react and change themselves).

Corrosion is the wearing away of a substance due to chemical reactions. It occurs whenever a gas or a liquid chemically attacks an exposed surface. This action is accelerated by heat, acids, and salts. Some materials naturally resist corrosion; others can be protected by painting, coatings, galvanizing, or anodizing.

Galvanizing involves the coating of zinc onto iron or steel to protect them against exposure to the atmosphere. If galvanizing is properly applied, it can protect the metals for 15–30 years or more.

Metals can be anodized for corrosion resistance, electrical insulation, thermal control, abrasion resistance, sealing, improving paint adhesion, and decorative finishing. Anodizing is a process that electrically deposits an oxide film from an aqueous solution onto the surface of a metal, often aluminum. During the process, dyes can be added to give the material a colored surface.

Hardness. The hardness of something describes its resistance to scratching. Hardening is a process that increases the hardness of a metal, deliberately or accidentally, by hammering, rolling, carburizing, heat treating, tempering, or other processes. All of these deform the metal by compacting the atoms or molecules to make the material denser.

Carburizing hardens the surface of steel with heat. It increases the hardness of the outer surface while leaving the core relatively soft. The combination of a hard surface and soft interior withstands very high stress. It also has a low cost and offers flexibility for manufacturing. To carburize, the steel parts are placed in a carbonaceous environment (with charcoal, coke, and carbonates, or carbon dioxide, carbon monoxide, methane, or propane) at a high temperature for several hours. The carbon diffuses into the surface of the steel, altering the crystal structure of the metal. Gears, ball and roller bearings, and piston pins are often carburized.

Heat treating changes the properties of a metal (including iron, steel, aluminum, copper, and titanium) by using heat. **Tempering** is the heat treating of metal alloys, particularly steel. For example, raising the temperature of hardened steel from 752°F (4000°C) and holding it for a time before quenching it in oil decreases its hardness and brittleness and produces strong steel.

Solids under Tension

The atoms of a solid are closely packed, so solids have a greater density than most liquids and gases. The rigidity results from the strong attraction between its atoms. A force pulling on a solid moves these atoms farther apart, creating an opposing force called tension. If a force pushes on a solid, the atoms move closer together, creating compression. These are the principles of how springs function. Springs are used in many automotive systems, the most obvious are those used in suspension systems.

An elastic substance is a solid that gets larger under tension, gets smaller under compression, and returns to its original size when no force is acting on it. Most solids show some elastic behavior, but there is usually a limit to the force that the material can face. When excessive force is applied, the material will not return to its original size and it will be distorted or will break. The limit depends on the material's internal structure, for example steel has a low-elastic limit and can only be extended about 1 percent of its length, whereas rubber can be extended to about 1000 percent. Another factor involved in elasticity is the cross-sectional area of the material.

Tensile strength is the ratio of the maximum load a material can support without breaking while being stretched. It is dependent on the cross-sectional area of the material. When stresses less than the tensile strength are removed, the material returns to its original size and shape. Greater stresses form a narrow, constricted area in the material, which is easily broken. Tensile strengths are measured in units of force per unit area.

Electrochemistry

Electrochemistry is concerned with the relationship between electricity and chemical change. Many spontaneous chemical reactions release electrical energy and some of these are used in batteries and fuel cells to produce electric power. The basis for electricity is the movement of electrons from one atom to another.

Electrolysis is an electrochemical process. During this process, electric current is passed through a substance causing a chemical change. This change causes either a gain or a loss of electrons. Electrolysis normally takes place in an electrolytic cell made of separated positive and negative electrodes immersed in an electrolyte.

An **electrolyte** is a substance that conducts current as a result of the breaking down of its molecules into positive and negative ions. The most familiar electrolytes are acids, bases, and salts, that ionize when dissolved in solvents such as water and alcohol. Ions drift to the electrode of the opposite charge and are the conductors of current in electrolytic cells.

ELECTRICITY AND ELECTROMAGNETISM

All electrical effects are caused by electric charges. There are two types of electric charge: positive and negative. These charges exert electrostatic forces on each other, due to the strong attraction of electrons to protons. An electric field is the area that these forces have an effect on. As discussed earlier in this chapter, in atoms, protons carry positive charge, while electrons carry negative charge. Atoms are normally neutral, meaning they have an equal number of protons and electrons, but an atom can gain or lose electrons, for example, by being rubbed. It then becomes a charged atom, or ion. Electricity has many similarities with magnetism. For example, the lines of the electric fields between charges take the same form as the lines of magnetic force, so magnetic fields can be said to be an equivalent to electric fields. Charges of the same type repel, while charges of a different type attract (Figure 2-26).

Electricity

An electric circuit is the path along which an electric current flows. Electrons carry negative charge and can be moved around a circuit by electrostatic forces. A circuit usually consists of a conductive material, such as a metal, in which the electrons are held very loosely to their atoms, thus making movement possible. The strength of the electrostatic force is the voltage. The resulting movement of the electric charge is called an electric current (Figure 2-27). The higher the voltage, the greater the current will be. But the current also depends on the thickness, length, temperature, and nature of the materials that conducts it. The resistance of a material is the extent to which it opposes the flow of electric current. Good conductors have low resistance, which means that a small amount of voltage will produce a large current. In batteries, chemical reactions in the metal electrode causes the freeing of electrons, resulting in their movement to another electrode and the formation of a current.

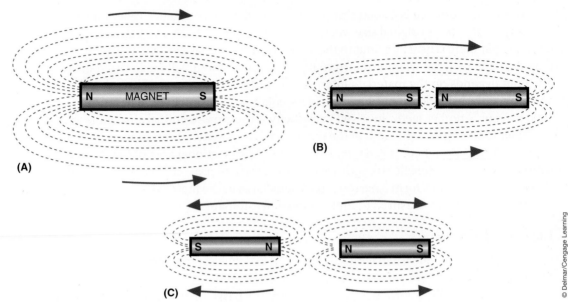

FIGURE 2-26 (A) In a magnet, lines of force emerge from the north pole and travel to the south pole before passing through the magnet back to the north pole. (B) Unlike poles attract, while (C) similar poles repel each other.

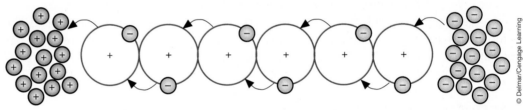

FIGURE 2-27 Electrical energy is the result of the controlled movement of electrons from one atom to another.

Magnets

Some materials are natural magnets; however, most magnets are produced. The materials typically used to make a permanent magnet are called *ferromagnetic* materials. These materials consist of mostly iron compounds that are heated. The heat causes the atoms to shift direction. Once they all point in the same direction the metal becomes a magnet. This sets up two distinct poles called the *north* and *south* poles. The poles are at the ends of the magnet and there is an attraction between the north pole and the south pole. This attraction or force set up by a magnet can be observed but the type of force is not known.

The lines of a magnetic field form closed lines of force from the north to the south. If another iron or steel object enters into the magnetic field, it is pulled into the magnet. If another magnet is introduced into the magnetic field, it will either move into the field or push away from it. This is the result of the natural attraction of a magnet from north to south. If the north pole of one magnet is introduced to the north pole of another, the two poles will oppose each other and will push away. If the south pole of a magnet is introduced to the north pole of another, the two magnets will join together because the opposite poles are attracted to each other.

The strength of the magnetic force is uniform all around the outside of the magnet. The force is strongest at the surface of the magnet and weakens with distance. If you double the distance from a magnet, the force is reduced by $\frac{1}{8}$.

The strength of a magnetic field is typically measured with devices known as magnetometers and in units of Gauss (G).

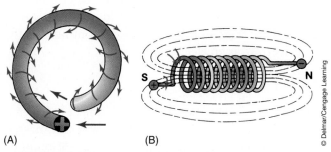

FIGURE 2-28 When current is passed through a conductor such as a wire, **(A)** magnetic lines of force are generated around the wire at right angles to the direction of the current flow. **(B)** These magnetic lines create a magnetic flux field around the entire electromagnet.

Electromagnetism

Any electrical current will produce magnetism that affects other objects in the same way as permanent magnets. The arrangement of force lines around a current-carrying conductor, its *magnetic field*, is circular. The magnetic effect of electrical current is increased by making the current-carrying wire into a coil.

When a coil of wire is wrapped around an iron bar, it is called an **electromagnet** (Figure 2-28). The magnetic field produced by the coil magnetizes the iron bar, strengthening the overall effect. A field like that of a bar magnet is formed by the magnetic fields of the wires in the coil. The strength of the magnetism produced depends on the number of coils and the size of the current flowing in the wires. Electromagnetic coils and permanent magnets are arranged inside an electric motor so that the forces of electromagnetism create rotation of the armature.

Producing Electrical Energy

There are many ways to generate electricity. The most common is to use coils of wire and magnets in a generator. Whenever a wire and magnet are moved relative to each other, a voltage is produced (Figure 2-29). In a generator, the wire is wound into a coil. The more turns in the coil and the faster the coil moves, the greater the voltage. The coils or magnets spin around at high speed, typically turned by steam pressure. The steam is usually generated by burning coal or oil, a process that creates pollution. Renewable sources of electricity, such as hydroelectric power, wind power, solar energy, and geothermal power, produce only heat as a pollutant. In automobiles, the generator is spun by a belt driven by the engine's crankshaft. In a generator, the kinetic energy of a spinning object is converted into electrical energy.

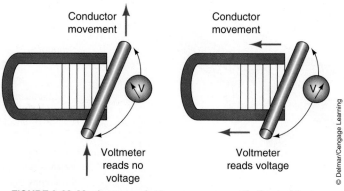

FIGURE 2-29 Moving a conductor across magnetic lines of force induces a voltage in the conductor.

A solar cell converts the energy of sunlight directly into electrical energy, using layers of semiconductors. Electricity is produced by causing electrons to leave the atoms in the semiconductor material. Each electron leaves behind a hole or gap. Other electrons move into the hole, leaving holes in their atoms. This process continues all the way around a circuit. The moving chain of electrons is an electrical current.

WAVES AND OSCILLATIONS

An **oscillation** is the back and forth movement of an object between two points. When that motion travels through matter or space, it becomes a **wave**. A mass suspended by a spring, for example, is acted upon by two forces: gravity and the tension in the spring. At the point of equilibrium, the resultant of these forces is zero. When the mass is given a downward push, the tension of the spring exceeds the weight of the mass. The resultant upward force accelerates the mass back up toward its original position and its momentum carries it farther upward. When the weight exceeds the spring's tension, the mass moves down again and the oscillation repeats itself until the mass is at equilibrium. As the mass oscillates toward the equilibrium position, the size of the oscillation decreases. As the mass oscillates, the air around it moves and becomes an air wave.

Vibrations

When an object oscillates, it vibrates (Figure 2-30). To prevent the vibration of one object from causing a vibration in other objects, the oscillating mass must be isolated from other objects. This is often a difficult task. For example, all engines vibrate as they run. To reduce the transfer of engine vibrations to the rest of the vehicle, the engine is held by special mounts. The materials used in the mounts must keep the engine in place and must be elastic enough to absorb the engine's vibrations (Figure 2-31). If the engine was mounted solidly, the vibrations would be felt throughout the vehicle.

Vibration control is also important for the reliability of components. If the vibrations are not controlled, the object could shake itself to destruction. Vibration control is the best justification for always mounting parts in the way they were designed to be mounted.

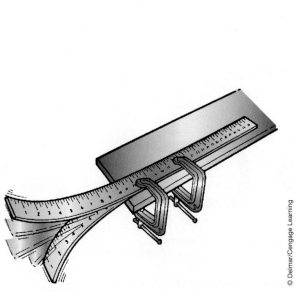

FIGURE 2-30 Vibrations happen in cycles.

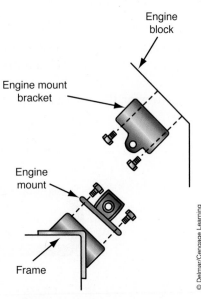

FIGURE 2-31 An engine mount holds the engine in place and isolates engine vibrations from the rest of the vehicle.

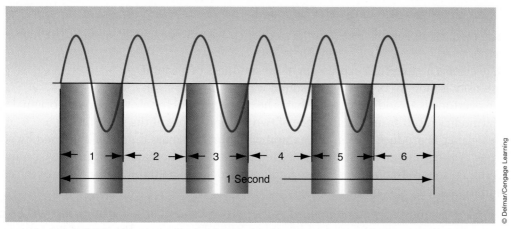

FIGURE 2-32 Frequency is a statement of how many cycles occur in one second.

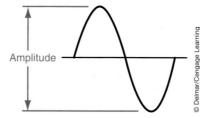

FIGURE 2-33 Amplitude is a measurement
of a vibration's intensity.

Unwanted and uncontrolled vibrations typically result from one component vibrating at a different frequency than another part. When two waves or vibrations meet, they add up or interfere. This is called the Principle of Superposition and is common to all waves. Making unwanted vibrations tolerable can be done by canceling them with equal and opposite vibrations.

How many times the vibration occurs in one second is called **frequency**. Frequency (Figure 2-32) is most often expressed in **hertz** (Hz). One hertz is equal to one cycle per second. The name is in honor of Heinrich Hertz, an early German investigator of radio wave transmission. The **amplitude** of a vibration is its intensity or strength (Figure 2-33). The velocity of a vibration is the result of its amplitude and its frequency. All materials have a unique resonant or natural vibration frequency.

AUTHOR'S NOTE: The manufacturers of vibration analyzers often measure the amplitude of the vibration in terms of "G-force" which represents the strength of the vibration.

Sound. Vibration results in the phenomenon of sound. In air, the vibrations that cause sound are transmitted as a wave between air molecules; many other substances transmit sound in a similar way. A vibrating object causes pressure variations in the surrounding air. Areas of high and low pressure, known as compressions and rarefactions, move through the air as sound waves. Compression makes the sound waves denser, whereas rarefaction makes them less dense. The distance between each compression of a sound wave is called its **wavelength**. Sound waves with a short wavelength have a high frequency and sound high-pitched.

When the rapid variations in pressure occur between about 20 Hz and 20 kHz, sound is audible. Audible sound is the sensation (as detected by the ear) of very small rapid changes in the air pressure above and below atmospheric pressure.

Certain terms are used to describe sound:

- The pitch of a sound is based on its frequency. The greater the frequency, the higher the pitch.
- A decibel is a numerical expression of the loudness of a sound.
- Intensity is amount of energy in a sound wave.
- An overtone is an additional tone that is heard because of the air waves of the original tone.
- Harmonics result from the presence of two or more tones at the same time.
- Resonance is produced when the natural vibration of a mass is greatly increased by vibrations at the same or nearly the same frequency of another source or mass. A cavity has certain resonant frequencies. These frequencies depend on the shape and size of the cavity and the velocity of sound within the cavity.

During diagnostics, you often need to listen to the sound of something. You will be paying attention to the type of sound and its intensity and frequency. The tone of the sound usually indicates the type of material that is causing the noise. If there is high pitch, you know the source of the sound is something that is vibrating quickly. This means the source is less rigid than something that vibrates with a low pitch. Although pitch is dependent on the sound's frequency, the frequency itself can identify the possible sources of the sound.

SUMMARY

- Matter is anything that occupies space. It exists as a gas, liquid, or solid.
- When a solid dissolves into a liquid, a solution is formed. Not all solids will dissolve in a liquid; rather the liquid will be either absorbed or adsorbed.
- Materials that *absorb* fluids are permeable substances. Impermeable substances *adsorb* fluids.
- Energy is the ability to do work and all matter has energy.
- The total amount of energy never changes; it can only be transferred from one form to another, not created or destroyed.
- When energy is released to do work, it is called kinetic energy. Stored energy may be called potential energy.
- Mass is the amount of matter in an object. Weight is a force and is measured in pounds or kilograms. Gravitational force gives the mass its weight.
- A force is a push or pull, can be large or small, and can be applied to objects by direct contact or from a distance.
- When an object moves in a circle, its direction is continuously changing and all changes in direction require a force. The forces required to maintain circular motion are called centripetal and centrifugal force.
- Pressure is a force applied against an opposing object and is measured in units of force per unit of surface area (pounds per square inch or kilograms per square centimeter).
- The greater the mass of an object, the greater the force needed to change its motion. Inertia is the tendency of an object at rest to remain at rest, or the tendency of an object in motion to stay in motion.
- When a force overcomes static inertia and moves an object, the object gains momentum. Momentum is the product of an object's weight and its speed.
- Speed is the distance an object travels in a set amount of time. Velocity is the speed of an object in a particular direction. Acceleration, which only occurs when a force is applied, is the rate of increase in speed. Deceleration is the reverse of acceleration, as it is the rate of decrease in speed.

TERMS TO KNOW

Acceleration

Acid

Alloy

Amplitude

Atoms

Base

Catalyst

Contraction

Curb weight

Deceleration

Density

Dynamic pressure

Electrolysis

Electrolyte

Electromagnet

Element

Equilibrium

Evaporate

Expansion

Force

Frequency

Friction

Gross weight

Heat

Hertz

Impermeable

Inertia

Kinetic energy

Latent heat

Lever

Load

Mass

Matter

- Newton's laws of motion state that an object at rest tends to remain at rest and an object in motion tends to remain in motion, unless some force acts on it; when a force acts on an object, the motion of the object will change; and for every action there is an equal and opposite reaction.
- Friction is a force that slows or prevents the motion of two moving objects that touch.
- Friction can be reduced in two main ways: by lubrication or by the use of rollers.
- When a force moves a certain mass a specific distance, work is done.
- A machine is any device that can be used to transmit a force and, in doing so, change the amount of force or its direction. Examples of simple machines are inclined planes, pulleys, levers, gears, and wheels and axles.
- Power is a measurement of the rate at which work is done. It is measured in watts.
- Horsepower is the rate at which torque is produced.
- Liquids and gases do not have a definite shape and will conform to the shape of their container.
- A gas will also readily expand or compress according to the pressure exerted on it. Liquids are basically incompressible, which gives them the ability to transmit force.
- A simple hydraulic system has liquid, a pump, lines to carry the liquid, control valves, and an output device.
- Pascal determined that force applied to liquid creates pressure (the transmission of the force through the liquid).
- Hydraulics are used to do work in the same way as a lever or gear does and can result in a mechanical advantage (increase) in the force available to do work. The multiplication of force through a hydraulic system is directly proportional to the difference in the piston sizes throughout the system.
- Heat is a form of energy caused by the movement of atoms and molecules and is measured in British thermal units (BTUs) and calories.
- Temperature is an indication of an object's kinetic energy and is measured with a thermometer, which has either a (F) Fahrenheit or (C) Celsius scale.
- As heat moves in and out of a mass, the size of the mass tends to change.
- Any electrical current will produce magnetism. When a coil of wire is wrapped around an iron bar, it is called an electromagnet.
- The most common way to produce electricity is to use coils of wire and magnets in a generator.

TERMS TO KNOW
(continued)

Mechanical advantage
Momentum
Oscillation
Oxidation
Permeable
pH scale
Potential energy
Power
Pressure
Redox
Reduction
Solution
Solvent
Specific gravity
Speed
Static pressure
Tempering
Tensile strength
Tension
Velocity
Wave
Wavelength
Weight
Work

REVIEW QUESTIONS

Short-Answer Essays

1. Describe Newton's first law of motion and give an application of this law in automotive theory.

2. Explain Newton's second law of motion and give an example of how this law is used in automotive theory.

3. Describe four different forms of energy.

4. Describe four different types of energy conversion.

5. What is the difference between speed and velocity?

6. What basic property of a liquid allows it to be used to transfer force?

7. How can a hydraulic system be used to create a mechanical advantage?

8. Why does the size of something change with a change in heat?

9. What is torque?

10. What is a solution and how is it formed?

Fill in the Blanks

1. The nucleus of an atom contains _____ and _____.

2. Work is calculated by multiplying _____ and _____.

3. Energy may be defined as the ability to do _____.

4. When one object is moved over another object, the resistance to motion is called _____.

5. Weight is the measurement of the Earth's _____ on an object.

6. Torque is a force that does work with a _____ action.

7. Torque is calculated by multiplying the applied force by the _____ from the center of the _____ to the point where the force is exerted.

8. Torque is measured in _____-_____ and _____-_____

9. A simple hydraulic system has _____, a _____, lines to carry the _____, _____ _____, and an output device.

10. One horsepower is equal to _____ pounds-foot per minute, or _____ pounds-foot per second.

MULTIPLE CHOICE

1. Which of the following statements about friction is not true?
   A. Friction can be reduced by lubrication.
   B. Bearings are a type of roller used to increase the friction between moving parts such as a wheel and its axle.
   C. The presence of oil or another fluid between two surfaces keeps the surfaces apart and thereby reduces friction.
   D. Friction can be reduced by the use of rollers.

2. While discussing different types of energy:
   Technician A says when energy is released to do work, it is called potential energy.
   Technician B says stored energy is referred to as kinetic energy.
   Who is correct?
   A. A only
   B. B only
   C. Both A and B
   D. Neither A nor B

3. While discussing friction in matter:
   Technician A says that friction is a force that slows or prevents motion of two moving objects or surfaces that touch.
   Technician B says that friction occurs in liquids, solids, and gases.
   Who is correct?
   A. A only
   B. B only
   C. Both A and B
   D. Neither A nor B

4. While discussing mass and weight:
   Technician A says that mass is the measurement of an object's inertia.
   Technician B says that mass and weight may be measured in cubic inches.
   Who is correct?
   A. A only
   B. B only
   C. Both A and B
   D. Neither A nor B

5. When applying the principles of work and force:
   A. Work is accomplished when force is applied to an object that does not move.
   B. In the metric system the measurement for work is cubic centimeters.
   C. No work is accomplished when an object is stopped by mechanical force.
   D. If a 50-pound object is moved 10 feet, 500 lbs-ft. of work is produced.

6. All these statements about energy and energy conversion are true, except:
   A. Thermal energy may be defined as light energy.
   B. Chemical-to-thermal energy conversion occurs when gasoline burns.
   C. Mechanical energy is defined as the ability to do work.
   D. Mechanical-to-electrical energy conversion occurs when the engine drives the generator.

7. Which of the following is not a conclusion Pascal made from his studies of liquids?
   A. When a force acts on an object, the motion of the object will change.
   B. Liquids can be used to achieve the same results as gears and levers.
   C. Force applied to liquid creates pressure.
   D. Force can be transmitted through liquids.

8. Which of the following statements about mass and weight is true?
   A. Mass is the amount of matter in an object.
   B. Weight is a measurement of mass expressed in pounds or kilograms.
   C. Gravitational force gives a weight its mass.
   D. Something that weighs a lot on Earth will weigh nothing once it is lifted from the Earth's surface.

9. A screw is a simple machine that operates as a(n):
   A. gear.
   B. pulley.
   C. inclined plane.
   D. lever.

10. Which of the following statements is not true?
   A. Materials that absorb fluids are permeable substances.
   B. When a solid dissolves into a liquid, its particles break away from this structure and mix evenly in the liquid, forming a solution.
   C. When most liquids are heated, they evaporate.
   D. Permeable substances adsorb fluids.

Chapter 3

GENERAL THEORIES OF OPERATION

UPON COMPLETION AND REVIEW OF THIS CHAPTER, YOU SHOULD BE ABLE TO:

- List the factors that determine when an automatic transmission will automatically shift.
- Describe the four basic systems of all automatic transmissions.
- Describe the operation and purpose of a torque converter.
- Explain Pascal's law and how it applies to the operation of an automatic transmission.
- Explain the basic operation of hydraulic machines.
- Identify the major components in a transmission's hydraulic circuit and describe how they provide fluid flow and pressurization.

- List the various types of reaction members commonly used in automatic transmissions.
- Describe the purpose of a transmission's valve body.
- List and describe the various load-sensing devices used in an automatic transmission's hydraulic system.
- Describe the basic construction of automatic transmission housings, including the purpose of the various mechanical and electrical connections.

A BIT OF HISTORY

The first fully automatic "gear boxes" were introduced by Mercedes of Germany in 1914. These were in very limited production cars built for high-ranking officials.

INTRODUCTION

Automatic transmissions are used in many rear-wheel-drive, all-wheel-drive, and four-wheel-drive vehicles. Automatic transaxles are used in most front-wheel-drive vehicles. The major components of a transaxle (Figure 3-1) are the same as those of a transmission, except the transaxle assembly includes the final drive and differential gears in addition to the transmission. Automatic transmissions and transaxles shift automatically without the driver moving a gearshift lever or depressing a clutch pedal.

Most of today's automatic transmissions and transaxles have electronic controls, in addition to hydraulics. This chapter is focused on all of the basic transmission systems. Electronic controls are touched on, but detailed discussions of them are in the following chapter. There are many different electronic systems, and they vary with the manufacturer and model of the transmission. The contents of this chapter apply to nearly all transmissions, regardless of the electronic control used with a particular transmission.

BASIC OPERATION

An automatic transmission receives engine power through a torque converter, which is driven by the engine's crankshaft. Hydraulic pressure (force) in the converter allows power to flow from the torque converter to the transmission's input shaft. The only time a torque converter is mechanically connected to the transmission is during the application of the torque converter clutch. This can only occur if the torque converter is fitted with a clutch.

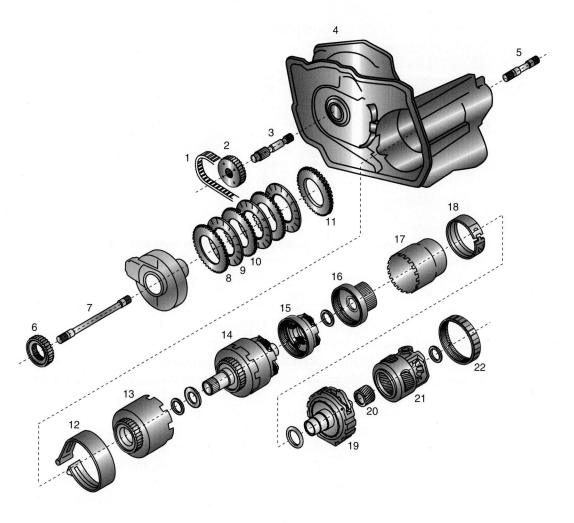

LEGEND

1. Drive link assembly
2. Drive sprocket
3. Turbine shaft
4. Transaxle case
5. Output stub shaft
6. Driven sprocket
7. Output shaft
8. 2nd clutch waved plate
9. 2nd clutch steel plate
10. 2nd clutch fiber plate
11. 2nd clutch backing plate
12. Intermediate 4th band
13. Reverse input clutch assembly
14. Direct and coast clutch assembly
15. Input carrier assembly
16. Input flange and forward clutch hub assembly
17. Forward clutch assembly
18. Low/Reverse band
19. Forward clutch support assembly
20. Final drive sun gear
21. Differential and final drive assembly
22. Final drive internal gear

FIGURE 3-1 **The planetary gears on a typical transaxle with their associated brakes and clutches.**

© Delmar/Cengage Learning

The input shaft drives a planetary gearset that provides the various forward gears, a neutral position, and one reverse gear. Power flow through the gears is controlled by multiple friction disc packs, one-way brakes and clutches, and brake bands. These hold a member of the gearset when force is applied to them. By holding different members of the planetary gearset (see Figure 3-1), different gear ratios are possible. In all gear positions, one member is driven, one is held, and the other becomes the output.

Hydraulic pressure is routed to the correct driving or holding element by the transmission's valve body, which contains many hydraulic valves and controls the pressure of the hydraulic fluid and its direction.

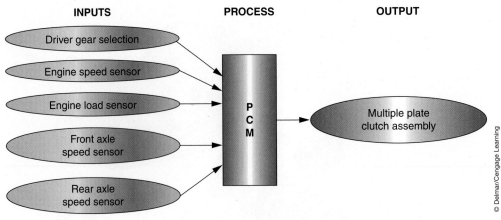

INPUTS **PROCESS** **OUTPUT**

Driver gear selection

Engine speed sensor

Engine load sensor

Front axle speed sensor

Rear axle speed sensor

P C M

Multiple plate clutch assembly

© Delmar/Cengage Learning

FIGURE 3-2 Up and downshifts are automatically made based on several different operating conditions.

Automatic transmission fluid is typically referred to as ATF.

An automatic transmission or transaxle selects gear ratios according to engine speed, engine load, vehicle speed, and other operating conditions (Figure 3-2). Both upshifts and downshifts occur automatically. The transmission can also be manually selected into a lower forward gear, reverse, neutral, or park. Depending on the forward gear range selected, the transmission may provide engine braking during deceleration.

In attempt to increase the efficiency of today's automobiles, manufacturers offer transmissions with a variety of forward speeds or gears. Most have at least four but can have as many as eight forward speeds. The additional speeds allow the engine to run at its most efficient speeds. The most common transmissions today have five or six speeds. These typically have two overdrive gears (fourth and fifth or fifth and sixth). While operating in overdrive, the transmission's input shaft rotates less than one full revolution for every one full revolution of the output shaft. This allows the engine to run at lower speeds during high vehicle speed, which increases engine life, reduces emissions, improves fuel economy, and reduces engine noise. Nearly all current automatic transmissions feature a torque converter clutch, which reduces the power lost through the operation of a conventional torque converter.

A transmission shift selector can have many positions (Figure 3-3). The selector lever is linked to the manual lever on the transmission to select the desired gear range. These ranges are:

P (park)—The transmission is in neutral with its output shaft locked to the case by engaging a parking pawl into the parking gear. The engine can be started in this selector position, and this is the only position in which the ignition key can be removed.

R (reverse)—The transmission is in reverse at a reduced or lower gear ratio.

N (neutral)—As in PARK, the transmission is in neutral and the engine can be started; however, the output shaft is not locked to the case and the parking brake should be applied.

OD (overdrive)—This is the normal driving gear range. Selection of this position provides for automatic shifting into all forward gears and allows for application and release of the converter clutch.

D (drive)—Selection of this gear range provides automatic shifting into all forward gears except overdrive gear, and allows for the application and release of the converter clutch. This position is selected when overdrive is not desired, such as when traveling on hilly or mountainous roads or when towing a trailer. Operating in overdrive during those conditions places

Shop Manual

Chapter 3, page 121

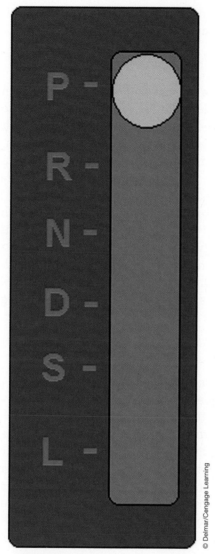

FIGURE 3-3 **A typical display for a shift range selector.**

© Delmar/Cengage Learning

an extraordinary amount of load on the engine and internal transmission components (such as gears and bearings), which can result in severe damage.

2 (manual second)—The selection of this gear position provides only second gear operation regardless of vehicle speed. Some transmissions, when MANUAL SECOND is selected, will start off in first gear, then upshift to second. Others will start and remain in second gear. The selection of this gear range is wise for acceleration on slippery surfaces or for engine braking while going down steep hills.

1 (manual low)—In most transmissions, this shift selection allows only first gear operation. Some transmissions are designed to shift into second gear when engine speed is high and manual low is selected. Selection of this gear at high speed results in a downshift into second gear. An automatic downshift to first will occur only when the vehicle's speed decreases to a predetermined level, normally about 30 mph. Once the transmission is in first gear, it will stay in low until the selector lever is moved to another position.

An automatic transmission automatically selects the gear ratio and torque output best suited for the existing operating conditions, vehicle load, and engine speed. A typical

A **torque converter** is a turbine-like device that uses hydraulic motion to transmit power from a driving source to a driven one.

Gear ratio is an expression of the number of revolutions made by a driving gear compared to the number of revolutions made by a driven gear.

A BIT OF HISTORY

In 1941, Chrysler introduced a four-speed, semi-automatic transmission with a hydraulic coupling.

The torque multiplication action of a torque converter is similar to adding another reduction gear to the transmission.

automatic transmission consists of four basic systems: the torque converter, planetary gearsets, hydraulic system, and **reaction members.**

AutoStick

Some late-model cars feature a manual shift option that allows the driver to control the shifting of the transmission. Chrysler calls this option "AutoStick." The benefit of this option is that it allows the driver to control gear changes without the use of a clutch pedal. When the shifter is moved into the AutoStick position, the transaxle remains in whatever gear it was using before AutoStick was activated. Moving the shifter toward the driver causes the transmission to upshift. Moving the shifter to the passenger side causes the transaxle to downshift. The selected gear is illuminated on the instrument panel.

The car can be launched in first, second, or third gear while in the AutoStick mode. Shifting into overdrive cancels the AutoStick option and the transmission resumes its normal overdrive operation. Although the driver has virtually full control of transmission shifting, there are limitations as to when shifts can be made. A shift into third or fourth gear cannot be made if the vehicle is going below 15 mph. Nor can the transmission be downshifted into first if the car is traveling at speeds greater than 41 mph. These safeguards prevent transmission and engine damage that could result if those gears were selected.

Torque Converters

Automatic transmissions use a **torque converter** to transfer engine torque from the engine to the transmission. The torque converter operates through hydraulic force provided by automatic transmission fluid.

The torque converter automatically transmits and disengages power from the engine to the transmission in relation to engine rpm. When the engine is running at the correct idle speed, there is an insufficient amount of fluid flow in the converter to allow for power transfer through the torque converter. As engine speed increases, fluid flow increases and creates a sufficient amount of force to transmit engine power through the torque converter to the input shaft of the transmission. This action connects the engine to the transmission's planetary gearset and provides some torque multiplication during some driving conditions.

The torque converter also absorbs the shock of the transmission while it is changing gears. The main components of a torque converter are the cover, turbine, impeller, and stator (Figure 3-4). The *converter cover* is connected to the flexplate and transmits power from the engine into the torque converter. The *impeller* is part of the converter cover. Its rear hub usually drives the transmission's oil pump. The *turbine* is splined to the *input shaft* and is driven by the force of the fluid from the impeller. The *stator* aids in redirecting the flow of fluid from the turbine to the impeller (Figure 3-5) and is equipped with a one-way clutch that allows it to remain stationary during maximum torque development. The one-way clutch will also be released and allowed to rotate with fluid flow once the fluid coupling point is obtained.

Most torque converters are fitted with a clutch mechanism that provides a mechanical connection between the transmission and the engine. The clutch essentially bypasses the normal operation of the T/C which means the turbine, stator, and impeller do not influence the transfer of power from the engine to the transmission when the clutch is engaged.

The clutch is only engaged when prescribed conditions are met. A computer controls the operation of the T/C clutch.

Many late-model transmissions are equipped with a modulated torque converter clutch. These allow for partial lockup during some conditions. Not only does it allow for a smooth engagement of the clutch, but it is also used to control transmission temperature by allowing some slip at the connection of the engine and transmission input shaft.

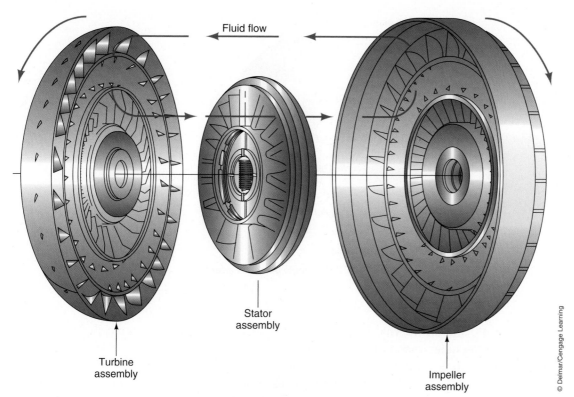

FIGURE 3-4 **The main components of a torque converter. The converter cover and impeller are shown as a single unit.**

Fluid flow

Stator assembly

Turbine assembly

Impeller assembly

© Delmar/Cengage Learning

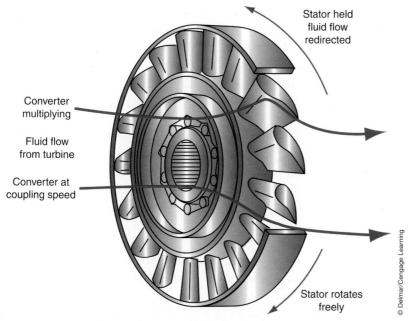

Stator held fluid flow redirected

Converter multiplying

Fluid flow from turbine

Converter at coupling speed

Stator rotates freely

© Delmar/Cengage Learning

FIGURE 3-5 **The stator aids in directing fluid flow from the turbine to the impeller.**

A BIT OF HISTORY

The transmission in a Model "T" Ford was based on a planetary gearset that was manually controlled.

Shop Manual
Chapter 3, page 88

PLANETARY GEARING

Planetary gearsets provide for the different gear ratios in an automatic transmission. The gear ratios are selected manually by the driver or automatically by the hydraulic control system, which engages and disengages the multiple-friction-disc packs and bands used to shift gears. A single planetary gearset consists of a sun gear, a planet carrier with three or more planet

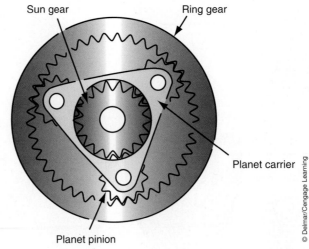

Sun gear   Ring gear

Planet carrier

Planet pinion

© Delmar/Cengage Learning

FIGURE 3-6 **A single planetary gearset.**

pinion gears, and an internal ring gear (Figure 3-6). These gears are typically helically cut gears that offer quiet operation. From a single planetary gearset, two reverse and five forward speeds are possible. The five forward speeds comprise two reduction ratios, a direct drive, and two overdrive ratios. These different speeds are obtained by holding one or more gearset members and driving another.

At the center of the planetary gearset is the sun gear. The placement of this gear in the gearset is the reason for its name; the Earth's sun is at the center of our solar system. Planet gears surround the sun gear, just like the Earth and other planets orbit the sun. These gears are mounted and supported by the planet carrier, and each gear spins on its own separate shaft. The planet gears are in constant mesh with the sun and ring gears. The ring gear is the outer gear of the gearset. It has internal teeth and surrounds the rest of the gearset. Its gear teeth are in constant mesh with the planet gears. The number of planet gears used in a planetary gearset varies according to the loads the transmission is designed to face. For heavy loads, the number of planet gears is increased to spread the workload over more gear teeth.

The planetary gearset can provide a gear reduction or overdrive, direct drive or reverse, or a neutral position. Because the gears are in constant mesh, gear changes are made without engaging or disengaging gears, as is required in a manual transmission. Rather, clutches and bands are used to drive, hold, or release different members of the gearset to get the proper direction of rotation or gear ratio.

Compound Planetary Gearsets

The input shaft provides power flow from the torque converter into the planetary geartrain.

A limited number of gear ratios are available from a single planetary gearset. To increase the number of available gear ratios, gearsets can be added. The typical automatic transmission has at least two planetary gearsets.

In automatic transmissions, the planetary gearsets (Figure 3-7) are connected together to provide the required gear ratios. There are two common designs of compound gearsets: the Simpson gearset, in which two planetary gearsets share a common sun gear, and the Ravigneaux gearset, which has two sun gears, two sets of planet gears, and a common ring gear. Some late-model six- and seven-speed transmissions use the Lepelletier system. This system connects a simple planetary gearset to a Ravigneaux gearset. In this arrangement, the ring gear of the simple gearset serves as the input to the gearsets and ring gear of the Ravigneaux serves as the output. Some Simpson-based transmissions are fitted with an additional single planetary gearset, which is used to provide an "add-on" overdrive gear. Although

The stator of a torque converter is called the reactor by some manufacturers.

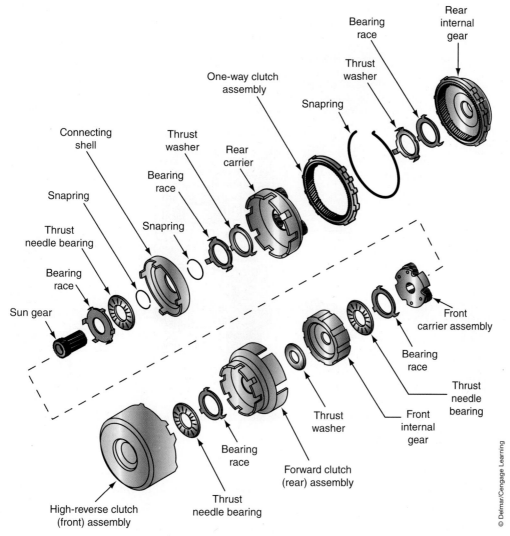

FIGURE 3-7 A Simpson planetary gearset.

many transmissions rely on compound planetary gearsets, some have two or three simple gearsets connected in tandem.

Planetary Gear–Based CVTs

Hybrid electric vehicles from Toyota and Ford rely on a planetary gearset to provide a continuously variable transmission (CVT). The transaxle contains two electric motor/generators, a differential, and a simple planetary gearset (Figure 3-8). The engine and the motor/generators are connected directly to the planetary gear unit. The planetary gearset is called the power-split device because it can transfer power among the engine, the motor/generators, and the drive wheels in nearly every possible combination. The power-split device splits power from the engine to different paths to drive one of the motor/generators, drive the car's wheels, or both. The other motor/generator can drive the wheels, assist the engine in driving the wheels, or be driven by the wheels. The speed ratios change in response to the torque applied to the various members of the gearset. In this arrangement, there are two basic sources of torque: the engine and an electric traction motor. Both rotate in the same direction but not at the same speed. Therefore, one can assist the rotation of the other or slow down the rotation of the other, or they can work together.

The input shaft is also called the turbine shaft.

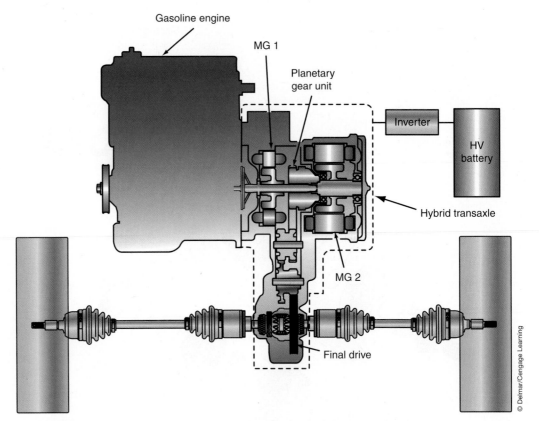

FIGURE 3-8 The transaxle assembly with MG1, MG2, and the power-split device.

CLUTCHES AND BANDS

Power flow through a planetary gearset depends entirely on what gearset members are the input or output and which member is held stationary. These actions are controlled by the driving and reaction members or devices.

Driving members connect the transmission's input shaft to a member of the planetary gearset. Therefore, they provide an input to the gearset. Driving devices are typically multiple-disc clutches, although in some cases, one-way driving clutches are used.

Reaction, or holding, members hold a gearset member to the transmission case in order to change gears. They act as a brake to hold a portion of a gearset in reaction. These devices are called reaction devices because they cause the other members of the gearset to react to the stationary member. The three types of reaction devices used today are multiple-disc clutches, bands, and one-way clutches.

One-way overrunning clutches are purely mechanical devices, whereas multiple-disc clutches and bands are hydraulically controlled mechanical devices.

Most automatic transmissions use two or more driving devices and two or more reaction devices in an automatic transmission.

Transmission Clutches

Transmission clutches are used as both driving and reaction devices.

Multiple-Disc Clutches. A multiple-disc clutch uses a series of friction discs to transmit torque or apply braking force. The discs have internal teeth that are sized and shaped to mesh with splines on the clutch assembly hub. Between each friction disc is a steel disc. These discs have external teeth that are sized and shaped to mesh with internal splines in the transmission housing or a clutch drum. The combination of friction and steel plates make up the clutch pack.

A **reaction member** of the planetary gearset is a component that is held stationary in a particular gear.

The area of a friction disc pack assembly with a 5-inch diameter is 19.6 square inches $(3.14 \times 2.5 \times 2.5)$.

74

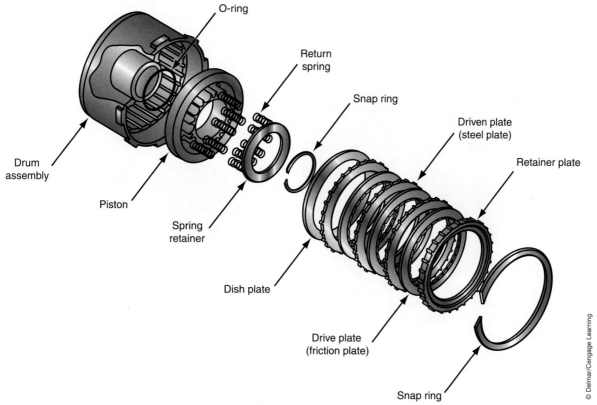

FIGURE 3-9 A multiple-disc clutch pack.

Labels in figure: O-ring, Return spring, Snap ring, Driven plate (steel plate), Retainer plate, Drum assembly, Piston, Spring retainer, Dish plate, Drive plate (friction plate), Snap ring

© Delmar/Cengage Learning

The clutch pack also has one very thick plate known as the pressure plate. The pressure plate has tabs around the outside diameter to mate with the channels in the clutch drum. Upon engagement, a clutch piston forces the clutch pack against the fixed pressure plate.

Clutch packs are enclosed in a large drum-shaped housing that can be either a separate casting or a part of the existing transmission housing (Figure 3-9). This drum housing also holds the other clutch components: cylinder, hub, piston, piston return springs, seals, pressure plate, friction plates, and snap rings. The hub is connected to the member of the planetary gearset that will receive the desired braking or transfer force when the clutch is applied or released.

One-way Clutches. A one-way clutch allows rotation in only one direction and operates at all times. One-way clutches can be either roller-type or sprag-type clutches. Both types of clutch can be used to hold or drive members of the planetary gearset. These clutches operate mechanically.

In a roller-type (Figure 3-10), rollers are held in place by springs to separate the inner and the outer race of the clutch assembly. One of the races is normally held by the transmission case and is unable to rotate. Around the inside of the outer race are several cam-shaped indentations. The rollers and springs are located in these pockets. Rotation of the race in one direction locks the rollers between the two races, preventing the race from moving. When the race is rotated in the opposite direction, the rollers move into the pockets and are not locked and the race is free to rotate.

A one-way sprag clutch (Figure 3-11) consists of a hub and a drum separated by figure-eight-shaped metal pieces called sprags. The sprags lock between the races when a race is turned in one direction. The sprags are longer than the distance between the two races. When a race rotates in one direction, the sprags float between the two races and allow the races to

One-way overrunning clutches, bands, and multiple-disc clutch packs are examples of apply devices used to provide input power to hold members of the gearset.

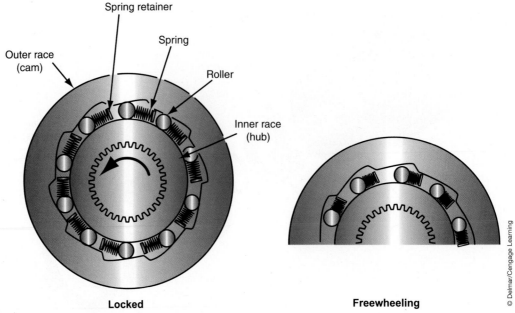

FIGURE 3-10 A roller-type one-way (overrunning) clutch assembly.

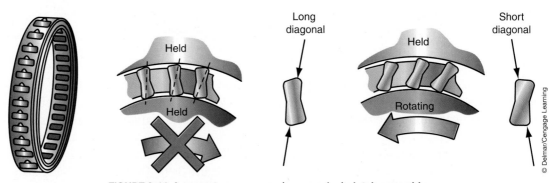

FIGURE 3-11 A sprag-type one-way (overrunning) clutch assembly.

move independently. When a race is moved in the opposite direction, the sprags straighten and lock the two races together.

Transmission Bands

A **band** is a braking assembly positioned around a drum. The band brings the drum to a stop by wrapping itself around it and holding it. The band is hydraulically applied by a servo assembly (Figure 3-12). Connected to the drum is a member of the planetary gearset. A band is a reaction member and causes the other members of the gearset to react as it holds one of the members stationary.

Bands provide excellent holding characteristics and require a minimum amount of space within the transmission housing. When a band closes around a rotating drum, a wedging action takes place to stop the drum from rotating. The wedging action is known as a self-energizing action. A typical band is designed to be larger in diameter than the drum it surrounds. This design promotes self-disengagement of the band from the drum when servo

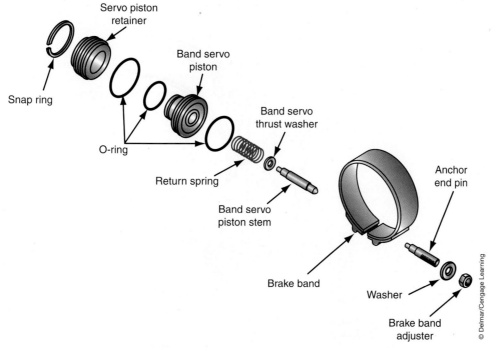

FIGURE 3-12 A brake band with its servo.

apply force is decreased to less than servo release spring tension. A friction material is bonded to the inside diameter of the band.

Band lugs connect the band with the servo through the actuating (apply) linkage or the band anchor (reaction) at the opposite end. The band's steel strap is designed with slots or holes to release fluid trapped between the drum and the applying band.

Bands used in automatic transmissions are rigid, flexible, single wrap, or double wrap types (Figure 3-13). Steel single wrap bands are used to hold geartrain components driven by

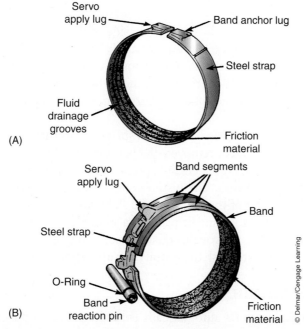

FIGURE 3-13 (A) Typical single wrap and (B) double wrap bands.

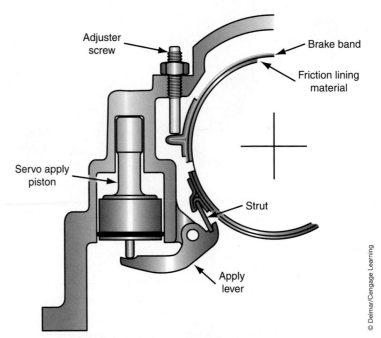

FIGURE 3-14 A servo for engaging and disengaging a brake band.

high-output engines. The double wrap band is a circular external contracting band normally designed with two or three segments. As the band closes, the segments align themselves around the drum and provide a cushion.

Transmission Servos. A servo converts hydraulic pressure into a mechanical force that applies a band to hold a drum stationary. Simple and compound servos are used in modern transmissions.

In a simple servo (Figure 3-14), the servo piston fits into the servo cylinder and is held in the released position by a coil spring. The piston is sealed with a rubber ring, which keeps fluid pressure confined to the apply side of the servo piston.

To apply a band, fluid pressure is to the apply side of the servo piston. The piston moves against the return coil spring and develops servo apply force. This force is applied to the band lug through the apply lever and strut. At the opposite end of the band is the anchor strut or end pin and adjustment screw. These hold that end of the band stationary as the band tightens around the rotating drum. The rotating drum comes to a stop and is held stationary by the band.

When servo apply force is released, the return spring forces the servo piston to move in the cylinder. With the servo apply force removed, the band springs free and permit drum rotation.

When the compound servo is applied, fluid pressure flows through the hollow piston pushrod to the apply side of the piston. The piston compresses the servo coil spring, and forces the pushrod to move one end of the band toward the adjusting screw and anchor. The band tightens around the rotating drum and brings it to a stop. Fluid pressure is then applied to the release side of the servo piston to provide equal pressure on both sides of the piston and allow the tension of the servo spring to push the piston back up its bore. This action releases the band.

TRANSMISSION CONTROLS

An automatic transmission shifts automatically through the gears during its forward range. The forward range is selected by the driver through the gearshift lever. The selection of park, neutral, and reverse is controlled by the driver as well. In order for the transmission to operate in the desired range, a series of other controls is needed. These controls may be mechanical, electrical (electronic), or hydraulic.

The mechanical controls comprise the gear selector linkage and other linkages that work with the electronic and hydraulic control devices.

Electronic Controls

Electrical and electronic circuits are used to monitor operating conditions, perform work, or control the operation of a device. An electrical switch is a simple electrical control that turns a circuit on or off, and consequently turns the electrical device connected to the circuit on or off. Switches can be opened or closed by the driver, mechanical linkages, or a predetermined condition. The latter typically is a low or high hydraulic pressure.

Electrical sensors are also a type of control; however, rather than merely opening and closing a circuit, sensors vary the flow of electricity through the circuit. Sensors are typically potentiometers or variable resistors that respond to changes in conditions. The sensor's resistance reflects the condition it is monitoring.

Solenoids are commonly used in transmissions. These devices convert electrical energy into mechanical energy. A solenoid provides for linear movement. Solenoids in automatic transmissions are used to control the direction and flow of hydraulic fluid pressure.

Until recently, all automatic transmissions were controlled by hydraulic circuits. However, nearly all transmissions now control the operation of the torque converter clutch and transmission through a computer and hydraulics (Figure 3-15). Based on information received from various electronic sensors and switches, a computer can control the operation of the torque converter and transmission's shift points. Computer-controlled electrical solenoids are typically used to control shifting and fluid pressures (Figure 3-16).

Shop Manual
Chapter 3, page 108

An understanding of the transmission's hydraulic control circuit is essential for proper diagnostics.

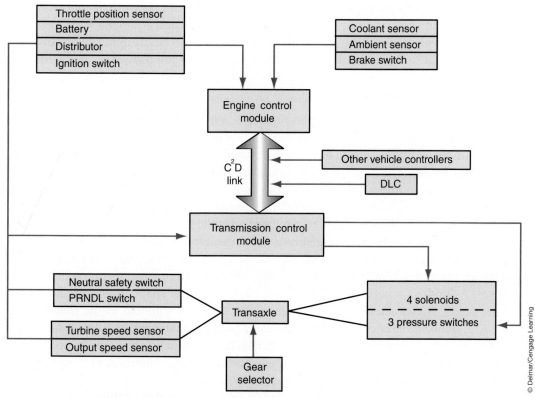

FIGURE 3-15 A basic look at the inputs and outputs for an electronically controlled transmission.

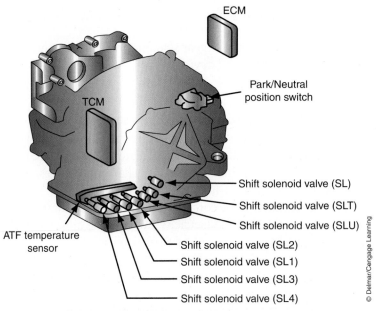

ECM

Park/Neutral
position switch

TCM

Shift solenoid valve (SL)

Shift solenoid valve (SLT)

Shift solenoid valve (SLU)

ATF temperature
sensor

Shift solenoid valve (SL2)

Shift solenoid valve (SL1)

Shift solenoid valve (SL3)

Shift solenoid valve (SL4)

© Delmar/Cengage Learning

FIGURE 3-16 An electronically controlled automatic transaxle.

Hydraulic Controls

ATF is a special oil designed to allow for proper transmission operation. Transmissions are equipped with a fluid cooler that prevents overheating of this fluid, which can result in damage to the transmission. The transmission's pump is the source of all fluid flow in the hydraulic system. It provides a constant supply of fluid under pressure to operate, lubricate, and cool the transmission. **Pressure-regulating valves** change the fluid's pressure to control the shift quality of a transmission and the shift points of the transmission equipped with a governor. **Flow-directing valves** direct the pressurized fluid to the appropriate apply device to cause a change in gear ratios. The hydraulic system also keeps the torque converter filled with fluid.

LAWS OF HYDRAULICS

An automatic transmission is a complex hydraulic circuit. To better understand how an automatic transmission works, a good understanding of how basic hydraulic circuits work is needed. A simple hydraulic system has liquid, a pump, lines to carry the liquid, control valves, and an output device. The liquid must be available from a continuous source, such as an oil pan or sump. An oil pump is used to move the liquid through the system. The lines to carry the liquid may be pipes, hoses, or a network of internal bores or passages in a single housing (Figure 3-17). Control valves are used to regulate hydraulic pressure and direct the flow of the liquid. The output device is the unit that uses the pressurized liquid to do work.

As can be seen, hydraulics involves the use of a liquid or fluid. Hydraulics is the study of liquids in motion. All matter in the universe, exists in three basic forms: solids, liquids, and gases. A fluid is something that does not have a definite shape; therefore, liquids and gases are fluids. A characteristic of all fluids is that they will conform to the shape of their container. A major difference between a gas and a liquid is that a gas will always fill a sealed container, whereas a liquid may not. A gas will also readily expand or compress according to the pressure exerted on it. A liquid will typically not compress, regardless of the pressure on it. Therefore, liquids are considered noncompressible fluids.

Liquids will, however, predictably respond to pressures exerted on them. Their reaction to pressure is the basis of all hydraulic applications. This fact allows hydraulics to do work.

Pressure-regulating valves change the fluid's pressure to control the shift quality of a transmission and the shift points of the transmission equipped with a governor.

Flow-directing valves direct the pressurized fluid to the appropriate apply device to cause a change in gear ratios.

To better understand how an automatic transmission works, a good understanding of how basic hydraulic circuits work is needed.

The operation of an automatic transmission is based on an important aspect of hydraulics, fluid power, which is a liquid's ability to transmit energy.

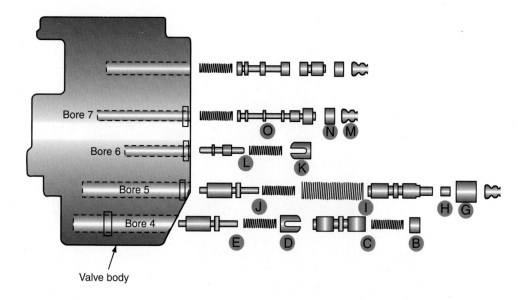

	Bore 4		Bore 5		Bore 6		Bore 7
A	Spring retainer plate	**F**	Clip	**K**	Retainer plate	**M**	Clip
B	Bore plug	**G**	Sleeve	**L**	TV limit valve and spring	**N**	Bore plug
C	Orifice control vale and spring	**H**	Plug			**O**	1–2 shift valve and spring
D	Spring retainer plate 1	**I**	3–4 shift valve and spring				
E	2–3 capacity modulator valve and spring	**J**	3–4 TV modulator valve and spring				

FIGURE 3-17 The various bores, valves, and passages in a valve body for a transmission's hydraulic system.

© Delmar/Cengage Learning

APPLICATION OF HYDRAULICS IN TRANSMISSIONS

A common hydraulic system within an automatic transmission is the servo assembly. The servo assembly is used to control the application of a band. The band must tightly hold the drum it surrounds when it is applied. The holding capacity of the band is determined by the construction of the band and the pressure applied to it. This pressure or holding force is the result of the action of a servo. The servo multiplies the force through hydraulic action (Figure 3-18).

If a servo has an area of 10 square inches and has a pressure of 70 psi applied to it, the apply force of the servo is 700 pounds. The force exerted by the servo is increased further by its lever-type linkage and the self-energizing action of the band. The total force applied by the band is sufficient to stop and hold the rotating drum connected to a planetary gearset member.

A multiple-disc assembly may also be used to stop and hold gearset members. This assembly also uses hydraulics to increase its holding force. If the fluid pressure applied to the clutch assembly is 70 psi and the diameter of the clutch piston is 5 inches, the force applying the clutch pack is 1374 pounds. If the clutch assembly uses a **Belleville spring** or piston spring (Figure 3-19), which adds a mechanical advantage of 1.25, the total force available to engage the clutch will be 1374 pounds multiplied by 1.25, or 1717 pounds.

A **Belleville spring** is a tempered spring steel cone-shaped plate used to aid the mechanical force in a pressure plate assembly.

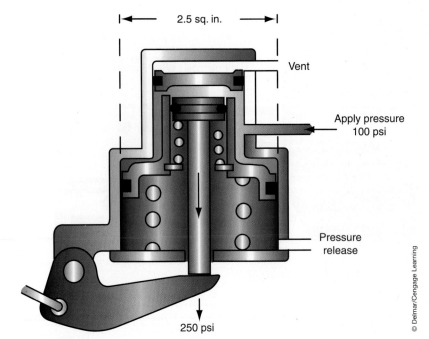

FIGURE 3-18 A servo uses hydraulics to increase the gripping power of a brake band.

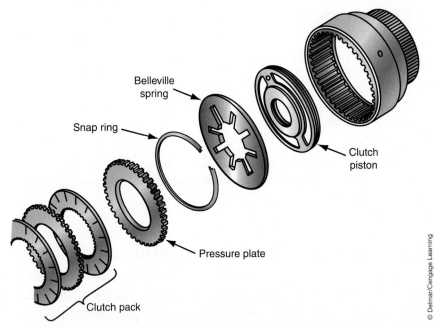

FIGURE 3-19 A clutch assembly with a Belleville (piston) spring.

ATF

The ATF circulating through the transmission and torque converter and over the parts of the transmission cools the transmission. The heated fluid typically moves to a transmission fluid cooler, where the heat is removed. As the fluid lubricates and cools the transmission, it also cleans the parts. The dirt is carried by the fluid to a filter where the dirt is removed.

Another critical job of ATF is its role in shifting gears. ATF moves under pressure throughout the transmission and causes various valves to move. The pressure of the ATF changes with changes in engine speed and load.

ATF is also used to operate the various apply devices (clutches and brakes) in the transmission. At the appropriate time, a switching valve opens and sends pressurized fluid to the

Shop Manual

Chapter 3, page 94

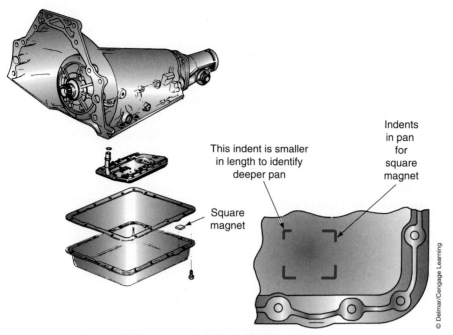

FIGURE 3-20 **A typical transmission oil pan.**

apply device that engages or disengages a gear. The valving and hydraulic circuits are contained in the valve body.

Reservoir

All hydraulic systems require a reservoir to store fluid and to provide a constant source of fluid for the system. In an automatic transmission, the reservoir is the pan, typically located at the bottom of the transmission case (Figure 3-20). Transmission fluid is forced out of the pan by atmospheric pressure into the pump and returned to it after it has circulated through the selected circuits. A transmission dipstick and filler tube are used to check the level of the fluid and to add ATF to the transmission. The tube and dipstick are normally located at the front of the transaxle housing and the right front of a transmission housing. Other transmissions have a side plug on the pan or the transmission to check and replenish fluid level.

Venting

In order to allow the fluid to be pumped through the transmission by the pump, all reservoirs must have an air vent that allows atmospheric pressure to force the fluid into the pump when the pump creates a low pressure at its inlet port. The pans of many automatic transmissions vent through the handle of the dipstick; others rely on a vent in the transmission case. Transmissions must also be vented to allow for the exhaust of built-up air pressure that results from heat and the moving components inside the transmission. The movement of these parts can force air into the ATF, which would not allow it to increase in pressure, cool, or lubricate the transmission properly.

Transmission Coolers

The removal of heat from ATF is extremely important to the durability of the transmission. Excessive heat causes the fluid to break down. Once broken down, ATF no longer lubricates well and has poor resistance to oxidation. Oxidized ATF may damage transmission seals. When a transmission is operated for some time with overheated ATF, varnish is formed inside the transmission. Varnish build up on valves can cause them to stick or move slowly. The result is poor shifting and glazed or burnt friction surfaces. Continued operation can lead to the need for a complete rebuilding of the transmission.

Shop Manual
Chapter 3, page 96

It is important to note that ATF is designed to operate at 175°F (80° C). At this temperature, the fluid should remain effective for 100,000 miles. However, when the operating temperature increases, the useful life of the fluid quickly decreases. A 20-degree increase in operating temperature will decrease the life of ATF by one-half!

Transmission housings are fitted with ATF cooler lines (Figure 3-21). These lines direct the hot fluid from the torque converter to the transmission cooler, normally located in the vehicle's radiator. The heat of the fluid is reduced by the cooler and the cool ATF returns to the transmission. In some transmissions, the cooled fluid flows directly to the transmission's bushings, bearings, and gears. Then, the fluid is circulated through the rest of the transmission. The cooled fluid in other transmissions is returned to the oil pan, where it is drawn into the pump and circulated throughout the transmission.

Some vehicles are equipped with an auxiliary fluid cooler (Figure 3-22), in addition to the one in the radiator. This cooler serves to remove additional amounts of heat from the fluid before it is sent back to the transmission. Auxiliary coolers are most often found on heavy-duty and performance vehicles.

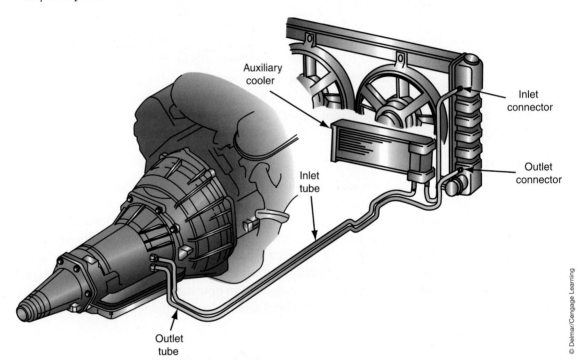

FIGURE 3-21 Routing of transmission cooler lines.

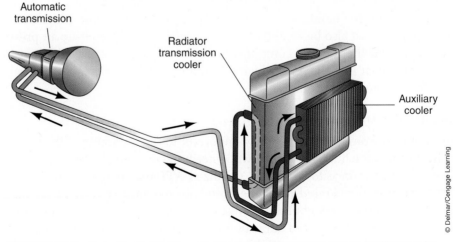

FIGURE 3-22 Routing of transmission cooler lines when the vehicle has an auxiliary cooler.

Valve Body

For efficient operation of the transmission, the bands and multiple-disc packs must be released and applied at the proper time. It is the responsibility of the hydraulic control system to control the hydraulic pressure being sent to the different hydraulic members. Central to the hydraulic control system is the valve body assembly (Figure 3-23). This assembly is made of two or three main parts: a valve body, *separator plate*, and *transfer plate* (Figure 3-24). These parts are bolted as a single unit to the transmission housing. The valve body is machined from aluminum or iron and has many precisely machined bores and fluid passages. Various valves are fit into the bores and the passages direct fluid to various valves and other parts of the transmission. The separator and transfer plates are designed to seal off some of these passages and to allow fluid to flow through specific passages.

Shop Manual
Chapter 3, page 114

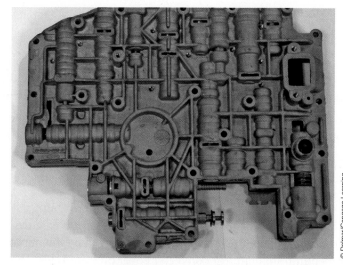

FIGURE 3-23 A valve body.

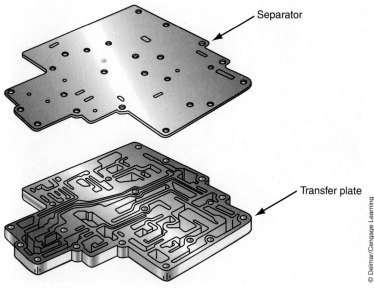

Separator

Transfer plate

FIGURE 3-24 The separator and transfer plates of a valve body.

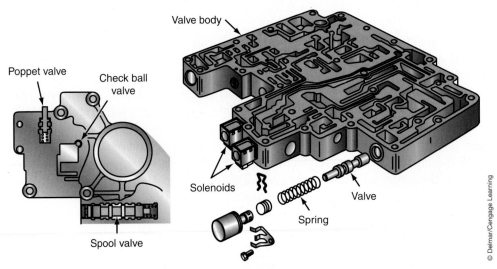

FIGURE 3-25 The various valves in a typical valve body.

The purpose of a valve body is to sense and respond to engine and vehicle load, as well as to meet the needs of the driver. Valve bodies are normally fitted with three different types of valves: spool valves, *check-ball valves*, and *poppet valves* (Figure 3-25). The purpose of these valves is to start, stop, or use movable parts to regulate and direct the flow of fluid throughout the transmission.

Oil Pump

The only source of fluid flow through the transmission is the oil pump. Three types of oil pumps are used in automatic transmissions: the *gear-type pump* (Figure 3-26), *rotor-type pump*, and *vane-type pump* (Figure 3-27). Oil pumps are driven by the pump drive hub of the torque converter or oil pump shaft and/or converter cover on transaxles. Therefore, whenever

FIGURE 3-26 A typical fixed displacement, gear-type oil pump.

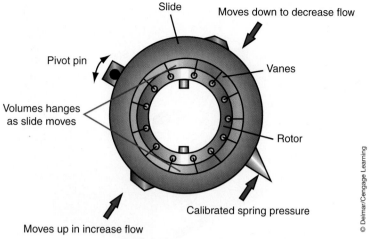

Slide

Moves down to decrease flow

Pivot pin

Vanes

Volumes hanges
as slide moves

Rotor

Calibrated spring pressure

Moves up in increase flow

© Delmar/Cengage Learning

FIGURE 3-27 A typical variable displacement, vane-type oil pump.

the torque converter cover is rotating, the oil pump is driven. The oil pump creates fluid flow throughout the transmission. The valve body regulates and directs the fluid flow to meet the needs of the transmission.

Transmissions are capable of creating excessive amounts of pressure, which may cause damage. Therefore, the transmission is equipped with a pressure-regulator valve.

Pressure-Regulator Valve

The *pressure-regulator valve* is normally located in the pump assembly. It maintains basic fluid pressure. Pressure-regulator valves are typically spool valves that toggle back and forth in their bore to open and close an exhaust passage. By opening the exhaust passage, the valve decreases the pressure of the fluid. As soon as the pressure decreases to a predetermined amount, the spool valve moves to close off the exhaust port and pressure again begins to build. The action of the spool valve regulates the fluid pressure.

Many late-model transmissions use an electronic pressure control (EPC or PC) solenoid to regulate system pressure.

Governor Assembly

The governor assembly is driven by the transmission's output shaft, senses road speed, and sends a fluid pressure signal to the valve body to either upshift or downshift. When vehicle speed is increased, the pressure developed by the governor is directed to the shift valve. As the speed (and therefore the pressure) increases, the spring tension on the shift valve is overcome and the valve moves. This action causes an upshift. Likewise, a decrease in speed will result in a decrease in pressure and a downshift.

Although the governor sends a signal that will force an upshift, engine load may cause a delay in the shift. This allows for operation in a lower gear when there is a heavy load and the vehicle needs the gear reduction. During heavy load operation, the governor pressure must be strong enough to overcome the high throttle pressure plus the spring tension on the shift valve before it can force an upshift. Because of this, the transmission will remain in a particular gear range until a higher than normal engine speed is reached.

PRESSURE BOOSTS

When the engine is operating under a heavy load, fluid pressure should be increased to improve the holding capacity of a multiple-disc clutch or band. Increasing fluid pressure reduces the chance of slipping while under heavy load. This is done by sending pressurized

Shop Manual
Chapter 3, page 121

Shop Manual
Chapter 3, page 120

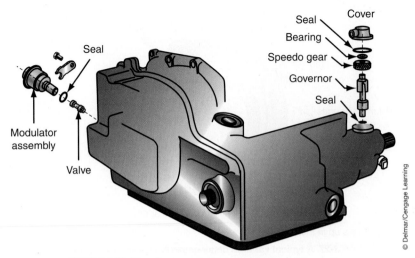

FIGURE 3-28 The placement of a vacuum modulator and governor assembly in a transaxle housing.

fluid to one side of the pressure regulator's spool valve. This pressure delays the exhausting of fluid by the valve which allows pressure to build to a higher than normal level.

Engine load is often electronically monitored by various electronic sensors that send information to an electronic control unit, which in turn controls the pressure at various points in the transmission. Load can also be monitored by throttle pressure. Throttle pedal movement, via a throttle cable, moves a **throttle valve** in the valve body. When the throttle pedal is depressed, the throttle valve opens and sends fluid under pressure to the pressure regulator. This delays the opening of the pressure regulator valve, which allows for an increase in pressure. When the driver lets off the throttle pedal, the pressure regulator valve is free to move and normal pressure is maintained. Many early transmissions were equipped with a **vacuum modulator**, which uses **engine vacuum** to monitor engine load (Figure 3-28). The vacuum modulator allowed for an increase in pressure when vacuum was low and a decrease in pressure when vacuum was high.

MAP Sensor

The most commonly used electronic load sensor is the **Manifold Absolute Pressure (MAP) sensor**. The MAP sensor (Figure 3-29) senses air pressure in the intake manifold. This sensor can be thought of as an electronic vacuum gauge since it responds to the low pressure in the manifold. When there is a great amount of load on the engine, the air pressure in the manifold will be high (low vacuum). Likewise, when there is little or no load on the engine, manifold pressure will be low (high vacuum). A pressure-sensitive ceramic or silicon element and electronic circuit in the sensor generates a voltage signal that changes in direct proportion to pressure. A MAP sensor measures manifold air pressure against a pre-calibrated absolute pressure; therefore, the readings from these sensors are not adversely affected by changes in operating altitudes or barometric pressures.

Vacuum Modulators

A vacuum modulator is a **load-sensing device** that regulates fluid pressure in response to a vacuum signal, which varies with throttle opening and vehicle load (Figure 3-30). Engine vacuum is low when there is a heavy load and high when the load is low. When engine vacuum is low, the modulator allows fluid flow to enter onto the spool valve in the pressure regulator, which allows for an increase in pressure.

> A **vacuum modulator** is a device that causes fluid pressure to increase when engine vacuum is low.

> **Engine vacuum** is the best indicator of the amount of load on an engine.

> A **load-sensing device** causes a change in operation in response to a change in engine load.

FIGURE 3-29 A MAP sensor.

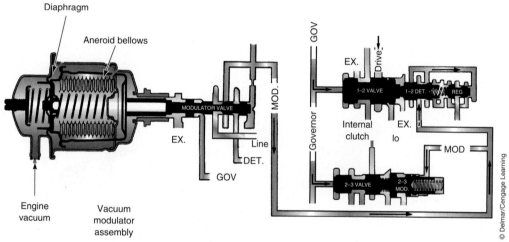

FIGURE 3-30 A vacuum modulator is a load-sensing device that regulates fluid pressure in response to a vacuum signal, which varies with throttle opening and vehicle load.

The vacuum modulator is normally a small round metal canister that is threaded or push fit into the transmission's housing. A vacuum line connects the vacuum modulator to the intake manifold of the engine. The vacuum modulator is divided into two chambers by a diaphragm. The chamber closest to the transmission is open to atmospheric pressure and the other chamber is closed and is connected to the vacuum source. A pushrod is connected to the open side of the diaphragm and is connected the **modulator valve**. Inside the closed chamber is a coil spring positioned between the diaphragm and the end of the canister. When engine load is low and the throttle closed, high manifold vacuum retracts the diaphragm and the pushrod against spring tension to reduce throttle pressure. As engine load increases and the throttle is opened wider, manifold vacuum drops and the diaphragm spring forces the pushrod into firmer contact with the modulator valve to increase fluid flow in the modulator circuit.

Low engine vacuum is an indication that the engine is under heavy load.

A vacuum diaphragm responds to differences in atmospheric pressure on one side and vacuum on the other side. Some vacuum modulators use an altitude-compensating vacuum valve with a spring-type bellows added to the atmospheric side of the valve. Inside the bellows is very low air pressure and the bellows collapses when it is exposed to atmospheric pressure. As altitude increases and atmospheric pressure decreases, the bellows unit expands because of the decrease in pressure. This allows the bellows unit to exert pressure against the atmospheric side of the diaphragm to compensate for the lower atmospheric pressure at higher altitudes.

Throttle Linkages

Some transmissions use mechanical linkages to control throttle valve pressure. The throttle cable is normally connected to a lever located at transmission. The throttle valve is moved by the throttle lever, which is moved by the throttle linkage at the throttle body (Figure 3-31). Some transmissions use a rod-type linkage assembly, not a cable, to move the throttle lever.

When the throttle pedal is depressed, the throttle valve opens to produce throttle pressure, which is directed to the pressure regulator valve. This pressure helps the pressure regulator valve spring hold the pressure regulator valve in position to close the exhaust port, which causes the pressure to increase. This increased pressure is used to hold the apply devices and causes a delay in the upshift. As the throttle pedal is released, throttle pressure is decreased. This decrease in pressure allows for a downshift when vehicle speed reaches a particular point.

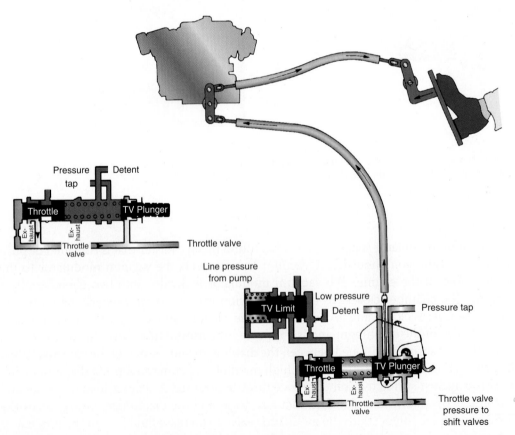

FIGURE 3-31 The throttle linkage moves the throttle valve against spring pressure, the more the throttle is opened, the higher the throttle pressure, until throttle pressure equals mainline pressure.

If the throttle pedal is quickly depressed, there is a sudden increase in throttle pressure and it becomes higher than governor pressure. The high throttle pressure and coil spring tension force the shift valve to move to the downshifted position against governor pressure. The transmission automatically downshifts to the next lower gear. The position of the shift valve blocks fluid at its inlet port, which prevents an upshift during this time.

Kickdown Valve

Often, the hydraulic circuit is fitted with a **kickdown** valve that applies additional pressure to the shift valve and provides for a quick and positive downshift. When the throttle pedal is quickly opened wide, throttle pressure rapidly increases and directs a large amount of pressure onto the kickdown valve. This moves the kickdown valve, which opens a port and applies pressure against the shift valve. The spring tension on the shift valve, the kickdown pressure, and throttle pressure will push on the end of the shift valve causing it to move (Figure 3-32). This forces a quick downshift to the next lower gear. The position of the shift valve blocks fluid at its inlet port, which prevents an upshift during this time.

A transmission can also automatically downshift when it is operating under a load, such as climbing a hill. During this time, throttle pressure exceeds governor pressure and forces a downshift.

Some transmissions use a solenoid kickdown valve operated electrically by a switch on the throttle linkage. This switch senses when the throttle plate is wide open and allows the kickdown valve to increase throttle pressure.

Electronic Pressure Control

Late-model transmissions use electronics to control transmission hydraulic pressures and operation. Sensors, such as the TP sensor, provide information on operating conditions to the electronic control unit. The TP sensor does the same thing as the throttle cable on earlier transmissions.

Based on the input signals, the control unit sends a varying amount of current to the pressure control solenoid located on the valve body. The control unit controls the resultant

Shop Manual
Chapter 3, page 127

Shop Manual
Chapter 3, page 120

A **MAP sensor** measures changes in the intake manifold pressure that result from changes in engine load and speed.

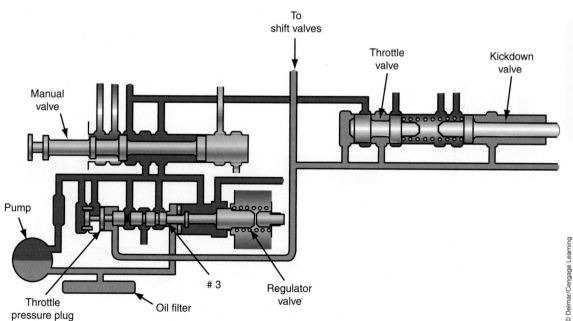

FIGURE 3-32 Fluid flow for the throttle and kickdown valves.

© Delmar/Cengage Learning

pressure by altering the duty cycle of the solenoid. Current flow and hydraulic pressure are proportional. Therefore, the throttle pressure signal increases as the solenoid duty cycle increases. When the pressure is applied to one end of the regulator valve, it pushes the valve against spring pressure and delays the shift.

SHIFT QUALITY

All transmissions are designed to change gears at the correct time, according to engine speed and load and driver intent. However, transmissions are also designed to provide for positive change of gear ratios without jarring the driver or passengers. If a band or clutch is applied too quickly, a harsh shift will occur. *Shift feel* is controlled by the pressure at which each hydraulic member is applied or released, the rate at which each is pressurized or exhausted, and the relative timing of the apply and release of the members.

To improve shift feel during gear changes, a band is often released while a multiple friction disc pack is being applied. The timing of these two actions must be just right or both components will be released or applied at the same time, causing engine flare-up or clutch and band slippage. Several other methods are also used to smooth gear changes and improve shift feel.

Multiple friction disc packs sometimes contain a wavy spring-steel separator plate that helps smoothen the application of the clutch. Shift feel can also be smoothed out by using a restricting **orifice** or an **accumulator** piston (Figure 3-33) in the band or clutch apply circuit. A restricting orifice or check ball (Figure 3-34) in the passage to the apply piston restricts fluid flow and slows the pressure increase at the piston by limiting the quantity of fluid that

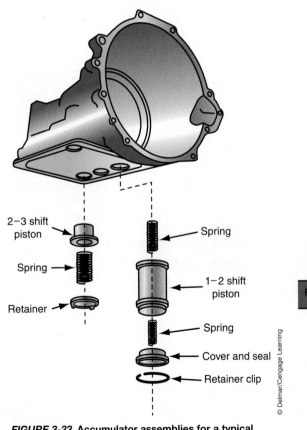

2–3 shift piston

Spring

Retainer

Spring

1–2 shift piston

Spring

Cover and seal

Retainer clip

© Delmar/Cengage Learning

FIGURE 3-33 Accumulator assemblies for a typical transmission.

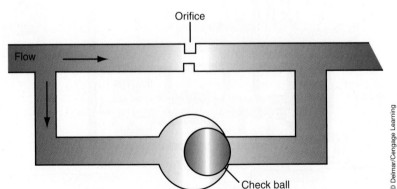

Orifice

Flow

Check ball

© Delmar/Cengage Learning

FIGURE 3-34 A check ball in the cylinder of a clutch pack.

can pass in a given time. An accumulator piston slows pressure buildup at the apply piston by diverting a portion of the pressure to a second spring-loaded piston in the same hydraulic circuit. This delays and smooths the application of a clutch or band.

Manufacturers have also applied electronics to get the desired shift feel. One of the most common techniques is the pulsing (turning on and off) of the shift solenoids. Doing this prevents the immediate engagement of a gear by allowing some slippage.

Shift Timing

Shift timing is determined by throttle pressure and governor pressure acting on opposite ends of the shift valve. When a vehicle is accelerating from a stop and the throttle is pressed down at least half way, the throttle pressure will be high and governor pressure will be low. As vehicle speed increases, the throttle pressure decreases and the governor pressure increases. When governor pressure overcomes throttle pressure and the spring tension at the shift valve, the shift valve moves to direct pressure to the appropriate apply device and the transmission upshifts.

TRANSMISSION AND TRANSAXLE HOUSINGS

Automatic transmission housings are typically aluminum castings. Aluminum housings are lightweight and are designed to allow for quick heat dissipation. The torque converter and transmission housings are normally cast as a single unit; however, in some designs the torque converter housing is a separate casting (Figure 3-35). By bolting the torque converter housing to the transmission housing, engineers can use the same transmission with different-sized torque converters to meet the specific needs of particular types of vehicles. Also, at the rear of some transmission housings is a bolted-on extension housing (Figure 3-36). This allows for the use of housings of different lengths; therefore, the transmission can be used on vehicles with different wheelbases.

Transaxle housings (Figure 3-37) are typically composed of two or three separate castings bolted together. One of these castings is the torque converter housing. Since FWD vehicles do not use a driveshaft to convey power to a drive axle, many transaxles are not fitted with an extension housing. Although they serve many of the same purposes as a transmission, transaxle housings are considerably different in appearance from transmission housings.

Shop Manual
Chapter 3, page 120

Extension housings conceal the transmission's output shaft.

The wheelbase of a vehicle is best defined as the distance between the front and rear axles.

A BIT OF HISTORY

The 1936 Imperia, which was built in Belgium, had a fully automatic transmission with a V-8 engine and front-wheel-drive.

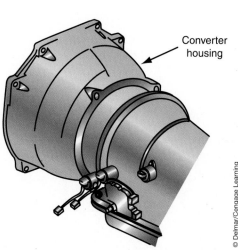

FIGURE 3-35 A torque converter housing cast separately from the transmission housing.

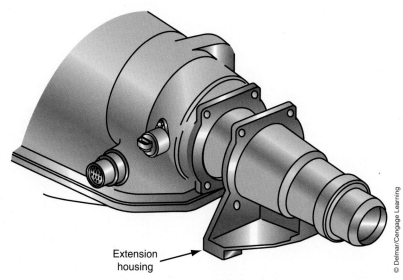

FIGURE 3-36 A transmission housing with a bolted-on extension housing.

© Delmar/Cengage Learning

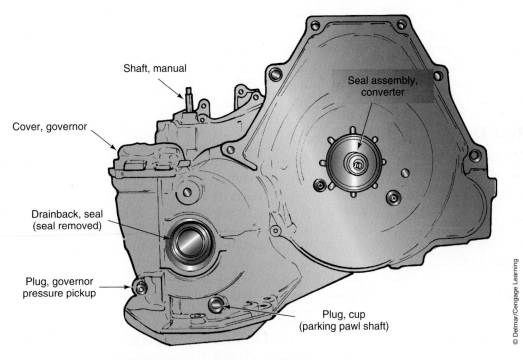

FIGURE 3-37 A side view of a transaxle housing. Note that the converter housing is an integral part of the housing.

Transmission housings are cast to secure and accommodate key components of the transmission. For example, linear keyways are cut into the inside diameter of the housing. These keyways are designed to hold a multiple-disc brake assembly in position in the housing (Figure 3-38). Some transmission housings have round structures projecting from the side. These are the cylinders of the servo and accumulator assemblies in which pistons, springs, and seals are fitted to operate the bands of the transmission. On some transaxles, these projections house the governor assembly and are sealed with a removable cover.

Some housings have band-adjusting screws located on the opposite side of the housing from the servo assemblies. These screws are adjusted to compensate for band wear. The band-adjusting screws can have square or hex heads and are locked in place with a jam nut.

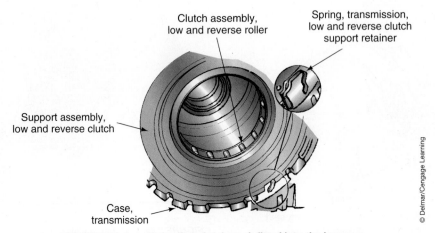

FIGURE 3-38 A multiple-disc clutch pack fitted into the keyways cast into the transmission housing.

Mechanical Connections

To monitor vehicle speed, vehicles are equipped with a speedometer. A speedometer can operate mechanically or electrically. On most new vehicles, it is an electrical device. A vehicle speed sensor (VSS) is most commonly found mounted in the housing near the transmission's **output shaft**. The shaft is fitted with a trigger tooth. As the output shaft rotates, the trigger rotates past the VSS. The VSS then sends an electrical signal to the processing unit. This unit then translates the signal into vehicle speed. Vehicles with a mechanical speedometer have a speedometer drive assembly geared to the output shaft. A flexible drive cable connects the drive assembly to the speedometer in the instrument panel.

Linkages

In addition to the throttle valve (TV) linkage, there may be another cable or rod linkage connected to the housing. This could be the gear selector linkage. As the gearshift selector is moved into its various positions, the linkage moves the manual shift lever, which in turn moves the manual shift valve in the transmission's valve body (Figure 3-39). Each gear position of the lever or shift valve is held in position by the internal linkage, which seats itself into the various seats or detents for the gear range selected.

Electrical Connections

Many different electrical switches, sensors, and connectors may be connected to the housing (Figure 3-40). As more electronics are used to control and monitor the functioning of the

Shop Manual

Chapter 3, page 121

Basically, an **output shaft** is the shaft that delivers the power from the transmission.

A detent is a half-round seat that a ball or roller fits into to hold a lever in position.

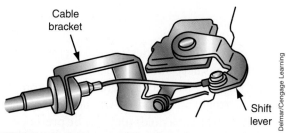

FIGURE 3-39 As the gearshift selector is moved into its various positions, the cable moves the manual shift lever at the transmission.

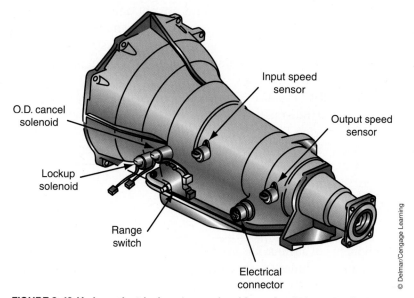

FIGURE 3-40 Various electrical sensors, solenoids, and switch connections to transmission housing.

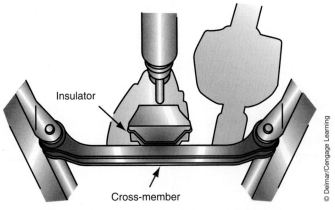

FIGURE 3-41 Typical rear transmission-to-frame mount.

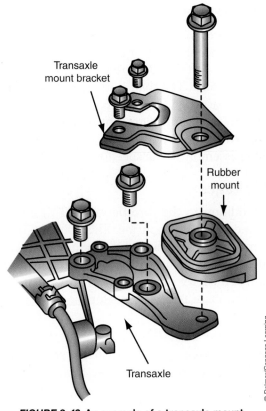

FIGURE 3-42 An example of a transaxle mount.

transmission, more connectors and electrical devices will be found at the housing. One of the more basic switches is the neutral safety switch, which allows the engine to start only when the transmission is in neutral or park. The backup light switch is sometimes incorporated with the neutral safety switch, as are other shift position switches. These switches may be individual switch assemblies. Connectors for shift solenoids may also be present on the outside of the case.

MOUNTS

Shop Manual
Chapter 3, page 105

The weight of the transmission or transaxle is supported by the engine and its mounts and by a transmission mount (Figure 3-41). These mounts are not only critical for proper operation of the transmission, but they also isolate transmission noise and vibrations from the passenger compartment. The mountings for the engine and transmission keep the powertrain in proper alignment with the rest of the driveline and help to maintain proper adjustment of the various linkages attached to the housing.

The transmission on a RWD vehicle is usually supported by a cross-member bolted with an insulator near the rear of the transmission housing. The engine sets on engine mounts bolted to the frame and sides of the engine. FWD vehicles have a number of different engine and transmission mounts (Figure 3-42). These mounts can be below or above the engine/transaxle assembly.

SUMMARY

- The major components of a transaxle are the same as those of a transmission, except the transaxle assembly includes the final drive and differential gears in addition to the transmission.
- The input shaft drives a planetary gearset, which provides the various forward gears, a neutral position, and one reverse gear.

- Power flow through the gears is controlled by multiple friction disc packs, one-way clutches, and friction bands. These hold a member of the gearset when force is applied to them, thereby making different gear ratios possible.
- An automatic transmission or transaxle selects gear ratios according to engine speed, engine load, vehicle speed, and other operating conditions.
- Gear ratio is an expression of the number of revolutions made by a driving gear compared to the number of revolutions made by a driven gear.
- The torque converter automatically engages or disengages power from the engine to the transmission in relation to engine rpm.
- The main components of a torque converter are the cover, turbine, impeller, and stator. The cover is connected to the flywheel and transmits power from the engine to the impeller, which drives the transmission's oil pump. The turbine is splined to the input shaft and is driven by the force of the fluid from the impeller. The stator redirects the flow of fluid from the turbine to the impeller.
- A single planetary gearset consists of a sun gear, a planet carrier with three or more planet pinion gears, and a ring gear.
- When two or more planetary gearsets are used in a transmission, they are collectively called a compound planetary gear system.
- Many transmissions rely on information sent from various electronic sensors and switches to a computer that controls the engagement of the torque converter clutch and transmission's shift points.
- Pressure-regulating valves change the pressure of the oil to control the shift points of the transmission. Flow-directing valves direct the pressurized oil to the appropriate reaction members, which cause a change in gear ratios.
- ATF is a hydraulic oil designed specifically for automatic transmissions. Its primary purpose is to transmit pressure to activate the transmission's brakes and clutches. It also serves as a fluid connector between the engine and the transmission, removes heat from the transmission, and lubricates the transmission's moving parts.
- To remove the heat from the fluid the ATF is directed through a cooler, which dissipates its heat.
- Typical reaction devices are bands, multiple-disc clutches, and overrunning clutches.
- Central to the hydraulic control system is the valve body assembly, which is typically made of two or three main parts: a valve body, separator plate, and transfer plate.
- The purpose of a valve body is to sense and respond to engine and vehicle load, as well as to meet the needs of the driver.
- Valve bodies are normally fitted with three different types of valves: spool valves, check-ball valves, and poppet valves. The purpose of these valves is to start, stop, or use movable parts to regulate and direct the flow of fluid throughout the transmission.
- The transmission's oil pump provides a constant supply of oil under pressure to operate, lubricate, and cool the transmission.
- Three types of oil pumps are used in automatic transmissions: gear-type, rotor-type, and vane-type pumps.
- The pressure-regulator valve is normally located in the pump assembly and maintains basic fluid pressure.
- When the engine is operating under heavy load conditions, fluid pressure is increased to increase the holding capacity of a reaction member.
- The governor assembly senses road speed and sends a fluid pressure signal to the valve body to either upshift or downshift.
- Throttle pressure increases the fluid pressure applied to the apply devices to hold them tightly, reducing the chance of slipping while the vehicle is operating under heavy load.
- The vacuum modulator is a load-sensing device that increases or decreases fluid pressure in response to the engine's vacuum. It provides an increase in pressure when vacuum is low and a decrease when vacuum is high.

TERMS TO KNOW

Accumulator

Belleville spring

Engine vacuum

Flow-directing valves

Kickdown

Load-sensing device

Manifold absolute
pressure (MAP) sensor

Modulator valve

Orifice

Output shaft

Pressure-regulating
valves

Reaction members

Throttle valve

Torque converter

Vacuum modulator

SUMMARY

- Transmissions not equipped with a vacuum modulator use a throttle cable or electronics to sense engine load and change fluid pressures.
- The throttle cable is connected to the throttle lever at the valve body and moves with the throttle pedal.
- The throttle valve converts mainline pressure into a variable throttle pressure based on the position of the throttle plate.
- Engine load can be monitored electronically through the use of a manifold absolute pressure (MAP) or mass air flow (MAF) sensor.
- As the gearshift selector is moved into its various positions, the manual lever moves a spool valve, called the manual shift valve, which is located in the transmission's valve body.
- The speedometer drive assembly is normally geared to the output shaft.
- Shift timing is determined by throttle pressure and governor pressure acting on opposite ends of the shift valve.
- Shift feel is controlled by the pressure at which each hydraulic member is applied or released, the rate at which each is pressurized or exhausted, and the relative timing of the apply and release of the members.
- Multiple friction disc packs sometimes contain a wavy spring-steel separator plate that helps smooth the application of the clutch.
- An accumulator is a hydraulic piston assembly that helps a servo apply a band or clutch smoothly. It does this by absorbing sudden pressure surges in the hydraulic circuit to the servo.

REVIEW QUESTIONS

Short-Answer Essays

1. How can shift feel be controlled?

2. What are the major components of all automatic transmissions?

3. What is the purpose of a vacuum modulator?

4. What determines the timing of the shifts in an automatic transmission?

5. How does engine load affect engine vacuum?

6. What is the purpose of a transmission fluid cooler?

7. Why must hydraulic line pressures increase when there is an increase of load on the engine?

8. Briefly explain how a torque converter works.

9. What are the primary purposes of the valves in the valve body?

10. What are purposes of the reaction members of a transmission?

Fill in the Blanks

1. Power flow through the gears is controlled by _____ - _____ _____ , _____ - _____ _____ , and _____ , which hold a member of the gearset when force is applied to them.

2. The main components of a torque converter are the _____ , _____ , _____ , and _____ .

3. The three most common types of oil pumps used in automatic transmission are _____ -type, _____ -type, and _____ -type pumps.

4. Oil pumps are driven by the _____ of the torque converter; therefore, whenever the torque converter _____ is rotating, the oil pump is driven.

5. The governor assembly senses _____ speed and sends a fluid pressure signal to the _____ to either upshift or downshift.

6. A transmission's oil pump provides a constant _____ of oil under _____ to _____ , _____ , and _____ the transmission.

7. _____ valves change the oil's pressure to control the shift points of governor type transmissions and shift quality, while _____ valves direct the pressurized oil to the appropriate reaction members to cause a change in gear ratios.

8. Engine load can be monitored _____ , by _____ _____ , or by _____ _____ .

9. The purpose of a valve body is to _____ and _____ to engine and vehicle load, as well as to meet the needs or desires of the _____ .

10. An automatic transmission or transaxle selects gear ratios according to engine _____ , engine _____ , vehicle _____ , and other operating conditions.

MULTIPLE CHOICE

1. While discussing stators:
   Technician A says a stator aids in directing the flow of fluid from the turbine to the impeller.
   Technician B says a stator is equipped with a one-way clutch that allows it to remain stationary under certain conditions.
   Who is correct?
   A. A only   C.   Both A and B
   B. B only   D.   Neither A nor B

2. *Technician A* says that one of the primary purposes of the pressure regulator valve is to fill the torque converter with fluid.
   Technician B says the pressure regulator valve directly controls throttle pressure.
   Who is correct?
   A. A only   C.   Both A and B
   B. B only   D.   Neither A nor B

3. While discussing impellers:
   Technician A says the torque converter's impeller is driven by the converter cover and its rear hub drives the transmission's oil pump.
   Technician B says the impeller is splined to the input shaft and is driven by the force of the fluid from the turbine.
   Who is correct?
   A. A only   C.   Both A and B
   B. B only   D.   Neither A nor B

4. While discussing oil pumps:
   Technician A says oil pumps are driven by the pump drive hub of the torque converter.
   Technician B says oil pumps are indirectly driven by the torque converter's stator.
   Who is correct?
   A. A only   C.   Both A and B
   B. B only   D.   Neither A nor B

5. While discussing load-sensing devices:
   Technician A says transmissions not equipped with a vacuum modulator use a throttle cable or electronics to sense engine load.
   Technician B says increasing the fluid pressure holds the planetary control units tightly to reduce the chance of slipping while under heavy load.
   Who is correct?
   A. A only   C.   Both A and B
   B. B only   D.   Neither A nor B

6. While discussing valve bodies:
   Technician A says a purpose of a valve body is to increase oil pressures in response to increases in engine speed.
   Technician B says valve bodies are normally fitted with three different types of valves that start or stop fluid flow, or use movable parts to regulate and direct the flow of fluid throughout the transmission.
   Who is correct?
   A. A only   C.   Both A and B
   B. B only   D.   Neither A nor B

7. While discussing engine load:
   Technician A says changes in engine load cause changes in hydraulic pressure inside the transmission.
   Technician B says engine load is monitored by throttle pedal movement or by engine vacuum.
   Who is correct?
   A. A only   C.   Both A and B
   B. B only   D.   Neither A nor B

8. Which of the following statements about planetary gears is *not* true?
   A. Gearsets designed for heavy-duty use will have more planet gears than units designed for light duty.
   B. The sun gear is often referred to as the internal gear.
   C. From a single gearset, five forward gear ratios are possible.
   D. Different gear ratios are obtained by driving one or more gearset members and holding another.

9. While discussing the hydraulic system:
   Technician A says the transmission's oil pump is the source of all pressures within the hydraulic system.
   Technician B says the hydraulic system also keeps the torque converter filled with fluid.
   Who is correct?
   A. A only   C. Both A and B
   B. B only   D. Neither A nor B

10. While discussing the levers that protrude from transmission housings:
    Technician A says the throttle lever is moved by the engine vacuum sent to the modulator in response to engine load.
    Technician B says that when the gearshift selector is moved it moves the manual lever, which moves the manual shift valve in the transmission's valve body.
    Who is correct?
    A. A only   C. Both A and B
    B. B only   D. Neither A nor B

Chapter 4

ELECTRONIC CONTROLS

UPON COMPLETION AND REVIEW OF THIS CHAPTER, YOU SHOULD BE ABLE TO:

- Explain basic electrical terms.
- Discuss the basic theories of electricity.
- Explain the value of Ohm's law.
- Explain how laws of magnetism are used to control automatic transmissions.
- Explain what a computer is and how it works.
- Describe the basic types of sensors used in electronically controlled transmission systems.
- Explain the advantages of using electronic controls for transmission shifting.

- Explain an elementary shift logic chart.
- Briefly describe what determines the shift characteristics of each selector lever position.
- Explain why adaptive learning is an important advancement for electronic control systems.
- Identify the input and output devices in a typical automatic transmission electronic control system and briefly describe the function of each.
- Describe the various types of transmissions used in hybird vehicles.

INTRODUCTION

Older automatic transmissions relied on hydraulic and mechanical devices that have been refined through the years. Much of the development was centered on increasing their efficiency and prolonging their useful life. These automatic transmissions waste a good amount of the torque produced by the engine through the generation of undesired heat by the moving fluid inside the transmission and torque converter. By using electronic controls for transmission operation, the amount of wasted power was reduced and the overall operation of the transmission is more responsive. However, the shifts are slightly slower, although smoother, because it takes time for the T/C clutch and the apply devices to release and apply.

Electronic controls are also more complicated transmissions due to the inclusion of various solenoids and sensors. However, the use of electronics eliminates many of the mechanical and vacuum circuits and valves in a typical valve body (Figure 4-1). With this comes increased reliability and increased responsiveness.

Electronics and electricity are based on the same theories and practices and are really the same thing. Electricity is a form of energy resulting from the movement of electrons. Electronics deals with the behavior and effects of electrons. Typically, very small electrical components are referred to as electronic components and devices larger than that are called electrical devices.

All automatic transmissions are equipped with some electrical sensors and switches. These switches primarily control the backup light circuits and prevent engine starting unless

A BIT OF HISTORY

In the mid-1980s, Toyota introduced the A140E transaxle. This was the first automatic transmission with electronic shift controls. This transmission began the trend in which, by 1990, all domestic manufacturers offered at least one electronic automatic transmission (EAT). Today, EATs are found in nearly all new vehicles.

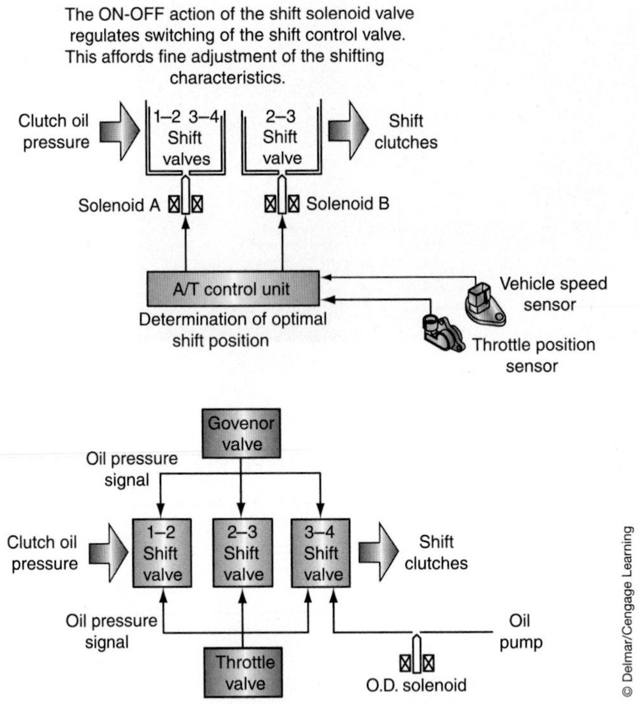

The ON-OFF action of the shift solenoid valve
regulates switching of the shift control valve.
This affords fine adjustment of the shifting
characteristics.

Clutch oil pressure → | 1–2 3–4 Shift valves | 2–3 Shift valve | → Shift clutches

Solenoid A ⊠⊠   ⊠⊠ Solenoid B

A/T control unit
Determination of optimal shift position

Vehicle speed sensor
Throttle position sensor

Govenor valve

Oil pressure signal

Clutch oil pressure → | 1–2 Shift valve | 2–3 Shift valve | 3–4 Shift valve | → Shift clutches

Oil pressure signal

Throttle valve

O.D. solenoid ⊠⊠

Oil pump

© Delmar/Cengage Learning

FIGURE 4-1 A comparison of electronic (TOP) and hydraulic (BOTTOM) transmission controls.

The computers used in electronic control systems are called controllers, micro-processors, micro-computers, or control modules.

the transmission selector is in park or neutral. Transmissions built within the last 15 years have both electrical and electronic devices. The electrical devices are still part of the starting and backup light circuits, while the electronic devices are used to control the operation of the transmission and torque converter.

BASIC ELECTRICITY

To understand the operation and purpose of electrical and electronic components, you must have a good understanding of electricity. A good understanding of electricity requires an understanding of the terms commonly used to describe it. *Electricity* is a form of energy that results from the movement of an electron from one atom to another (Figure 4-2). Since everything is made up of atoms and all atoms have at least one electron, everything has the potential of having electricity in it. However, in some things, the movement of electrons occurs more easily than in others. Materials in which electrons easily move are often referred to as *conductors*. Materials in which electrons have a very difficult time moving from one atom to another are called *insulators*.

Wires are the most commonly used and most recognizable conductors used in an automobile. They allow electricity to flow to designated parts of the car and provide the

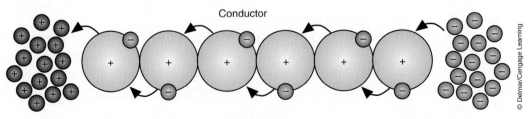

Conductor

© Delmar/Cengage Learning

FIGURE 4-2 Electricity is the flow of electrons from one atom to another.

path for electricity to follow through the car. These wires are covered with insulation, which prevents the electricity from leaking out of the wire. Because insulating materials do not readily allow for the movement of electrons, they are said to have very high electrical resistance.

Electrical Flow

Electrical *resistance* is the ability to resist the flow of electricity. The principles of resistance are not only applied to the insulated covering on wires, but are also used within electrical circuits. Using resistance within an electrical circuit controls the flow of electricity within that circuit. Materials used within a circuit do not have extremely high resistance like an insulator. Rather, they make it difficult for electricity to flow and cause the flow to lose or give up some of its energy. The amount of resistance in a circuit determines the amount of electricity that will flow through the circuit.

An electrical system can be compared to an automobile's cooling system. Coolant must be able to move throughout the cooling system and return to the radiator. Hoses are used to connect the components together and to keep the flow of coolant contained within the system. A water pump forces the coolant through the system, while valves and the thermostat control the amount of coolant that will flow.

The same is true for electricity: it flows from the battery through its circuit and then back to the battery. Conductors serve the same purpose as the hoses of the cooling system; they keep electrical flow contained and directed throughout the system. The battery accomplishes the work of the water pump; it forces current through the circuit. This force is electrical pressure, which is measured in volts. Resistors or loads control the amount of current flow in the same way that the thermostat and valves do in the cooling system.

If the pressure is increased in the cooling system, more coolant will flow through it. Likewise, if the valves are nearly closed, little coolant will be able to flow. An electrical circuit behaves in the same way. When resistance is increased, current will decrease and when voltage is increased, current will also increase.

A light bulb is able to shine because the filament inside the bulb resists the flow of electricity. The bulb shines from the heat given off, as energy is lost while electricity flows through the filament or resistor. Any device that uses electricity for its operation has some resistance. This resistance converts the electrical energy into another form of energy for use by the device. The resistance of a device is measured in ohms by using an ohmmeter or the resistance–measuring function of a multimeter.

The random movement of electrons in a conductor is not current flow. Electrical *current* is the directed movement of electrons from atom to atom within a conductor. In an automobile, electrons flow from one post of the battery through a circuit, then back to the other post of the battery. The amount of resistance in the circuit determines how many electrons or how much current will flow through the circuit. If the resistance is high, current will be low. If a wire is cut and the ends placed a distance apart, the air gap between the two ends will create a great enough resistance to stop all current flow. The rate of current is measured in amperes by using an ammeter or the current-measuring function of a multimeter.

The amount of current flow in a circuit is determined by the amount of resistance in the circuit and the amount of pressure pushing the current through. This pressure is set up by the attraction of the negative side of the battery to the positive side of the battery. **Voltage** is the amount of electrical pressure or force present at a particular point within a circuit (Figure 4-3). This electrical pressure is measured in units of volts. Voltage is also referred to as electromotive force, which can be defined as something that causes current to flow. A normal automotive battery has 12 volts present across its positive

Shop Manual
Chapter 4, page 156

Energy cannot be destroyed nor can it be created. It is merely changed to another form of energy when it is used.

Amperes are commonly referred to as **amps**.

Voltage is best described as the electrical pressure forced by the attraction of electrons to protons.

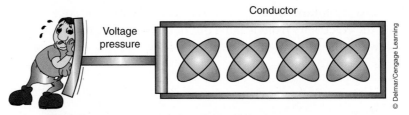

FIGURE 4-3 Voltage is the force that causes electrons to move.

and negative posts. Voltage pushes current through the circuit's resistors or loads. The amount of voltage that is available decreases as the current passes through the load. By the time the current passes through the last bit of resistance in a circuit, all of the voltage is used.

The amount of voltage used is called the *voltage drop*. Voltage drop is the amount of voltage required to cause an amount of current to flow through a certain resistance. Voltage drop indicates the amount of energy or force that was required to push the current through.

Ohm's Law

Shop Manual

Chapter 4, page 156

In a complete electrical circuit, voltage, current, and resistance are present. The amounts of each depend on a number of variables. However, most often the amount of resistance is the factor that determines the amount of current and voltage. *Ohm's law* describes the relationship between volts, amperes, and resistance in a circuit. Ohm's law states that one volt of electrical pressure is needed to push one ampere of electrical current through one ohm of resistance. Using Ohm's law, the value of current, voltage, or resistance can be calculated. The amount of voltage (E for electromotive force) needed equals the amount of current (I for intensity of current) multiplied by the amount of resistance (R) or $E = I \times R$ (Figure 4-4).

This law can be used to calculate unknown values in a circuit. To calculate the resistance that would cause a particular amount of current in a fixed voltage circuit, the formula $R = E/I$ is used. To calculate the current of a circuit, $I = E/R$ is used.

> **AUTHOR'S NOTE:** The important thing for an automatic transmission technician to learn about Ohm's law is not how to calculate values, but rather, how changes in resistance, voltage, and current will affect the operation of a circuit.

The most practical use of this law is to explain and predict the effects of a change in a circuit. In a light circuit with 12 volts available from the battery and the resistance of the bulb being 3 ohms, 4 amps of current will flow through the circuit (Figure 4-5). However, if there

Voltage (E) = current (I) times resistance (R), therefore

$$E = I \times R.$$

Current (I) = voltage (E) divided by resistance (R), therefore

$$I = E/R.$$

Resistance (R) = voltage (E) divided by current (I), therefore

$$R = E/I.$$

FIGURE 4-4 Ohm's Law.

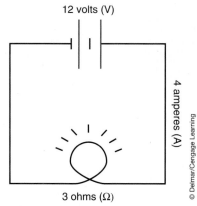

12 volts (V)

4 amperes (A)

3 ohms (Ω)

© Delmar/Cengage Learning

FIGURE 4-5 A simple 12-volt series circuit.

is a bad connection at the bulb that creates an additional 1 ohm of resistance, the current will decrease to 3 amps. As a result of the added resistance and decreased current, the bulb will lose about half of its brightness.

This lack of brightness is also caused by the decreased amount of circuit current and the amount of voltage dropped by the bulb. In the normal circuit, the bulb drops 12 volts. However, in the circuit with the added resistance, 3 volts will be dropped by the bad connection and the bulb will drop the remaining 9 volts.

The amount of power or heat given off by the bulb, or any electrical device, can be determined by multiplying the current flowing through the load by the amount of voltage dropped by the load. This formula is expressed as $P = I \times E$. The power used to light the bulb in the above normal circuit is equal to 4 amps times 12 volts or 48 watts. The wattage used by the bulb in the bad circuit is only 27. Electrical power is measured and expressed in *watts*.

The way resistance affects a circuit depends on the placement of the resistance in the circuit. If the resistance, wanted or unwanted, is placed directly into the circuit, the resistances are said to be in series and there is a *series circuit*. If a resistor is placed so that it allows an alternative path for current, it is a parallel resistor or there is a *parallel circuit*.

A series circuit allows current to follow only one path and the amount of current that flows through the circuit depends upon the total resistance of the circuit. To determine the total resistance in a series circuit, all resistance values are added together ($R_1 + R_2 + R_3 + \ldots = R_T$). At each resistor, voltage will be dropped. The total amount of voltage dropped in a series circuit is equal to the sum of the drops throughout the circuit. Regardless of the possible differences and resistance values of the loads, current in a series circuit is always constant and is the same at any point within the circuit.

Parallel circuits are designed to allow current to flow in more than one path. This allows one power source to power more than one circuit or load. All of a car's accessories and other electrical devices are powered by the battery and can be individually controlled through the use of parallel circuits. Within a parallel circuit there is a common path from and back to the power source. In an automobile, the path goes from the battery through the individual legs of the circuit and back to the battery through a common ground. This common ground is normally the frame of the car (Figure 4-6). Each branch or leg of a parallel circuit behaves as if it were an individual circuit. Current will flow only through the individual circuits when each is closed or completed. All legs of the circuit do not need to be complete in order for current to flow through one of them. In parallel circuits, the total amperage of the circuit is equal to the sum of the amperages in all of the legs of the circuit (Figure 4-7). There is no voltage lost when the circuit splits into its branches; therefore, equal amounts of voltage are applied to each branch of the circuit. The total resistance of a parallel circuit is always less than the resistance of the leg with the smallest amount of resistance. The total resistance of two resistors in parallel can be calculated by dividing the product of the two by their sum.

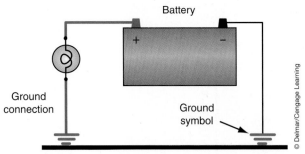

FIGURE 4-6 A simple light circuit using the vehicle as the negative conductor for the circuit.

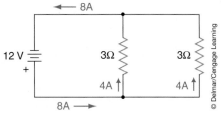

FIGURE 4-7 A simple parallel circuit.

The legs of a parallel circuit can contain a series circuit. To determine the resistance of that leg, the resistance values are added together. The resistance values of each leg are used to calculate the total resistance of a parallel circuit. Total circuit current flows only through the common power and ground paths; therefore, a change in a branch's resistance will not only affect the current in the branch but will also affect total circuit current.

Electrical Problems

Shop Manual

Chapter 4, page 156

Motors attempting to move a heavier than normal load or an immovable object, such as a binding window, will rotate very slowly or not at all and will draw excessive amounts of current.

Normally, parallel circuits in an automobile have a circuit protection device placed in a common path from the positive side of the battery. This protection device is usually a fuse, fusible link, or a circuit breaker. They are designed to protect the wires and components from damage due to excessive current flow. When a great amount of current flows through the fuse or breaker, an element will burn out or open, causing the circuit to be opened and stopping current flow. This action prevents the high current from burning up the wires or components in the protected circuit.

Low resistance causes high current. A decrease in the amount of resistance is typically the result of a short. A *short* is best defined as an additional and unwanted path to ground. Most shorts, such as a bare wire contacting the frame of the car, create an extremely low resistance parallel branch. Low resistance and high current can also be caused by a slow-turning motor.

A short to ground can be present before the load in the circuit or internally within the load or component. A short can also connect two or more circuits together, causing additional parallel legs and uncontrolled operation of components. An example of a possible result from a wire-to-wire short would be the horn blowing each time the brake pedal was depressed (Figure 4-8). This could be caused by a wire-to-wire short between the horn and brake light circuits. Shorts are one of the three common types of electrical problems.

Another common electrical fault is the *open*. An open causes an incomplete circuit and can result from a broken or burned wire, loose connection, or a faulty component. If a circuit is open, there is no current flow and the component will not operate. If there is an open in one leg of a parallel circuit, the remaining part of that parallel circuit will operate normally.

Excessive resistance at a connector, internally in a component, or within a wire is also a common electrical problem. High, unwanted resistance will cause low current flow and the component will not be able to operate normally, if at all.

AUTHOR'S NOTE: High resistance problems always cause low circuit current and lower than normal voltage drops across the intended load. Opens always result in zero current flow and therefore there are no measurable voltage drops in the circuit. A short always increases current flow but the voltage drops will be about normal. Voltage drops will be normal because a short is an unwanted parallel branch in the circuit.

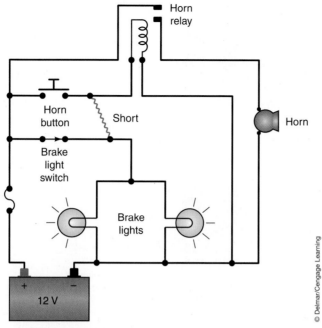

FIGURE 4-8 This wire-to-wire short will cause the horn to sound whenever the brake pedal is depressed.

Magnetism

Magnetism is a form of energy that can cause current flow in a conductor and, just like electricity, results from the movement of electrons. The movement of electrons causes the atoms of some materials to align and set up magnetic lines of flux. These *flux lines* establish an attraction of the north pole of the magnet to its south pole.

Current flowing through a conductor normally causes a weak magnetic field to form around the conductor. To strengthen the magnetic field, the conductor can be wound into a coil. This concentrates the flux lines into a smaller area and gives the effect of a large magnet. As long as current is flowing through the wound conductor, a magnetic field will be present. To increase the concentration of the magnetic field, a strip of soft iron can be inserted into the center of the windings. By controlling the current flow through the windings, this magnet can be switched on and off and is referred to as an *electromagnet*. Many switching devices, including relays, use this principle.

A *relay* is both a protection device and a switch (Figure 4-9). A low current circuit is used to close and open a high current circuit. Relays use a low current to set up a magnetic field

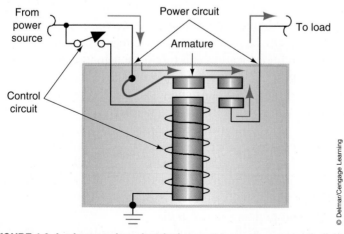

FIGURE 4-9 A relay uses low electrical current to create a magnetic field to draw the contacts closed, which energizes a high-current circuit.

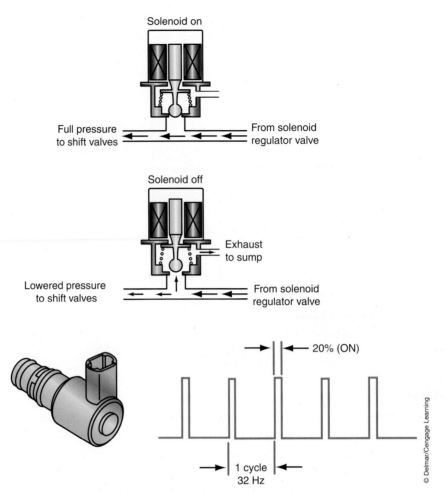

FIGURE 4-10 The action of a typical duty-cycled solenoid.

that pulls a set of contacts closed. The closing of the contacts completes the high current circuit and allows it to operate.

The electromagnetic device most commonly used with automatic transmissions is the *solenoid* (Figure 4-10). A solenoid is a magnetic switch, like a relay. However, rather than moving contacts external to the magnetic field, the soft iron core in the center of the solenoid's windings moves to open or close a circuit. The movement of the core may also be used to mechanically perform a function, such as moving a lever or opening and closing a hydraulic valve or circuit.

Soft iron can also become a permanent magnet. A *permanent magnet* is one that retains its magnetic field after current has ceased to flow. Once the electrons of the material are aligned, they tend to stay in their location and retain their magnetic qualities.

When a conductor is passed through lines of magnetic flux, a voltage is induced in the conductor. The amount of induced voltage depends on the strength of the magnetic field and how quickly the conductor is moved through the field. This principle is used in generators, which recharge the battery, and in many engine control devices.

As the conductor moves into the magnetic field, a positive voltage is produced. When the conductor is moved out of the field, a negative voltage is induced. Equal amounts of movement in and out of the field will produce equal amounts of positive and negative voltages. Many computer inputs or sensors rely on this change of voltage to inform the computer of the location of a moving component (Figure 4-11).

In a generator, a magnetic field is rotated past a number of conductors. As the field rotates, the stationary conductors continuously pass through the field and voltage is induced. The use of multiple conductors increases the amount of induced voltage.

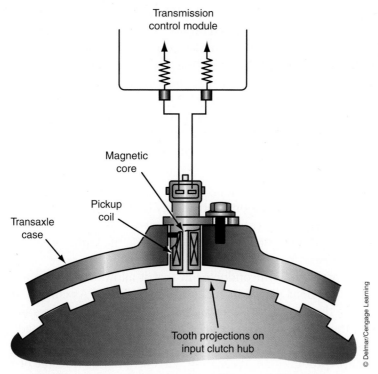

FIGURE 4-11 **This speed sensor is a voltage-generating device.**

Principles of magnetism are also used in motors. Magnetic poles of the same polarity will oppose each other, while opposite poles are attracted to one another. Motors are constructed of two magnets, one inside the other. Rotation is caused by the constant attraction and opposition of the poles. The speed of the rotation is controlled by the strength of the magnetic fields, which can be controlled by the amount of current flow through the electromagnets. Some motors use two electromagnets, while others use a permanent magnet and an electromagnet.

Solenoids. A solenoid is a device which converts electrical energy into linear motion. The electricity is sent to a coil which creates an electromagnetic field. In the center of the magnetic field is a metal cylinder that acts as a plunger. One end of the plunger is a spring. The spring holds the plunger partially out of the electromagnetic coil. When current is sent to the coil, the plunger is pulled into the coil and compresses the spring. The movement of the plunger results in some action. Either end of the plunger can do something. One end of the plunger is moved toward the center of the coil and the other end is moved toward the outside of the coil. This results in a pull on one end and a push on the other.

In automatic transmissions, the plunger opens a pathway for hydraulic fluid. When current stops flowing in the coil, the magnetic field is no longer present and the spring forces the plunger back to its original position.

Circuit Controls

Relays are often used to control a motor circuit. The use of a relay allows other circuits to be activated along with the motor. An example of this is the starter motor relay, which not only completes the circuit from the battery to the starter motor but also completes the circuit from the battery to the ignition circuit. Both of these actions allow the engine to start.

Circuits are controlled by switches that open and close the circuit and by variable resistors that control the amount of current flow in the circuit. These switches and controls can be manually operated or controlled by electricity and magnetism. Manual switches and variable resistors are either controlled by the driver or through mechanical linkages that move in response to operating conditions.

The different models of vehicles have many different automatic switches. These switches open or close a circuit depending on the conditions set up for the switches. Often the driver is unaware of the action of the switches unless one of them fails. These switches can be controlled by temperature or pressure and will activate or deactivate a circuit whenever the conditions cause them to close or open. Automobiles equipped with electronic engine controls rely heavily on this type of switch for the activation of circuits as well as for sending information to the computer. Many manual switches in a torque converter clutch (TCC) circuit are controlled by fluid pressure. When pressure is present at the diaphragm of the switch, the diaphragm moves and either opens or closes the switch.

Semiconductors

Computerized engine and transmission controls rely heavily on semiconductors, both in the computer and in the sensing devices. A *semiconductor* is a material that conducts electricity only when the conditions are right. Two basic types of semiconductor are used in the automobile: diodes and transistors.

A *diode* is the most basic semiconductor in that it allows current to flow through it in one direction only. A diode is placed in series within a circuit. As current attempts to flow through one side of a diode, it is met with a great amount of resistance and little current, if any, flows through the circuit. When current flows toward the opposite side, it is met with very little resistance and the current flows through it.

An AC generator produces an alternating current (AC) as a result of the conductors passing in and out of the magnetic field. AC is a type of electricity in which current flows both from positive to negative and negative to positive within a circuit. The normal current flow in an automobile is direct current (DC), in which current flows in one direction only. In order to recharge a car's battery, the AC from an AC generator must be changed to DC. This is accomplished by diodes.

Diodes are also used in a number of other circuits. They are used in some warning circuits to light an indicator lamp, activate a buzzer or chimes, or activate a light and buzzer. Specially designed diodes are used to control more than the direction of current flow. *Zener diodes* will allow current to flow in both directions when the voltage is above a particular level. *Clamping diodes* are often used to protect systems from voltage surges or to precisely control the voltage in a circuit (Figure 4-12). They are commonly found in electronic control modules, voltage regulators, and instrumentation displays.

A light-emitting diode (LED) is often used in ignition systems, instrumentation displays, and in many measuring instruments. When the proper amount of current is introduced to

AC is commonly used to represent alternating current, whereas **DC** is used to represent direct current.

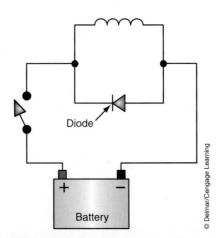

FIGURE 4-12 A clamping diode is connected in parallel with a coil to prevent voltage spikes that normally occur when the circuit's switch is opened.

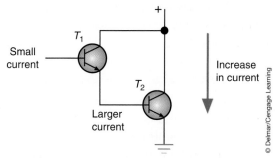

FIGURE 4-13 This arrangement is called a Darlington pair and is used to amplify a signal. T1 acts as a pre-amplifier for T2.

these diodes, they release some electrical energy in the form of light. Through careful arrangement of these diodes it is possible to light a display of numbers and letters.

Transistors are semiconductors that are used as amplifiers and switches. Activated by low amounts of voltage and current, they can amplify current and voltage and serve as a relay to switch a high-current circuit on or off. When used in combination with diodes, transistors can quickly and precisely open and close a high-current circuit. The advantage of using a transistor to serve as a relay is that a transistor has no moving parts that could wear out or break.

Amplifying transistors are commonly used in radios and electronic ignitions to strengthen weak signals. They provide for a high current in response to low-voltage signals (Figure 4-13).

The main switching component of a transistor is its base. When the emitter and collector are connected into a circuit, the transistor will not function until the correct amount of voltage is applied to the base. At that time, the transistor will either serve a switch or will amplify the signal it receives.

Combinations of diodes and transistors are the basis of the solid-state electronics used in today's electronic control modules and computers. Often they are compactly constructed to form an integrated circuit (IC). A single IC chip can perform the functions of thousands of semiconductors at one-millionth of their size.

Semiconductor materials are also used in many sensors to monitor the speed of shafts. The most common of these sensors are called Hall-effect sensors (Figure 4-14). The action of a semiconductor when it is moved through a magnetic field is similar to that of a conductor. However, the semiconductor must have current flowing through it before the passing of the magnetic field will induce a voltage in it. As the magnetic field moves toward the semiconductor material, the voltage potential at the semiconductor begins to change. Throughout

Amplifying transistors are commonly referred to as power transistors.

An integrated circuit is commonly referred to as an **IC**.

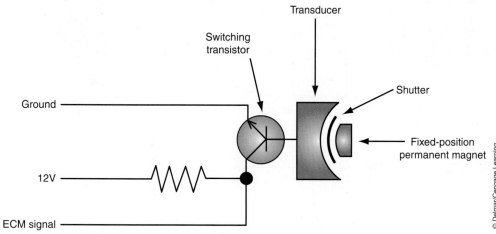

FIGURE 4-14 A Hall-effect switch and basic circuit.

the movement of the field, the voltage will respond to the position of the field. This change in voltage serves as an information signal to the computer or module, indicating the exact position and speed of the rotating object.

BASICS OF ELECTRONIC CONTROLS

The basic part of all electronic control systems is a computer. A *computer* is an electronic device that receives, stores, processes, and communicates information (Figure 4-15). All of the information it works with is really nothing more than electricity. To a computer, certain voltage and current values mean something and, based on these values, the computer becomes informed.

Computers receive information from a variety of input devices, which send voltage signals to the computer. These signals tell the computer the current condition of a particular part or the conditions that a particular part is operating in. After the computer receives these signals, it stores them and interprets the signals by comparing the values to data it has in its memory. By processing this data, the computer knows what conditions the input data represents. It also can search its memory to identify any actions it should take in response to those conditions. If an action is required, the computer will send out a voltage signal to the device that should take the action, causing it to respond and correct the situation.

This entire process describes the operation of an electronic control system. Information is received by a microprocessor from input sensors. The computer processes the information, then sends commands to output devices (Figure 4-16). Most control systems are also designed to monitor their own work and will check to see if their commands resulted in the expected results. If the result is not what was desired, the computer will alter its command until the desired outcome is achieved.

Inputs

The input devices used in electronic control systems vary with each system; however, they can be grouped into distinct categories: *reference voltage sensors* and *voltage generators*. Voltage generation devices are typically used to monitor rotational speeds. The most common of these devices is the PM generator used as a vehicle speed sensor. A *speed sensor* is a magnetic pick up that senses and transmits low-voltage pulses (Figure 4-17). These pulses are generated each time a small magnet, normally located on the circumference of a shaft or

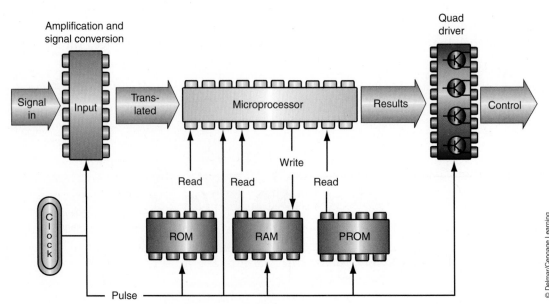

FIGURE 4-15 The basic workings and components of a typical automotive computer.

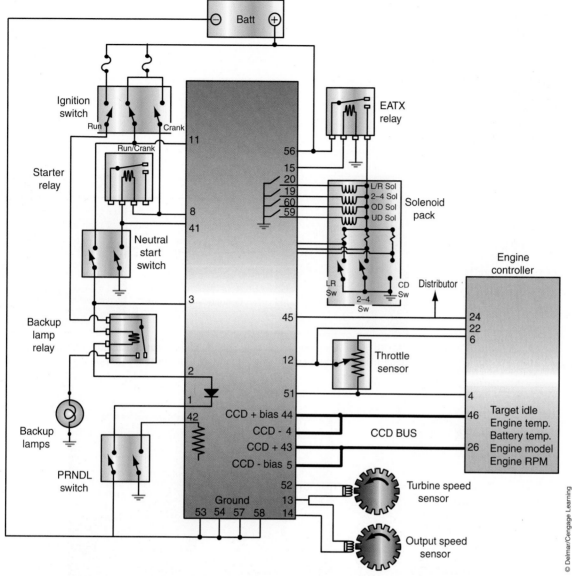

FIGURE 4-16 The electric circuit for a typical electronically controlled transmission. Note: The CCD data BUS is the data source for other inputs in this multiplexed circuit.

rotating component, passes by the wire coil of the pick up unit. The pulse of voltage is the signal sent to the computer. The computer then compares the signal to its clock or counter, which converts the signal into speed.

Common voltage reference sensors include *on/off switches, potentiometers, thermistors,* and *pressure sensors.* These sensors normally complete a circuit to and from the computer. The computer sends a reference voltage to them and reads the voltage sent back to it by them. The change in voltage represents a condition that the computer can again identify by looking at its program and comparing values. Potentiometers (Figure 4-18), thermistors, and pressure sensors are designed to change their electrical resistance in response to something else changing. Normally a potentiometer is linked to devices such as the throttle linkage. As the throttle is moved, the resistance of the potentiometer in the throttle position (TP) sensor changes. This causes a varied return signal to the computer as the throttle opening changes.

Thermistors (Figure 4-19) also change their resistance values in response to conditions (Figure 4-20). However, they respond to changes in heat. Pressure sensors respond to pressure

Shop Manual
Chapter 4, page 185

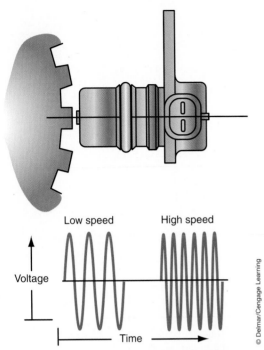

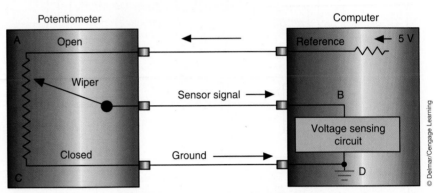

FIGURE 4-17 (TOP) A typical speed sensor. (BOTTOM) The signal a speed sensor sends to the computer.

FIGURE 4-18 A potentiometer sensor circuit measures the amount of voltage drop to determine the position of the device attached to the sensor's wiper.

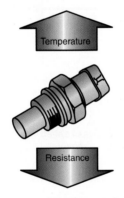

Thermistor temperature to resistance valves		
°F	°C	Ohms
210	100	185
160	70	450
100	38	1800
40	4	7500
0	−18	25,000
−40	−40	100,700

FIGURE 4-19 A thermistor's resistance decreases as the temperature rises.

Shop Manual
Chapter 4, page 185

applied to a movable diaphragm in the sensor. As pressure increases, so does the movement of the diaphragm and the amount of resistance in the sensor (Figure 4-21).

On/off switches are simple in operation. When a particular condition exists, the switch is either forced closed or open. When the switch is closed, the computer receives a return signal from the switch. When it is open, there is no return signal. The return signal from a switch represents circuit on and off times. If a switch is cycled on and off very rapidly, the return signal is a rapid on/off one. This describes a *digital signal*: a series of on/off pulses.

A computer is a digital device that processes information through a series of on/off signals. In order for a computer to process information, it must receive its data digitally. The inputs from voltage reference sensors arrive at the computer as a change in voltage, not as a digitized signal. Therefore, in order for the computer to analyze the data, it must be first changed to a digital signal (Figure 4-22). This is the first processing task of a computer: to translate the analog signals into a digital signal.

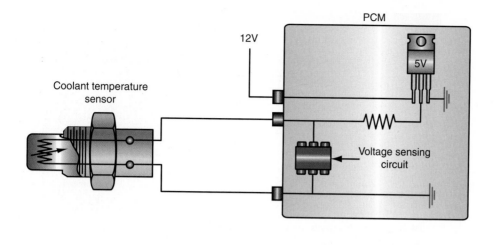

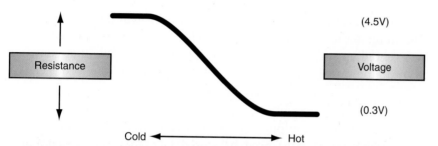

FIGURE 4-20 The resistance of a thermistor changes with changes in temperature, therefore the return voltage from the sensor to the computer changes according to changes in temperature.

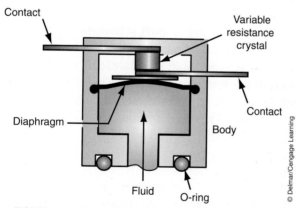

FIGURE 4-21 The working parts of a pressure switch.

Outputs

After the information has been processed, the computer sends out commands to its output devices. The typical output devices are solenoids and motors, which cause something mechanical or hydraulic to change. The movement of these outputs is controlled by the commands of the computer. Sometimes the command is merely the application of voltage to operate or energize the device. Other times it is a variable signal that causes the device to cycle in response to the signal. Most often, the solenoids are controlled on their ground side. Battery voltage is applied to the solenoids whenever the ignition is on. The computer controls the ground of the solenoid.

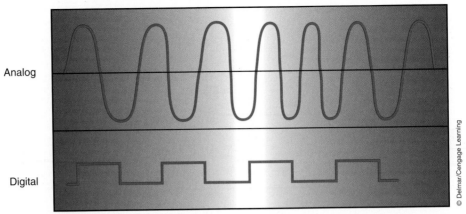

Analog

Digital

© Delmar/Cengage Learning

FIGURE 4-22 Analog signals are constantly variable, whereas digital signals are either on or off, or high or low.

The typical electronically controlled transmission uses hydraulically operated clutches and brakes. The computer receives information from various inputs and controls a solenoid assembly. The solenoid assembly consists of two to seven solenoids that control hydraulic pressure and flow to the various apply devices and to the clutch of the torque converter.

Processing

The primary purpose of using a computer is to process information. It does so by comparing all data it receives against data it has stored. All stored data remains in the computer's memory and is used as needed to analyze and correct operating conditions. The computers that control transmission functions rely on programming stored in their memory to provide gear shifting at the optimum time. The decision to shift or not to shift is based on shift schedules and logic programmed into the memory of the computer.

In order for a computer in any electronic control system to determine when to initiate a gearshift change, it must be able to refer to **shift schedules** that it has stored in its memory. A shift schedule contains the actual shift points to be used by the computer according to the input data it receives from the sensors. Shift schedule logic chooses the proper shift schedule for the current conditions of the transmission. It uses the shift schedule to select the appropriate gear, and then determines the correct shift schedule or pattern that should be followed.

The first input a computer looks at to determine the correct shift logic is the position of the gearshift lever. All shift schedules are based on the gear selected by the driver. The choices of shift schedules are limited by the type and size of engine that is coupled to the automatic transmission. Each engine/transmission combination has a different set of shift schedules. These schedules are coded by selector lever position and current gear range, and use throttle angle and vehicle speed as primary determining factors. The computer also looks at different temperature, load, and engine operation inputs for more information.

The basic shift logic of the computer allows the releasing apply device to slip slightly during the engagement of the engaging apply device. Once the apply device has engaged and the next gear is driven, the releasing apply device is pulled totally away from its engaging member and the transmission is fully into its next gear. This allows for smooth shifting into all gears.

The shift schedules set the conditions that need to be met for a change in gears. Since the computer frequently reviews the input information, it can make quick adjustments to the schedule if needed and as needed. The result of the computer's processing of this information and commanding outcomes according to a logical program is optimum shifting of the automatic transmission. This results in improved fuel economy and overall performance.

The electronic control systems used by the manufacturers differ with the various transmission models and the engines they are attached to. The components in each system and

A **shift schedule** is a three-dimensional graph that plots engine speed and load, as well as other operating conditions. Certain parts of the graph have designated gear ranges. When the conditions fall into a range, the computer causes the transmission to shift into that gear.

Current operating conditions are typically referred to as *real-time* conditions.

Shop Manual
Chapter 4, page 171

Range	Gear	A solenoid	B solenoid	4th clutch	Reverse band	2nd clutch	3rd clutch	3rd roller clutch	Input clutch	Input sprag	Forward band	1–2 roller clutch	2–1 band
P - N		On	On						*	*			
D	1st	On	On						Appl	Hold	Appl	Hold	
	2nd	Off	On			Appl			*	Orun	Appl	Hold	
	3rd	Off	Off			Appl	Appl	Hold			Appl		
	4th	On	Off	Appl		Appl	*	Orun			Appl		
D	3rd	@Off	@Off			Appl	Appl	Hold	Appl	Hold	Appl		
	2nd	@Off	@On			Appl			*	Orun	Appl	Hold	
	1st	@On	@On						Appl	Hold	Appl	Hold	
2	2nd	@Off	@On			Appl			*	Orun	Appl	Hold	Appl
	1st	@On	@On						Appl	Hold	Appl	Hold	Appl
1	1st	@On	@On				Appl	Hold	Appl	Hold	Appl	Hold	Appl
R	Rev	On	On		Appl				Appl	Hold			

★ Applied but not effective

On - Solenoid energized

Off - Solenoid de-energized

@ The solenoid's state follows a shift pattern which depends upon vehicle speed and throttle opening. It does not depend upon the selected gear.

FIGURE 4-23 Range reference chart for a 4T60-E transaxle.

the overall operation of the system also vary among different transmissions (Figure 4-23). However, all operate in a similar fashion and use basically the same parts.

ELECTRONICALLY CONTROLLED TRANSMISSIONS

Electronic transmission control has become increasingly more common on today's cars. These controls provide automatic gear changes when certain operating conditions are met. Through the use of electronics, transmissions have better shift timing and quality. As a result, the transmissions contribute to improved fuel economy, lower exhaust emission levels, and improved driver comfort. Although these transmissions function in the same way as earlier hydraulically based transmissions, a computer determines their shift points. The computer uses inputs from several different sensors and matches this information to a predetermined schedule.

Hydraulically controlled transmissions relied on signals from a governor and throttle pressure device to force a shift in gears. Electronically controlled transmissions typically don't have governors or throttle pressure devices. Hydraulically controlled transmissions relied on pressure differentials at the sides of a shift valve to hold or change a gear. Electronic transmissions still do. However, the pressure differential is caused by the action of shift solenoids that allow for changes in pressure on the side of a shift valve. The computer controls these solenoids. The solenoids do not directly control the transmission's clutches and bands. These are engaged or disengaged in the same way as hydraulically controlled units. The solenoids simply control the fluid pressures in the transmission and do not perform a mechanical function.

Most electronically controlled systems are complete computer systems. There is a central processing unit, inputs, and outputs. Often the central processing unit is a separate computer designated for transmission control. This computer may be the transmission control module (TCM), powertrain control module (PCM), or body control module (BCM). When transmission control is not handled by the PCM, the controlling unit communicates with the PCM. In this chapter, the controlling computer for the transmission will be referred to as the TCM regardless of whether it is a separate computer or is integrated into another computer.

Inputs

The inputs include transmission operation monitors plus some of the sensors used by the PCM. Input sensors, such as the throttle position (TP) sensor, supply information for many different control systems. Various control modules may share this information.

Shop Manual
Chapter 4, page 183

By putting information on the bus bar, the information is made available for a number of circuits; therefore, this is a **multiplexed** system.

Communication Networks. The computer can receive information from two different sources: directly from a sensor, or through a bus circuit, which connects the various vehicle computer systems together (Figure 4-24). This type of modulated bidirectional bus system allows the various computers in the vehicle to share information. This is called **multiplexing**. Multiplexing describes the activity of a system that allows for the transmission of information from several sources through a single circuit. Many of the separate computers use information from the same sensors; multiplexing eliminates the need to run separate circuits from the sensor to the computers that need that input. Each input is connected to only one computer, which reads and distributes the sensor data to the other computers over the data bus. All of this information is transmitted at the same time.

There are many different types of multiplex systems used in today's vehicles. These vary in communication speed, data signals, and the wires used as the bus. These different systems, commonly called protocols of communication, are rules or operating parameters that allow for the exchange of information. Most vehicles use different protocols for different computers. The quicker protocols are used for the most important computer systems, such as engine control. When different protocols are used in the same vehicle, the vehicle is equipped with a Gateway ECU (electronic control unit). This ECU changes the arrangement of data so that information can be shared between the different protocols. Multiplex communication relies on serial communication, which changes the data from many different sources into serial data.

The two most common protocols for serial communication are the **Programmable Controller Interface Data Bus (PCI Bus)** and the **CAN (Controller Area Network) Bus**. The PCI data bus system is a single wire multiplex system and the serial data is comprised on bits and frames. A bit is the data and is expressed in binary values, which are digital or on-and-off voltage signals. The combination of these signals is interpreted by the computer. The frame surrounds the group of binary signals. The beginning of the frame represents the beginning of the data and an end message indicates the end. This is how the computers recognize the origin of the signals (Figure 4-25). Communication between computers takes place through a single wire by the transmission of **variable pulse width modulation (VPWM)** signals (Figure 4-26).

The communication wire in a CAN system is a twisted-pair wire. Each of the wires of the pair has a different applied voltage. The differing voltages allow for communication (Figure 4-27) while the computer is looking at the difference between the two voltages. The bus wires are twisted to reduce the chance of the signals being disrupted by radio frequency interference. This interference can cause the sensitive voltage signals to be altered and send false information

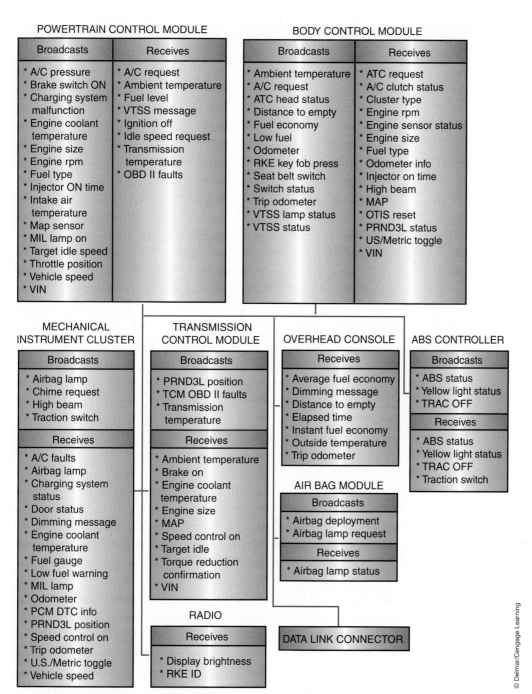

POWERTRAIN CONTROL MODULE

Broadcasts	Receives

* A/C pressure
* Brake switch ON
* Charging system
  malfunction
* Engine coolant
  temperature
* Engine size
* Engine rpm
* Fuel type
* Injector ON time
* Intake air
  temperature
* Map sensor
* MIL lamp on
* Target idle speed
* Throttle position
* Vehicle speed
* VIN
	* A/C request
	* Ambient temperature
	* Fuel level
	* VTSS message
	* Ignition off
	* Idle speed request
	* Transmission
	  temperature
	* OBD II faults

BODY CONTROL MODULE

Broadcasts	Receives
* Ambient temperature * A/C request * ATC head status * Distance to empty * Fuel economy * Low fuel * Odometer * RKE key fob press * Seat belt switch * Switch status * Trip odometer * VTSS lamp status * VTSS status	* ATC request * A/C clutch status * Cluster type * Engine rpm * Engine sensor status * Engine size * Fuel type * Odometer info * Injector on time * High beam * MAP * OTIS reset * PRND3L status * US/Metric toggle * VIN

MECHANICAL INSTRUMENT CLUSTER

Broadcasts
* Airbag lamp * Chime request * High beam * Traction switch

Receives

* A/C faults
* Airbag lamp
* Charging system
  status
* Door status
* Dimming message
* Engine coolant
  temperature
* Fuel gauge
* Low fuel warning
* MIL lamp
* Odometer
* PCM DTC info
* PRND3L position
* Speed control on
* Trip odometer
* U.S./Metric toggle
* Vehicle speed

TRANSMISSION CONTROL MODULE

Broadcasts
* PRND3L position * TCM OBD II faults * Transmission   temperature

Receives

* Ambient temperature
* Brake on
* Engine coolant
  temperature
* Engine size
* MAP
* Speed control on
* Target idle
* Torque reduction
  confirmation
* VIN

RADIO

Receives
* Display brightness * RKE ID

OVERHEAD CONSOLE

Receives
* Average fuel economy * Dimming message * Distance to empty * Elapsed time * Instant fuel economy * Outside temperature * Trip odometer

AIR BAG MODULE

Broadcasts
* Airbag deployment * Airbag lamp request

Receives
* Airbag lamp status

DATA LINK CONNECTOR

ABS CONTROLLER

Broadcasts
* ABS status * Yellow light status * TRAC OFF

Receives
* ABS status * Yellow light status * TRAC OFF * Traction switch

© Delmar/Cengage Learning

FIGURE 4-24 The information available on the bus, which is identified by the colored lines.

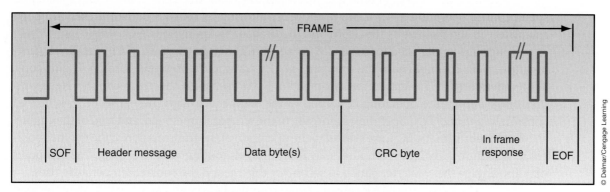

FRAME

SOF | Header message | Data byte(s) | CRC byte | In frame response | EOF

© Delmar/Cengage Learning

FIGURE 4-25 A typical PCI bus message.

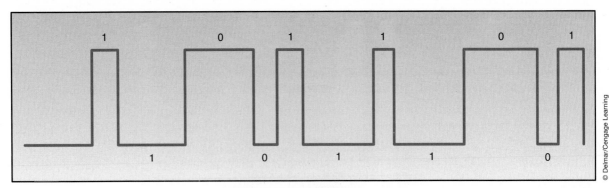

FIGURE 4-26 A VPWM signal.

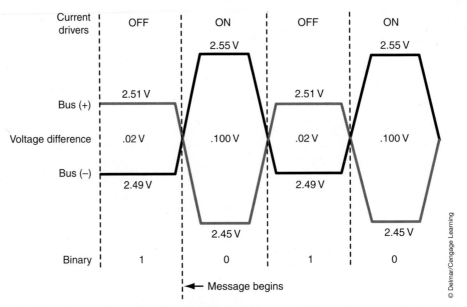

FIGURE 4-27 In a CAN system, the computer looks at the difference in the voltage values on the two bus wires to interpret the values encoded in the data stream.

to the PCM. The computers that share the bus have unique frequencies that serve as identification. The frequency may also be altered by radio frequency interferences. Radio frequency interference is a form of electromagnetic interference or electrical noise caused by secondary ignition, high current flow, and the operation of devices such as motors and solenoids.

Primary Inputs. Engine speed, throttle position, engine temperature, engine load, and other typical engine-related inputs are used by the computer to determine the best shift points. Many of these inputs are available through and are inputted from the common bus. Other information, such as engine and body identification, the TCM's target idle speed, and speed control operation are not the result of monitoring by sensors; rather, these have been calculated or determined by the TCM and made available on the bus.

Typical bus inputs used by the TCM come from the mass airflow (MAF), intake air temperature (IAT), manifold absolute pressure (MAP), barometric pressure (BARO), engine coolant temperature (ECT), and crankshaft position (CKP) sensors. These provide the TCM with information about the operating condition of the engine. Through these, the TCM is able to control shifting and TCC operation according to the temperature, speed, and load of the engine.

The inputs from the ECT are critical to the operation of the transmission. If the engine's coolant temperature is cold, the computer may delay upshifts to improve driveability. The computer may also engage the converter clutch in second or third gear if the coolant temperature rises.

The TCM receives the MAP sensor signal from the PCM and uses it to calculate the load on the engine. By monitoring the MAP values, the TCM can change the shift schedule to

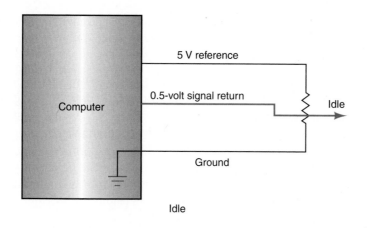

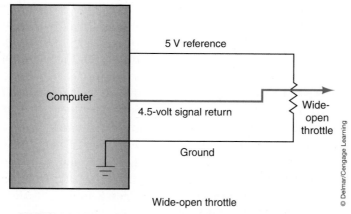

FIGURE 4-28 The voltage signal from a typical TP sensor varies from a closed throttle to a wide-open throttle.

reduce gear hunting by the transmission during times of high-load operation, such as going up a steep hill. The MAP signal, combined with the signals from the TP sensor, allows the TCM to determine the load on the engine and the driver's intent. The TP sensor sends a voltage signal to the TCM in response to throttle position (Figure 4-28). TP signals are used by the TCM to help determine the best shift points. As the throttle pedal is pressed or released, a wiper inside the sensor moves sweeps across a resistor inside the sensor. The positive side of the resistor is fed 5 volts by the PCM and the negative is grounded. The wiper is connected to an output wire and, as the wiper moves along the resistor, it picks up varying amounts of voltage. This voltage is signal sent back to the PCM to reflect on throttle position. Typically, the voltage ranges between approximately 0.5 volts when the throttle is closed to approximately 4.5 volts when the throttle is fully open.

The TCM may use the signal from the IAT to calculate the temperature of the battery. The TCM then uses this temperature calculation to estimate transmission fluid temperature.

The CKP sensor provides engine speed to the TCM. Although engine speed information is available at the bus, the computer receives this signal directly from the distributor pick up coil or CKP sensor. With the direct feed, any time delay at the bus circuit is avoided and the computer is aware of current engine speeds. This input is used to determine shift timing and TCC apply and release.

The signals from the BARO are used by the TCM to adjust line pressures according to changes in altitude. This sensor input may not be used; its use depends on the type of intake air monitoring system the vehicle is equipped with. On those vehicles using the BARO sensor as an input, the sensor may be integrated on the PCM circuit board or mounted externally.

The direct inputs are those sensors that provide information to the TCM and do not use the bus circuit. Many of these sensors produce an analog signal that must be changed to a digital signal before the TCM can respond. This conversion is handled by an analog-to-digital (A/D) converter, the PCM, or a digital radio adapter controller (DRAC). These typically convert an analog AC signal to a digital 5-volt square wave.

Shop Manual
Chapter 4, page 185

On-Off Switches. The brake switch is used to disengage the torque converter clutch when the brakes are applied. Its input has little to do with the up and down shifting of gears, except in some systems where it signals a need for engine braking. An A/C request switch informs the TCM that the A/C has been turned on. The TCM then changes line pressure and shift timing to accommodate the extra engine load created by A/C compressor operation.

The transmission range (TR) sensor informs the TCM of the gear selected by the driver. This sensor normally also contains the neutral safety switch and the reverse light switch. The TR sensor is typically a multiple pole-type on/off switch (Figure 4-29).

Transmission Fluid Temperature (TFT) Sensor. The TFT sensor monitors the temperature of the transmission's fluid. When the signal from this sensor is normal, the transmission will operate within its normal range. However, when the signal indicates that the fluid is overly hot, the TCM will allow the transmission to only operate in such a way that will allow the transmission to cool down. This prevents damage to the transmission. When the TFT signal indicates that the fluid is cooler than normal, the TCM will alter the shift schedule.

The TFT sensor is a thermistor located in the valve body. Its electrical resistance varies with ATF temperature. The TCM integrates this input with others to control TCC operation. Typically the TCM prevents TCC engagement until fluid temperatures reach about 68°F (20°C). If fluid temperature reaches about 250°F (122°C), the TCM applies the TCC in second, third, or fourth gears. If mechanically connecting the engine to the input shaft of the transmission does not reduce fluid temperature, and it reaches 300°F (150°C), the TCM will release the TCC to prevent damage to the converter clutch from excessive temperatures. If the fluid reaches about 310°F (154°C), the TCM sets a fluid temperature trouble code and uses a fixed value as the fluid temperature input signal.

The TFT is sometimes referred to as the TOT (transmission oil temperature) sensor.

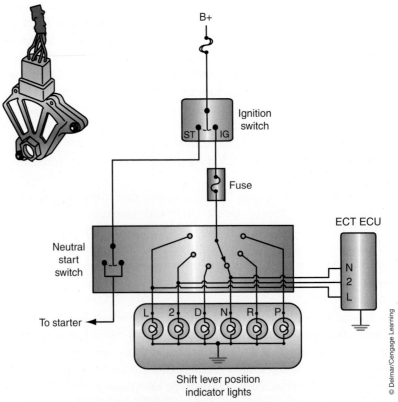

FIGURE 4-29 A neutral start switch (TR sensor) and immediate circuit.

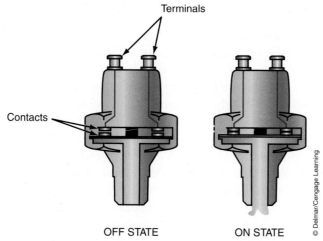

FIGURE 4-30 The action of a pressure switch.

Transmission Pressure Switches. Various transmission pressure switches can be used to keep the TCM informed as to which hydraulic circuits are pressurized and which clutches and brakes are applied. These input signals can serve as verification to other inputs and as self-monitoring or feedback signals (Figure 4-30).

Voltage-Generating Sensors. The vehicle speed (VSS) and **output shaft speed (OSS)** sensors are used to monitor output of the transmission or vehicle speed. In some electronic control systems, only one of these sensors is used. When a vehicle has both sensors, the OSS signal is used as a verification signal for the VSS by the engine control system. The VSS is used as a verification signal for the OSS by the TCM. Some transmissions use these speed-related inputs in place of a governor. These signals are used to regulate hydraulic pressure and shift points and to control TCC operation. Four-wheel-drive vehicles may use a third speed sensor installed in the transfer case. The TCM determines vehicle speed from this sensor, rather than the OSS.

Some transmissions have an **input shaft speed (ISS)** sensor. This sensor and its operation are identical to the OSS and its signal is used by the TCM to calculate converter turbine speed. Input and output speeds provided by the two sensors are used by the TCM to help determine line pressure, shift patterns, and TCC apply pressure and timing (Figure 4-31).

Some manufacturers call the OSS a **TOSS (transmission output speed sensor)**.

Shop Manual
Chapter 4, page 190

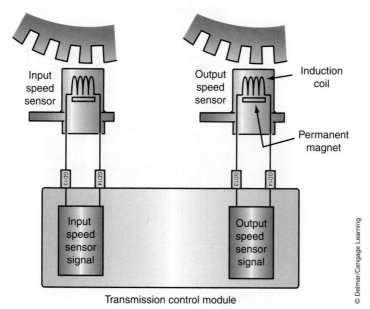

FIGURE 4-31 The difference between the signals from the input shaft speed sensor and the output shaft speed sensor is used by the TCM to determine many different operating parameters.

Adaptive Controls

Many late-model transmissions have systems that allow the computer to change transmission behavior in response to the operating conditions and the habits of the driver. The system monitors the condition of the engine and compensates for any changes in the engine's performance. It also monitors and memorizes the typical driving style of the driver and the operating conditions of the vehicle. With this information, the computer adjusts the timing of shifts and converter clutch engagement to provide good shifting at the appropriate time.

These systems are constantly learning about the vehicle and driver. The computer adapts its normal operating procedures to best meet the needs of the vehicle and the driver. When systems are capable of doing this, they are said to have **adaptive learning** capabilities. To store this information, the computer includes a long-term adaptive memory.

The computer also learns the characteristics of the transmission and changes its programming accordingly. It learns the release and application rates of various transmission components during various operating conditions. Adaptive learning allows the computer to compensate for wear and other events that might occur and cause the normal shift programming to be inefficient. Doing this, the adaptive learning capability of the transmission computer allows for this smooth shifting throughout the life of the transmission. As components wear and shift overlap times increase, the TCM adjusts line pressure to maintain proper shift timing calibrations.

The adaptive learning takes place as the TCM reads input and output speeds more than 140 times per second. The computer responds to each new reading. This learning process allows the TCM to make adjustments to its program so that quality shifting always occurs.

Clutch Volume Indexes

As part of their adaptive strategy, Chrysler has incorporated a monitoring system that detects and adjusts to wear. It does this by monitoring **clutch volume indexes (CVIs)** whenever the vehicle is driven. The CVI is the measurement of the physical amount of fluid and time required to fill the clutch and stroke the piston. As friction elements wear, it takes more fluid volume and time to apply the element. The system uses inputs from the input shaft and output shaft speed sensors to determine the time, and calculates the required fluid volume based on a CVI value stored in its memory. When a clutch pack is worn or there is a leak at a piston seal, the TCM will see an increase in the time it takes to fully apply the clutch (Figure 4-32). In

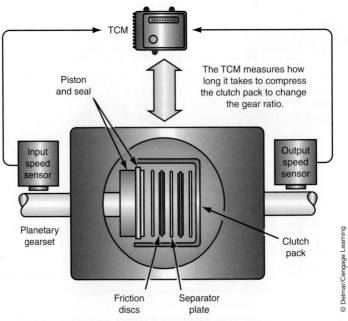

FIGURE 4-32 The calculation of the CVI.

response to this, the TCM will send a greater volume of fluid to the piston to allow the clutch pack to fully engage in the same time it would require if it was not worn or leaking.

Outputs

Common outputs used with EATs are indicator lamps and solenoids. The indicator lamps show the selected gear range and may also be a MIL designated for the transmission or this warning light may be incorporated into the circuit for the engine's MIL. Shift, pressure control, and T/C clutch solenoids are used in all modern EATs. All of these are controlled by the TCM. The solenoids are typically located inside the transmission and are mounted to the valve body.

A transmission control system uses several solenoids (Figure 4-33), controlled by the TCM, to regulate shift timing, feel, and TCC application. The number and purpose of each depends on the model of transmission. However, all electronic transmissions have solenoids to control gear shift timing, throttle pressure control (shift quality), accumulator back-pressure, and torque converter lockup. Many of a transmission's solenoids are housed together in a solenoid pack.

SHIFT SOLENOID OPERATION CHART					
Transaxle range selector lever position	Powertrain control module Gear commanded	Eng braking	AX4N solenoids		
			SS 1	SS 2	SS 3
P/N[a]	P/N	NO	OFF[b]	ON[b]	OFF
R (REVERSE)	R	YES	OFF	ON	OFF
Overdrive	1	NO	OFF	ON	OFF
	2	NO	OFF	OFF	OFF
	3	NO	ON	OFF	ON
	4	YES	ON	ON	ON
D (DRIVE)	1	NO	OFF	ON	OFF
	2	NO	OFF	OFF	OFF
	3	YES	ON	OFF	OFF
Manual 1	2[c]	YES	OFF	ON	OFF
	3[c]	YES	OFF	OFF	OFF
		YES	ON	OFF	OFF

[a] When transmission fluid temperature is below 50°F then SS 1 = OFF, SS 2 = ON, SS 3 = ON to prevent cold creep.

[b] Not contributing to powerflow.

[c] When a manual pull-in occurs above calibrated speed the transaxle will downshift from the higher gear until the vehicle speed drops below this calibrated speed.

FIGURE 4-33 Shift solenoid operation for an AX4N transaxle.

The TCM relies on engine and transmission sensors to detect such things as throttle position, vehicle speed, engine speed, engine load, brake pedal position, and so on to determine the shift points and how soft or firm the shift should be. To control shifting, the solenoids redirect hydraulic fluid to the appropriate clutch or servo.

Two types of solenoids are used in automatic transmissions: On/off or linear solenoids. On/off solenoids are normally connected to the TCM by a single wire and are turned on when voltage is applied to them. These are the most commonly used solenoids and are used to control a shift valve. Shift solenoids can be two-way solenoids that either allow or block fluid flow or they can be three-way solenoids that switch the flow of hydraulic fluid from one path to another.

Linear solenoids normally have two wires from the TCM. One lead is the power and the other is a ground that is controlled by the TCM. The action of this type of solenoid is controlled by current flow to it. The TCM rapidly switches the solenoid off and on. Linear solenoids are normally used to control line pressure, apply pressure, accumulator back-pressure, and torque converter clutch pressure.

Shift Solenoids. Shift solenoids control the delivery of fluid to the manual shift valve. Each shift solenoid offers two possible on/off combinations to control fluid to the shift valves. These solenoids are on/off solenoids that are normally off and are either normally vented or normally applied. A normally vented solenoid (Figure 4-34) does not allow fluid pressure to reach the shift valve until it is energized. When they are turned off, the fluid to the solenoid is exhausted or vented. Normally applied solenoids (Figure 4-35) allow fluid pressure to reach the shift valve until it is energized, then it blocks the fluid flow and vents the pressure.

The number and purpose of each depends on the model of transmission. In a typical four-speed unit, there are two shift solenoids. The solenoids offer four possible on/off combinations to control fluid to the shift valves. This provides the engagement of the four forward gears. Transmissions with additional forward gears rely on additional shift solenoids. The on/off combinations of these provide the additional gears. It is important to note that the TCM will stop current flow to a shift solenoid if it detects a fault in the solenoid or its circuit.

Pressure Control Solenoids. An EPC (Electronic Pressure Control) solenoid is used to control hydraulic pressures throughout the transmission. This solenoid operates on a duty cycle controlled by the TCM. Its purpose is to regulate the operating hydraulic pressures to the clutches and brakes to provide smooth and precise shifting. The pressure control solenoid is installed on the valve body.

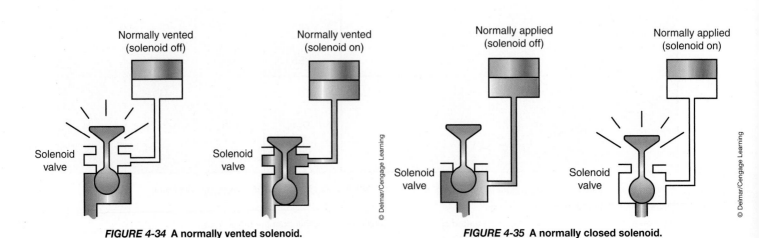

FIGURE 4-34 A normally vented solenoid.

FIGURE 4-35 A normally closed solenoid.

The EPC solenoid, or Pressure Control Solenoid (PCS), replaces the conventional TV cable to provide changes in pressure in response to engine load (Figure 4-36). This solenoid is a pulse width modulated (PWM) linear solenoid (Figure 4-37) or a **variable force solenoid** (**VFS**) and contains a spool valve and spring. To control fluid pressure, the TCM sends a varying signal to the solenoid. This varies the amount the solenoid will cause the spool valve to move.

When the solenoid is off, the spring tension keeps the valve in place to maintain maximum pressure. As more current is applied to the solenoid, the solenoid moves the spool valve more, which moves to uncover the exhaust port, thereby causing a decrease in pressure (Figure 4-38). The EPC solenoid controls line pressure, at all times, based on the programming

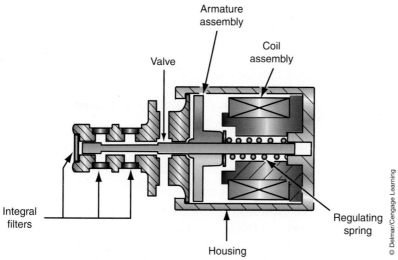

FIGURE 4-36 A typical variable force solenoid or motor.

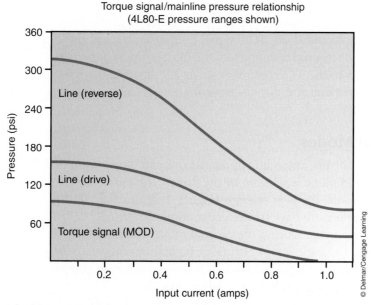

FIGURE 4-37 The EPC solenoid controls hydraulic pressures throughout the transmission by controlling the current going to the solenoid.

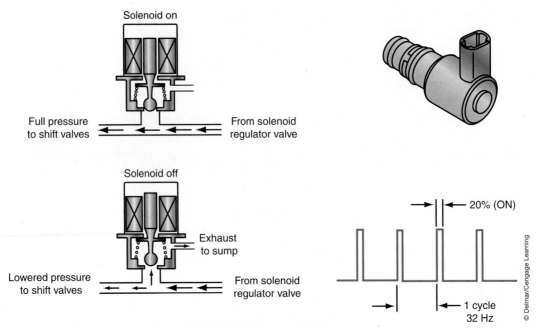

Solenoid on

Full pressure
to shift valves

From solenoid
regulator valve

Solenoid off

Exhaust
to sump

Lowered pressure
to shift valves

From solenoid
regulator valve

20% (ON)

1 cycle
32 Hz

© Delmar/Cengage Learning

FIGURE 4-38 (TOP) A typical PWM solenoid. (BOTTOM) The signal representing the control or ordered duty cycle from the computer.

of the system's computer and is able to match shift timing and feel with the current needs of the vehicle.

Torque Converter Clutch Solenoid. The torque converter clutch (TCC) solenoid is used to control the application, modulation, and release of the torque converter clutch. The operation of the converter clutch is also totally controlled by the TCM. The exception to this is during first gear and reverse gear operation when the clutch is disabled hydraulically to prevent engagement regardless of the commands by the computer. Normally, the converter clutch is hydraulically applied and electrically controlled through a PWM solenoid, which is controlled by the TCM. Modulating the pressure to the converter clutch allows for smooth engagement and disengagement and also allows for partial engagement of the clutch.

The PWM solenoid is installed in the valve body. It controls the position of the TCC apply valve. When the solenoid is off, TCC signal fluid exhausts and the converter clutch remains released. Once the solenoid is energized, the plunger moves the metering ball to allow TCC signal fluid to pass to the TCC regulator valve. Modulating the pressure to the clutch allows for smooth engagement and disengagement and also allows for partial engagement of the clutch.

Operational Modes

With electronic controls, automatic transmissions can be programmed to operate in different modes. The desired mode is selected by the driver. The mode selection switch can be located on the center console, the gear selector (Figure 4-39), or the instrument panel. Most transmissions with this feature have two selective modes, normally called "Normal" and "Power" or Manual or Automatic. During the normal mode, the transmission operates according to the shift schedule and logic set for normal or regular operation. In the power mode, the TCM uses different logic and shift schedules to provide for better acceleration and performance with heavy loads. Normally this means delaying upshifts.

Limp-In Mode. When the TCM detects a serious transmission problem or a problem in its circuit (or in some cases an engine control or data bus problem), it may switch to a default, limp-in, or fail-safe mode. Limp-in may also be initiated if the TCM loses its battery power

FIGURE 4-39 The gear selector for a manual shifting BMW automatic transmission.

feed. This mode allows for limited driving capabilities and is designed to prevent further transmission damage while allowing the driver to drive with decreased power and efficiency to a service facility for repairs.

The capabilities of a transmission while it is in limp-in depend on the extent of the fault, the manufacturer, and the model of transmissions. When the TCM moves into the limp-in mode, a DTC is set and the transmission will only operate in this mode until the problem is corrected. Examples of operating characteristics during limp-in include the following:

- The transmission locks in third gear when the gear selector is in the Drive position or second gear when it is in a lower position.
- The transmission will remain in whatever gear it was in but will shift into third or second gear and stay there as soon as the vehicle slows down.
- The transmission will only use first and third gears while in Drive.
- The transmission will operate only in Park, Neutral, Reverse, and one forward gear and will not upshift or downshift. The allowed forward gear depends on the transmission; most restrict driving to second gear only while others allow operation in only third or fourth gear.

Overdrive Cancel. Many transmissions have an overdrive cancel button that is used to disable the overdrive gear. The button may be labeled "OD off" or "OD cancel". When overdrive is canceled in a four-speed transmission, the transmission will shift up and down through first, second, third gears but never shift into fourth gear. Canceling overdrive prevents the transmission from rapid up-down-up shifting that can occur when the vehicle is moving in stop and go traffic, on rolling hills, or climbing a medium grade. By preventing overdrive operation, the resulting locking and unlocking of the torque converter clutch, due to the changing load, is also prevented. Preventing the transmission from shifting into overdrive can extend the life of the transmission by reducing the number of unnecessary shifts and torque converter lockup cycles.

Tow/Haul Mode. Some late-model vehicles have a "Tow/Haul" switch rather than an overdrive cancel switch. This operating mode delays upshifts and increases operating pressures when the vehicle is operating with a heavy load. The mode also provides quicker downshifts during deceleration, which allows for more engine braking. The torque converter's lockup clutch is applied sooner and in more gears than normal during acceleration. This helps to keep the temperature of the fluid in the torque converter down. Also, the clutch remains engaged for a longer period of time during deceleration to improve engine braking. In this mode, the transmission is able to shift into overdrive when the vehicle's load is overcome.

Manual Shifting

One of the most publicized features of electronically controlled transmissions is the availability of manual shift controls. Although not all electronic transmissions have this feature, they all could. Basically, these systems allow the driver to manually upshift and downshift the transmission at will, much like a manual transmission. Unlike a manual transmission, the driver does not need to depress a clutch pedal. All the driver does is move a gear selector or hit a button and the transmission changes gears. If the driver doesn't change gears and engine speed is high, the transmission shifts on its own. If the driver elects to let the transmission shift automatically, a switch disconnects the manual control and the transmission operates automatically.

Marketed as a sport option and a combination of a manual and automatic transmission, these transmissions are still based on an automatic transmission with a torque converter. Therefore, performance numbers are not quite as good as if the vehicle was equipped with a manual transmission. In fact, manual control of an automatic transmission often results in slower acceleration times than when the transmission shifts by itself.

All manually shifted automatics do not behave in the same way, nor do they control the same things. Actually the behavior of the transmission depends on the car it is installed in. Some of these cars are pure high-performance cars, while others are moderate-performance family sedans. What follows are basic descriptions of some of the manually shifted automatic transmissions available. There are many other similar systems available, these are mentioned to only serve as examples.

BMW's Steptronic. Steptronic systems are based on five-or six-speed transmissions and offer the option of shifting gears manually using the selector lever or through steering wheel mounted buttons. The driver can select to shift the gears manually or let the transmission shift automatically. The system relies on shift-by-wire technology, which replaced mechanical linkages to the transmission. These transmissions also are equipped with adaptive transmission controls (ATC), which responds to the driver's style as well as the operating conditions. ATC looks at the travel and movement patterns of the accelerator, deceleration rates during braking, and lateral acceleration in curves. ATC then selects the most suitable shift schedule. Upshifts are delayed on uphill stretches to allow better use of the engine's power. On downhill grades, ATC downshifts when you are forced to brake to counteract undesired acceleration.

In the manual mode, ATC works to avoid over speeding the engine by upshifting just before the engine reaches its automatic cutoff speed. At low speeds, it downshifts automatically without input from the driver. In the kickdown mode, the system downshifts to the lowest gear possible without overrevving the engine.

Chrysler Autostick. In this system, manual shifting is performed by moving a control on the console. Moving the selector to the right provides for an upshift. A movement to left allows for a downshift.

The transaxle is not modified for this option; rather, it is fitted with a special gear selector and switch assembly. The driver can either manually shift the gears or allow the transaxle to shift automatically. The selected gear is displayed on the instrument panel to keep the driver informed of the selected gear.

Although the driver has control of the shifting, the TCM will override the controls during some conditions. Sometimes, the TCM will force upshifts or prevent manual shifting. These conditions relate to the projected engine speed if a gear change does or does not take place. The Autostick feature will also be deactivated if the TCM senses problems and/or sets a trouble code that relates to the TR sensor or Autostick switch or when there is a high engine and transmission temperature code.

Honda's Sequential Sportshift. Manual shifting is performed by moving a control on the console or buttons on the steering wheel. Moving the selector forward provides for an upshift. A movement down allows for a downshift. Some models of this transmission are unique in that they will not automatically upshift if the driver brings the engine's speed too high. All other transmissions of this design will upshift automatically at a predetermined engine speed to prevent damage to the engine. Honda vehicles use a rev limiter that prevents the engine from overspeeds. When some Acura models are operated in the manual shifting mode, they will start out in first gear and shift automatically to second gear. The transmission can only be shifted manually between gears.

Tiptronic. This is a five- or six-speed transmission available in a few European cars, such as Porsches and Audis. Although the concept of all Tiptronic units is the same, the driver control varies quite a bit. Manual shifting on the Audi is performed by moving a control on the console. Moving the selector forward provides for an upshift. A movement down allows for a downshift. To shift Porsche's transmission, the driver moves the gear selector into the manual gate, next to the automatic ranges, and depresses buttons on the steering wheel. These systems also are typically controlled by a computer that controls the change of gears and tries to mimic the action of a manual transmission. While operating in the manual shift mode, these systems study the driving habits of the driver and select the optimum driving range, reducing the driver's workload particularly in stop-and-go traffic.

Toyota/Lexus Systems. Toyota has a series of high-performance cars that feature five-speed transmissions that can operate in either of two modes providing fully automatic shifting or electronic manual control. The TCM is programmed to allow for rapid shifts in response to the driver's commands. It will also prevent shifting during conditions that may cause engine or transmission failure. The transmission may also be shifted by its gated console-mounted shift lever (Figure 4-40). The shift lever allows the driver to select individual gear ranges as well as the full-automatic mode.

Manual shifting may also be controlled by fingertip shifting buttons located on both horizontal spokes of the steering wheel (Figure 4-41). The buttons are located so that thumbs can be used to easily change the gears.

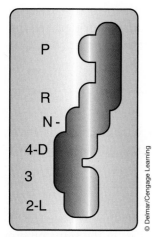

FIGURE 4-40 The gated shifter for Lexus' automatic/manual transmission.

FIGURE 4-41 Fingertip controls for manually shifting Lexus' automatic/manual transmission.

CONTINUOUSLY VARIABLE TRANSMISSIONS (CVTs)

A continuously variable transmission (CVT) may be regarded as an automatic transmission in that the driver does nothing to gears. However, the design is also more mechanical than most automatic transmissions; therefore, it is sort of a manual transmission. A CVT can automatically select any desired drive ratio within its operating range. It automatically and continuously selects the best overall ratio for the operating conditions. During the drive ratio changes, there is no perceptible shift. The controls of this type of transmission attempt to keep the engine operating at its most efficient speed. This decreases fuel consumption and exhaust emissions. When driving a vehicle with a CVT, shifts are unnoticeable and the transmissions are referred to as "stepless or non-stepped."

During maximum acceleration, the drive ratio is adjusted to maintain peak engine horsepower. At a constant vehicle speed, the drive ratio is set to obtain maximum fuel mileage while maintaining good driveability.

Many late-model CVTs are equipped with a feature that stimulates the activity of a manual shifting automatic transmission. Transmissions with this feature are typically called stepped CVTs. They have five or six predetermined areas that the pulleys stop in when the driver selects manual control. While in manual, these steps provide the feel and shift effect of distinct gear ratios.

To change the direction of the vehicle, two wet clutches may be used. One for forward and one for reverse. A planetary gearset is used in conjunction with the clutches to change direction. These clutches are applied when the shifter is placed into drive or reverse.

CVT Controls

The size of the pulleys is electronically controlled to select the best overall drive ratio based on throttle position, vehicle speed, and engine speed. Different speed ratios are available any time the vehicle is moving. Since the size of the drive and driven pulleys can vary greatly, vehicle loads and speeds can be responded to without changing the engine's speed.

The electronic control system for a typical CVT consists of a TCM, various sensors, linear solenoids, and an inhibitor solenoid. Input from the various sensors determines which linear solenoid the TCM will activate (Figure 4-42). Activating the shift control solenoid changes the shift control valve pressure, causing the shift valve to move. This changes the pressures applied to the driven and drive pulleys, which changes the effective pulley ratio.

If the car is continuously driven at full throttle acceleration, the TCM causes an increase in pulley ratio. This reduces engine speed and maintains normal engine temperature while not adversely affecting acceleration. After the car has been driven at a lower speed or not accelerated for a while, the TCM lowers the pulley ratio. When the gear selector is placed into reverse, the TCM sends a signal to the PCM. The PCM then turns off the car's air conditioning and causes a slight increase in engine speed.

HYBRID TRANSMISSIONS

Perhaps the most complex transmissions are those used in many hybrid vehicles. However, some hybrids rely on conventional CVTs or manual or automatic transmissions. This section covers the common nontraditional transmissions used in hybrids. CVTs are desirable in hybrids because they allow the engine to run at its most efficient speed, which means increased fuel economy and decreased exhaust emissions. Honda and others use a conventional CVT with pulleys. Toyota, Ford, and Nissan hybrids use a planetary gear based CVT. These CVTs do not rely on belts and pulleys; rather the electric motors in the transmission not only help propel the vehicle, but also provide a constantly variable drive ratio. Other manufacturers have electric motors connected in series with planetary gearsets. This arrangement allows for variable and stepped gear ratios.

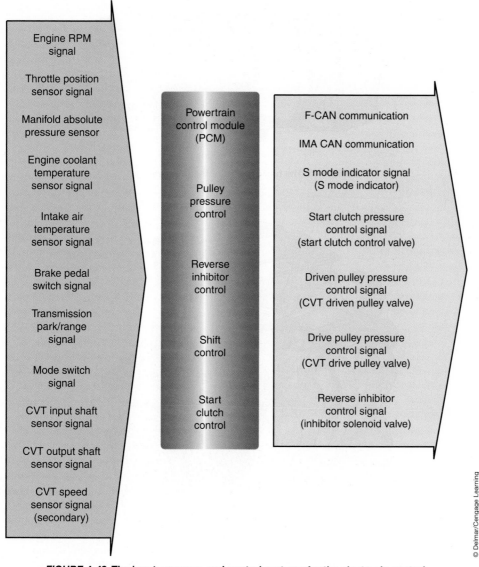

FIGURE 4-42 The input, process, and control systems for the electronic control system of a Honda CVT.

© Delmar/Cengage Learning

Hybrid systems allow for regenerative braking which decreases the amount of load placed on the engine to turn the generator. Regenerative braking recharges the batteries when the vehicle slows down or coasts. This means the engine driven generator works less hard, thereby allowing the engine to work less hard. Also, hybrids have a stop–start feature. During braking, when vehicle speed drops below 13 mph, the engine is shut off and remains off until the driver releases the brake pedal.

Honda IMA System

Honda hybrids are available, depending on model, with a CVT, five-speed manual, or a five-speed automatic transmission. The important thing, other than changes to the automatic transmission, relates directly to the **Integrated Motor Assist (IMA)** system used in these hybrid vehicles.

The IMA system combines a modified gasoline engine with a thin permanent magnet electric motor (Figure 4-43). This 144-volt synchronous AC motor has a three-phase coil stator and a permanent magnet rotor that is directly connected to the engine crankshaft. The motor is used to assist the engine, when extra power is needed. The motor also serves as a generator.

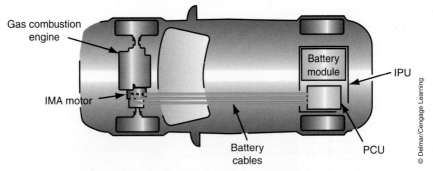

FIGURE 4-43 The basic layout of the IMA system.

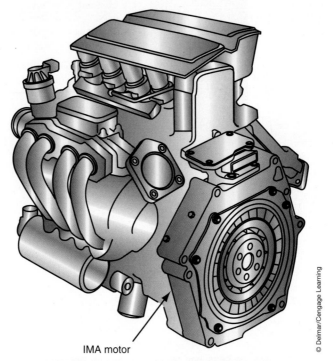

FIGURE 4-44 The placement of the IMA motor.

All transaxles used in a Honda hybrid are modified and are made more compact so they can fit behind the electric motor mounted at the rear of the engine (Figure 4-44) and occupy the same amount of space as the transaxle in a nonhybrid vehicle.

The transmission's torque converter is slightly smaller and has a stronger lockup clutch. The transmission is also fitted with a stronger input shaft and different gear ratios to provide for better acceleration, fuel economy, and regenerative braking. The transmission is fitted with an electric auxiliary oil pump. The pump is used to provide for sufficient line pressure in the transmission when the engine is restarted during the stop–start sequence.

Flywheel Alternator Starter Hybrid System

Similar to the Honda IMA system, the flywheel/alternator/starter assembly, sometimes called an **integrated starter alternator damper (ISAD)**, is positioned between the engine and the transmission, although it can be mounted to the side of the transmission. It does not require the very high voltages required by Honda and it typically does not assist the engine.

The ISAD system replaces the conventional starter, generator, and flywheel with an electronically controlled compact 42-volt AC asynchronous induction electric motor. This unit is

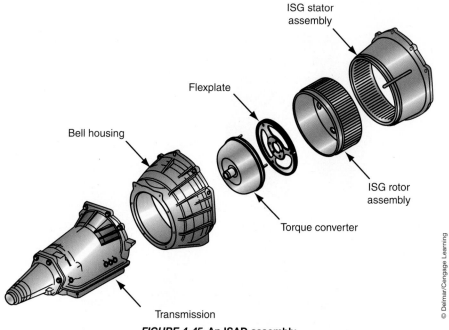

ISG stator assembly

Flexplate

Bell housing

ISG rotor assembly

Torque converter

Transmission

© Delmar/Cengage Learning

FIGURE 4-45 An ISAD assembly.

housed in the transmission's bell housing. The stator of the starter/generator is mounted to the engine block. The rotor is attached to the end of the engine's crankshaft (Figure 4-45). As the crankshaft rotates, so does the rotor, or vice versa. Current is sent to the stator when the unit is functioning as a motor. When functioning as a generator, current flows from the stator.

Belt Alternator Starter Hybrid System

A **Belt Alternator Starter (BAS)** system (Figure 4-46) replaces the traditional starter and generator in a conventional vehicle. The unit is located where the generator would normally be and is connected to the engine's crankshaft by a drive belt. This unit, an electric motor, serves as the starting motor and generator. When the engine is running, a drive belt spins the rotor of the motor and the motor acts as a generator to charge the batteries. To start the engine, the motor's rotor spins and moves the drive belt, which in turn cranks the engine. These systems have the capability of providing stop–start, regenerative braking, and high-voltage generation.

A typical BAS includes the motor/generator, electronic controls, and a 42-volt battery. The ability to start the engine quietly and quickly is important to the operation of the stop–start feature, therefore the system uses a rather robust motor. The motor can be either a permanent magnet or induction motor. The required electronic controls for the system depend on the type of motor used. Some systems are also equipped with a conventional starting motor. These are used during extremely cold temperatures. Once the engine is warm, the BAS takes over.

General Motors. GM uses the BAS system in their Saturn VUE and Chevrolet Malibu Hybrids. The system is based on a dual-voltage architecture of a 42-/14-battery. A 42-volt NiMH battery pack provides power for the motor and some accessories, while the traditional 12-volt battery powers typical auxiliary devices, such as lights. The system's electronic circuitry monitors many vehicle operating conditions and controls the operation of the motor/generator and the engine. The electronics must synchronize the activity of the motor/generator with engine systems, such as the fuel injection system. Without precise control, early fuel shutoff during deceleration and quick restarts would not be possible. Nor could there be regenerative braking.

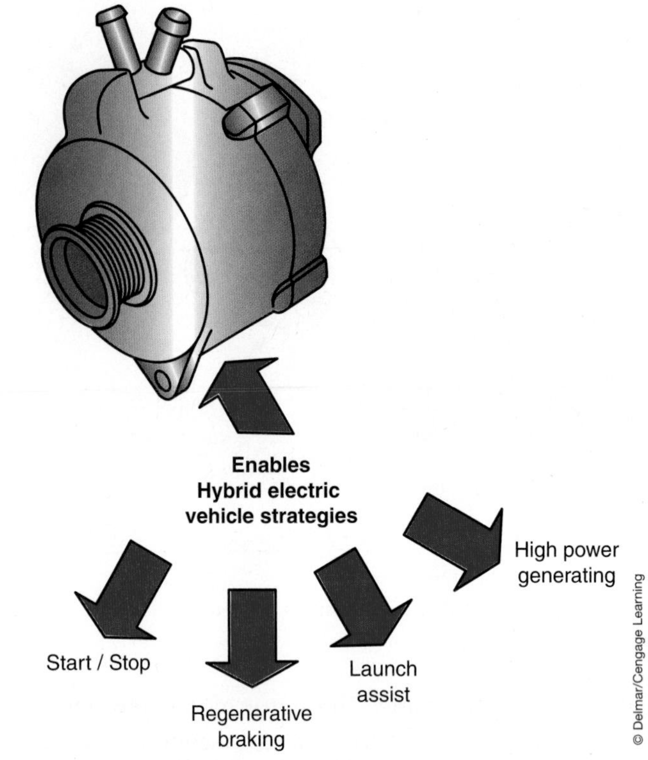

Belt-driven integrated
starter alternator

Enables
**Hybrid electric
vehicle strategies**

High power
generating

Start / Stop

Launch
assist

Regenerative
braking

© Delmar/Cengage Learning

FIGURE 4-46 **A BAS system.**

When working as a generator, the electric motor/generator provides more than twice the output of a typical generator. It is capable of providing 3000 watts of continuous power. The generator's output of 42-volts AC is converted to 42-volts DC and is used to charge the battery pack. A DC–DC converter is used to convert the 42-volt output to 14 volts to charge the conventional 12-volt battery and power most of the vehicle's electrical accessories. An inverter takes the 42-volts DC from the batteries and converts it to 42-volts AC to power the motor.

Toyota and Lexus Hybrids

The power-split device used in Toyota and Lexus hybrids operates as a continuously variable transaxle, although it does not use the belts and pulleys. The variability of this transaxle depends on the action of a motor/generator, referred to as MG1, and the torque supplied by another motor/generator referred to as MG2, and/or the engine. The transaxle assembly contains the following:

Differential assembly
Reduction unit
Motor Generator 1 (MG1)
Motor Generator 2 (MG2)
Transaxle damper
Planetary gearset

Using a motor as a generator or vice versa is accomplished by changing the electrical connections to the rotor and stator.

A conventional differential unit is used to allow for good handling when the car is making a turn. The reduction unit increases the final drive ratio so that ample torque is available to the drive wheels. MG1, which generates energy and serves as the engine's starter, is connected to the planetary gearset. So is MG2, which is also connected to the differential unit by a drive chain (Figure 4-47). This transaxle does not have a torque converter or clutch. Rather a damper is used to cushion engine vibration and the power surges that result from the sudden engagement of power to the transaxle.

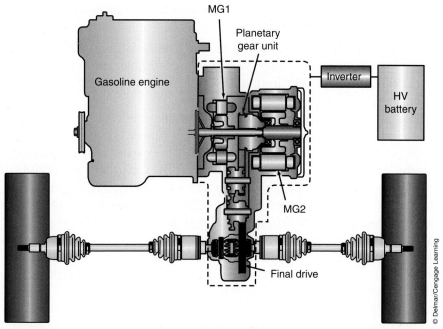

FIGURE 4-47 Toyota's hybrid system uses a combination of two basic types of motive forces: an engine and a motor (MG2). The engine drives a generator (MG1) or is started by MG1.

The engine, MG1, and MG2 are mechanically connected at the planetary gearset. The gearset can transfer power between the engine, MG1, MG2, and drive wheels in nearly any combination of these. The unit splits power from the engine to different paths: to drive MG1, drive the car's wheels, or both. MG2 can drive the wheels or be driven by them.

In the **power-split device**, the sun gear is attached to MG1. The ring gear is connected to MG2 and the final drive unit in the transaxle. The planetary carrier is connected to the engine's output shaft (Figure 4-48). The key to understanding how this system splits power is to realize that when there are two sources of input power, they rotate in the same direction but not at the same speed. Therefore, one can assist the rotation of the other, slow down the rotation of the other, or work together. Also, keep in mind the rotational speed of MG2 largely depends on the power generated by MG1. Therefore, MG1 basically controls the CVT

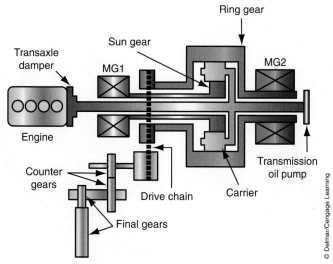

FIGURE 4-48 The layout of the main components connected to Toyota's power-splitting device.

function of the transaxle. Here is a summary of the action of the planetary gearset during different operating conditions.

- To start the engine, MG1 is energized and the sun gear becomes the drive member of the gearset (Figure 4-49). Current is sent to MG2 to lock or hold the ring gear. The carrier is driven by the sun gear and walks around the inside of the ring gear to crank the engine at a speed higher than that of the sun gear.
- After the engine is started, MG1 becomes a generator. The ring gear remains locked by MG2 and the carrier now drives the sun gear, which spins MG1.
- When the car is driven solely by MG2 (Figure 4-50), the carrier is held because the engine is not running. MG2 rotates the ring gear and drives the sun gear in an opposite direction. This causes MG1 to spin in the opposite direction without generating electricity.
- If more torque is needed while operating on MG2 only, MG1 is activated to start the engine. There are now two inputs to the gearset, the ring gear (MG2) and the sun gear (MG1). The carrier is driven by the sun gear and walks around the inside of the rotating ring gear. This cranks the engine at a faster speed than when the ring gear is held.
- After the engine is started, MG2 continues to rotate the ring gear and the engine rotates the carrier to drive the sun gear and MG1, which is now a generator.
- When the car is operating under light acceleration and the engine is running, some engine power is used to drive the sun gear and MG1 and the rest is sent to the ring gear to move the car (Figure 4-51). The energy produced by MG1 is fed to MG2. MG2 is also rotating the ring gear and the power of the engine and MG2 combine to move the vehicle.
- This condition continues until the load on the engine or the condition of the battery changes. When the load decreases, such as during low-speed cruising, the HV ECU increases the generation ability of MG1, which now supplies more energy to MG2. The increased power at the ring gear allows the engine to do less work while driving the car's wheels and do more work driving the sun gear and MG1.
- During full throttle acceleration, battery power is sent to MG2, in addition to the power generated by MG1. This additional electrical energy allows MG2 to produce more torque. This torque is added to the high output of the engine at the carrier.
- When the car is decelerating and the transmission is in Drive, the engine is shut off which effectively holds the carrier (Figure 4-52). MG2 is now driven by the wheels and acts as a generator to charge the battery pack. The sun gear rotates in the opposite direction and MG1 does not generate electricity. If the car is decelerating from a high speed, the engine is kept running to prevent damage to the gearset. The engine, however, merely keeps the carrier rotating within the ring gear.

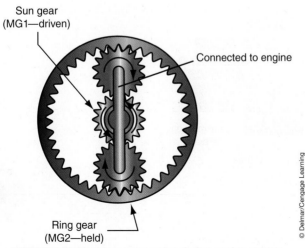

FIGURE 4-49 Planetary gear action during engine startup.

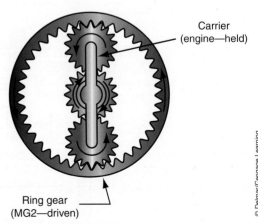

FIGURE 4-50 Planetary gear action when MG2 is propelling the vehicle.

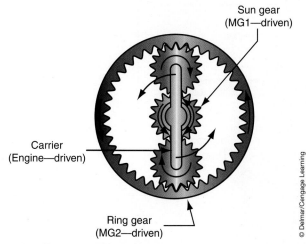

FIGURE 4-51 Planetary gear action when the engine and MG2 are propelling the vehicle.

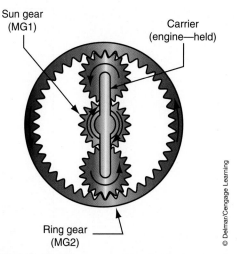

FIGURE 4-52 Planetary gear action during deceleration. MG2 is driven by the wheels of the vehicle.

- When the car is decelerating and the transmission is moved into the B range, MG2 acts as a generator to charge the battery pack and to supply energy to MG1. MG1 rotates the engine, which is not running at this time, to offer some engine braking.
- During normal deceleration with the brake pedal depressed, the engine is off and the skid control ECU calculates the required amount of regenerative brake force and sends a signal to the HV ECU. The HV ECU, in turn, controls the generative action of MG2 to provide a load on the ring gear. This load helps to slow down and stop the car. The hydraulic brake system does the rest of the braking.
- When reverse gear is selected, only MG2 powers the car. MG2 and the ring gear rotate in the reverse direction. Since the engine is not running, the carrier is effectively being held. The sun gear is rotating in its normal rotational directional but slowly and MG1 is not acting as a generator. Therefore, the only load on MG2 is the drive wheels.

It is important to remember that at anytime the car is powered only by MG2, the engine may be started to correct an unsatisfactory condition, such as low battery SOC, high battery temperature, and heavy electrical loads.

4WD Hybrids. On Toyota and Lexus hybrid SUVs, the front transaxle assembly has been modified to include a speed reduction unit. This unit is a planetary gearset coupled to the power-split planetary gearset. This compound gearset has a common or shared ring gear that drives the vehicle's wheels. The sun gear of the power-split unit is driven by MG1 and the carrier is driven by the engine. In the reduction gearset, the carrier is held and the sun gear is driven by MG2. Because the sun gear is driving a larger gear, the ring gear, its output speed is reduced and its torque output is increased proportionally. High torque is available because MG2 can rotate at very high speeds. The rotational speed of MG1 essentially controls the overall gear ratio of the transaxle. The torque of the engine and MG2 flows to a common ring gear and to the final drive gear and differential unit.

The transaxle has three distinct shafts: a main shaft that turns with MG1, MG2, and the compound gear unit, a shaft for the counter driven gear and final gear, and a third shaft for the differential. Since a clutch or torque converter is not used, a coil spring damper is used to absorb torque shocks from the engine and the initiation of MG2 to the driveline.

At the rear axle, an additional motor/generator (MGR) is placed in its own transaxle assembly to rotate the rear drive wheels. Unlike conventional 4WD vehicles, there is no physical connection between the front and the rear axles (Figure 4-53). The aluminum housing of the rear transaxle contains the MGR, a counter drive gear, counter driven gear, and a differential. The unit has three shafts: MGR and the counter drive gear are located on the main

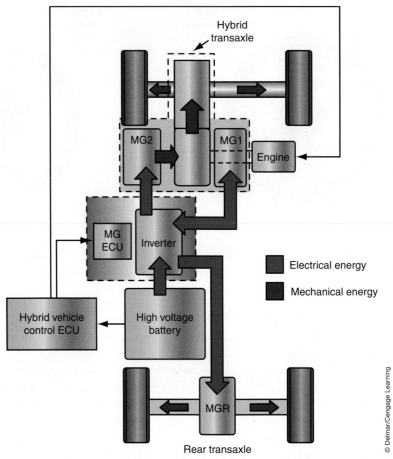

FIGURE 4-53 The rear axle is driven by an electric motor (MGR) and is not mechanically linked to the front drive axles.

shaft (MGR drives the counter drive gear), the counter driven gear and the differential drive pinion gear are located on the second shaft, and the third shaft holds the differential.

Ford Motor Company Hybrids

Ford's hybrids are equipped with an electronically controlled continuously variable transmission (eCVT). Based on a simple planetary gearset, like the Toyotas, the overall gear ratios are determined by the motor/generator. Ford's transaxle is different in construction from that found in a Toyota. In a Ford transaxle, the traction motor is not directly connected to the ring gear of the gearset. Rather it is connected to a transfer gear assembly (Figure 4-54). The transfer gear assembly has three gears, one connected to the ring gear of the planetary set, a counter gear, and the drive gear of the traction motor.

The effective gear ratios are determined by the speed of the members in the planetary gearset. This means the speed of the motor/generator, engine, and traction motor determines the torque that moves to the final drive unit in the transaxle. The operation of these three is controlled by the TCM. Based on information from a variety of inputs, the TCM calculates the amount of torque required for the current operating conditions. A motor/generator control unit then sends commands to the inverter. The inverter, in turn, sends phased AC to the stator of the motors. The timing of the phased AC is critical to the operation of the motors as is the amount of voltage applied to each stator winding.

Angle sensors (resolvers) at the motors' stator track the position of the rotor within the stator. The signals from the resolvers are also used for the calculation of rotor speed. These calculations are shared with other control modules through CAN communications. The TCM also monitors the temperature of the inverter and transaxle fluid.

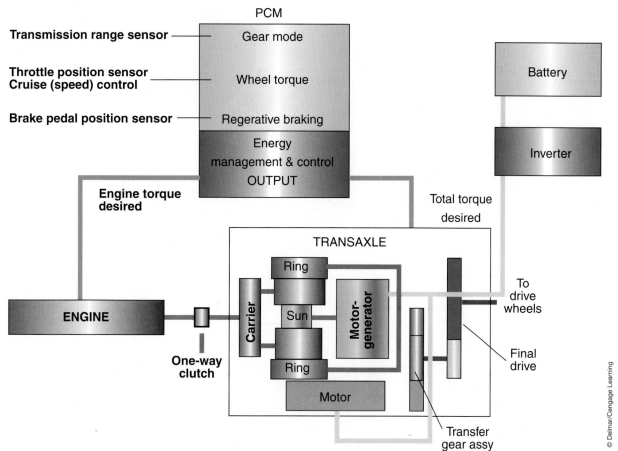

FIGURE 4-54 Ford's hybrid system is similar to Toyota's except they use a transfer gear to connect the traction motor to the output.

4WD Hybrids. Unlike Toyotas with 4WD, the Escape and Mariner do not have a separate motor to drive the rear wheels. Rather these wheels are driven in a conventional way with a transfer case, a rear driveshaft, and a rear axle assembly. This 4WD system is fully automatic and has a computer-controlled clutch that engages the rear axle when traction and power at the rear is needed. The system relies on inputs from sensors located at each wheel and the accelerator pedal. By monitoring these inputs, the control unit can predict and react to wheel slippage. It can also make adjustments to torque distribution when the vehicle is making a tight turn; this eliminates any driveline shutter that can occur when a 4WD vehicle is making a turn.

> The acronym "CAN" stands for Controller Area Network. CAN is a network protocol used to interconnect a network of electronic control modules.

Two-Mode GM, DCX, and BMW Hybrids

GM, BMW, and DaimlerChrysler have developed a two-mode full hybrid system that can be used with gasoline or diesel engines. The **two-mode hybrid system** relies on advanced hybrid, transmission, and electronic technologies to improve fuel economy and overall vehicle performance. It is claimed that the fuel consumption of a full-size truck or SUV is decreased by at least 25 percent when it is equipped with this hybrid system.

The system fits into a standard transmission housing and is basically three planetary gearsets coupled to two electronically controlled electric motors, which are powered by a 300-volt battery pack (Figure 4-55). This combination results in four forward speeds plus continuously variable gear ratios at low speeds, and motor/generators for hybrid operation.

Operation. One motor is used to restart the engine after it shuts down at a traffic light or stop sign. It also assists the engine during low-speed acceleration. The other motor provides all propulsion power when reverse gear is selected and assists the engine during low speeds with a heavy load and when cruising at high speeds. During light-load operation up to

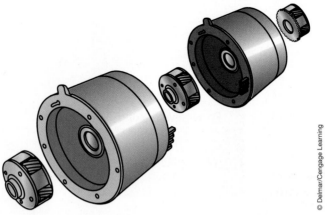

FIGURE 4-55 GM, Chrysler LLC, BMW of North America, LLC, two-mode hybrid electric motors, and planetary gearsets.

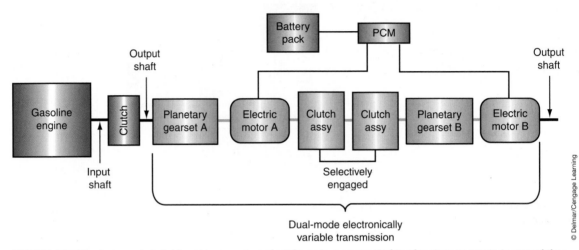

FIGURE 4-56 The two-mode hybrid system uses two electric motors connected to planetary gearsets to propel the vehicle or assist the engine during propulsion.

30 mph, the motor can propel the vehicle without the assistance of the engine. Both motors are used as generators to charge the battery pack when the vehicle is decelerating and braking.

The transmission uses clutch-to-clutch technology. The variable gear ratios are available through a mixing of power from the electric motors with the engine's power (Figure 4-56). When the motors are not providing power, the transmission operates like a conventional four-speed automatic.

The hybrid system has two distinct modes of operation. It operates in the first mode during low speed and low load conditions and the second mode is used while cruising at highway speeds. The system can operate solely on electric or engine power or by a combination of the two. Typically, when one or both of the motors are not providing propulsion power, they are working as generators driven by the engine or by the drive wheels for regenerative braking.

The first mode of operation is called the input split and second is the compound split (Figure 4-57). During the input split mode, the vehicle can be propelled by battery power, engine power or both. This is the normal mode of operation when the vehicle is slowly accelerating from a stop and when it is cruising at slow speeds. When the control unit determines that battery power is sufficient for the current conditions, the engine shuts off. During this time, one motor is working to move the vehicle, while the other may be working as a generator to supply power for the traction motor or to recharge the battery. If the engine is commanded to start, the traction motor may shut down but the second motor can continue to operate as a generator if needed.

During normal driving under light loads, the vehicle is powered solely by the engine. Depending on the load and other conditions, the engine may switch off some of its cylinders.

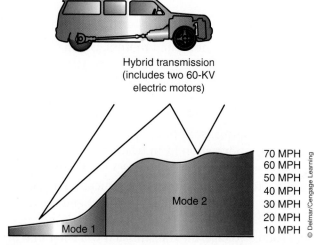

FIGURE 4-57 The two operating modes for a two-mode hybrid system.

When vehicle speed increases or when a heavy load is introduced to the vehicle, such as hard acceleration, climbing a hill, or towing, the system switches to the compound split mode.

In this mode, the control unit can order both motors to supply assist to the engine or require one of them to operate as a generator. The goal of the control unit is to maximize fuel economy while meeting the needs of the current operating conditions. The control unit also works with engine controls to determine if the other fuel-saving features, such as cylinder deactivation and late intake valve closing (Atkinson cycle), should be initiated. It is important to realize that the deactivation of cylinders and the initiation of the Atkinson cycle reduce the power output of the engine. These fuel-saving features do not hurt the performance of the vehicle because the engine's output is supplemented by the electric motor. This feature distinguishes a two-mode hybrid from other full hybrids. Typically, electric assist is available only when there is a high demand for power. In the two-mode system, the motors make it possible to reduce the work of the engine, even during light and moderate loads.

SUMMARY

- A computer is an electronic device that receives, stores, processes, and communicates information.
- The input devices used in electronic control systems vary with each system; however, they can be grouped into distinct categories: reference voltage sensors and voltage generators.
- Voltage generation devices are typically used to monitor rotational speeds. The most common of which is the PM generator used as a vehicle speed sensor. A speed sensor is a magnetic pick-up that senses and transmits low-voltage pulses.
- Common voltage reference sensors include on/off switches, potentiometers, thermistors, and pressure sensors.
- Potentiometers, thermistors, and pressure sensors are designed to change their electrical resistance in response to something else changing.
- If a switch is cycled on and off very rapidly, the return signal is a rapid on/off one. A digital signal is a series of on/off pulses.
- A TCM can be either a separate computer designated for transmission operation or part of the PCM. The PCM can be a multifunction computer that controls all engine and transmission operations. It receives information on engine operating conditions from different sensors and switches.

SUMMARY

- The typical output devices are solenoids and motors, which cause something mechanical or hydraulic to change.
- The decision to shift or not to shift is based on shift schedules and logic programmed into the memory of the computer.
- Adaptive learning allows the computer to compensate for wear and other events that might occur and cause the normal shift programming to be inefficient.
- The computer may receive information from two different sources: directly from a sensor, or through a bus circuit, which connects all of the vehicle computer systems.
- The two most common protocols for serial communication in a multiplex system are the Programmable Controller Interface Data Bus (PCI Bus) and the Controller Area Network (CAN) Bus.
- If the computer loses source voltage, the transmission will enter into default (limp-in) mode. The transmission will also enter into default mode if the computer senses a transmission failure. While in the default mode, the transmission will operate only in PARK, NEUTRAL, REVERSE, and second gears. The transmission will not upshift or downshift. This allows the vehicle to be operated, although its efficiency and performance is hurt.
- Fluid flow to the various apply devices is directly controlled by the solenoids.
- Pressure switches, which give inputs to the transmission computer, are all located within the solenoid assembly.
- Engine speed, throttle position, temperature, engine load, and other typical engine-related inputs are also used by the computer to determine the best shift points. Many of these inputs are available through multiplexing and are input from the common bus.
- Adaptive learning takes place while the computer reads input and output speeds over 140 times per second.
- The CVI is the measurement of the physical amount of fluid and time required to fill the clutch and stroke the piston.
- The TCM relies on information from the engine control system, transmitted through such sensors as the MAP and TP sensor, as well as information from the transmission to determine the optimum shift timing.
- The transmission control system uses many solenoids for control of operation. One solenoid is used for modulated converter clutch control. Another, the EPC solenoid, is used to control hydraulic pressures throughout the transmission. The remaining solenoids are shift solenoids.
- The EPC solenoid replaces the conventional TV cable setup to provide changes in pressure in response to engine load.
- The pulse width modulated (PWM) solenoid is a normally closed valve installed in the valve body. It controls the position of the TCC apply valve.
- On some models, an operational mode selector switch is located on the center console or instrument panel. This switch allows the driver to select different modes to change transmission upshift characteristics.
- Hybrid vehicles are equipped with special automatic transmissions that allow the engine to run at its most efficient speed, which means increased fuel economy and decreased exhaust emissions.
- Honda hybrids use the IMA system, which places an electric motor between the engine and the transmission.
- Some lower voltage hybrid systems use an integrated starter alternator damper (ISAD) between the engine and the transmission to provide for the stop–start feature.
- The two-mode hybrid system fits into a standard automatic transmission housing and is comprised of two planetary gearsets coupled to two electric motor/generators. This arrangement allows for two distinct modes of hybrid drive operation.

TERMS TO KNOW

Adaptive learning

Belt Alternator Starter (BAS)

Controller Area Network (CAN) Bus

Clutch volume indexes (CVIs)

Integrated Motor Assist (IMA)

Integrated starter alternator damper (ISAD)

Input shaft speed (ISS)

Multiplexing

Output shaft speed (OSS) sensor

Power-split device

Programmable Controller Interface Data Bus (PCI Bus)

Shift schedule

Two-mode hybrid system

Variable pulse width modulation (VPWM)

Variable force solenoid (VFS)

Voltage

REVIEW QUESTIONS

Short-Answer Essays

1. A computer relies on many different reference voltage sensors. These can be divided into types. Name the types and give a brief description of their operation.

2. Although computers receive different information from a variety of sensors, the decisions for shifting are actually based on more than the inputs. What are they based on?

3. What is the purpose of a protection device in an electrical circuit?

4. Some transmissions receive information through multiplexing. How does this work?

5. How does a solenoid in an electronically controlled transmission work?

6. Most late-model transmission control systems have adaptive learning. What does this mean?

7. What inputs are of prime importance to a computer in deciding when to shift gears?

8. What are the advantages of using electronic controls rather than relying on conventional hydraulic controls in a transmission?

9. What is so unique about a Honda transmission?

10. What is the major difference between the Honda IMA system and a typical ISAD system?

Fill-in the Blanks

1. Amperage in a series circuit is always _____.

2. The voltage required to push amperage through a resistance is called _____ _____.

3. A VSS and an OSS are examples of _____ _____ sensors.

4. A circuit that has more than one path for current to flow through is a _____ _____.

5. A computer is an electronic device that _____, _____, _____, and _____ information.

6. The input devices used in electronic control systems vary with each system; however, they can be grouped into distinct categories: _____ _____ _____ and _____ _____.

7. Common voltage reference sensors include _____ switches, _____, _____, and _____ sensors.

8. In an electronic control system, the typical output devices are _____ and _____ that cause something _____ or _____ to change.

9. Most electronically controlled transmissions rely on _____ _____ solenoids to control all forward gears.

10. Voltage generation devices are typically used to monitor _____ _____.

MULTIPLE CHOICE

1. Voltage generation devices are typically used to monitor rotational speeds. Which of the following is NOT a voltage-generating type sensor?
   A. Vehicle speed sensor   C. MAP
   B. OSS   D. ISS

2. While discussing transmission solenoids:
   Technician A says an EPC solenoid replaces the conventional TV cable setup to provide changes in fluid pressure in response to engine load.
   Technician B says a TCC solenoid can be modulated to smooth the engagement of the torque converter clutch.
   Who is correct?
   A. A only   C. Both A and B
   B. B only   D. Neither A nor B

3. *Technician A* says throttle position is an important input in most electronic shift control systems.

Technician B says vehicle speed is an important input for most electronic shift control systems.

Who is correct?

A. A only   C. Both A and B

B. B only   D. Neither A nor B

4. Some transmission control systems monitor the speed of the input shaft and compare it to the speed of the output shaft to determine the _____.

A. Gear ratio

B. CVI

C. Amount of internal wear

D. All of the above

5. While discussing various sensors used with an electronic transmission control system:

Technician A says potentiometers are typically used to measure temperature changes.

Technician B says vacuum modulators are used to measure engine load.

Who is correct?

A. A only   C. Both A and B

B. B only   D. Neither A nor B

6. *Technician A* says shift solenoids direct fluid flow to and away from the various apply devices in the transmission.

Technician B says shift solenoids are used to mechanically apply a brake band or multiple disc clutch assembly.

Who is correct?

A. A only   C. Both A and B

B. B only   D. Neither A nor B

7. While discussing transmission adaptive learning:

Technician A says the transmission learns to respond according to the current condition of the engine and transmission.

Technician B says the TCM learns to respond to the driver's driving habits.

Who is correct?

A. A only   C. Both A and B

B. B only   D. Neither A nor B

8. Which of the following statements about electrical problems is NOT true?

A. A short will cause the components in the circuit to operate poorly because of decreased current flow to them.

B. In an open circuit, there will be no current flow.

C. If there is an open in one leg of a parallel circuit, the remaining part of that parallel circuit will operate normally.

D. Excessive resistance at a connector, internally in a component, or within a wire will cause low current flow and the component will not be able to operate normally, if at all.

9. *Technician A* says multiplexing allows information to be shared with many computers.

Technician B says multiplexing reduces the number of wires and components needed in current vehicles.

Who is correct?

A. A only   C. Both A and B

B. B only   D. Neither A nor B

10. While discussing valve body assemblies in late-model transmissions:

Technician A says the valve body is no longer needed in some electronically controlled transmissions.

Technician B says an EPC solenoid maintains constant fluid pressure through the valve body regardless of the vehicle's operating condition.

Who is correct?

A. A only   C. Both A and B

B. B only   D. Neither A nor B

Chapter 5

TRANSMISSION DESIGNS

UPON COMPLETION AND REVIEW OF THIS CHAPTER, YOU SHOULD BE ABLE TO:

- Explain the differences between a transaxle and a transmission.
- Describe the total driveline for FWD, RWD, and 4WD vehicles.
- Explain how to identify the features of a transmission by its model number.
- Describe the major internal parts of an automatic transmission/transaxle.
- Describe the different designs of compound planetary gearsets.

- Name the major types of planetary gear controls used on automatic transmissions and explain their basic operating principles.
- Explain the purposes of seals and gaskets that are found in an automatic transmission/transaxle.
- Explain the purposes of engine and transmission/ transaxle mounts.

INTRODUCTION

Throughout the years, the automatic transmission has evolved into a complex machine with electronic, hydraulic, and mechanical components. It is safe to say that this evolution will continue for years to come. The focus of this chapter is on the mechanical components of an automatic transmission.

This chapter also looks at the different designs of automatic transmissions and the major subassemblies that are part of a typical transmission. Before you can thoroughly understand the purpose of each and every part of a transmission, you must have an understanding of the roles and construction of the major subassemblies. Each of these subassemblies is covered in detail in later chapters.

Transmission designs vary from manufacturer to manufacturer and within each manufacturer. These variations may be simply different electronic controls for shifting or for the torque converter or the number of forward speeds a transmission has. Variations also result from the way a particular model of transmission is constructed and the vehicle it will be installed in. These topics are the focus of this chapter.

TRANSMISSION VERSUS TRANSAXLE

One of the primary variations in the design of an automatic transmission is based on the driveline of the vehicle the transmission was designed for. As you know, automobiles are propelled in one of three ways: by the rear wheels, by the front wheels, or by all four wheels. The type of driveline helps determine whether a transmission or a transaxle will be used.

A BIT OF HISTORY

In 1939, the big bit of automotive news was the Hydra-Matic Drive introduced by Oldsmobile. The Hydra-Matic was a combination of a liquid flywheel and a fully automatic transmission. "The '40 Olds shifts without a clutch? What will they think of next?"

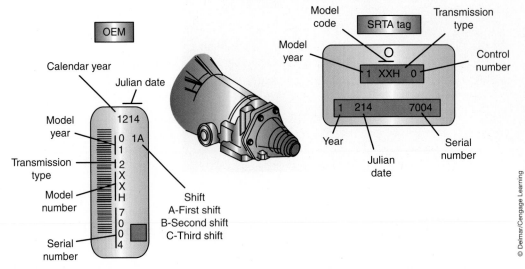

FIGURE 5-1 A RWD transmission with its identification numbers and production information highlighted.

Longitudinally mounted engines are sometimes said to have "north-south" orientation, whereas transverse engines are oriented "east-west."

Vehicles propelled by the rear wheels normally use a transmission. Transmission gearing is located within an aluminum or iron casting called the transmission case assembly (Figure 5-1). The transmission case assembly is attached to the rear of the engine, which is normally located in the front of the vehicle. A driveshaft links the output shaft of the transmission with the differential and drive axles located in a separate housing at the rear of the vehicle. The differential splits the driveline power and redirects it to the two rear drive axles, which then pass it on to the wheels.

The front wheels propel front-wheel-drive (FWD) vehicles. For this reason, they must use a drive design different from that of a rear-wheel-drive vehicle. FWD vehicles are typically equipped with a transaxle and have no need for a separate differential and drive axle housing. A transaxle is a compact unit that combines the transmission gearing, differential, and drive axle connections into an aluminum housing located in front of the vehicle. This design has two primary advantages: good traction on slippery roads due to the weight of the powertrain components being directly over the driving axles of the vehicle, and transverse engine and transaxle configurations that allow for lower hood lines, thereby improving the vehicle's aerodynamics.

Not all FWD vehicles have a transversely mounted engine. Some have the engine mounted longitudinally and use a transaxle that looks like a conventional transmission design modified to drive the front wheels directly.

Four-wheel-drive vehicles typically use a transmission and transfer case. The transfer case mounts on the side or back of the transmission. A chain or gear drive inside the transfer case receives power from the transmission and transfers it to two separate driveshafts. One driveshaft connects to a differential on the front drive axle. The other driveshaft connects to a differential on the rear drive axle.

AUTHOR'S NOTE: There are a few mid-engined and rear-engined cars out there. These can use either a transmission or a transaxle. To simplify things, you can look at the driveline and tell whether or not the vehicle has a transmission or a transaxle. If there is a separate drive axle unit with a differential, a transmission is used. If the drive axles extend from the transmission unit, it is a transaxle.

BASIC DESIGNS

All automotive transmissions and transaxles are equipped with a varied number of forward speed gears, a neutral gear, and one reverse speed. Transmissions can be divided into groupings based on the number of forward speed gears they have. Current designs have four or five forward speeds. A growing number have six and few have seven or eight. Of course, there is a wide variety of continuously variable transmissions, as well.

Transmissions are designed and built for particular applications. Things that must be considered in the design of a transmission are the power output of the engine, the weight of the vehicle the transmission will be installed in, and the typical workload of the vehicle. All of these play a part in the size and strength of the parts and materials used in a transmission. Obviously, high-powered and hard-working vehicles need larger and stronger materials.

Overall transmission/transaxle size is also an important design consideration. This often dictates the placement of the various shafts inside the transmission and the size of the components. Transmission designs also vary according to the type of compound planetary gearsets used.

Transaxle design is most affected by the intended application. Since these units attach directly to the engine and that combination must fit between the two drive wheels, size is important. A variety of shaft arrangements can be found in today's transaxles. Long shafts may be divided into two parts and the ends of each connected by gears or chains. Final drive units vary in design as well. Common final drive setups for FWD vehicles use helical gears, planetary gears, hypoid gears, and/or a chain drive.

It is important to know that no two transmission models are exactly alike, regardless of their outward appearance. Keep in mind that these variations are based on external and internal components.

Although there are many different designs, all transmissions and transaxles rely on gears, shafts, bearings, and apply devices to function. All automatic transmissions, except CVTs, have a torque converter to connect and disconnect the engine's power to the transmission. All have an input shaft to transfer power from the torque converter to the internal gearsets and drive members. They also have gearsets to provide the different gear ratios, a reverse gear, and a neutral position. All have an output shaft to transfer power from the transmission to the final drive unit. They also have a hydraulic system that includes the pump, valve body, pistons, servos, brakes, and clutches.

Model Numbers

In the past, manufacturers identified their transmission models with internal codes that represented the design of the transmission. Although technicians who worked on transmissions learned to know the differences between the models, there was little logic used to denote the features of a particular design.

Recently, manufacturers began using more definitive codes for model identification. Most use an alphanumeric code and the features of each transmission model are explained in their service manuals. Let's take a look at the transmission model codes for some manufacturers.

Interpreting a General Motors transmission code will reveal the work capacity of the transmission, the number of forward gears, its directional placement in the vehicle, and if it has electronic controls. A 4T40-E transmission is a four-speed, transversely mounted, light-to-medium duty, electronically controlled unit. The "4" in the model number designates the number of forward gears, the "T" shows that it is a transversely mounted unit, the "40" is the product series, and the "E" means it has electronic controls.

Other commonly used GM transmissions are the 4L60-E and the 4L80-E. Both of these are found in pickup trucks and SUVs. Both of these models are longitudinally mounted four-speed units with electronic controls. The internal construction of these units is different, as noted by the different product series numbers. The 4L80-E unit is designed for heavier duty than the 4L60-E.

Chrysler uses a similar system (Figure 5-2). The first character denotes the number of forward gears. The second character is the duty rating. In contrast to GM's use of a two-digit code for duty rating, Chrysler uses a single-digit code. Using a 45RFE transmission as an example of Chrysler's transmission code, the "4" indicates there are four forward speeds, the "5" is the duty rating, "R" indicates the transmission is designed for rear-wheel-drive, and "FE" indicates that the unit is fully electronic.

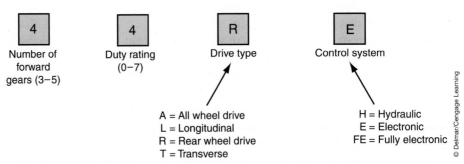

TRANSMISSION CODE

4	4	R	E
Number of forward gears (3–5)	Duty rating (0–7)	Drive type	Control system

Drive type:
A = All wheel drive
L = Longitudinal
R = Rear wheel drive
T = Transverse

Control system:
H = Hydraulic
E = Electronic
FE = Fully electronic

© Delmar/Cengage Learning

FIGURE 5-2 Transmission model designation code interpretation for Chrysler vehicles.

Ford Motor Company uses similar logic for model designations; however, its duty rating is more specific. The 4F27E transmission is used in the Focus. The model code is broken into four character groups. The "4" indicates that the transmission is a four-speed unit. The "F" denotes that this is a transaxle for FWD vehicles. The "27" is the duty rating and represents the maximum input torque. In this case the "27" represents 270 lbs.-ft. (365 Nm). The "E" means the transaxle is fully electronically controlled.

The importance of using the service manual to decipher the codes is apparent when considering the model code for another commonly used Ford transmission. The 4R70W is a four-speed transmission for RWD pickups and SUVs. This transmission is electronically controlled, but the model doesn't contain an "E"; rather, there is a "W." The W indicates this heavy-duty (maximum input torque of 700 lbs.-ft.) unit has wide-ratio gears.

Toyota has two primary classifications of transmissions/transaxles, the A and U series. The A-series are two to eight-speed automatic transmissions for front-wheel-drive, all-wheel-drive, or rear-wheel-drive vehicles built by Aisin-Warner. The U-series are automatic transmissions for front-wheel-drive applications built primarily by Toyota.

The designations of individual transmissions are defined by an alphanumeric system. This system is used for both automatic and manual transmissions. The first character is the series of transmission. The second numerical digit states the number of forward gears. The last numeric digit denotes the version of the base transmission. This number typically changes according to the application. These numbers may be followed by a letter which denotes special features of the transmission. Examples of these letters are as follows: L = Lockup torque converter, E = Electronic control, F = Four-wheel-drive, and H = AWD with a transversely mounted engine.

To show the designations define the transmission, let's look at the AA80E. This eight-speed automatic transmission is used in current Lexus models, including the LS460, GS460, and IS-F. This transmission is built by Aisin-Warner and is electronically controlled. The 0 after the 8 shows this is the first version of this transmission.

MOUNTS

The weight of the transmission or transaxle is supported by the engine and its mounts and by a transmission mount. These mounts are not only critical for proper operation of the transmission, but also isolate transmission noise and vibrations from the passenger compartment. The mountings for the engine and transmission keep the powertrain in proper alignment with the rest of the drivetrain and help to maintain proper adjustment of the various linkages attached to the housing.

An engine oscillates (vibrates) as it runs. Since the transmission is directly connected to the engine, those vibrations carry through to the transmission. Faulty mounts will not only allow these vibrations and the resulting noise to transfer into the vehicle, but can also cause internal transmission problems. When an engine is mounted transversely, the inherent vibrations of the engine are easily transmitted to the vehicle's suspension and wheels. Also, most FWD vehicles use compact in-line 4-cylinder or V-6 engines that do not run as smoothly as larger engines. For these reasons, the manufacturers of FWD vehicles have developed many different mounting systems for their engines and transmissions.

The typical mount bolts to the transaxle and is connected to a plate or to the surface on the transaxle housing. A bolt passes through a rubber insulator and connects the mount to the transaxle (Figure 5-3). This type of mount is used on the side and toward the top of the engine/transaxle assembly. With this mount is a lower mount located between the subframe and the transaxle (Figure 5-4). These two mounts keep the assembly in place, but do little to control noise and vibration; therefore, additional mounts are used.

Another common way to suppress vibrations and noise is the use of an engine mount strut. This strut limits the rocking motion of the engine and connects the top of the engine to the frame of the vehicle (Figure 5-5). Again, the connecting points between the two parts of the mount are made through an insulator. Some vehicles use more than one of these struts or have an additional strut mounted to the side of the engine.

Some models have a lower engine mount that connects the lower front of the engine to the frame of the vehicle (Figure 5-6) and an upper engine mount of the rear of the assembly (Figure 5-7).

A few models use an adjustable mount that responds to vibrations (Figure 5-8). This type of system relies on the action of a solenoid on a hydraulic mount. The solenoid

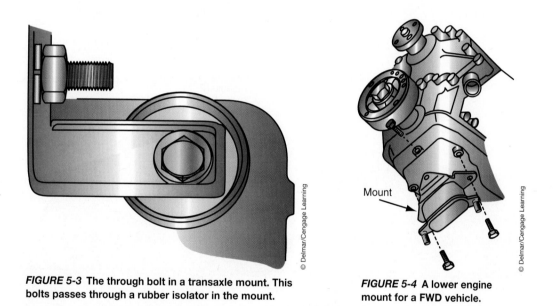

FIGURE 5-3 The through bolt in a transaxle mount. This bolts passes through a rubber isolator in the mount.

FIGURE 5-4 A lower engine mount for a FWD vehicle.

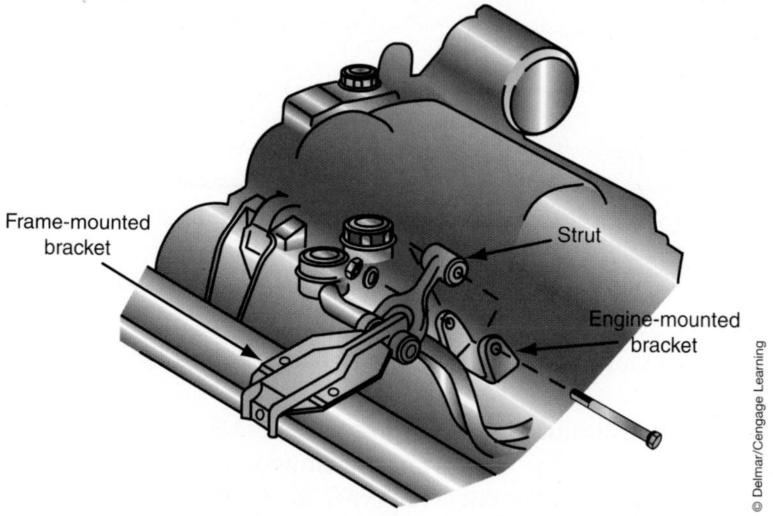

FIGURE 5-5 An engine mount strut.

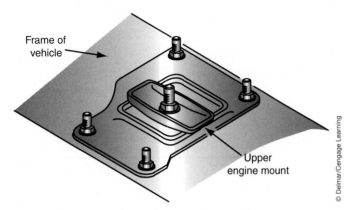

FIGURE 5-6 A lower engine mount.

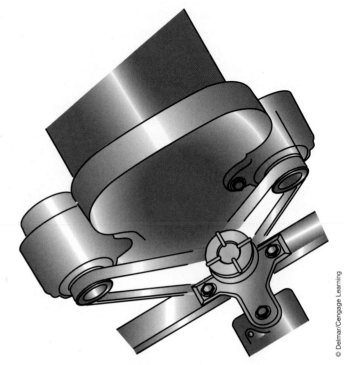

FIGURE 5-7 An upper engine mount.

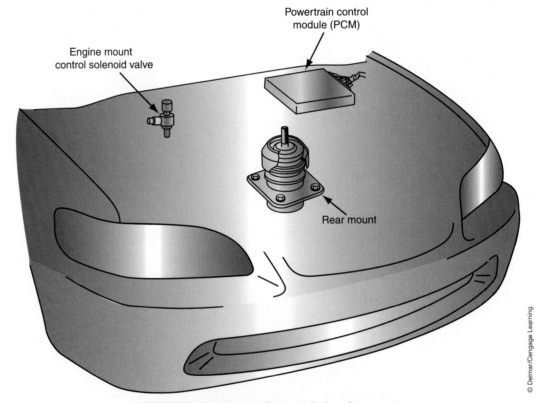

FIGURE 5-8 An electronically controlled engine mount.

responds to the commands from the PCM and decreases and increases the fluid pressure at the mount.

Isolation of vibrations is not as much of a requirement for FWD vehicles with a longitudinally mounted powertrain. The engine and transmission in these models are mounted in much the same way as in a RWD vehicle (Figure 5-9). Although these mounts are less complex, they are still critical to the overall operation of the vehicle.

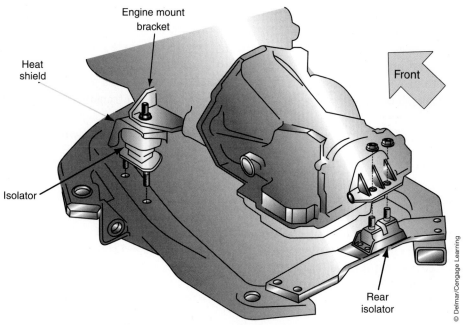

FIGURE 5-9 Mounts for a longitudinally mounted transaxle.

HOUSINGS

The basic shape and size of a transmission/transaxle reflects its design to some degree. Automatic transmission housings are typically aluminum castings. The torque converter and transmission housings are normally cast as a single unit; however, in some designs the torque converter housing is a separate casting. Transaxle housings are typically comprised of two or three separate castings bolted together; one of these castings is the torque converter housing (Figure 5-10).

Shop Manual
Chapter 5, page 240

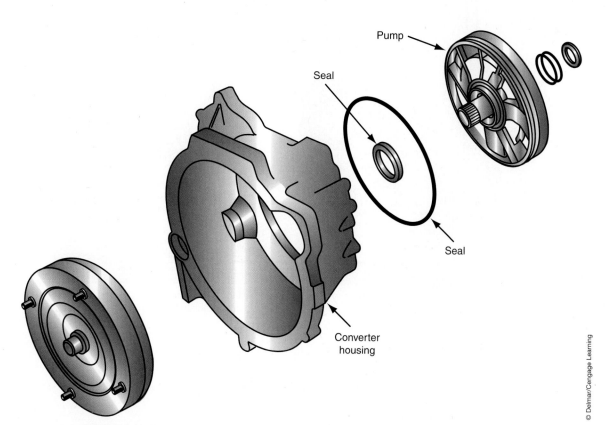

FIGURE 5-10 A converter housing with associated components separated from the transaxle.

Although they serve many of the same purposes as a transmission, transaxle housings are considerably different in appearance from transmission housings.

The surfaces at which the sections of a transmission/transaxle mate are critical to the operation of the unit. Proper mounting surfaces are necessary to keep the various shafts aligned and to provide a good seal. It is important to remember that a poor seal will not only cause fluid leaks, but will also allow dirt to enter the unit. Dirt is a transmission's most feared enemy.

Transmission housings are cast to secure and accommodate the following components:

- Multiple friction-disc assemblies—Inside the housing are linear keyways designed to hold a multiple-disc clutch assembly in position in the housing.
- Fluid passages (Figure 5-11).
- Threaded bores and/or studs to fasten mounts.
- Bores for the fluid level dipstick and filler tube.
- Threaded bores for a variety of sensors and switches (Figure 5-12).
- Bores for gear selector attachment (Figure 5-13).
- Fittings for the fluid cooling lines.

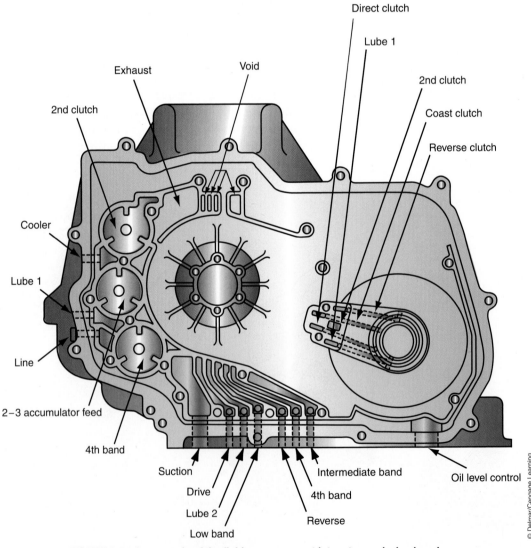

FIGURE 5-11 An example of the fluid passages cast into a transmission housing.

© Delmar/Cengage Learning

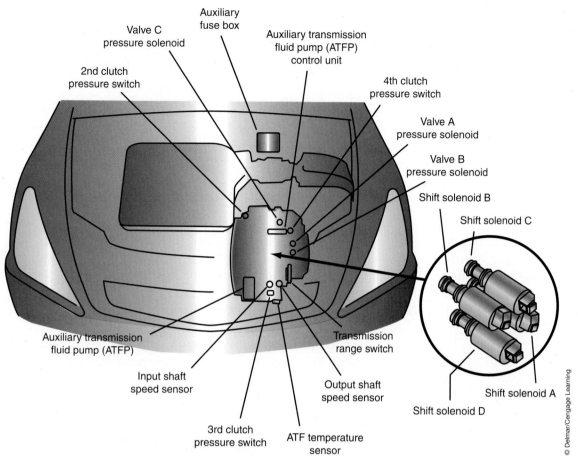

FIGURE 5-12 Various sensors and switches attached to a late-model transaxle housing.

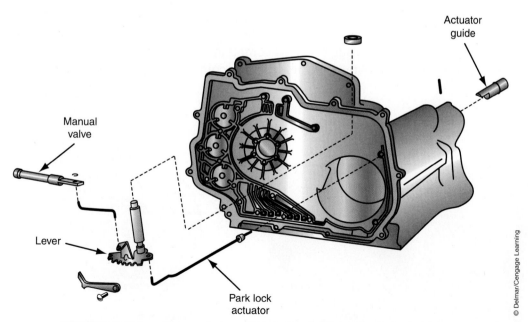

FIGURE 5-13 This parking pawl assembly is just one of the many shafts and linkage parts that pass through or into a typical transmission housing.

- Mounting points for the pump and valve body.
- Mounting for the oil pan.
- Round structures projecting from the side to serve as the cylinders for the servo and accumulator assemblies. In some designs there are also projections to house the governor assembly.

In addition, some housings have bores for the band adjusting screws. A transmission/transaxle case assembly is a precisely machined and designed unit. It has many more important roles than simply housing the internal components.

GEARS

Nearly all automatic transmissions rely on planetary gearsets (Figure 5-14) to transfer power and multiply engine torque to the drive axle. A simple planetary gearset consists of three parts: a sun gear, a carrier with planetary pinions mounted to it, and an internally toothed ring gear or *annulus*. The sun gear is located in the center of the assembly and meshes with the teeth of the planetary pinion gears. **Planetary pinion** gears are small gears fitted into a framework called the **planetary carrier**. The planetary carrier can be made of cast iron, aluminum, or steel plate and is designed with a shaft for each of the planetary pinion gears (Figure 5-15).

Planetary pinion gears rotate on needle bearings positioned between the planetary carrier shaft and the planetary pinions. The carrier and pinions are considered one unit—the mid-size gear member (Figure 5-16).

The planetary pinions are surrounded by the annulus or ring gear, which is the largest part of the simple gearset. The ring gear holds the entire gearset together and provides great strength to the unit.

Each member of a planetary gearset can spin (revolve) or be held at rest. Power transfer through a planetary gearset is only possible when one of the members is held at rest, or if two of the members are locked together.

Any one of the three members can be used as the driving or input member. At the same time, another member might be kept from rotating and thus becomes the held or stationary member. The third member then becomes the driven or output member. Depending on which member is the driver, which is held, and which is driven, either a torque increase or a speed increase is produced by the planetary gearset. Output direction can also be reversed through various combinations. Refer to Figure 5-17 for a summary of the possible outcomes of the different combinations available in a planetary gearset.

FIGURE 5-14 A single planetary gearset.

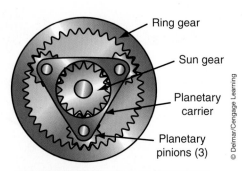

Ring gear

Sun gear

Planetary carrier

Planetary pinions (3)

FIGURE 5-15 Planetary gear configuration is similar to the solar system, with the sun gear surrounded by the planetary pinion gears. The ring gear surrounds the complete gearset.

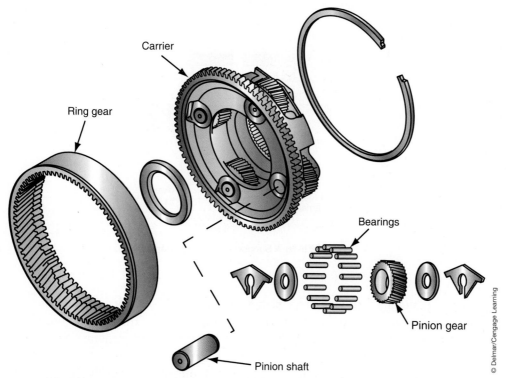

FIGURE 5-16 The pinion gears and carrier assembly for a typical front planetary unit.

Sun Gear	Carrier	Ring Gear	Speed	Torque	Direction
1. Input	Output	Held	Maximum reduction	Increase	Same as input
2. Held	Output	Input	Maximum reduction	Increase	Same as input
3. Output	Input	Held	Maximum increase	Reduction	Same as input
4. Held	Input	Output	Maximum increase	Reduction	Same as input
5. Input	Held	Output	Reduction	Increase	Reverse of input
6. Output	Held	Input	Increase	Reduction	Reverse of input
7. When any two members are held together, speed and direction are the same as input. Direct 1:1 drive occurs.					
8. When no member is held or locked together, output cannon occur. The result is a neutral condition.					

FIGURE 5-17 The basic laws of planetary gear operation.

Compound Planetary Gearsets

Compound gearsets combine simple planetary gearsets so load can be spread over a greater number of teeth for strength and also to obtain the largest number of gear ratios possible in a compact area. A limited number of gear ratios are available from a single planetary gearset. To increase the number of available gear ratios, gearsets can be added. The typical automatic transmission with four forward speeds has at least two planetary gearsets. Some transmissions are fitted with an additional single planetary gearset which is used to provide additional forward gear ratios.

Simpson Gearset. The Simpson geartrain is an arrangement of two separate planetary gearsets with a common sun gear, two ring gears, and two planetary pinion carriers (Figure 5-18). One half of the Simpson gearset or one planetary unit is referred to as the front planetary

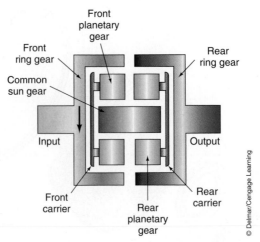

FIGURE 5-18 A Simpson planetary gearset.

and the other planetary unit is the rear planetary. The two planetary units do not need to be the same size or have the same number of teeth on their gears. The size and number of gear teeth determine the actual gear ratios obtained by the compound planetary gear assembly.

The different gear ratios and direction of rotation are the result of applying torque to one member of either planetary unit, holding at least one member of the gearset, and using another member as the output. For the most part, automobile manufacturer use the same parts of the planetary assemblies as input, output, and reaction members; therefore, they have similar power flows.

Ravigneaux Gearset. A Ravigneaux gearset has two sun gears, two sets of planet gears, and a common ring gear (Figure 5-19). This gearset provides forward gears with a reduction, direct drive, overdrive, and a reverse operating range. The Ravigneaux offers some advantages over a Simpson geartrain. It is very compact. It can carry large amounts of torque because of the great amount of tooth contact. It can also have three different output members. However, it has a disadvantage to students and technicians, it is more complex and therefore its actions are more difficult to understand.

The Ravigneaux geartrain is designed to use two sun gears, one small and one large. They also have two sets of planetary pinion gears, three long pinions, and three short pinions. The planetary pinion gears rotate on their own shafts that are fastened to a common planetary carrier. A single ring gear surrounds the complete assembly.

The small sun gear is meshed with the short planetary pinion gears. These short pinions act as idler gears to drive the long planetary pinion gears. The long planetary pinion gears mesh with the large sun gear and the ring gear.

Lepelletier Gearset. The Lepelletier design connects a simple planetary gearset to a Ravigneaux gearset. In this design, the input shaft is connected to the ring of the simple planetary gear and can simultaneously be connected to the carrier and large sun gear of the

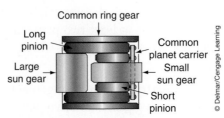

FIGURE 5-19 A Ravigneaux gearset.

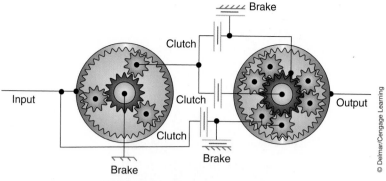

FIGURE 5-20 A Lepelletier gearset.

Ravigneaux gearset using separate clutches (Figure 5-20). The output shaft is connected to the ring of the Ravigneaux gear.

Planetary Gears in Tandem. Rather than rely on a compound gearset, some automatic transmissions use two (Figure 5-21) or three simple planetary units in series. In this type of arrangement, gearset members are not shared instead the holding devices are used to lock different members of the planetary units together.

The combination of the planetary units functions much like a compound unit. The tandem units do not share a common member, rather certain members are locked together or are integral with each other. The front planetary carrier is locked to the rear ring gear and the front ring gear is locked to the rear planetary carrier.

Helical Gear-Based Units. Honda and Saturn nonplanetary based transaxles are rather unique in that they use constant-mesh helical and square-cut gears (Figure 5-22) in a manner similar to a manual transmission. These transaxles have a mainshaft and countershaft on which the gears ride. To provide the forward gear ratios and a reverse gear, different pairs of gears are locked to the shafts by hydraulically controlled clutches. Reverse gear is obtained

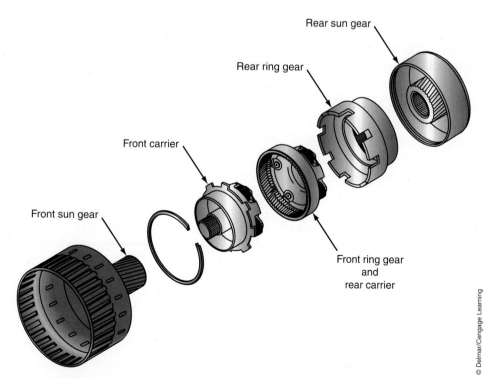

FIGURE 5-21 Two planetary units connected in tandem with the ring gear of one gearset connected to the planet carrier of the other.

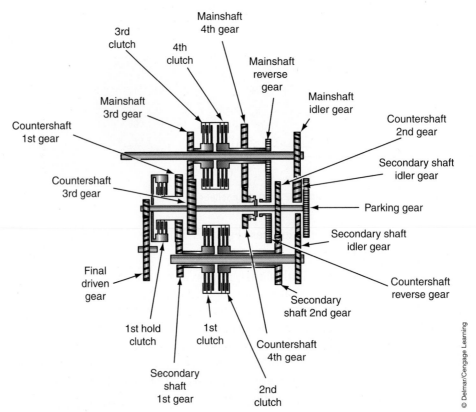

3rd clutch

4th clutch

Mainshaft 4th gear

Mainshaft reverse gear

Mainshaft idler gear

Countershaft 2nd gear

Mainshaft 3rd gear

Countershaft 1st gear

Secondary shaft idler gear

Countershaft 3rd gear

Parking gear

Secondary shaft idler gear

Final driven gear

Countershaft reverse gear

1st hold clutch

1st clutch

Secondary shaft 2nd gear

Countershaft 4th gear

Secondary shaft 1st gear

2nd clutch

© Delmar/Cengage Learning

FIGURE 5-22 A cutaway of a constant-mesh helical gear automatic transmission.

through the use of a shift fork that slides the reverse gear into position. The power flow through these transaxles is also similar to that of a manual transaxle.

Planetary Gear Controls

Certain parts of the geartrain must be held while others must be driven to provide the needed torque multiplication and direction for vehicle operation. "Planetary gear controls" is the general term used to describe transmission bands, servos, and clutches.

Transmission Bands. A band is used as a brake and is positioned around a rotating drum. The band brings a drum to a stop by wrapping itself around the drum and holding it. The band is applied hydraulically by a servo assembly. Connected to the drum is a member of the planetary geartrain. The band holds a member of the planetary gearset by preventing the drum from rotating and the attached planetary gear member becomes the reaction gear for the gearset. Bands have excellent braking characteristics and require a minimum amount of space within the transmission housing.

The *servo* assembly converts hydraulic pressure into a mechanical force that applies a band to hold a drum stationary. Simple and compound servos are used in modern transmissions.

Overrunning Clutches. In an automatic transmission, both sprag and roller overrunning (one-way) clutches are also used to brake members of the planetary gearset. These clutches operate mechanically. An overrunning clutch allows rotation in only one direction and does not require hydraulic activation.

Multiple-Disc Clutches. In contrast to a band or one-way clutch, which only brake a planetary gear member, multiple-disc clutches are capable of both holding and driving gearset members.

Shop Manual
Chapter 5, page 259

A multiple-disc clutch has a series of friction discs that have internal teeth that are sized and shaped to mesh with splines on the clutch assembly hub. In turn, this hub is connected to a member of the planetary gearset that will receive the desired braking or transfer force when the clutch is applied or released.

Multiple-disc clutches are enclosed in a large drum-shaped housing that can be either a separate casting or part of the existing transmission housing. This drum housing also holds the other clutch components: cylinder, hub, piston, piston return springs, seals, pressure plate, friction plates, and snap rings.

INTERNAL COMPONENTS

The housing contains the basic parts that transfer power and change gears. Fitted into the housing is a key component, the torque converter. A torque converter uses hydraulic fluid flow to connect and disconnect the power of the engine from the transmission. The converter is a doughnut-shaped device filled with ATF. Inside the converter are the parts that make it work.

Like most parts of an automatic transmission, this component has many different designs and shapes. A torque converter is bolted to the engine and not the transmission. However, the transmission is designed to support the unbolted end. This end rides on a bearing or bushing in the transmission.

The torque converter not only transfers engine power to the transmission, it also drives the transmission pump. The pump is a critical component because it supplies fluid flow for the entire transmission, including the torque converter.

The valve body contains the valves and orifices that control the operation of a transmission. The valve body is typically bolted to the bottom of the transmission housing and is covered by the oil pan. In response to the movement of the valves, fluid is routed to the different hydraulic apply devices. This action causes a change in gears.

Although the pump is the main source of fluid flow, there are many devices involved in the flow and pressurization of the fluid. All of these work together to enable the transmission to shift at the correct time and with the correct feel.

Shafts

Transmissions have at least two shafts, an input shaft and an output shaft. The input shaft connects the output of the torque converter to the driving members inside the transmission. Each end of the input shaft is externally splined to fit into the internal splines of the torque converter's turbine and the driving member in the transmission. Normally, the front clutch pack's hub is the driving member. In many transmissions there is a tube, called the stator shaft, that surrounds the input shaft. The stator shaft is splined to the torque converter's stator and is a stationary shaft.

The output shaft connects the driven members of the gearsets to the final drive gearset. The rotational torque and speed of this shaft varies with input speed and the operating gear. The output shaft may be splined to any member of each planetary gearset. For example, in a Simpson gearset, the carrier of the input gearset and the ring gear of the reaction set are splined to the output shaft.

Some transaxles have additional shafts. These shafts are actually a continuation of the input and output shafts. They are placed in parallel where the rotating torque can be easily transferred from one shaft to another. The shafts are divided to keep the transaxle unit compact.

Bearings, Bushings, and Thrust Washers

When a component slides over or rotates around another part, the surfaces that contact each other are called bearing surfaces. A gear rotating on a fixed shaft can have more than one bearing surface; it is supported and held in place by the shaft in a radial direction. The gear tends to move along the shaft in an axial direction as it rotates, and is therefore held in place by some other components. The surfaces between the sides of the gear and the other parts are bearing surfaces.

Shop Manual
Chapter 5, page 244

Shop Manual
Chapter 5, page 251

A bearing is a device placed between two bearing surfaces to reduce friction and wear. Most bearings have surfaces that either slide or roll against each other. In automatic transmissions, sliding bearings are used where one or more of the following conditions prevail: low rotating speeds, very large bearing surfaces compared to the surfaces present, and low use. Rolling bearings are used in circumstances including high-speed applications, high loads with relatively small bearing surfaces, and high use.

Transmissions use sliding bearings that are composed of a relatively soft bronze alloy. Many are made from steel with the bearing surface bonded or fused to the steel. Those that take radial loads are called bushings and those that take axial loads are called thrust washers (Figure 5-23).

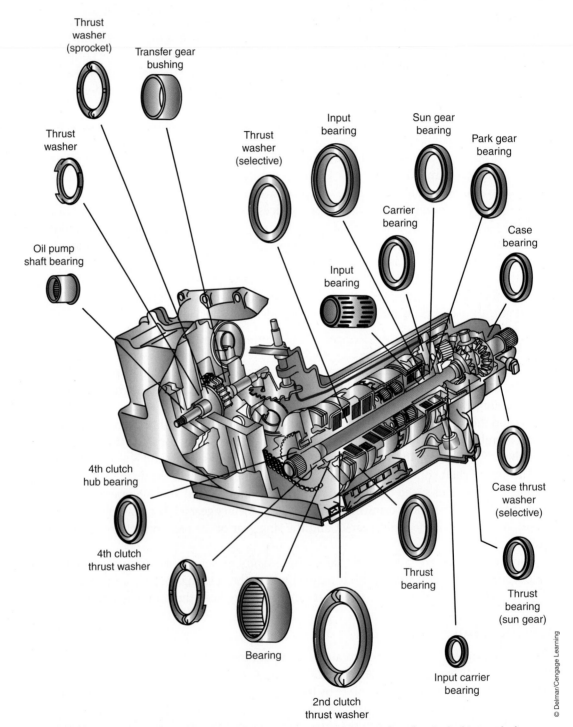

FIGURE 5-23 Locations of the various bushings, bearings, and thrust washers in a typical transmission.

The bearing's surface usually runs against a harder surface, such as steel, to produce minimum friction and heat wear characteristics.

Bushings are cylindrically shaped and usually held in place by press fit. Since bushings are made of a soft metal, they act like a bearing and support many of the transmission's rotating parts. They are also used to precisely guide the movement of various valves in the transmission's valve body. Bushings can also be used to control fluid flow. Some restrict the flow from one part to another while others are made to direct fluid flow to a particular point or part in the transmission.

Often serving both as a bearing and a spacer, thrust washers are made in various thicknesses. They may have one or more tangs or slots on the inside or outside circumference that mate with the shaft bore to keep them from turning. Some thrust washers are made of nylon or Teflon, which are used when the load is low. Others are fitted with rollers to reduce friction and wear.

Thrust washers normally control free axial movement or endplay. Since some endplay is necessary in all transmissions because of heat expansion, proper endplay is often accomplished through selective thrust washers. These thrust washers are inserted between various parts of the transmission. Whenever endplay is set, it must be set to manufacturer's specifications. Thrust washers work by filling the gap between two objects and become the primary wear item because they are made of softer materials than the parts they protect. Normally, thrust washers are made of copper-faced soft steel, bronze, nylon, or plastic.

Shop Manual
Chapter 5, page 258

Torrington bearings are thrust washers fitted with roller bearings. These thrust bearings are primarily used to limit endplay but also to reduce the friction between two rotating parts. Most often Torrington bearings are used in combination with flat thrust washers to control endplay of a shaft or the gap between a gear and its drum.

The bearing surface is greatly reduced through the use of roller bearings. The simplest roller bearing design leaves enough clearance between the bearing surfaces of two sliding or rotating parts to accept some rollers. Each roller's two points of contact between the bearing surfaces are so small that friction is greatly reduced. The bearing surface is more like a line than an area.

If the roller length to diameter is about 5:1, or more, the roller is called a needle and such a bearing is called a needle bearing. Sometimes the needles are loose or they can be held in place by a steel cylinder or by rings at each end. Often the latter are drilled to accept pins at the ends of each needle that act as an axle. These small assemblies help prevent the agony of losing one or more loose needles and the delay caused by searching for them.

Many other roller bearings are designed as assemblies. The assemblies consist of an inner and outer race, the rollers, and a cage. Roller bearings are designed for radial loads. Tapered roller bearings are designed to accept both radial and axial loads and are rarely used in automatic transmissions. Ball bearings are constructed similarly to a roller bearing, except that the races are grooved to accept the balls. Ball bearings can withstand heavy radial loads, as well as light axial loads.

Snap Rings

Many different sizes and types of snap rings are used in today's transmissions. External and internal snap rings are used as retaining devices throughout the transmission. Internal snap rings are used to hold servo assemblies and clutch assemblies together. In fact, snap rings are also available in several thicknesses and may be used to adjust the clearance in multiple-disc clutches. Some snap rings for clutch packs are waved to smooth clutch application. External snap rings are used to hold gear and clutch assemblies to their shafts.

Shop Manual
Chapter 5, page 257

FINAL DRIVES

The last set of gears in the drive train is the final drive. In most RWD cars, the final drive is located in the rear axle housing. Most FWD vehicles have the final drive located within the transaxle. Some FWD cars with longitudinally mounted engines locate the differential and final drive in a separate case that bolts to the transmission.

Types of Final Drives

A transaxle's final drive gears provide a way to transmit the transmission's output to the differential section of the transaxle. There are four common configurations used as the final drives on FWD vehicles: helical gear, planetary gear, hypoid gear, and chain drive. The helical, planetary, and chain final drive arrangements are found with transversely mounted engines. Hypoid final drive gear assemblies are normally found in vehicles with a longitudinally placed engine.

Hypoid Gears. RWD final drives normally use a hypoid gearset that turns the power flow 90 degrees from the driveshaft to the drive axles. On FWD cars with a transversely mounted engine, the power flow axis is naturally parallel to that of the drive axles; therefore the power doesn't need to be turned. Simple gear connections can be made to connect the output of the transmission to the final drive to the drive wheels.

A hypoid assembly in a transaxle is basically the same unit as would be used on RWD vehicles and is mounted directly to the transmission. The drive pinion gear is connected to the transmission's output shaft and the ring gear is attached to the differential case. The pinion and ring gearset provides for a multiplication of torque.

The teeth of the ring gear usually meshes directly with a gear on the transmission's output shaft. However on some transaxles, an intermediate shaft is used to connect the transmission's output to the ring gear.

The differential case rotates with the final drive output gear and is supported by tapered roller bearings on each side. The differential case contains four bevel gears: two pinion gears and two side gears. The pinion gears are installed on a pinion shaft that is retained by a pin in the differential case. The pinion gears transmit power from the case to the side gears. Each side gear is splined to one of the halfshafts that transmit power to the front wheels.

Helical Gear Drives. Helical gears are gears with teeth that are cut at an angle or are spiral to the gear's axis of rotation. Some transaxles route power from the transmission's output shaft through two helical-cut gears to a transfer shaft (Figure 5-24). A helical-cut pinion gear attached to the opposite end of the transfer shaft drives the differential ring gear and carrier (Figure 5-25). The differential assembly then drives the axles and wheels.

Helical final drive gearsets require that the centerline of the pinion gear is at the centerline of the ring gear. The pinion gear is cast as part of the main shaft and is supported by tapered-roller bearings. The pinion gear is meshed with the ring gear to provide the required torque multiplication. Because the ring is mounted on the differential case, the case rotates in response to the pinion gear.

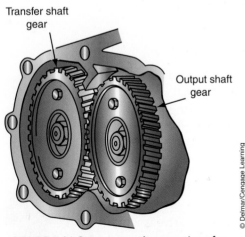

FIGURE 5-24 Some transaxles use a transfer shaft and gear to move the output to the final drive unit.

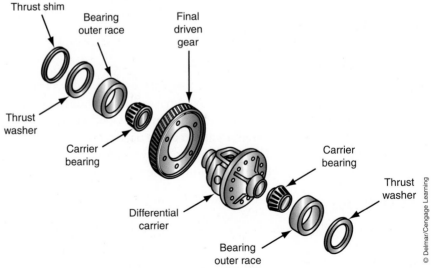

FIGURE 5-25 A helical-gear differential for a transaxle.

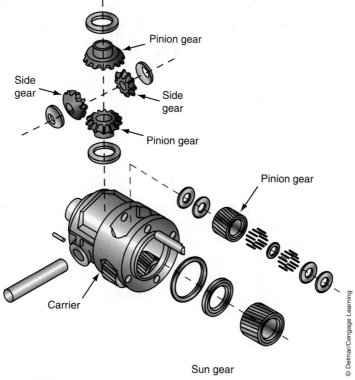

FIGURE 5-26 A final drive unit that utilizes a planetary gearset.

Planetary Gear Drives. Rather than use helical-cut gears in the final drive assembly, many transaxles use a simple planetary gear (Figure 5-26). The sun gear of this planetary unit is driven by the output shaft of the transaxle. The final drive sun gear meshes with the final drive planetary pinion gears, which rotate on their shafts in the planetary carrier. The planetary carrier is part of the differential case, which contains typical differential gearing: two pinion gears and two side gears.

The final drive pinion gears mesh with the ring gear, which has lugs around its outside diameter. These lugs fit into grooves machined inside the transaxle housing. The lugs and grooves hold the ring gear stationary. The final drive pinion gears walk around the inside of the stationary ring gear and drive the planetary carrier and differential case. This combination provides maximum torque multiplication from a simple planetary gearset.

When the ring gear is held and input is sent to the sun gear, forward gear reduction takes place. This gear reduction is the final drive gear ratio.

Chain and Gear Drives. Chain-drive final drive assemblies use a multiple-link chain to connect a drive sprocket, connected to the transmission's output shaft, to a driven sprocket (Figure 5-27), which is connected to the differential case. This design allows for remote positioning of the differential within the transaxle housing. Final drive gear ratios are determined

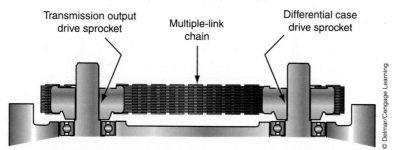

FIGURE 5-27 The chain setup for a final drive unit in a transaxle.

by the size of the driven sprocket compared to the drive sprocket. The driven sprocket is attached to the differential case, which provides differential action for the drive wheels.

GASKETS AND SEALS

The gaskets and seals of an automatic transmission help contain the fluid within the transmission and prevent the fluid from leaking out of the various hydraulic circuits. Different types of seals are used in automatic transmissions; they can be made of rubber, metal, or Teflon (Figure 5-28). Transmission gaskets are made of rubber, cork, paper, synthetic materials, or plastic.

Gaskets

Gaskets are used to seal two parts together or to provide a passage for fluid flow from one part of the transmission to another (Figure 5-29). Gaskets are easily divided into two separate groups, hard and soft, depending on their application. **Hard gaskets** are used whenever the surfaces to be sealed are smooth. This type of gasket is usually made of paper. A common application of a hard gasket is the gasket used to seal the valve body and oil pump against the transmission case. Hard gaskets are also often used to direct fluid flow or seal off some passages between the valve body and the separator plate.

Gaskets that are used when the sealing surfaces are irregular or in places where the surface may distort when the component is tightened into place are called **soft gaskets**. A typical

Shop Manual
Chapter 5, page 242

A **gasket** is typically used to seal the space between two parts that have irregular surfaces.

A **hard gasket** is one that will compress less than 20 percent when it is tightened in place.

A **soft gasket** is one that will compress more than 20 percent when it is tightened in place.

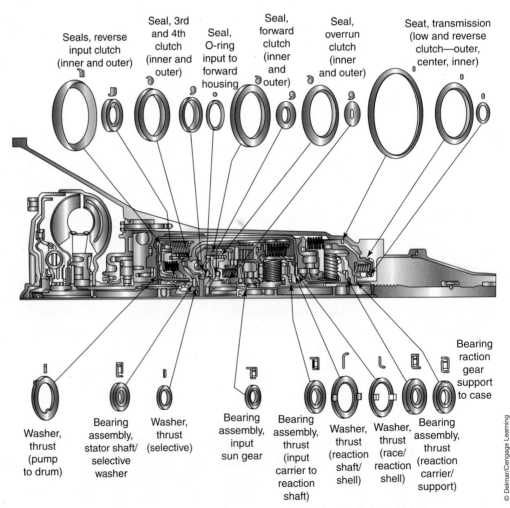

FIGURE 5-28 Locations of various seals and gaskets in a typical transmission.

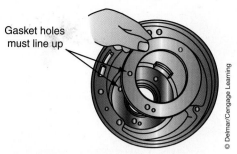

FIGURE 5-29 A typical application of a gasket in an automatic transmission.

A **composition gasket** is one that is made from two or more different materials.

RTV stands for **room temperature vulcanizing**, which means this sealant will begin to solidify at room temperature and form a seal prior to being heated by the operation of the transmission.

Shop Manual
Chapter 5, page 255

A **static seal** prevents fluid from passing between two or more parts that are always in the same relationship with each other.

A **dynamic seal** is one that prevents leaks between two or more parts that do not have a fixed position and move in relation to each other.

Synthetic rubber is made from neoprene, nitrile, silicone, fluoroelastomers, and polyacrylics.

location of a soft gasket is the oil pan gasket, which seals the oil pan to the transmission case. Oil pan gaskets are typically a **composition gasket** made with rubber and cork. However, some transmissions use a **RTV** sealant instead of a gasket to seal the oil pan.

Seals

Because valves and transmission shafts move within the transmission, it is essential that the fluid and pressure be contained within its bore. Any leakage would decrease the pressure and result in poor transmission operation. Seals are used to prevent leakage around valves, shafts, and other moving parts. Rubber, metal, or Teflon materials are used throughout a transmission to provide for **static seals** and **dynamic seals**. Both static and dynamic seals can provide for positive and nonpositive sealing. A definition of each of the different basic classifications of seals follows.

Static—A seal used between two parts that do not move in relationship to each other.

Dynamic—A seal used between two parts that do move in relationship to each other. This movement is either a rotating or reciprocating (up and down) motion.

Positive—A seal that prevents all fluid leakage between two parts.

Nonpositive—A seal that allows a controlled amount of fluid leakage. This leakage is typically used to lubricate a moving part.

Three major types of rubber seals are used in automatic transmissions: the O-ring, the lip seal, and the lathe-cut seal or square-cut seal (Figure 5-30). Rubber seals are made from synthetic rubber rather than natural rubber.

O-rings are round seals with a circular cross section. Normally an O-ring is installed in a groove cut into the inside diameter of one of the parts to be sealed. When the other part is inserted into the bore and through the O-ring, the O-ring is compressed between the inner part and the groove. This pressure distorts the O-ring and forms a tight seal between the two parts (Figure 5-31).

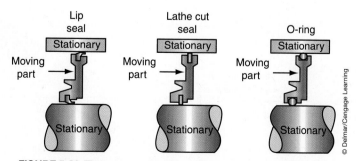

FIGURE 5-30 Three types of seals shown in their typical position and mountings.

167

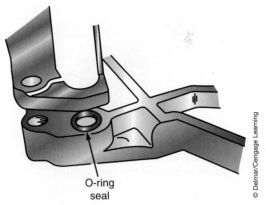

FIGURE 5-31 A typical application of an O-ring seal.

O-rings can be used as dynamic seals but are most commonly used as static seals. An O-ring can be used as a dynamic seal when the parts have relatively low amounts of axial movement. If there is a considerable amount of axial movement, the O-ring will quickly be damaged as it rolls within its groove. O-rings are never used to seal a shaft or part that has rotational movement.

Lip seals are used to seal parts that have axial or rotational movement. They are round in order to fit around a shaft and into a bore. The primary sealing part is a flexible lip (Figure 5-32). The flexible lip is normally made of synthetic rubber and shaped so that it is flexed when it is installed to apply pressure at the sharp edge of the lip. Lip seals are used around input and output shafts to keep fluid in the housing and dirt out. Some seals are double-lipped.

When the lip is around the outside diameter of the seal, it is used as a piston seal (Figure 5-33). Piston seals are designed to seal against high pressures and the seal is positioned so that the lip faces the source of the pressurized fluid. The lip is pressed firmly against the cylinder wall; as the fluid pushes against the lip a tight seal is formed. The lip then relaxes its seal when the pressure on it is reduced or exhausted.

Lip seals are also commonly used as shaft seals. When used to seal a rotating shaft, the lip of the seal is around the inside diameter of the seal and the outer diameter is bonded to the inside of a metal housing. The outer metal housing is pressed into a bore. To help maintain good sealing pressure on the rotating shaft, a garter spring is fitted behind the lip. This **toroidal** spring pushes on the lip to provide for uniform contact on the shaft. Shaft seals are not designed to contain pressurized fluid; rather, they are designed to prevent fluid from leaking over the shaft and out of the housing. The tension of the spring and of the lip is designed to allow an oil film of about 0.0001 of an inch. This oil film serves as a lubricant for the lip. If the tolerances increase, fluid will be able to leak past the shaft, and if the tolerances are too small, excessive shaft and seal wear will result.

> The word **toroidal** infers that the spring is doughnut shaped.

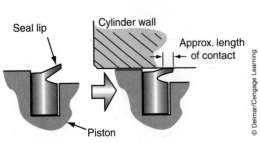

FIGURE 5-32 Sealing action of a lip seal.

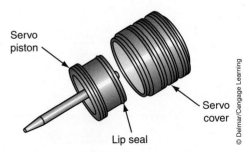

FIGURE 5-33 A typical application of a lip seal.

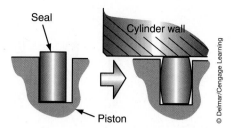

FIGURE 5-34 Sealing action of a rubber seal as a piston moves in its bore.

A **square-cut seal** is similar to an O-ring; however, a square-cut seal can withstand more axial movement than an O-ring can. Square-cut seals are also round seals but have a rectangular or square cross section. They are designed this way to prevent the seal from rolling in its groove when there are large amounts of axial movement. Added sealing comes from the distortion of the seal during axial movement. As the shaft inside the seal moves, the outer edge of the seal moves more than the inner edge, causing the diameter of the sealing edge to increase, which creates a tighter seal (Figure 5-34).

A **square-cut seal** is often called a lathe-cut seal.

Metal Sealing Rings

There are some parts of the transmission that do not require a positive seal and where some leakage is acceptable. These components are sealed with ring seals, which fit into a groove on a shaft (Figure 5-35). The outside diameter of the ring seals slide against the walls of the bore that the shaft is inserted into. Most ring seals in a transmission are placed near pressurized fluid outlets on rotating shafts to help retain pressure. Ring seals are made of cast iron, nylon, or Teflon.

Three types of metal seals are used in automatic transmissions: butt-end seals, open-end seals, and hook-end seals. In appearance, butt-end and open-end seals are much the same; however, when an *open-end seal* is installed, there is a gap between the ends of the seal. When a *butt-end seal* is installed, the square-cut ends of the seal touch or butt against each other. **Hook-end seals** (Figure 5-36) have small hooks at their ends, which are locked together during installation to provide better sealing than the open-end or butt-end seals provide.

Shop Manual
Chapter 5, page 256

Metal seals are often called steel rings.

Hook-end seals are also referred to as locking-end seals.

Teflon Seals

Some transmissions use Teflon seals instead of metal seals. Teflon provides for a softer sealing surface, which results in less wear on the surface that it rides on and therefore a longer-lasting seal. Teflon seals are similar in appearance to metal seals except for the hook-end type. The ends of locking-end Teflon seals are cut at an angle (Figure 5-37) and the locking hooks are somewhat staggered.

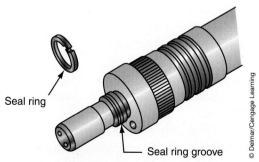

FIGURE 5-35 Metal sealing rings are fit into grooves on a shaft.

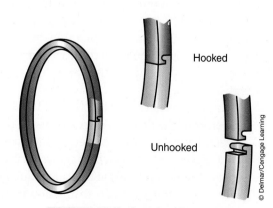

FIGURE 5-36 Hook-end sealing rings.

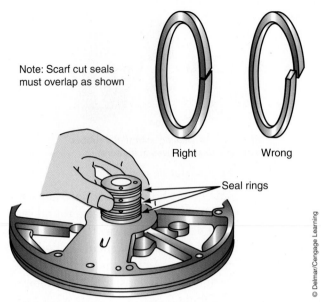

Note: Scarf cut seals must overlap as shown

Right   Wrong

Seal rings

© Delmar/Cengage Learning

FIGURE 5-37 Scarf-cut seals; notice that the ends of the seal are cut at opposing angles.

> Teflon locking-end seals are normally called scarf-cut rings.

Many late-model transmissions are equipped with solid one-piece Teflon seals. Although the one-piece seal requires some special tools for installation, it provides for a nearly positive seal. These Teflon rings seal much better than other metal sealing rings.

GM uses a different type of synthetic seal on some late-model transmissions. The material used in these seals is Vespel, which is a flexible but highly durable plastic-like material. Vespel seals are found on 4T60-E and 4T80-E transaxles.

SUMMARY

- Transmissions are designed and built for particular applications. Things that must be considered in the design of a transmission are the power output of the engine, the weight of the vehicle, and the typical workload of the vehicle.
- Most manufacturers use an alphanumeric code that indicates the features of the transmission model.
- The weight of the transmission or transaxle is supported by the engine and its mounts and by a transmission mount.
- Transmission mounts isolate transmission noise and vibrations from the passenger compartment.
- Proper mounting surfaces of a transmission/transaxle are critical in order to keep the various shafts aligned and to provide a good seal.
- The valve body contains the valves and orifices that control the operation of a transmission.
- Transmissions have at least two shafts. The input shaft connects the output of the torque converter to the driving members inside the transmission. The output shaft connects the driven members of the gearsets to the final drive gearset.
- Certain parts of the planetary geartrain must be held, while others must be driven, to provide the needed torque multiplication and direction for vehicle operation. Planetary gear controls include the transmission bands, servos, and clutches.

- A band is a braking assembly positioned around a stationary or rotating drum that is connected to a member of the planetary gearset.
- A servo assembly hydraulically applies the band. It converts hydraulic pressure into a mechanical force that applies a band to hold a drum stationary.
- In an automatic transmission operation, both sprag and roller overrunning clutches are used to hold or drive members of the planetary gearset. These clutches operate mechanically.
- A multiple-disc pack uses a series of friction discs to transmit torque or apply braking force.
- Multiple-disc clutches are enclosed in a large drum-shaped housing that can be either a separate casting or part of the existing transmission housing.
- Nearly all automatic transmissions rely on planetary gearsets to transfer power and multiply engine torque to the drive axle.
- Compound gearsets combine simple planetary gearsets so that the load can be spread over a greater number of teeth for strength and also to obtain the largest number of gear ratios possible in a compact area.
- Any one of the three members of a planetary gearset can be used as the driving or input member. At the same time, another member might be the held or stationary member. The third member then becomes the driven or output member. Depending on which member is the driver, which is held, and which is driven, either a torque increase or a speed increase is produced by the planetary gearset.
- A bearing is a device placed between two bearing surfaces to reduce friction and wear.
- Bushings act like a bearing and support many of the transmission's rotating parts.
- Thrust washers normally control free axial movement or endplay.
- Torrington bearings are thrust washers fitted with roller bearings.
- Internal snap rings are used to hold servo assemblies and clutch assemblies together.
- External snap rings are used to hold gear and clutch assemblies to their shafts.
- A transaxle uses helical gears, planetary gears, or a chain-type final drive unit when the engine is transversely mounted. Hypoid final drive gear assemblies are normally found in vehicles with a longitudinally placed engine.
- The gaskets and seals of an automatic transmission help to contain the fluid within the transmission and prevent the fluid from leaking out of the various hydraulic circuits. Different types of seals are used in automatic transmissions; they can be made of rubber, metal, or Teflon.
- Transmission gaskets are made of rubber, cork, paper, synthetic materials, or plastic.
- Three major types of rubber seals are used in automatic transmissions: the O-ring, the lip seal, and the square-cut seal.
- Three types of metal seals are used in automatic transmissions: butt-end seals, open-end seals, and hook-end seals.

TERMS TO KNOW

Band

Composition gasket

Dynamic seal

Gasket

Hard gasket

Hook-end seal

Planetary carrier

Planetary pinion

Room temperature vulcanizing (RTV)

Soft gasket

Square-cut seal

Static seal

Toroidal

REVIEW QUESTIONS

Short-Answer Essays

1. The coding used to identify the transmission model typically includes what kind of information?

2. Transmission housings are cast to secure and accommodate many important components. List at least five of them.

3. What are the two common designs of compound planetary gearsets. How do they differ?

4. What are the input and output shafts connected to in a typical transmission?

5. What is the purpose of a band in a transmission?

6. Why are Torrington bearings used in automatic transmissions?

7. Why might a chain drive be used in a transaxle's final drive unit?

8. An automatic transmission uses specially designed snap rings. What is so special about them?

9. What is the purpose of a servo?

10. When a transmission is described as having two planetary gearsets in tandem, what does that mean?

Fill in the Blanks

1. Multiple-disc clutches are enclosed in a large drum-shaped housing that can be either a _____ _____ or part of the _____ _____ .

2. Since _____ are made of a soft metal, they act like a bearing and support many of the transmission's rotating parts.

3. In an automatic transmission operation, both _____ and _____ overrunning clutches are used to hold or drive members of the planetary gearset.

4. The four common configurations used as the final drives on FWD vehicles are the _____ gear, _____ gear, _____ gear, and _____ _____ .

5. The major components of a planetary gearset are the: _____ _____ , _____ _____ , and _____ _____ .

6. When a small gear drives a larger gear, torque is _____ while speed is _____ .

7. The three major types of rubber seals used in automatic transmissions are the _____ , the _____ , and the _____ seal.

8. The three types of metal seals used in automatic transmissions are the _____ , _____ , and _____ seals.

9. Transmission gaskets are made of _____ , _____ , _____ , _____ _____ , or _____ .

10. A Ravigneaux gearset has two _____ gears, two sets of _____ gears, and a common _____ gear.

MULTIPLE CHOICE

1. Which of the following is used all automatic transmissions?
   A. Input shaft
   B. Planetary gearset
   C. Brake bands
   D. All of the above

2. Overrunning clutches are capable of _____ .
   A. Holding a planetary gear member stationary
   B. Driving a planetary gear member
   C. Both A and B
   D. Neither A nor B

3. While discussing current manufacturer transmission coding:
   Technician A says Ford model numbers do not indicate the duty rating of the transmission.
   Technician B says Chrysler model numbers indicate the number of planetary units used in the transmission.
   Who is correct?
   A. A only
   B. B only
   C. Both A and B
   D. Neither A nor B

4. While discussing transmission housings:
   Technician A says good mating surfaces are necessary to provide a good seal between the mating sections or parts.
   Technician B says proper mating surfaces are necessary to keep shafts properly aligned.
   Who is correct?
   A. A only
   B. B only
   C. Both A and B
   D. Neither A nor B

5. *Technician A* says a multiple-friction-disc assembly is used to transfer power from one member to another.
   Technician B says a multiple-friction-disc assembly is used to brake or hold a member of the gearset.
   Who is correct?
   A. A only
   B. B only
   C. Both A and B
   D. Neither A nor B

6. *Technician A* says a nonpositive static seal will allow some fluid leakage between two parts that do not move in relationship to each other.

Technician B says a positive dynamic seal allows a controlled amount of leakage between two parts that move in relationship to each other.

Who is correct?
A. A only
B. B only
C. Both A and B
D. Neither A nor B

7. Which of the following statements is *not* true?
A. A torque converter uses hydraulic fluid flow to connect and disconnect the power of the engine to and from the transmission.
B. The transmission pump is driven by the input shaft.
C. The valve body contains the valves and orifices that control the operation of a transmission.
D. The transmission pump is the main source of fluid flow in a transmission.

8. *Technician A* says the output shaft connects the output of a driving member to the final drive unit.

Technician B says the input shaft connects the impeller of the torque converter to the front driven member in the transmission.

Who is correct?
A. A only
B. B only
C. Both A and B
D. Neither A nor B

9. *Technician A* says engine/transaxle mounts isolate the inherent vibrations of the assembly from the passenger compartment.

Technician B says improper mounting of an engine and/or transmission can cause problems with the transmission.

Who is correct?
A. A only
B. B only
C. Both A and B
D. Neither A nor B

10. *Technician A* says a Simpson gearset is two planetary gearsets that share a common sun gear.

Technician B says a Ravigneaux gearset has two sun gears, two sets of planet gears, and a common ring gear.

Who is correct?
A. A only
B. B only
C. Both A and B
D. Neither A nor B

Chapter 6

TORQUE CONVERTERS AND PUMPS

UPON COMPLETION AND REVIEW OF THIS CHAPTER, YOU SHOULD BE ABLE TO:

- Identify the major components in a torque converter and explain their purpose.
- Understand the fluid flows that occur in a torque converter.
- Explain the necessity of curved blades to enhance torque converter efficiency.
- Discuss stator design and operation.
- Explain how fluid flows in and out of a torque converter.
- Explain torque converter operation in the stall and coupling phases.

- Explain the basic design and operation of standard and torque converters with a clutch.
- Describe the design and operation of a centrifugal converter clutch.
- Describe the design and operation of a piston-type converter clutch.
- Explain the operation of typical electronic controls for a converter clutch.
- Name and explain the different designs of pumps used in modern transmissions.

A BIT OF HISTORY

In 1939, Chrysler introduced the Fluid Drive, a three-speed Synchromesh transmission with a fluid coupling between the engine and a conventional friction clutch. In 1941, this was replaced by the Vacamatic, a manually shifted two-speed transmission in series with a vacuum-actuated, automatic overdrive. The result was four forward

INTRODUCTION

A torque converter uses fluid to smoothly transfer engine torque to the transmission. The torque converter is a doughnut-shaped unit located between the engine and the transmission that is filled with ATF (Figure 6-1). Internally, the torque converter has three main parts: the impeller, turbine, and stator. Each of these has blades that are curved to increase torque converter efficiency.

The *impeller* is driven by the engine and directs fluid flow against the turbine blades, causing them to rotate and drive the turbine shaft, which is the transmission's input shaft. The *stator* is located between the impeller and the turbine and returns fluid from the turbine to the impeller, so that the cycle can be repeated.

During certain operating conditions, the torque converter multiplies torque. It provides extra reduction to meet the driveline needs while under a heavy load. When the vehicle is operating at cruising speeds, the torque converter operates as a *fluid coupling* and transfers engine torque to the transmission. It also absorbs the shock from gear changing in the transmission. Not all of the engine's power is transferred through the fluid to the transmission; some is lost. To reduce the amount of power lost through the converter, especially at cruising speeds, manufacturers equip most of their current transmissions with a torque converter clutch.

The engagement of the converter clutch is based on both engine and vehicle speeds and the clutch is normally controlled by transmission hydraulics and onboard computer electronic controls. When the clutch engages, a mechanical connection exists between the engine and the drive wheels. This improves overall efficiency and fuel economy.

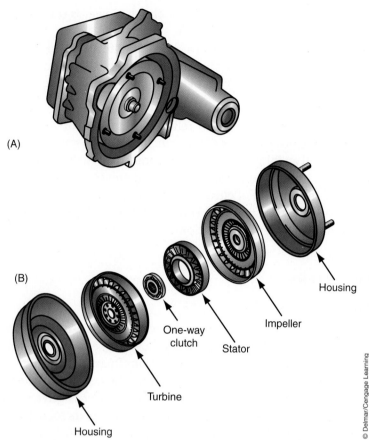

(A)

(B)

Housing

Impeller

One-way
clutch

Stator

Turbine

Housing

© Delmar/Cengage Learning

FIGURE 6-1 (A) The location of a torque converter in an automatic transaxle, (B) A cutaway of a typical late-model torque converter.

A BIT OF HISTORY

(continued)

speeds, only half of which were engaged automatically. Ford launched its similar Liquamatic Drive on 1942 Lincolns and Mercurys. Few were sold, and many owners had standard transmissions retrofitted during the war, when replacement parts were hard to find.

It also eliminates the heat that is produced by the converter, which is the largest amount of heat produced by an automatic transmission; this extends the life of the ATF. Consider this: at 90 percent converter efficiency, there is a 10 percent loss of energy at cruising speed. This loss of energy is converted to unwanted heat!

All automatic transmissions, except the many different designs of the constant variable transmission (CVT) and some concept transmissions, use a torque converter to transfer power from the engine to the transmission. Some concept vehicles are fitted with automatic transmissions that use a wet clutch assembly and a dual-mass flywheel rather than a torque converter. These units work just like an automatic version of the driver-operated clutch in a manual transmission and are primarily controlled by electronics.

Continuously Variable Transmission (CVT) Units

Most vehicles (hybrid and nonhybrid) that are equipped with a CVT do not use a torque converter. Rather they use an internal start clutch that allows the engine to maintain an idle speed while it is in gear and at a stop. The start clutch is designed to slip just enough to get the car moving without stalling or straining the engine. The start clutch can be electrically or hydraulically controlled.

Many units have an electromagnetic clutch controlled by a clutch control unit. The clutch control unit switches current to energize and deenergize the electromagnetic clutch in response to inputs from various sensors. When the electromagnetic clutch is energized, engine torque is transferred to the transmission's input shaft and the drive pulley.

Other units have hydraulically controlled start clutches. Activating the start clutch control solenoid moves the start clutch valve. This valve allows or disallows pressure to the start

A BIT OF HISTORY

The use of fluid couplings actually began in the 1900s with steamship propulsion and later was used to help dampen the vibrations of large diesel engines. Just before WWII, Chrysler was the first American automobile manufacturer to use a fluid coupling. The fluid coupling was added to a manual transmission driveline, which allowed the engine to idle in gear, but a foot-operated clutch was still needed to shift gears.

The impeller and turbine are sometimes called the torus halves because of their convex shape.

The impeller is commonly referred to as the drive torus or the primary pump. Likewise, the turbine is referred to as the driven torus.

clutch assembly. When pressure is applied to the clutch, power is transmitted from the pulleys to the final drive gearset.

Nissan's CVT operates with a torque converter, one clutch and a simple planetary gearset. The action of the converter is controlled by the TCM. Once the vehicle begins to move, the torque converter locks to allow the CVT's belt and pulleys to provide all of the drive ratios without slippage.

Primary Inputs for Control. The TCM relies on many sensors to control the action of the start clutch or torque converter. These inputs include signals sent by a brake switch, accelerator pedal switches, and an inhibitor switch. The brake switch deenergizes the clutch when the vehicle is slowing or coming to a stop. The control unit uses inputs from the accelerator pedal switches to vary the amount of current to the clutch and to signal ratio changes to the adjustable pulleys inside the transmission. The inhibitor switch prevents clutch engagement when the gear selector is in the P or N position.

TORQUE CONVERTERS

Fluid Couplings

A simple fluid coupling is comprised of three basic members: a housing, impeller, and turbine (Figure 6-2). The impeller and the turbine are shaped like two halves of a doughnut. The impeller and turbine have internal vanes radiating from their centers. Both the impeller and turbine are enclosed in the housing. ATF is forced into the housing by the transmission's oil pump. The housing is sealed and filled with fluid. The impeller is driven by the engine. The impeller acts like a pump and moves the fluid in the direction of its rotation. The moving fluid hits against the turbine vanes. This causes the turbine to rotate, which brings an input into the transmission via the input shaft.

When the transmission is in gear, the fluid coupling allows the vehicle to stop without stalling the engine because the impeller doesn't rotate fast enough to drive the turbine. As engine speed increases, the force from the fluid flow off the impeller increases and forces the turbine to rotate. Based on the fluid flow, a fluid coupling can transmit from very little to very much force. Maximum efficiency is approximately 90 percent, but that figure depends on a number of factors: fluid type, impeller and turbine vane design, vehicle load, and engine torque.

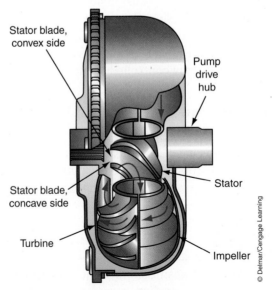

Stator blade, convex side

Pump drive hub

Stator blade, concave side

Stator

Turbine

Impeller

© Delmar/Cengage Learning

FIGURE 6-2 A torque converter's major internal parts are its impeller, turbine, and stator.

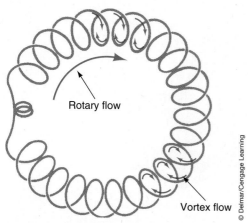

FIGURE 6-3 Difference between rotary and vortex flow. Note that vortex flow spirals its way around the converter.

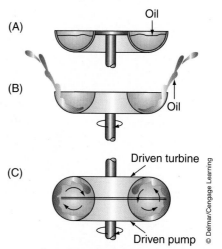

FIGURE 6-4 Fluid travel inside the torque converter: (A) Fluid at rest in the impeller/pump; (B) fluid thrown up and outward by the spinning pump; and (C) fluid flow harnessed by the turbine and redirected back into the pump.

Fluid Flow. The operation of a fluid coupling is totally based on the flow of fluid inside the coupling. Two types of fluid flow take place inside the fluid coupling: rotary and vortex flow (Figure 6-3). These different flows complement each other, depending on the difference in speed between the impeller and the turbine. **Rotary flow** is the movement of the fluid in the direction of impeller rotation and results from the paddle action of the impeller vanes against the fluid. As this rotating fluid hits against the blades of the slower turning or stationary turbine, it exerts a turning force onto the turbine (Figure 6-4).

Vortex flow is the fluid flow circulating between the impeller and turbine as the fluid moves from the impeller to the turbine and back to the impeller. This type of fluid flow is only present when there is a difference in rotational speeds between the impeller and the turbine.

Both types of flow can occur within a fluid coupling at the same time. When the impeller is rotating fast but turbine speed is restricted because of a load, most of the fluid within the housing moves with a vortex flow. However, as the load is overcome and turbine speed increases, more rotary flow is occurring.

BASIC TORQUE CONVERTERS

The torque converter is a type of fluid coupling or connector that connects the engine's crankshaft to the transmission's input shaft. The torque converter, working as a fluid connector, smoothly transmits engine torque to the transmission. The torque converter allows some slippage between the engine and the transmission so that the engine will remain running when the vehicle is stopped while it is in gear. The torque converter also multiplies torque when the vehicle is under load to improve performance. Torque converters are more efficient than fluid couplings because they allow for increased fluid flow, which multiplies the torque from the engine when it is operating at low speeds.

Torque Converter Mountings

On all RWD and most FWD vehicles, the torque converter is mounted in line with the transmission's input shaft. Some transaxles use a drivechain to connect the converter's output shaft with the transmission's input shaft (Figure 6-5). Offsetting the input shaft from the centerline of the transmission gives the manufacturers more flexibility in the positioning of the transaxle.

Rotary flow occurs when the fluid path is in the same circular direction as the rotation of the impeller.

Vortex flow occurs when the fluid path is at a right angle to the rotary oil flow and the rotation of the impeller.

An impeller is a finned wheel-like device that is turned by the engine to pump transmission fluid into the turbine, which drives the transmission's input shaft.

Shop Manual
Chapter 6, page 283

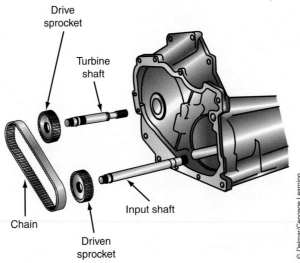

Drive
sprocket

Turbine
shaft

Chain

Driven
sprocket

Input shaft

© Delmar/Cengage Learning

FIGURE 6-5 Typical drivechain setup for offsetting the torque converter and the input shaft.

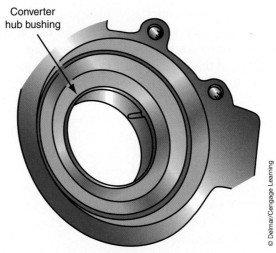

Converter
hub bushing

© Delmar/Cengage Learning

FIGURE 6-6 The converter's hub rides on a bushing in the transmission housing.

A Bit of History

The most important improvement to automatic transmissions was the result of Buick's development of a torque converter for tanks during the war. The converter was incorporated into the Dynaflow transmission in 1948, and eventually into other automatics.

Shop Manual

Chapter 6, page 303

The torque converter fits between the transmission and the engine and is normally supported by the transmission housing's oil pump bushing and the engine's crankshaft. The rear hub of the converter's cover fits into the transmission's pump (Figure 6-6). The hub or a shaft drives the pump whenever the engine is running.

The front of a torque converter is mounted to a flexplate, which is bolted to the end of the engine's crankshaft. The converter cover is normally fitted with studs that are used to tighten the cover to the flexplate. The flexplate is designed to be flexible enough to allow the front of the converter to move forward or backward if it expands or contracts because of heat or pressure. The centerlines of the converter and crankshaft are matched and this alignment is maintained by a pilot on the converter cover that fits into a recess in the crankshaft.

An externally toothed ring gear is normally pressed or welded to the outside diameter of the flexplate. The starter motor's drive gear meshes with the ring gear to crank the engine for starting.

The transmission's input shaft is supported by bushings in the stator support inside the torque converter. There is no mechanical link between the output of the engine and the input of the transmission. The fluid connects the power from the engine to the transmission. The combined weight of the fluid, torque converter, and flexplate serves as the flywheel for the engine.

Torque Converter Construction

A typical torque converter consists of three elements sealed in a single housing: the impeller, the turbine, and the stator. The impeller is the drive member of the unit and its fins are attached directly to the converter cover. Therefore, the impeller is the input device for the converter and always rotates at engine speed.

The turbine is the converter's output member and is coupled to the transmission's input shaft (Figure 6-7). The turbine is driven by the fluid flow from the impeller and always turns at its own speed. The fins of the turbine face toward the fins of the impeller. The impeller and the turbine have internal fins, but the fins point toward each other.

The stator is the reaction member of the converter (Figure 6-8). This assembly is about one-half the diameter of the impeller or turbine and is positioned between the impeller and turbine. The stator is not mechanically connected to either the impeller or turbine; rather, it fits between the turbine outlet and the inlet of the impeller. All of the fluid returning from the turbine to the impeller must pass through the stator. The stator redirects the fluid leaving the turbine back to the impeller (Figure 6-9). By redirecting the fluid so that it is flowing in the same direction as engine rotation, it allows the impeller to rotate more efficiently, creating torque multiplication.

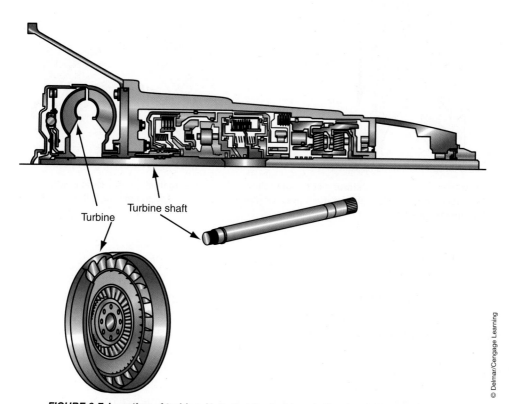

© Delmar/Cengage Learning

FIGURE 6-7 Location of turbine. Note that the turbine shaft extends from the turbine to the transmission and serves as the input shaft of the transmission.

A BIT OF HISTORY

Variations on the basic three-element torque converter have been used. The 1948 Buick Dynaflow had two impellers, two stators, and one turbine. In 1953 the Twin-Turbine Dynaflow was released, with two turbines, one impeller, and one stator. In 1956, Buick introduced a multiple-turbine torque converter that had a variable pitch stator. By the late 1960s, the industry, including Buick, had returned to the basic three-element converter.

A turbine is a finned wheel-like device that receives fluid from the impeller and forces it back to the stator. The turbine transmits engine torque to the transmission's input shaft.

A stator is sometimes called a reactor.

Shop Manual
Chapter 6, page 294

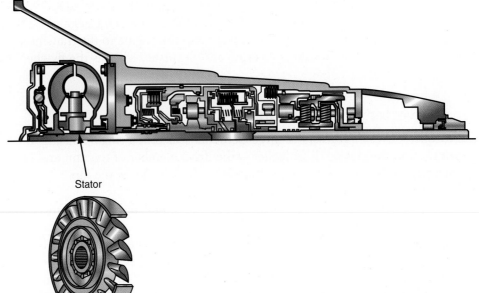

Stator

FIGURE 6-8 Location of stator assembly. Note that the stator is not attached to the turbine or the impeller.

© Delmar/Cengage Learning

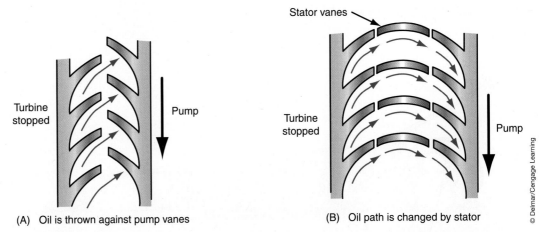

Turbine stopped — Pump

(A) Oil is thrown against pump vanes

Stator vanes

Turbine stopped — Pump

(B) Oil path is changed by stator

FIGURE 6-9 (A) Without a stator, fluid leaving the turbine works against the direction in which the impeller is rotating. (B) With a stator in its lock mode, fluid is directed to help push the impeller in its rotating direction.

© Delmar/Cengage Learning

TORQUE CONVERTER OPERATION

The impeller rotates at engine speed whenever the engine is running. The impeller is composed of many curved vanes radiating out of an inner ring, which form passages for the fluid. As soon as the converter begins to rotate, the vanes of the impeller begin to circulate fluid.

The fluid in the torque converter is supplied by the transmission's oil pump. It enters through the converter's hub, then flows into the passages between the vanes. As the impeller rotates, the fluid is moved outward and upward through the vanes by centrifugal force because of the curved shape of the impeller. The faster the impeller rotates, the greater the centrifugal force becomes.

The fluid moves from the outer edge of the vanes into the turbine. As the fluid strikes the curved vanes of the turbine, it attempts to push the turbine into a rotation. Because the

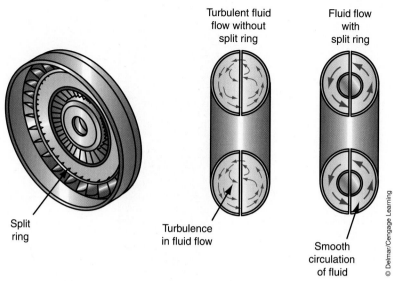

Turbulent fluid flow without split ring

Fluid flow with split ring

Split ring

Turbulence in fluid flow

Smooth circulation of fluid

© Delmar/Cengage Learning

FIGURE 6-10 Location and action of the split guide rings.

impeller is turning in a clockwise direction (as viewed from the front), the fluid also rotates in a clockwise direction as it leaves the vanes of the impeller. However, the turbine vanes are curved in the opposite direction of the impeller and the fluid turns the turbine in the same direction as the impeller.

The higher the engine speed, the faster the impeller turns, and the more force is transferred from the impeller to the turbine by the fluid. This explains why a torque converter allows the engine to idle in gear. When engine speed is low, the fluid does not have enough force to turn the turbine against the load on the drivetrain. The movement of the fluid in the converter is very weak and it is just circulating from the impeller to the turbine and back to the impeller. Therefore, little if any power is transmitted through the torque converter to the transmission.

As engine speed increases, the fluid is thrown at the turbine with a greater force, causing the turbine to rotate. Once the turbine begins to turn, engine power is transmitted to the transmission. However, the force from the fluid must be great enough to overcome the load of the vehicle before the turbine can rotate. Some of the energy in the moving fluid returns to the impeller as the torque converter responds to the torque requirements of the vehicle and to vortex flow. At low speeds, most of the energy in the fluid is lost as the fluid moves through the curvature of the turbine vanes.

Vortex flow is a continuous circulation of the fluid, outward in the impeller and inward in the turbine, around the split *guide rings* attached to the turbine and the impeller (Figure 6-10). The guide rings direct the vortex flow to provide for a smooth and turbulence-free fluid flow.

As the vortex flow continues, the fluid leaving the turbine to return to the impeller is moving in the opposite direction as crankshaft rotation. If the fluid were allowed to continue in this direction, it would enter the impeller as an opposing force and some of the engine's power would be used to redirect the flow of fluid. To prevent this loss of power, torque converters are fitted with a stator.

Torque Multiplication

The stator receives the fluid thrown off by the turbine and redirects the fluid so that it reenters the impeller in the same direction as crankshaft rotation (Figure 6-11). The redirection of the fluid by the stator not only prevents a torque loss, but also provides for a multiplication of torque.

Torque multiplication in a torque converter occurs when the vortex flow is redirected through the stator to increase flow to the turbine.

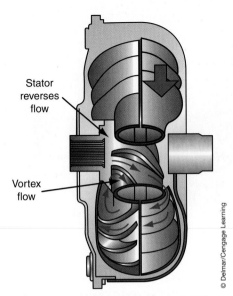

FIGURE 6-11 Action of the stator on fluid flow.

The stator is attached to a circular hub, which is mounted on a one-way clutch (Figure 6-12). This clutch assembly has an inner and outer race separated by spring-loaded roller bearings or sprags. The inner race is splined to the stator support; therefore, it cannot turn. The outer race is fitted into the stator. The rollers or sprags are fitted between the two races and will allow the outer race to rotate in one direction only. The stator locks in the opposite direction of turbine rotation. When the stator is turned in the same direction as turbine rotation, the rollers are free between the races and the stator is able to turn.

The fluid leaving the turbine has to pass through the stator blades before reaching the impeller. In passing through the stator, the direction of fluid flow is reversed by the curvature of the stator blade. The fluid now moves in a direction that aids in the rotation of the impeller. The impeller accelerates the movement of the fluid and it now leaves with nearly twice the energy and exerts a greater force on the turbine. This action results in a torque multiplication.

It is vortex flow that allows for torque multiplication. Torque multiplication occurs when there is high impeller speed and low turbine speed. Low turbine speed and the stator cause the returning fluid to have a high-velocity vortex flow. This allows the impeller to rotate more efficiently and increase the force of the fluid pushing the turbine in rotation. When an increase in torque is required, the driver depresses the throttle, which increases engine and impeller speed. This creates a greater speed difference between the turbine and the impeller causing vortex flow and an increase in torque multiplication.

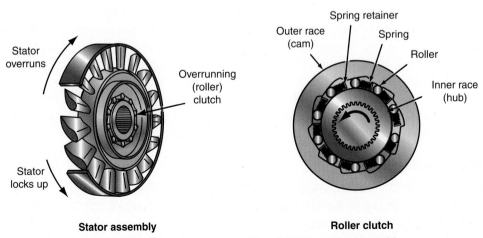

FIGURE 6-12 The one-way overrunning clutch in a stator assembly.

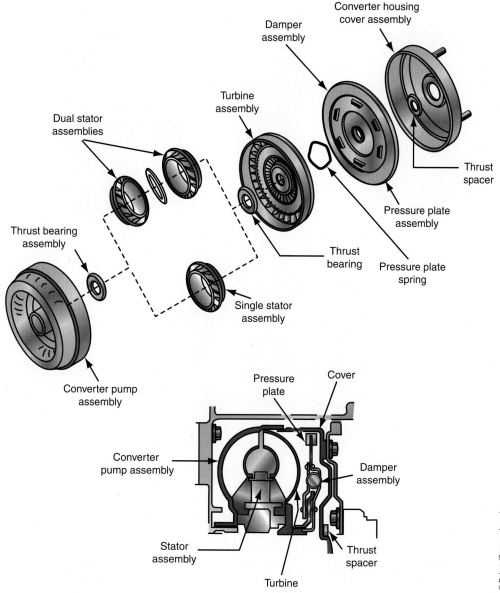

FIGURE 6-13 Typical construction of a torque converter. Note that this converter may be equipped with two stator assemblies for additional torque multiplication.

As the vortex flow slows down, torque multiplication is reduced. Torque multiplication is obtained anytime the turbine is turning at less than 90 percent of impeller speed. Most automotive torque converters are capable of maximum torque multiplication factors ranging from 1.7:1 to 2.8:1 (Figure 6-13).

Coupling Phase

Torque multiplication occurs because of the redirection of the fluid flow by the stator. This only takes place when the impeller is rotating faster than the turbine. As the speed of the turbine increases, the direction of the flow changes and there is less multiplication of torque. When the speed of the turbine nearly equals the speed of the impeller, fluid flows against the stator vanes in the same direction as the fluid from the impeller (Figure 6-14). This releases the one-way clutch and allows the stator to rotate freely. At this point, there is little vortex flow and the engine's torque is carried through the converter by the rotary flow of the fluid. This is the *coupling phase* of the torque converter and no torque multiplication takes place.

The amount of torque multiplication a converter will produce is sometimes called the converter ratio.

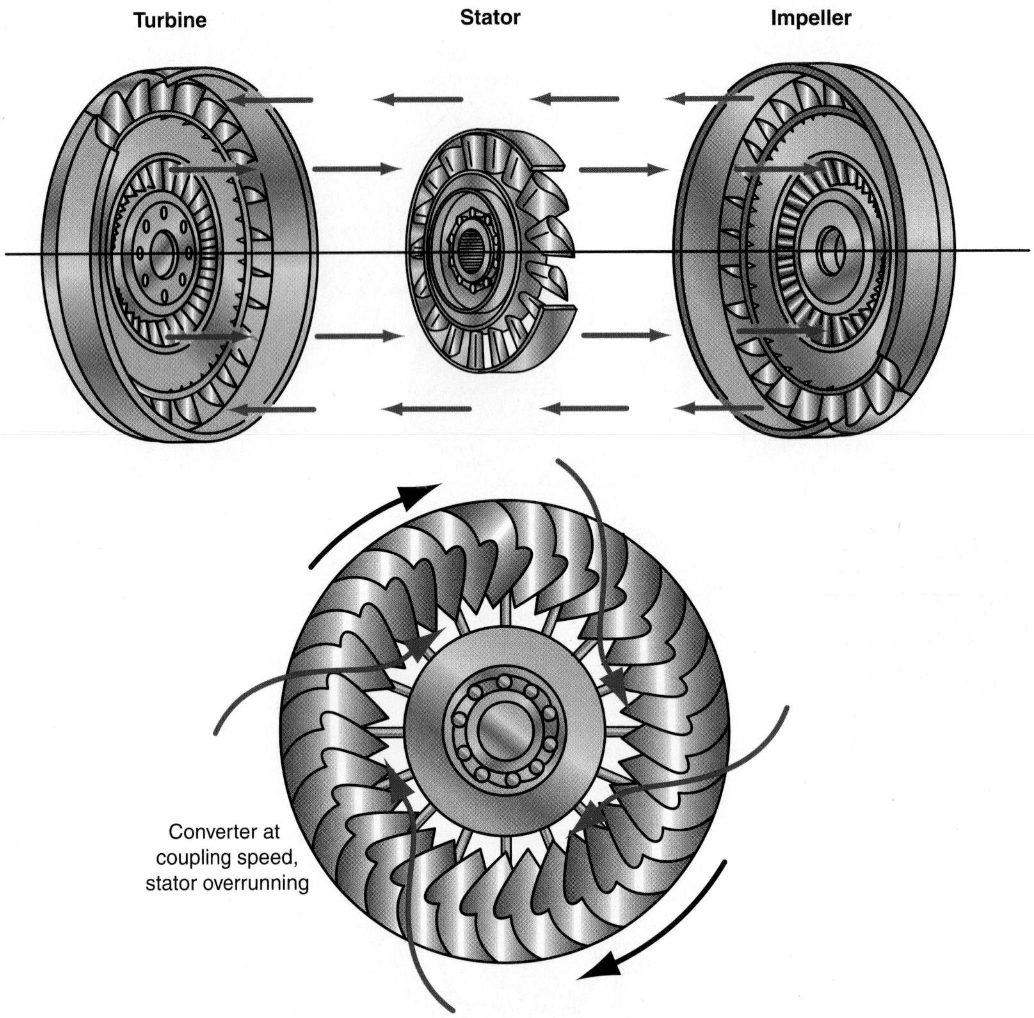

Turbine   Stator   Impeller

Converter at
coupling speed,
stator overrunning

FIGURE 6-14 **Fluid flow through the converter when the coupling phase has been achieved and the stator is overrunning.**

When the speed of the turbine reaches approximately 90 percent of the impeller's speed, coupling occurs. During coupling, the converter is acting like a fluid coupling transmitting engine torque to the transmission. The coupling phase does not occur at a specific speed or condition. It occurs whenever the speed of the turbine nearly equals the speed of the impeller.

Stall speed

The condition called stall occurs when the turbine is held stationary and the impeller is spinning. **Stall speed** is the fastest speed an engine can reach while the turbine is at stall. Some stall occurs every time the vehicle begins to move forward or backward, as well as each time the vehicle is brought to a stop. Today, most torque converters have a stall speed of 1200–2800 rpm. Torque converters with a high stall speed are normally used with the less powerful engines. These converters allow the engine to operate at higher speeds when they are making the most power. Low stall speed converters are normally used with engines that produce a great amount of torque at low speeds.

A torque converter is not a very efficient device. This is especially true at stall speeds when all of the engine torque that enters the torque converter is lost as heat and no power is inputted into the transmission. From stall speed to the coupling speed, the efficiency of a converter increases to approximately 90 percent.

Stall speed is dictated by the design of the torque converter. The diameter of the converter, the angles of the impeller, turbine, and/or stator vanes, and the clearance between the impeller and the turbine fins will affect the stall speed.

Shop Manual
Chapter 6, page 292

Stall speed
represents the
highest engine
speed attained with
the turbine stopped
(at stall).

© Delmar/Cengage Learning

The diameter of the converter affects stall speed because it determines the distance the fluid travels inside the converter. Also, the speed of the fluid inside a large converter is higher, therefore the converter will lockup at lower speeds.

Small diameter torque converters have high stall speeds and multiply torque at high engine speeds, but they do not couple until high engine speeds. Large diameter converters offer low stall speeds and torque multiplication at lower speeds. It is said that a one-inch increase or decrease in the diameter of a converter will change the stall speed by 30 percent.

The angle of the impeller and turbine vanes also influence the stall speed of a converter. The vanes can be angled forwards (positive angle) or backwards (negative angle). A forward angle produces higher fluid speeds; therefore, the stall speed tends to be lower. When the vanes have a backward angle, the stall speed increases.

The angle of the stator vanes also changes capacity and stall of the converter. As the angle of the vanes increases, more fluid will return to the impeller and stall speed will increase. Changing the vane angle of an impeller, turbine, or stator is a common way to change torque converter capacity.

A converter with a high stall speed is sometimes referred to as a loose converter and one with a low stall speed as a tight converter.

Converter Capacity

The capacity of a torque converter is an expression of the converter's ability to absorb and transmit engine torque with a limited amount of slippage. It also reflects the fluid volume of the converter. A low-capacity converter has a high stall speed but allows for a relatively large amount of slippage during the coupling phase. However, they also provide for more torque multiplication and better acceleration. A high-capacity converter allows for less slippage but has a low stall speed. This type of converter is very efficient at highway speeds but offers low torque multiplication. The maximum torque capacity of a converter is determined by the converter's ability to dissipate heat and the materials used in its construction.

Using a torque converter beyond its capacity will cause it to slip. When the converter continuously experiences large amounts of slippage, the converter may overheat. Overheating can cause a number of problems, such as destroyed seals, seized or broken stator clutch, fin deformation or separation, or converter ballooning.

Performance Torque Converters

Changing the torque converter to gain performance is a common modification. Choosing the correct converter should only be done after evaluating the vehicle and its normal operation. Ideally, the torque converter should reach its stall speed at the same time the engine is producing a large amount of torque. The engine's torque curve changes whenever the engine has been modified. When changes have been made to the fuel system, camshaft, displacement, and engine control systems, a higher stall speed is required to take advantage of the performance gains. The higher stall speed will allow for more torque multiplication before the converter couples. The higher stall speed will also allow the converter to slip more at idle which means engines with radical camshafts will idle better.

The correct converter is also determined by the transmission's gear ratios, the final drive ratio, tire diameter, engine size, and the vehicle's weight.

Remember, the higher the stall speed, the higher the torque multiplication and the better the acceleration. In fact, the heavier the vehicle, the more it will benefit from a higher stall speed. But with a higher stall speed, the engine's speed must be higher before the converter can couple and the vehicle can move. This can be annoying in every day driving and will increase fuel consumption.

High-performance cars are often fitted with a small diameter torque converter to raise the stall speed and to use the power of high engine speeds.

Most vehicles come with a 12–13 inch diameter torque converter from the factory. These provide a stall speed of 1600–1800 rpm. They are designed to provide good driveability and fuel economy. Performance torque converters are normally grouped by diameter and stall speed. The most common sizes are 11-, 10-, 9-, and 8-inch converters. As the diameter decreases, the stall speed increases. The 8- and 9-inch converters are designed for race only vehicles with highly modified engines. For street performance, the best choice is a 10- or 11-inch converter. This will provide increased acceleration and reasonable drive-ability. They will increase the stall speed from 200 to 2000 rpm. Again, the stall speed should be matched to the engine's torque curve.

A slight increase (200–300 rpm) in stall speed is also better for vehicles that tow. The converter will allow slightly more slip and torque multiplication.

Vehicles built for off-the-road use benefit from a slightly lower stall speed. But the converter of choice should also provide more torque multiplication.

Performance and heavy duty converters also are built to withstand loads. Their impeller and turbine are often strengthened by a process called furnace brazing. Furnace brazing is a process that forces molten brass into seams and joints. This provides a strong bond between the internal parts of the converter.

> **AUTHOR'S NOTE:** Torque converters generate most of the heat that is inside a transmission. When the stall speed is increased, there is a proportional increase in the heat that is produced. To dissipate the heat and protect the transmission, an auxiliary fluid cooler should be added whenever the stall speed is changed.

Cooling the Torque Converter

Slippage in a torque converter results from a loss of energy. Most of these losses are in the generation of frictional heat. Heat is produced as the fluid hits and pushes its internal members. To maintain the efficiency that it has, the fluid must be cooled. Excessive fluid heat would result in even more inefficiencies.

The transmission's oil pump continuously delivers ATF into the torque converter through a hollow shaft in the center of the torque converter assembly (Figure 6-15). A seal is used to

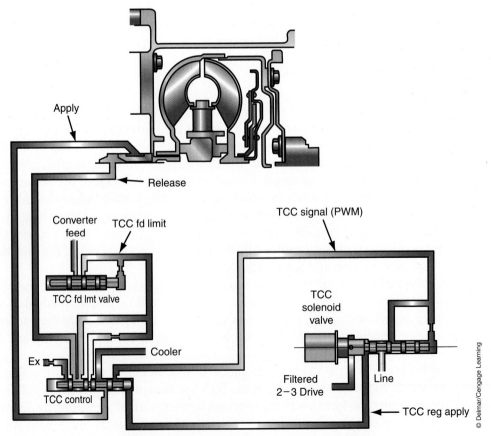

FIGURE 6-15 Typical oil circuit for converter and oil cooler.

prevent fluid from leaking at the point where the shaft enters the converter. The fluid is circulated through the converter and exits past the turbine through the turbine shaft, which is located within the hollow fluid feed shaft. From there, the fluid is directed to an external oil cooler and back to the transmission where it is distributed to lubrication bores in the input and output shafts. The cooled fluid is now ready to efficiently lubricate and cool all of the transmission's components. The cooling circuit also ensures that the fluid flowing into the converter is cool enough to maintain the converter's normal efficiency.

DIRECT DRIVE

Up to 10 percent of the engine's energy is wasted in a pure fluid connection. This wasted energy is in the form of heat. Torque converter slippage or the speed difference between the impeller and turbine during the coupling phase serves as evidence of the wasted energy. To eliminate this slippage, most late-model vehicles have a torque converter clutch that mechanically links the engine to the input of the transmission during some operating conditions. The results of using a converter clutch are improved fuel economy and reduced transmission fluid temperatures.

Manufacturers have adapted automatic transmissions to provide a direct mechanical drive for input through the use of centrifugally applied clutches, hydraulically operated clutches, and planetary gearsets in the torque converter. The use of planetary gearsets is not as common as the use of clutches. The most common converter clutch systems rely on electronics to control a hydraulically operated clutch.

When applied, the converter clutch connects the turbine to the cover of the converter. Application of the clutch occurs at various operating speeds, depending on the model of the vehicle and the driving conditions. Most converters equipped with a clutch consist of the three basic elements: impeller, turbine, and stator, plus a piston and clutch plate assembly, special thrust washers, and roller bearings.

The piston and clutch plate assembly has frictional material on the outer portion of the plate with a spring-cushioned **damper** assembly in the center. The clutch plate is splined to the turbine shaft and when the piston is applied, the plate locks the turbine to the converter (Figure 6-16). The thrust washers and roller bearings control the movements of and provide bearing surfaces for the components of the converter.

A BIT OF HISTORY

Packard's Ultramatic of 1949 featured a torque converter clutch, but it was discontinued in 1957 due to its jerky action.

The converter clutch assembly is commonly called a lockup clutch.

Shop Manual
Chapter 6, page 286

A **damper** is a device that absorbs vibrations and noise.

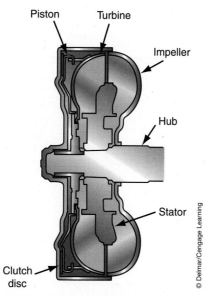

FIGURE 6-16 Piston-type converter clutch assembly.

Converter apply or encouragement only occurs when the powertrain control module (or other control) determines that conditions are right for direct drive. The converter clutch should not connect the engine to the transmission when torque multiplication from the torque converter is needed, such as during acceleration. The clutch should be disengaged during braking; this prevents the engine from stalling. Likewise, the clutch should be disengaged when the vehicle is at a standstill while the transmission is in gear. Normally, during deceleration, the clutch should not be applied. If the clutch is applied during deceleration, fuel may be wasted and there may be high exhaust emissions. The clutch should be applied only when the engine is warm enough and is running at a great enough speed to prevent a shudder or stumble when it engages. The control systems use inputs from many different sensors to determine when the conditions are suitable for clutch engagement.

Hydraulic Converter Clutch

Shop Manual

Chapter 6, page 283

The most common way to provide for a mechanical link between the engine and the transmission is through the use of a hydraulically operated clutch. Introduced in the late 1970s, this design of converter clutch was controlled and operated totally by hydraulics and provided a mechanical link only when the transmission was in high gear. Today's transmissions apply the clutch in more than one forward gear and rely on computer-controlled electrical solenoids to control the hydraulic pressure to the clutch assembly (Figure 6-17).

The operation of the hydraulic clutch is rather simple. However, the systems that control it can be fairly complex. The converter clutch is usually applied when the fluid flow through the torque converter is reversed by a valve. When the torque converter clutch control valve moves, the fluid begins to flow in a reversed direction (Figure 6-18). This forces the clutch disc or pressure plate against the front of the torque converter's cover. The position of the clutch now blocks the fluid flow through the converter and a mechanical link exists between the impeller and the turbine.

Normally, when the converter clutch is not engaged, fluid flows down the turbine shaft, past the clutch assembly, through the converter, and out past the outside of the turbine shaft.

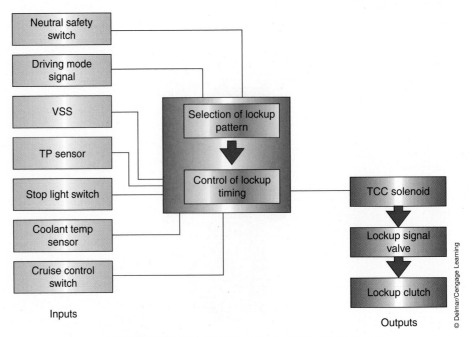

FIGURE 6-17 A basic T/C lockup control system.

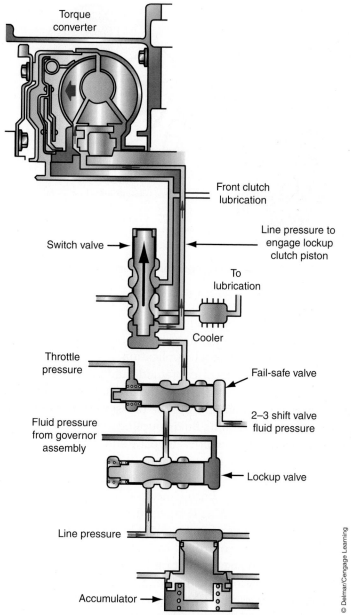

FIGURE 6-18 Lockup control valve operation to produce total clutch engagement in a piston-type converter clutch assembly.

Normal torque converter pressures keep the clutch firmly against the turbine. There is no mechanical link and only a fluid coupling exists.

To engage the clutch, mainline pressure is directed between the plate and the turbine. This forces the plate into contact with the front inner surface of the cover and locks the turbine to the impeller. With the engagement of the clutch, the fluid in front of the clutch is squeezed out before the clutch is totally engaged. The presence of this fluid softens the engagement of the clutch and acts much like an accumulator.

Viscous Clutch

A viscous converter clutch (VCC) is used on some older automobiles to provide torque converter clutch engagement. This design allows the clutch to engage in a very smooth manner with no engagement shock. It operates in the same way as a hydraulically applied clutch

The viscous converter clutch system from General Motors is referred to as the VCC system.

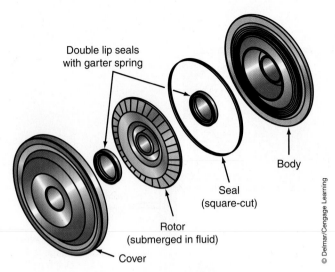

Double lip seals
with garter spring

Body

Seal
(square-cut)

Rotor
(submerged in fluid)

Cover

© Delmar/Cengage Learning

FIGURE 6-19 A viscous-type converter clutch assembly.

Late-model
Cadillacs with 4T60
transmissions are
the most common
vehicles equipped
with a VCC.

except that it uses a *viscous clutch* assembly to lock the impeller to the turbine. The viscous clutch assembly consists of a rotor, body, clutch cover, and silicone fluid (Figure 6-19). The *silicone fluid* is sealed between the cover and the body of the clutch assembly. It is this viscous silicone fluid that cushions the feel of clutch application.

The viscous converter clutch is a self-contained fluid coupling with a built-in friction-faced pressure plate. When the clutch is engaged, the pressure plate is forced against the converter cover. Engine power is transmitted from the pressure plate through the fluid coupling to the transaxle's input shaft. The clutch's fluid coupling uses the viscous properties of thick silicone fluid between the closely spaced pressure plate and cover plate to transmit the power.

When the clutch is applied, there is a constant but minor amount of slippage in the viscous unit, about 40 rpm at 60 mph. However, this slippage is nothing compared to a conventional torque converter without lockup. Engagement of the viscous clutch allows for improved fuel economy and reduced fluid operating temperatures. When the clutch is disengaged, the assembly operates in the same way as a conventional torque converter.

The viscous converter clutch is controlled by a solenoid, which is controlled by a computer. The computer bases clutch engagement on conditions such as vehicle speed, throttle angle, transmission gear, transmission fluid temperature, engine coolant temperature, outside temperature, and barometric pressure. When the computer determines it is time to engage the clutch, it completes the ground circuit to the viscous clutch solenoid.

Clutch Assembly

Shop Manual
Chapter 6, page 291

A typical hydraulic converter clutch has a plate that acts as a clutch piston and is splined to the front of the turbine. Friction material is usually bonded to either the forward face of the clutch plate or the inner front surface of the converter cover. A large ring of friction paper is the typical friction material used for a converter clutch.

A converter clutch disc is splined to the turbine so that it can drive the turbine when the frictional material is forced against the torque converter's cover. Most converter clutch assemblies are fitted with a damper assembly that directs the power flow through a group of coil springs. The damper springs are either located at the center of the disc or they are grouped around the outer edge of the disc. These springs are designed to absorb the normal torsional vibrations of an engine, which vary with changes in speed and load.

Torsional vibrations
are slight speed
fluctuations that
occur during normal
crankshaft revolutions. These vibrations can produce
gear noise in a
transmission, as
well as a noticeable shaking of the
vehicle.

The *isolator springs* are placed evenly from the center of the disc and are sandwiched between two steel plates and riveted together. One of the plates is attached directly to the clutch assembly's hub and the other to the clutch disc. The two plates move as a single unit after the plates have moved against the tension of the isolator springs. During clutch engagement, the sudden application of torque to one plate is absorbed by the springs as they compress and start the other plate turning. The damper assembly acts as a shock absorber to the pulses of the engine's vibrations and softens the engagement of the clutch. When the converter clutch is not applied, the fluid inside the torque converter absorbs the torsional vibrations. Therefore, a damper assembly is not needed until the clutch mechanically connects the engine to the input of the transmission.

CONVERTER CONTROL CIRCUITS

Many different systems have been used to control the application of the converter clutch. The systems vary from simple hydraulic controls at the valve body to complex computer-controlled solenoids (Figure 6-20). Simple systems limit the engagement of the clutch in high gear only when the vehicle is traveling above a particular speed. More complex systems allow the clutch to engage or partially engage whenever they decide efficiency would be improved by engaging the clutch.

Shop Manual
Chapter 6, page 284

Most of the computer-controlled systems base clutch engagement on inputs from various sensors, which give information about the engine's fuel system, ignition system, vacuum, operating temperature, and vehicle speed. This information allows the computer to engage the clutch at exactly the right time according to operating conditions.

The clutch is typically engaged by a clutch piston controlled by an electric solenoid and one or more spool valves. The solenoid controls pressurized fluid that moves the spool valve to move the clutch piston. The movement of the piston engages or disengages the converter clutch. When oil pressure moves the piston against the clutch disc, lockup occurs. As the piston moves away from the disc, the clutch unlocks.

A quick look at some of the systems used by different manufacturers shows the similarities in the systems. In spite of the different systems, they all function to provide a mechanical link between the engine and the transmission.

Chrysler Converter Controls

Early Chrysler electronically controlled lockup converter systems used an electric solenoid mounted on the valve body to engage and disengage the converter clutch. The operation of this solenoid is controlled by the engine control computer through a relay. The computer determines when to engage or disengage the torque converter clutch based on information from the coolant temperature sensor, vacuum transducer, neutral safety switch, vehicle speed sensor, and carburetor or throttle ground switch. This information allows for precise converter clutch control, which leads to improved fuel economy, reduced transmission temperature, and reduced engine speeds.

Electronic control for the transmission was incorporated into the existing electronic combustion computer program for engine control. The system uses sensors already developed for

A solenoid is an electric device capable of converting electrical energy into mechanical force.

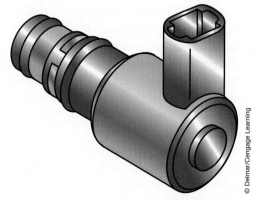

© Delmar/Cengage Learning

FIGURE 6-20 A typical torque converter clutch solenoid valve.

use with the PCM to provide inputs for control of the torque converter clutch. Prior to controlling the clutch electronically, Chrysler relied on hydraulic pressure to control the clutch. These hydraulic controls allowed clutch engagement when the transmission was in third gear only and operating above a preset speed, regardless of load.

With electronic control, clutch equipment is still preset to the optimum speed, but the solenoid-operated hydraulic valve is controlled by the computer; therefore, the actual time of clutch engagement is controlled by the computer.

Prior to engaging the converter clutch, the computer looks at engine coolant temperature. If the temperature is above 113°F, the computer scans other information from its sensors. If the temperature is below 113°F, the computer keeps the TCC solenoid off until the temperature is reached. Once this temperature is reached, the computer looks at vehicle speed, engine vacuum, and throttle information to determine if the clutch should be engaged.

The vehicle speed sensor located on the speedometer cable sends an electronic signal to the computer, which reports that vehicle speed is above or below 40 mph. If the vehicle has reached the preset engagement speed, the computer will make sure the throttle is open. This prevents converter clutch engagement during coasting.

If the throttle is closed, regardless of other conditions, the clutch will remain off. When the throttle is open, the computer will wait a short time after it has been opened before it proceeds to engage the clutch.

The final factor that is looked at by the computer is engine vacuum. A *vacuum transducer* relays load information to the computer. Since engine vacuum relates to engine load, the computer will allow clutch engagement only during no-load or low-load conditions. The engine's vacuum must be above 4 inches and below 22 inches.

If all conditions are met, the computer energizes the clutch relay, which grounds the TCC solenoid. The activation of the solenoid moves the solenoid check ball against its seat. This prevents the solenoid valve from exhausting line pressure. Line pressure increases and forces the switch valve to move against coil spring tension. Pressure is then directed to the impeller drive hub and stator support to fill the torque converter with fluid. Line pressure flows from the impeller and turbine to fill the space behind the torque converter clutch piston. This high pressure forces the engagement of the clutch.

To disengage the clutch, the solenoid is turned off. This moves the solenoid check ball away from its seat and fluid is exhausted. This action redirects the pressure inside the converter and relaxes the force on the clutch piston, thereby allowing the clutch to disengage.

Chrysler added a three-valve module to the transmission's valve body to provide for control of the lockup clutch's piston. This is controlled by governor pressure. When the vehicle's speed reaches a particular point, governor pressure forces the valve to move against the tension of its spring. The movement of the valve permits pressure to move to the fail-safe valve. When the transmission is in third gear, fluid flows to the **fail-safe valve** and moves it. Line pressure is now directed between the turbine shaft and the stator support to fill the torque converter and engage the clutch.

If the throttle is quickly opened during clutch apply, throttle pressure increases, causing the fail-safe valve to move and block the line pressure passage to the converter clutch valve. The clutch is disengaged as the fluid is redirected to the other side of the clutch's piston assembly.

Later-model Chrysler converter clutch systems operate in much the same way. However, as engine control computers became more complex with the addition of new sensors, so did converter clutch control. Slightly different conditions must be met before the computer will engage the clutch. The determinative factor, however, is the relationship between throttle position and vehicle speed.

The TCC solenoid is constantly receiving pressurized fluid. When it is deenergized, it exhausts the pressure, which prevents a buildup of pressure in the converter. When the computer completes the ground circuit for the solenoid, the exhaustion of fluid is stopped and the pressure builds. This buildup of pressure causes the engagement of the clutch.

Five conditions must be present before the computer will order clutch engagement.

1. The temperature of the ATF must be at least 113°F.
2. The park/neutral switch must indicate that the transmission is in a forward (3, 4, or D) gear.
3. The brake switch must indicate that the brakes are not applied.
4. The vehicle must be traveling above 35 mph with a steady throttle.
5. The signals from the throttle position sensor must indicate an off idle condition.

The computer also looks at the rate of change from the TP sensor. This rate indicates the driver's intent and helps the computer define the actual operating conditions. For example, if the vehicle is traveling at 55 mph and the voltage from the TP sensor changes less than 1.74 volts while holding that speed, the solenoid will be activated and the converter clutch engaged. If the vehicle is traveling above 45 mph, the clutch will be disengaged whenever the TP sensor rate of change decreases by more than 0.040 volts per 11 milliseconds.

Most late-model Chrysler transmissions and transaxles have a torque converter clutch that can engage in second, third, and overdrive gear. The Electronic Modulated Converter Clutch (EMCC) system allows the converter clutch to partially engage during certain operating conditions. This feature limits the amount of converter slippage and serves to limit the amount of engine vibration that is transferred to the transmission. It is also used to control transmission temperature. The TCM can duty cycle the solenoid to provide a smooth partial or full application of the clutch. Full torque converter engagement occurs when the input shaft is within 60 rpm of the crankshaft.

Partial engagement will occur only when the TCM sees that the vehicle is operating in conditions that are suitable. To determine the condition, the TCM looks at the position of the gear selector, the current gear range, transmission fluid temperature, engine coolant temperature, input shaft speed, engine speed, and throttle opening.

When the TCM determines that partial engagement would be beneficial, it modulates the solenoid. Partial engagement is maintained until full engagement is initiated by the TCM. Partial engagement normally occurs at low speeds, low load, and light throttle conditions. During partial engagement some slip is allowed.

Full engagement is initiated once certain conditions are met, one of which is the amount of slip that resulted from partial engagement. When the difference between engine and input shaft speeds is within a specified range, the TCM increases the duty cycle of the solenoid to cause full engagement.

The system includes a logic-controlled solenoid torque converter clutch control valve (Figure 6-21). This valve makes sure the converter clutch is never engaged during first-gear operation. The valve also redirects the fluid for first gear to the converter clutch valve. In order to downshift into first gear, the EMCC must receive information that the clutch is disengaged.

A TP sensor is a sensor that monitors throttle position.

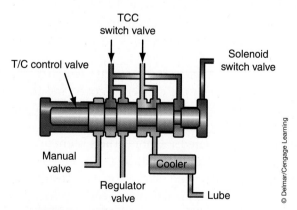

FIGURE 6-21 Controls for the fluid flow to and from the converter clutch assembly.

Ford Converter Clutch Controls

The electronically controlled hydraulic converter clutch circuits in Ford Motor Company products are controlled by the PCM. This is the main processing device for the engine control system. The PCM receives information from various sensors and switches and generates output signals to control air/fuel mixture, emission controls, idle speed, ignition timing, and transmission operation. In addition, the system controls A/C compressor clutch operation, idle speed, transmission converter clutch operation, and third to fourth gear shifting on some models. Ford has used four slightly different system designs to control converter clutch engagement, each based on particular transmission models. The main sensors for each of these designs are basically the same, however the outputs are different.

During certain conditions, the PCM sends the appropriate signal to the TCC solenoid, which allows fluid pressure within the torque converter to force the piston plate and damper assembly against the cover, creating a mechanical link between the engine's crankshaft and the input shaft of the transaxle.

The TCC solenoid is a **pulse width** modulated (PWM) style solenoid. The solenoid is used to control the apply and release of the bypass clutch in the torque converter. By modulating the pulse width of the solenoid, the pressure in the clutch circuit varies, thereby modulating the apply and release of the bypass clutch in the torque converter. This allows for partial clutch engagement during some operating conditions and for smooth engagement.

Some converter clutches are controlled by only a converter clutch control valve and a solenoid. The converter clutch is controlled by a converter clutch bypass solenoid, which vents line pressure to the oil pan when the converter clutch is released. The solenoid also redirects line pressure into an apply passage when the solenoid is energized to apply the clutch. The solenoid is energized by the completion of its ground circuit. Battery voltage is always applied to it when the ignition is on. If the required operating conditions exist and the vehicle is traveling at approximately 35 mph (56 km/h), the clutch is energized. The converter clutch cannot be applied when the transmission is in first or second gear.

Engine coolant temperature and throttle position are monitored by the PCM and are used to determine if the converter clutch should be engaged. Before the converter clutch can be engaged, regardless of other conditions, the engine temperature must be at least 75°F (24°C). After the engine is warmed up and when the transmission is in third gear, the clutch will only be engaged if the throttle opening is greater than 8 percent and less than 59 percent. While in fourth gear, the throttle opening must be between 5 and 45 percent before the clutch is engaged. The PCM will also not allow clutch engagement, in any gear, if the throttle is closed, quickly opened, or opened wide.

When the converter clutch is disengaged, no current flows through the solenoid preventing application of the converter clutch. When the solenoid is off, converter fluid moves from the converter regulator valve through the converter clutch control valve into a release passage. The release fluid then moves between the impeller and turbine shafts pushing the converter clutch piston plate away from the converter cover. This releases the converter clutch (Figure 6-22).

The converter clutch is applied when the converter clutch bypass solenoid is energized. This causes the release fluid between the converter cover and the clutch piston plate to be exhausted and allows fluid pressure to be applied to the back of the piston plate causing it to contact the converter cover, thus locking the clutch to the cover (Figure 6-23).

Other transmissions use modulated solenoid. The system uses a pulse width–modulated solenoid allows the PCM to vary clutch operation from full release to controlled slip to full apply. The PCM turns the solenoid on and off at a constant **frequency**. If greater output pressure is needed, the pulse width is lengthened during each cycle. When less output pressure is needed, the pulse width is shortened. By varying the pulse width, the PCM can permit clutch slippage to obtain the best combination of performance and economy. The solenoid is often referred to as the Modulated Converter Clutch Control (**MCCC**) solenoid.

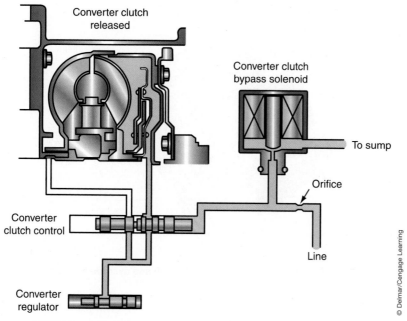

FIGURE 6-22 Fluid flow through clutch circuit when it is disengaged.

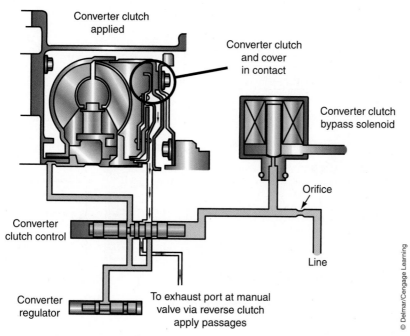

FIGURE 6-23 Fluid flow through clutch circuit when it is engaged.

The converter clutch in other transmissions is applied and released hydraulically but can be overridden electronically. The override solenoid (Figure 6-24) is normally energized to prevent fluid flow from disengaging the converter clutch.

A check valve is positioned in the circuit between the override solenoid and the converter clutch shift valve (Figure 6-25). This check valve operates according to line pressure and spring tension. When governor pressure is low, line pressure from the check valve keeps the converter clutch in the downshift position and the torque converter remains disengaged.

As governor pressure increases, it moves the converter shift valve up which exhausts fluid from the spring end of the check valve. This directs line pressure through the converter apply circuit. If governor pressure decreases enough, the converter clutch shift valve moves back down and the check valve returns to its normal position.

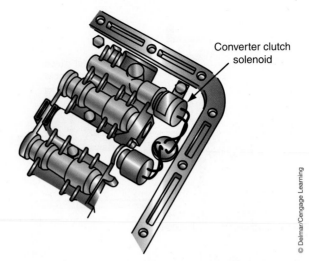

Converter clutch
solenoid

FIGURE 6-24 Location of converter clutch and override solenoids.

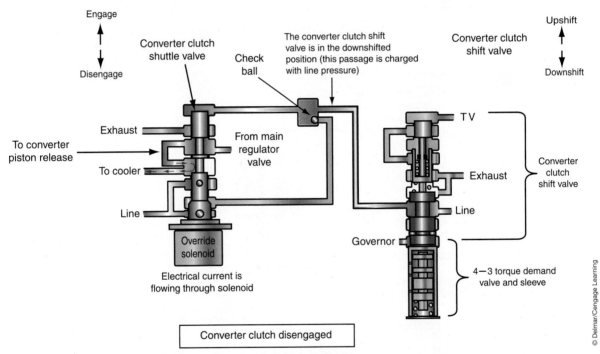

Engage ↑ Disengage

Converter clutch
shuttle valve

Check
ball

The converter clutch shift
valve is in the downshifted
position (this passage is charged
with line pressure)

Converter clutch
shift valve

Upshift ↑ Downshift

Exhaust

From main
regulator
valve

To converter
piston release

To cooler

Line

Override
solenoid

Electrical current is
flowing through solenoid

TV

Exhaust

Line

Governor

Converter
clutch
shift valve

4—3 torque demand
valve and sleeve

Converter clutch disengaged

FIGURE 6-25 Oil circuit and flows when the clutch is released or not connected to the converter's cover.

The override solenoid is energized when the PCM detects a condition unfit for converter engagement. To release the converter clutch, the PCM deenergizes the override solenoid. This allows line pressure to flow through the solenoid and into the TCC inhibition circuit. Pressure in the circuit moves the two-way check ball installed between the check valve and the converter shift valve to block the passageway. This allows the converter clutch shift valve to remain in the upshifted position, but it is ineffective. The check valve returns to its released position and the converter apply fluid is exhausted. These actions release the clutch.

General Motors' Converter Clutch Controls

The TCC assembly allows for a mechanical link between the impeller and the turbine during all gear ranges except: Park, Reverse, Neutral, and Drive Range-First Gear. On some late-model vehicles, the converter clutch is engaged and disengaged by a solenoid (Figure 6-26). This solenoid is controlled by the brake switch, third gear clutch pressure switch, and the

Shop Manual
Chapter 6, page 288

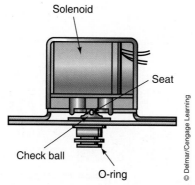

FIGURE 6-26 Typical clutch solenoid valve assembly.

computer command control system. The computer controls the application of the clutch by providing a ground circuit for the TCC solenoid circuit. This solenoid assembly is energized by the PCM to redirect ATF to the clutch apply valve in the converter clutch control valve assembly. In some older vehicles, converter clutch operation is controlled by the converter clutch shift valve and the solenoid, both of which are located together.

When the TCC solenoid's ground circuit is completed by the PCM, fluid pressure is exhausted from between the converter's pressure plate and the converter cover, and converter clutch apply pressure from the converter clutch regulator valve pushes the converter pressure plate against the converter cover to apply the converter clutch. The apply feel of the clutch is controlled by the converter clutch regulator valve and the converter clutch accumulator (Figure 6-27).

When the TCC solenoid is deactivated, fluid pressure is applied between the converter cover and the pressure plate. Converter feed pressure from the pressure regulator valve passes through the converter clutch apply valve into the release passage. This fluid is directed between the pump shaft and the turbine shaft and pushes the pressure plate away from the converter's cover to release the converter clutch (Figure 6-28).

Various sensors provide input to the PCM for the control of the converter clutch (Figure 6-29). Not all models have the same sensors, but all systems operate in a similar way.

Computer command control is the common name given to GM's early engine control computer system.

The converter clutch shift valve is called the TCC shift valve and it controls the application and release of the clutch according to the postion of the TCC regulator valve.

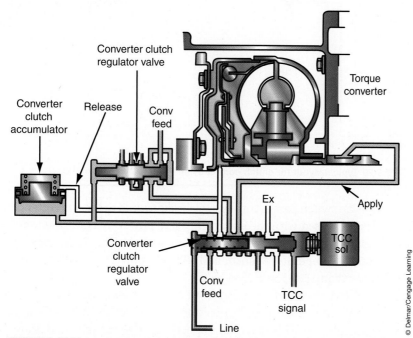

FIGURE 6-27 The converter clutch apply feel is controlled by the converter clutch regulator valve and the converter clutch accumulator.

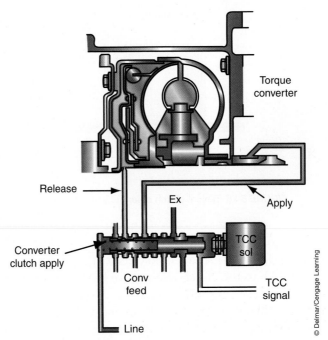

FIGURE 6-28 Oil circuit for a typical GM hydraulic converter clutch system.

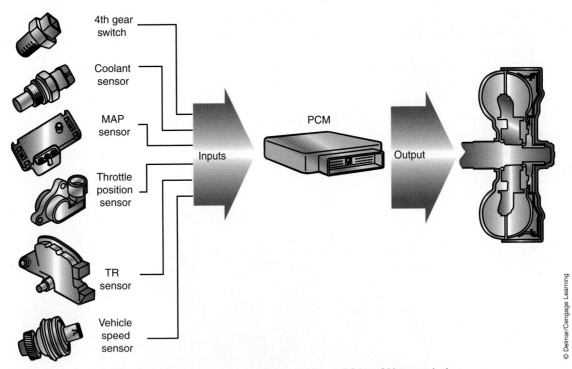

FIGURE 6-29 Typical layout for clutch control by an PC in a GM transmission.

Power from the ignition switch passes through the brake switch to the converter clutch solenoid. If the brake pedal is depressed with the clutch engaged, power to TCC solenoid is interrupted. This releases the clutch to prevent the engine from stalling. The PCM receives information from the engine coolant temperature sensor and will not allow TCC operation until the coolant temperature is higher than 130–150°F (54–66°C).

The computer also receives information from other sensors, such as the TP sensor, which provides the PCM with information on the position of the throttle. TCC operation is prevented when the throttle position signal is less than a specified value. Engine load is also looked at by the computer through inputs received from vacuum sensors, such as a MAP sensor.

Vehicle speed is a critical input, especially when the speed is compared to the other inputs. This combination of inputs can clearly define the exact operating conditions of the vehicle. Vehicle speed must be greater than a certain value before the TCC can be applied.

To prevent converter clutch engagement in first and second gears, some systems have a third and fourth gear switch. This switch prevents TCC operation until the transmission is operating in third or fourth gear. The position of this switch may be monitored by the PCM or the switch may be wired in series with the power feed to the TCC solenoid. When the switch is open, no power will flow to the solenoid.

Most late-model GM transmissions feature their Electronic Controlled Capacity Clutch (ECCC) system. The ECCC allows for a variable amount of clutch slippage. The amount of slippage allowed depends on what is needed to dampen engine vibrations and allow for smooth gear changes. The transmission and torque converter in GM hybrid vehicles have been modified to work in this application. The torque converter was made smaller in order to fit the transmission housing and uses a dual-plate torque converter clutch. The controls for the clutch also were changed to accommodate the action of the electric motors.

Honda/Acura Converter Clutch Controls

Late-model Hondas and Acuras use an electronically controlled five-speed transmission. The torque converter clutch may be off or partially or fully engaged, depending on conditions. The TCC is controlled by the transmission's hydraulic system and the TCM (Figure 6-30).

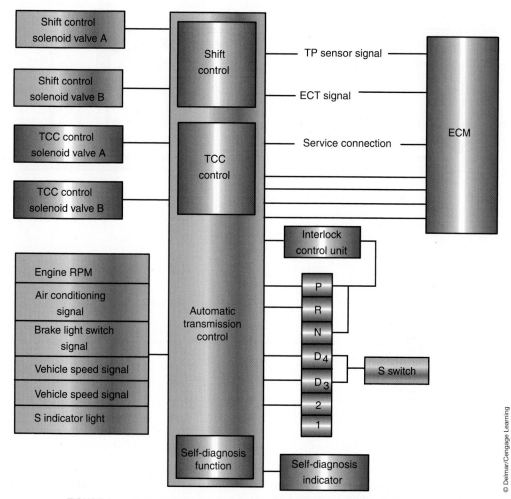

FIGURE 6-30 Typical layout for clutch control in a late-model Honda Transaxle.

The clutch can lockup when the gear selector is in the D position and the transmission is operating in second, third, fourth, or fifth gear.

During lockup the TCM energizes the TCC solenoid and the clutch pressure control solenoid. The clutch pressure control solenoid valve regulates and applies hydraulic pressure to the lockup control valve to control the amount of the lockup. When the clutch is applied, pressurized fluid is discharged from behind the torque converter and the TCC piston is pushed against the torque converter cover (Figure 6-31). The TCC solenoid valve is mounted on the torque converter housing and the clutch pressure control solenoid valve is mounted on the transmission housing. Both are directly controlled by the PCM.

When the clutch is disengaged, the fluid in the converter flows in the opposite direction. As a result, the TCC piston is moved away from the converter cover. The pressurized fluid regulated by the modulator is present at both ends of the TCC shift valve and at one side of the clutch control valve. With equal pressure present at both ends of the TCC shift valve, the shift valve is moved to the right by the tension of the valve spring alone. This causes fluid to flow to the backside of the pressure plate so the clutch is pushed away from the converter's cover.

As the vehicle's speed reaches a specified point, the TCM turns on the TCC solenoid. This action releases the modulator pressure from one side of the clutch shift valve and sends pressurized fluid to the other side of the valve. This action causes the valve to move and engage the clutch. The movement of the valve determines how much the clutch is engaged.

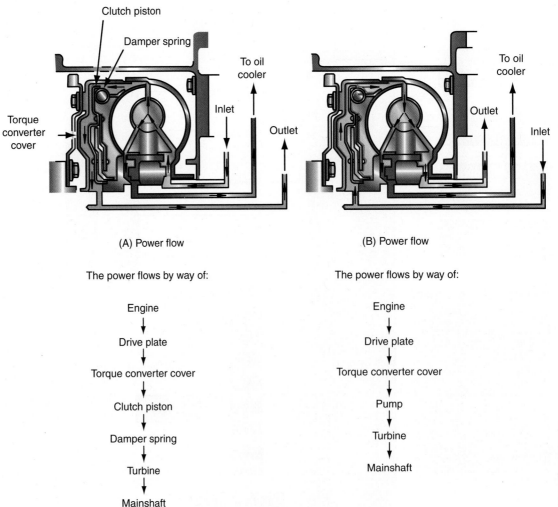

(A) Power flow

The power flows by way of:

Engine
↓
Drive plate
↓
Torque converter cover
↓
Clutch piston
↓
Damper spring
↓
Turbine
↓
Mainshaft

(B) Power flow

The power flows by way of:

Engine
↓
Drive plate
↓
Torque converter cover
↓
Pump
↓
Turbine
↓
Mainshaft

© Delmar/Cengage Learning

FIGURE 6-31 Power flow diagram and fluid flow in a Honda torque converter when the clutch is engaged (A) and when the clutch is released (B).

Modulator pressure is divided into two separate passages: one for the torque converter's inner pressure, which enters into the clutch assembly to engage the clutch, and one for torque converter backpressure, which enters between the pressure plate and the cover to disengage the clutch (Figure 6-32). Converter backpressure is regulated by the TCC control

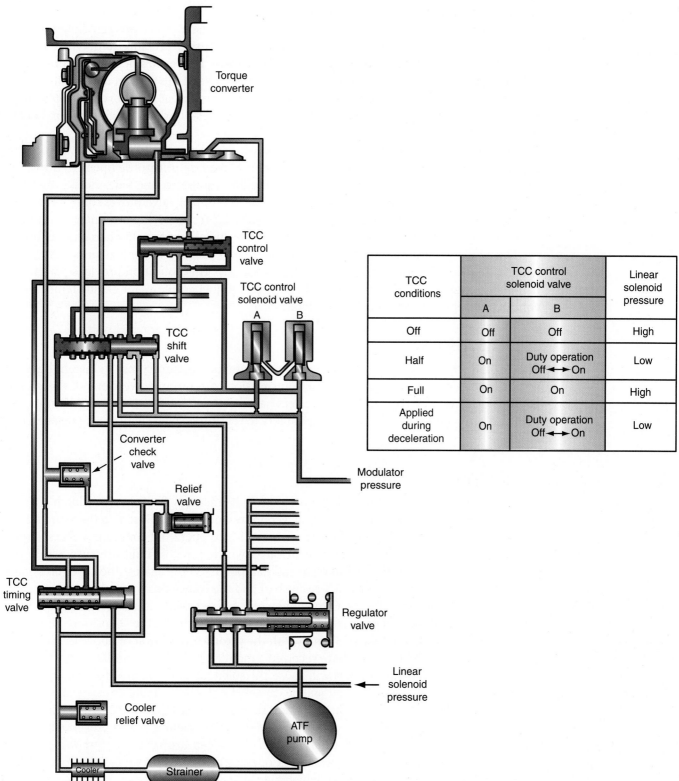

TCC conditions	TCC control solenoid valve		Linear solenoid pressure
	A	B	
Off	Off	Off	High
Half	On	Duty operation Off ◄──► On	Low
Full	On	On	High
Applied during deceleration	On	Duty operation Off ◄──► On	Low

FIGURE 6-32 Hydraulic system for a torque converter clutch.

valve, whereas the position of the TCC timing valve is determined by the throttle pressure, the tension of the valve spring, and the pressure regulated by the modulator. The position of the TCC control valve is determined by the backpressure of the clutch control valve and torque converter pressure, which is regulated by a check valve.

When modulator pressure is maintained at one end of the clutch control valve, the valve moves slightly to one side. This movement causes backpressure to slightly decrease. This results in a partial engagement of the clutch.

When both solenoids are turned on by the computer, the torque converter clutch is in the half-applied position. Modulator pressure is released, causing pressure to decrease on one side of the clutch control valve and the TCC timing valve. However, since throttle pressure is still low, the TCC timing valve remains to one side because of the tension of its spring. Since backpressure is also low, a greater amount of pressure is present at the converter clutch and it attempts to engage. However, because some backpressure still exists, the clutch is prevented from fully engaging.

Full engagement occurs when both solenoids are on (Figure 6-33). However, before this happens, vehicle speed must increase along with an increase in throttle opening. The TCC timing valve overcomes the tension of its spring and moves to one side, closing the fluid port leading to the converter's check valve. The clutch control valve moves to the other side because the throttle pressure is greater than modulator pressure. The movement of the control valve exhausts all of the converter's backpressure and allows the clutch to engage fully.

During deceleration, one solenoid remains on and the other operates on a duty cycle, which causes it to rapidly turn on and off (Figure 6-34). This condition provides partial and half-clutch engagement to occur in a sequence. This sequence allows for engine temperature control and engine braking during deceleration.

Honda Hybrids

Some Honda hybrid vehicles are available with automatic transmissions. To match the characteristics of a hybrid powertrain, these transmissions are modified versions of those used in nonhybrid vehicles. The torque converters have been designed to improve shift response time and to maximize the electric motor's ability to regenerate electricity during regenerative braking.

Toyota Transmissions

The PCM in Toyota's TCC system has a program in its memory for converter clutch engagement in each of its driving modes (normal, economy, and power). Based on these programs, the PCM turns the clutch solenoid on or off according to operating conditions. The solenoid controls the position of the clutch control valve, which controls fluid pressure within the torque converter to force the piston plate and damper assembly against the cover creating a mechanical link between the engine's crankshaft and the transaxle's input shaft.

The PCM will energize the TCC solenoid when three conditions are present. The vehicle must be traveling in second gear or higher with the gear selector in the D position. The speed of the vehicle must be at or above 37 mph with the throttle opening at or above the specified value. Finally, the PCM must not have received a mandatory lockup cancellation signal.

Mandatory cancellation signals will cause the solenoid to be deenergized when certain conditions exist. Clutch engagement will be canceled if the brake pedal is depressed and the brake light is turned on or the throttle position sensor indicates that the throttle is closed. By preventing clutch engagement during these times, engine stalling is also prevented. Clutch engagement will also be canceled when the vehicle's speed drops about 10 mph below the set speed while the cruise control system is operating. Doing this allows the torque converter to serve as a torque multiplier, which aids in acceleration or overcoming heavy loads. If the engine's coolant temperature falls below a specified temperature, clutch apply will be

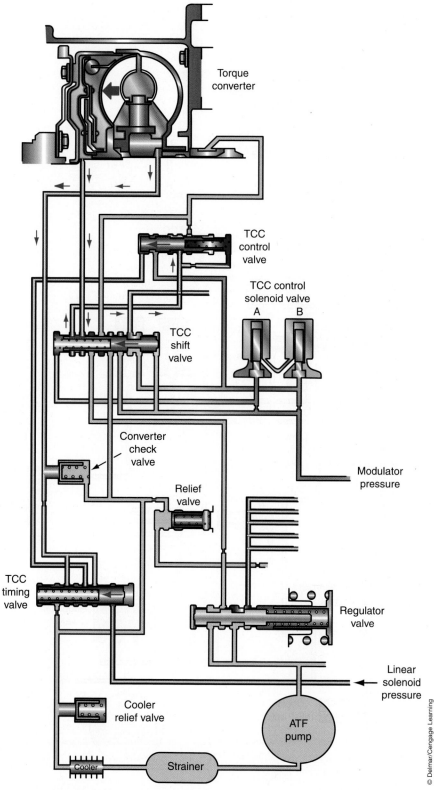

FIGURE 6-33 Hydraulic system for the converter clutch during full engagement.

© Delmar/Cengage Learning

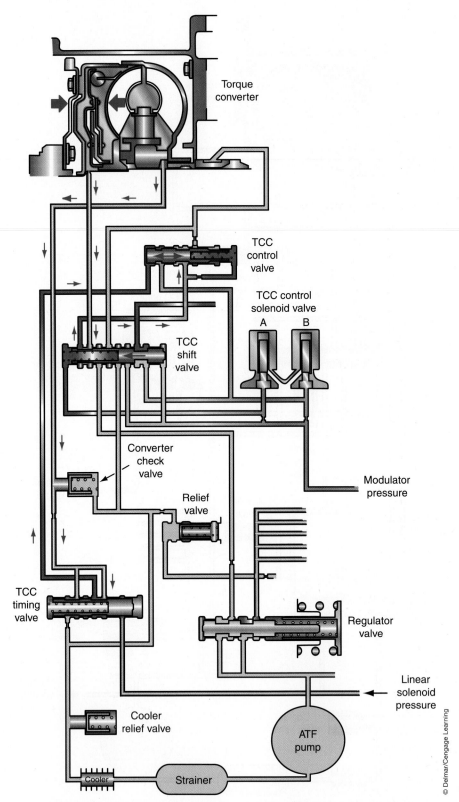

Torque converter

TCC control valve

TCC control solenoid valve
A   B

TCC shift valve

Converter check valve

Relief valve

Modulator pressure

TCC timing valve

Regulator valve

Cooler relief valve

Linear solenoid pressure

ATF pump

Cooler

Strainer

© Delmar/Cengage Learning

FIGURE 6-34 Hydraulic system for the converter clutch during deceleration.

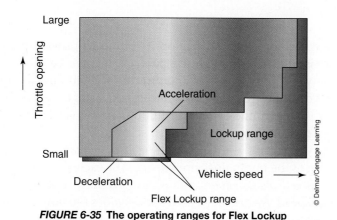

FIGURE 6-35 The operating ranges for Flex Lockup and Normal Lockup.

canceled and prevented. This allows for improved driveability, reduced emissions, and quicker transmission warming. The PCM will also temporarily turn the TCC off during upshifting or downshifting to decrease the harshness of the shifts.

When the lockup clutch is actuated, it rotates with the torque converter's impeller and turbine. Engaging and disengaging of the clutch is dependent on where the pressurized fluid enters the torque converter. Fluid can enter in the front or behind the lockup clutch. The pressure difference between the front and back of the clutch determines the engagement of the clutch. The flow of the pressurized fluid is controlled by a relay valve and a signal valve. Both valves are spring-loaded and are normally in the clutch-disengaged position.

When the vehicle is running at speeds less than 37 mph, pressurized fluid flows to the front of the clutch. There is now equal pressure at the front and rear of the clutch and the clutch is disengaged. Once the vehicle increases in speed, the pressurized fluid is sent to the rear of the clutch. The relay valve opens and exhausts the fluid at the front of the clutch and the pressure from behind the clutch causes the clutch to engage. Again, the pressure differential determines how much the clutch is engaged and this is determined by the modulation of the solenoid.

Flex Lockup. In addition to a typical On/off clutch lockup system, many late-model vehicles have a Flex Lockup system. In this system, the clutch may be partially or fully engaged during acceleration and deceleration (Figure 6-35). Toyota calls this the intermediate stage of clutch lockup. Its basic purpose is to reduce T/C slippage in lower gears and at lower vehicle speeds. Doing this increases fuel economy and improved driveability. The flex lockup also works during up and downshifts, this helps smoothen the shifts.

In most Toyota transmissions with Flex lockup, a linear solenoid is used. During acceleration, once the transmission has shifted into a higher gear, the solenoid allows the clutch to partially lock. This improves acceleration and reduces the amount of fuel used. Once the vehicle reaches a specified speed during deceleration, the clutch is locked. This adds to the effects of engine braking and also reduces fuel consumption.

BMW

BMW's 6HP26 six-speed transmission is based on the Lepelletier planetary gear design and relies on an electronic control module called the Mechatronik module. This module offers electronic shift control and adaptive learning, plus a feature designed to reduce fuel consumption—the Standby Control (SBC) system. This system disconnects the transmission's input shaft from the torque converter whenever the vehicle is stationary and the brakes are applied. This action reduces the amount of power and energy lost when the vehicle is sitting and the engine is working only to spin the torque converter.

This six-speed transmission controlled by the Mechatronik system is also called the "myTronic6."

Pumps

The torque converter and transmission rely on its pump to provide the circulation of the ATF through the transmission. Although the pump is the source of all fluid flow through the transmission, the valve body regulates and directs the fluid flow to provide for the gear changes. The pump is driven by the hub of the torque converter or a shaft from the converter; therefore, it operates whenever the engine is running. Three types of pumps are commonly used: *gear-type*, *rotor-type*, and *vane-type* pumps.

Gear- and rotor-type pumps are considered *fixed-displacement pumps*. Fixed displacement means the inside of the pump does not change. For each shaft rotation, there is a fixed amount of fluid that a fixed-displacement pump causes to flow. The flow rate of these pumps depends on rotational speed. As the speed increases, so does the flow rate. If, at a given speed, the flow rate requirements decrease, the pump cannot decrease the flow, thereby wasting energy and horsepower.

Modern transmissions commonly use vane-type pumps because they are usually variable in displacement. If, at a given speed, the flow rate requirements decrease, the inside of the pump has the ability to decrease the displacement, causing a decrease in pressure. This saves energy. With a variable-displacement pump, decreasing fluid pressure can happen without a decrease in speed.

Pump Drives

Automatic transmission pumps are driven by the engine's crankshaft through the torque converter housing (Figure 6-36). The pumps in all rear-wheel-drive and some front-wheel-drive

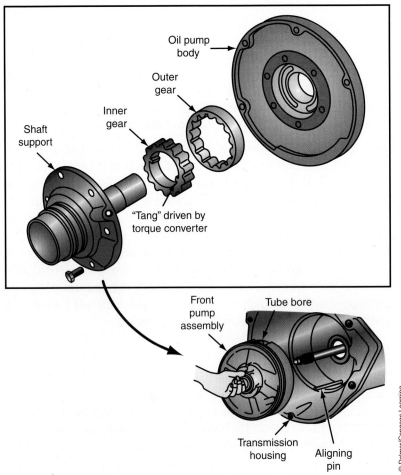

FIGURE 6-36 The pump is driven by the torque converter.

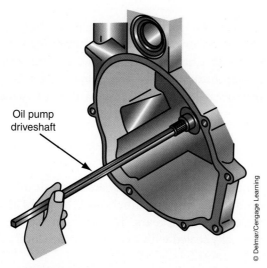

FIGURE 6-37 A typical oil pump driveshaft used on some transmissions and transaxles.

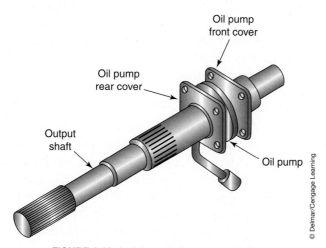

FIGURE 6-38 An integral oil pump assembly.

transmissions are driven externally by the torque converter's drive hub. The hub has two slots or flats machined into it, which fit into the pump drive. As the converter rotates, the pump is driven directly by the hub. Many front-wheel-drive transaxles use a hex-shaped or splined shaft fitted in the center of the converter cover to drive the pump (Figure 6-37). This system is called an internal drive. The Honda CVT uses a chain drive to drive the pump. Many older and a few late-model transmissions use a second pump mounted at the rear of the transmission case and driven by the output shaft. This second pump allows for pump operation whenever the output shaft is rotating, such as when the vehicle is being push-started (Figure 6-38).

Operation

While the pump is operating, the pump creates a low-pressure area at its inlet port. The oil pan is vented to the atmosphere and atmospheric pressure forces the fluid into the inlet port of the pump. The fluid is now at atmospheric pressure and is moved by the pump through the outlet port, causing fluid flow.

During pump operation, ATF moves from an area of high pressure to an area of low pressure.

Atmospheric pressure is approximately 14.7 psi at sea level.

AUTHOR'S NOTE: Transmission pumps run at very close tolerances. They are very likely to wear if the ATF is dirty or broken down. If the pump wears, fluid movement and pressure will decrease. This could cause poor transmission performance. This is just another reason why an automatic transmission needs to be taken care of by its owner and why you need to work on these things in a clean area.

Shop Manual

Chapter 6, page 308

Some gear-type pumps use a wear plate instead of a pump cover to seal the gears in the housing.

A **crescent** is a half-moon shaped part that isolates the inlet side of the pump from the outlet side of the pump.

Gear-Type Pumps

External tooth gear-type pumps consist of two gears in mesh to cause fluid flow. The gears rotate on their own shafts and in opposite directions. The gears are assembled in a housing that surrounds and totally encloses the gears. The shafts are sealed in the housing by bushings and the housing is normally sealed with a cover. In order for the pump to create low inlet pressures, the pump housing must be sealed.

One pump gear may be driven by the torque converter and drives the other gear. As the gears rotate, the gear teeth move in and out of mesh. As the teeth move out of mesh, inlet oil is trapped between the gear teeth and the walls of the housing. The trapped oil is carried around with the teeth until the gears again mesh. At this time the meshing of the teeth forces the oil out. The continuous release of trapped oil provides for flow as the fluid is pushed out of the pump's outlet port. The meshing of the gears also forms a seal that stops the fluid from moving out of the inlet port. Atmospheric pressure on the fluid ensures that the gap between the gear teeth will be refilled with fluid.

This type of pump is made with close tolerances. Excessive wear or play between the teeth of the gears or between the gears and the housing or pump cover will reduce the output and efficiency of the pump.

Another common type of gear pump is the **crescent** pump, which also uses two gears. One gear of this pump has internal teeth and the other has external teeth. The smaller gear with external teeth is in mesh with one part of the larger gear. In the gap where the teeth are not meshed is a crescent-shaped separator (Figure 6-39). The small gear is driven by the torque converter and it drives the larger gear. The gears' teeth mesh tightly together. As the gears rotate, the teeth mesh and then separate. As they separate, a low pressure is created between the gear teeth. The fluid is pushed by atmospheric pressure into this void until it is full.

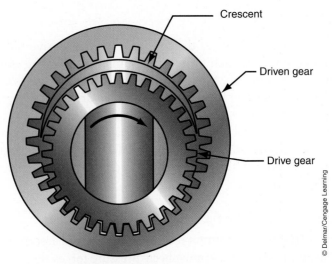

Crescent

Driven gear

Drive gear

© Delmar/Cengage Learning

FIGURE 6-39 Typical gear-type pump.

As the gears rotate, oil is trapped between the teeth and the crescent. The fluid is then carried around the pump housing toward the outlet port. As the gear teeth get close to the outlet and the gear teeth begin to mesh, the gap between the gear teeth begins to narrow. This narrowing of the gap continues until the gear teeth fully mesh. The narrowing of the gap squeezes the fluid and forces it through the outlet port as the gears move into mesh. A continuous flow of oil toward the outlet port pushes the fluid out into the transmission's hydraulic circuit. The crescent blocks the pressurized fluid and prevents it from leaking back toward the pump's inlet port.

Crescent pumps are made with close tolerances and can deliver the same amount of fluid each time the gear makes one complete revolution. The rate of fluid delivery changes with changes in engine speed.

Rotor-Type Pump

A rotor-type pump is a variation of the gear-type pump. However, instead of gears, an inner and outer rotor turns inside the housing (Figure 6-40). A rotor utilizes rounded lobes instead of teeth. The lobes of one rotor mesh with the recess area between the lobes of the other rotor. The torque converter drives the inner rotor, which rotates inside the outer rotor or rotor ring. The inner rotor has one fewer lobe than the ring, so that only one lobe is engaged with the outer ring at any one time.

Fluid is carried in the recess between the lobes and squeezed out toward the outlet port as the lobe moves into the recess of the outer ring (Figure 6-41). A constant supply of fluid being forced out of outlet port supplies fluid for the operation of the transmission. Fluid is prevented from backing up into the inlet port by the action of the lobes sliding over the lobes of the outer ring. The lobe-to-lobe contact causes them to seal against each other and creates small fluid chambers whose volumes increase as the lobes separate on the inlet side. This creates low pressure that allows fluid to be pushed into the pump inlet. As the lobes move into the recesses of the outer ring, oil is squeezed out.

Shop Manual
Chapter 6, page 309

Gear and rotor-type pumps are some times referred to as IX pumps, IX is an abbreviation that refers to the internal/external design of the gears or lobes used in these pumps.

The rotor-type pump is sometimes referred to as the gerotor-type pump.

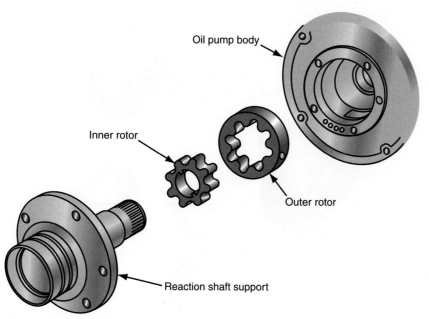

© Delmar/Cengage Learning

FIGURE 6-40 **Typical rotor-type pump.**

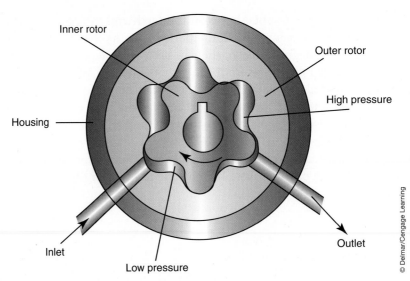

FIGURE 6-41 Action of rotors to move the fluid.

Vane-Type Variable-Capacity Pumps

Shop Manual
Chapter 6, page 309

Many late-model automatic transmissions are equipped with vane-type pumps (Figure 6-42). This type of pump is a *variable-capacity pump* or variable-displacement pump whose output can be reduced when high fluid pressures are not necessary. To monitor and control output flow, a sample of the output is applied to the back of the pump's slide. The pressure moves the slide against spring pressure and changes its position in relation to the rotor. This action controls the size variations between the chamber at the inlet and the outlet ports, which in turn controls the output pressure of the pump.

The rotor and vanes of the pump are contained within the bore of a movable slide that is able to pivot on a pin. The position of the slide determines the output of the pump

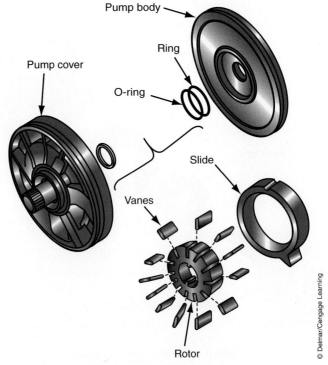

FIGURE 6-42 Typical vane-type pump.

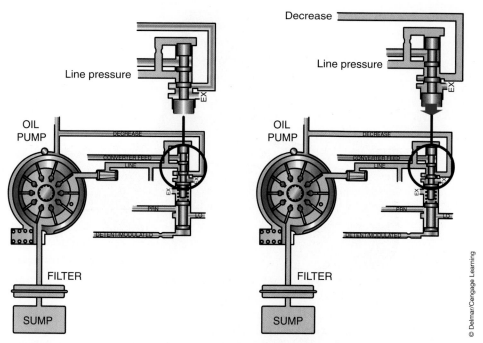

FIGURE 6-43 Action of vane-type pump, maximum output on the left and minimum output on the right.

(Figure 6-43). When the slide is in the fully extended position, the slide and rotor vanes are at maximum output. As the rotor and vanes rotate within the bore of the slide, the changes in the chamber size form low and high pressure areas. Fluid trapped between the vanes at the inlet port is moved to the outlet side. As the slide pivots toward the center or away from the fully extended position, a greater amount of fluid is allowed to move from the outlet side back to the inlet side. When the slide is centered, there is no output from the pump and a neutral or no-output condition exists. Since the slide moves in response to the sample of pressure sent to it, the slide can be at an infinite number of positions.

The output of a variable-capacity pump will vary according to the needs of the transmission, not according to the engine's speed. Therefore, it does not waste energy like a positive-displacement pump does. A variable-capacity pump can deliver a large volume of fluid when the need is great, especially at low pump speeds. At high pump speeds, the pressure requirements are normally low and the variable-capacity pump can reduce its output accordingly. Once the needs of the transmission are met, the pump delivers only the amount of fluid that is needed to maintain the regulated pressure.

Dual-Stage Oil Pumps

Some late-model transmissions have a **dual-stage** positive-displacement **oil pump**. These pumps have two driven gears (primary and secondary) driven by the torque converter hub through a central drive gear (Figure 6-44). At low speeds, both gears supply fluid to the transmission. As speed increases, so does the output from the primary stage so it is able to meet the demands of the transmission by itself. Once enough fluid pressure is developed, a regulator valve closes and shuts down the secondary circuit. When this happens, the fluid from the secondary gear is recirculated through the main pressure regulator valve, causing a check valve in the pump outlet to close and the primary gear supplies all fluid under pressure.

This design allows the pump to deliver the pressure and fluid flow of a large displacement pump at low speeds, and the economy of a small displacement pump at higher engine speeds.

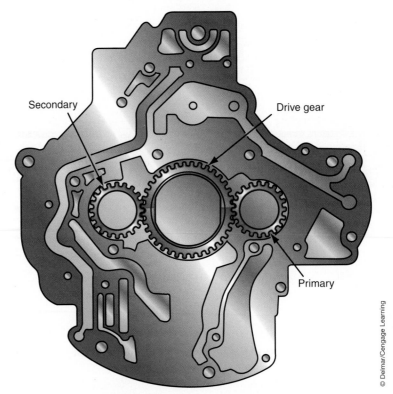

FIGURE 6-44 A dual-stage oil pump.

The dual-stage oil pump of some Chrysler transmissions also houses some of the valves that would normally be found in the valve body (Figure 6-45). These valves (the converter clutch switch valve, converter regulator valve, converter limit valve, and pressure regulator valve) control or limit hydraulic pressure throughout the transmission and torque converter.

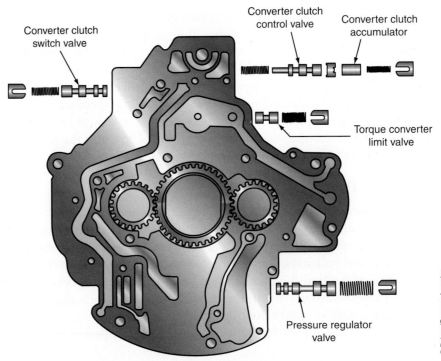

FIGURE 6-45 The valves in a Chrysler dual-stage oil pump.

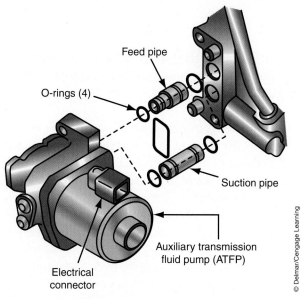

FIGURE 6-46 An auxiliary oil pump.

Auxiliary Electric Oil Pumps

Hybrid vehicles utilize a stop–start feature. This turns off the engine when it is at a stop and instantly restarts it when the driver removes pressure on the brake pedal or when the electric motor can no longer propel the vehicle by itself. When this technology is used with an automatic transmission, there can be a delay in the action of the transmission because it takes a while for the pressure to build up inside it. To overcome this, many hybrid transmissions are fitted with an electric oil pump (Figure 6-46). These pumps will maintain pressure within the transmission even when the engine is not running. Some hybrid transmissions are designed with only an electric oil pump, and others have an electric pump that works only when the engine is not running.

SUMMARY

- A torque converter uses fluid to smoothly transfer engine torque to the transmission.
- All automatic transmissions, except some designs of constant variable transmission, use a torque converter to transfer power from the engine to the transmission.
- A simple fluid coupling is composed of three basic members: a housing, an impeller, and a turbine.
- The impeller acts like a pump and moves the fluid in the direction of its rotation. The moving fluid hits against the turbine vanes. This causes the turbine to rotate, which brings an input into the transmission.
- The fluid coupling allows the vehicle to stop while in gear without stalling the engine, because at low engine speeds, the impeller does not rotate fast enough to drive the turbine.
- Two types of fluid flow take place inside the fluid coupling: rotary flow and vortex flow.
- Rotary flow is in the direction of impeller rotation and results from the paddle action of the impeller vanes against the fluid.
- Vortex oil flow is the fluid flow circulating between the impeller and turbine as the fluid moves from the impeller to the turbine and back to the impeller.
- On all RWD and most FWD vehicles, the torque converter is mounted in line with the transmission's input shaft. Some transaxles use a drivechain to connect the converter's output shaft with the transmission's input shaft.

SUMMARY

- The torque converter fits between the transmission and the engine and is supported by the transmission's oil pump housing and the engine's crankshaft. The rear hub of the converter's cover fits into the transmission's pump. The hub drives the oil pump whenever the engine is running.
- The front of a torque converter is mounted to a flexplate. The flexplate is designed to be flexible enough to allow the front of the converter to move forward or backward if it expands or contracts because of heat or pressure.
- A typical torque converter consists of three elements sealed in a single housing: the impeller, the turbine, and the stator.
- The stator is the reaction member of the converter. This assembly is about one-half the diameter of the impeller or turbine and is positioned between the impeller and turbine.
- The stator redirects the fluid leaving the turbine back to the impeller. By redirecting the fluid so that it is flowing in the same direction as engine rotation, it drives the impeller to rotate more efficiently.
- A stator is supported by a one-way clutch that is splined to the stator support shaft and allows the stator to rotate only in the same direction as the impeller.
- The fluid in the torque converter is supplied by the transmission's oil pump and enters through the converter's hub, then flows into the passages between the vanes. As the impeller rotates, the fluid is moved outward and upward through the vanes by centrifugal force because of the curved shape of the impeller.
- The higher the engine speed, the faster the impeller turns, and the more force is transferred from the impeller to the turbine by the fluid.
- The redirection of the fluid by the stator not only prevents a torque loss, but also provides for a multiplication of torque. The fluid now moves in a direction that aids in the rotation of the impeller and the fluid now leaves with nearly twice the energy and exerts a greater force on the turbine.
- Vortex flow allows for torque multiplication. Torque multiplication occurs when there is high impeller speed and low turbine speed.
- When there is little vortex flow and the engine's torque is carried through the converter by the rotary flow of the fluid, the coupling phase of the torque converter occurs and no torque multiplication takes place.
- The condition called stall occurs when the turbine is held stationary and the impeller is spinning. Stall speed is the fastest speed an engine can reach while the turbine is at stall.
- The transmission's pump continuously delivers pressurized ATF into the torque converter through a hollow shaft in the center of the torque converter assembly.
- The fluid is circulated through the converter and exits past the turbine through the turbine shaft, which is located within the hollow fluid feed shaft. From there, the fluid is directed to an external oil cooler and back to the transmission's oil circuit.
- When applied, the converter clutch locks the turbine to the cover of the converter.
- The piston and clutch plate assembly have frictional material on the outer portion of the plate with a spring-cushioned damper assembly in the center. The clutch plate is splined to the turbine shaft and when the piston is applied, the plate connects the turbine to the converter.
- The converter clutch is applied when the fluid flow through the torque converter is reversed by a valve. When the torque converter clutch control valve moves, the fluid begins to flow in a reversed direction. This forces the clutch disc or pressure plate against the front of the torque converter's cover.
- A viscous converter clutch assembly consists of a rotor, body, clutch cover, and silicone fluid.

- Frictional material is normally bonded to either the forward face of the clutch plate or the inner front surface of the converter cover.
- A converter clutch disc is splined to the turbine so that it can drive the turbine when the frictional material is forced against the torque converter's cover.
- All converter clutch assemblies are fitted with a damper assembly that directs the power flow through a group of coil springs. These springs are designed to absorb the normal torsional vibrations of an engine, which vary with changes in speed and load.
- Early Chrysler electronic controlled converter clutch systems use an electrical solenoid mounted on the valve body to engage and disengage the converter clutch. The operation of this solenoid is controlled by the engine control computer through a relay. The computer determines when to engage or disengage the torque converter clutch based on information from the coolant temperature sensor, vacuum transducer, vehicle speed sensor, and carburetor or throttle ground switch.
- Late-model Chryslers with the 41TE transaxle have a torque converter that can lock up in second, third, and overdrive gear. The electronic modulated converter clutch (EMCC) system allows the converter clutch to partially engage between 23 and 47 mph. This feature limits the amount of converter slippage but serves to cushion the engine from the driveline.
- Electronically controlled hydraulic converter clutch circuits in Ford Motor Company products are controlled by the PCM.
- An AX4S uses a pulse width–modulated solenoid to allow the PCM to vary clutch operation from full release to controlled slip to full apply.
- The converter control system in a 4R70W also uses a pulse width–modulated solenoid called the modulated converter clutch control (MCCC) solenoid. It operates in the same way as the AX4S.
- The converter clutch in an A4LD transmission is applied and released hydraulically, but it can be overridden electronically. The A4LD override solenoid is normally energized to prevent fluid flow from unlocking the converter clutch.
- The torque converter clutch assembly's computer controls the application of the clutch by providing a ground circuit for the torque converter clutch (TCC) solenoid circuit.
- When the torque converter clutch solenoid's ground circuit is completed by the PCM, fluid pressure is exhausted from between the converter's pressure plate and the converter cover, and converter clutch apply pressure from the converter clutch regulator valve pushes the converter pressure plate against the converter cover to apply the converter clutch.
- A Honda torque converter can operate in various degrees of clutch engagement: zero, partial, half, full, and cycling. The TCC is controlled by the transmission's hydraulic system and the transmission control module.
- Three types of pumps are commonly used: gear type, rotor type, and vane type.
- The pumps in all rear-wheel-drive and some front-wheel-drive transmissions are driven externally by the torque converter's drive hub. The hub has two slots or flats machined into it, which fit into the pump drive.
- Many front-wheel-drive transaxles use a hex-shaped or splined shaft fitted in the center of the converter cover to drive the pump.
- While the pump is operating, it creates a low-pressure area at its inlet port. The oil pan is vented to the atmosphere and atmospheric pressure forces the fluid into the inlet port of the pump. The fluid is now moved by the pump and forced out at a higher pressure through the outlet port.
- Gear-type pumps consist of two gears set in mesh to move the fluid.
- Another common type of gear pump is the crescent pump, which also uses two gears. One gear of this pump has internal teeth and the other has external teeth. The smaller gear with external teeth is in mesh with one part of the larger gear.

TERMS TO KNOW

Crescent

Damper

Dual-stage oil pump

Fail-safe valve

Frequency

TERMS TO KNOW

(continued)

Pulse width

Rotary flow

Stall speed

Vortex flow

SUMMARY

- A rotor-type pump is a variation of the gear-type pump. However, instead of gears, an inner and outer rotor turns inside the housing.
- Vane-type pumps are variable-displacement pumps whose output can be reduced when high fluid pressures are not necessary.
- Automatic transmissions used in hybrid vehicles may have an electric oil pump that is used to maintain pressure throughout the transmission when the engine is not running.

REVIEW QUESTIONS

Short-Answer Essays

1. What does a CVT use instead of a torque converter? Briefly describe how that assembly works.

2. What is the difference between rotary and vortex oil flow?

3. Name the three major components of a torque converter and describe the purpose of each.

4. What drives a transmission pump?

5. What is meant by stall speed?

6. What is the benefit of modulating the TCC solenoid?

7. How is the fluid in the torque converter cooled?

8. What is the purpose of the damper in a converter clutch?

9. What is the purpose of the stator in a torque converter?

10. What determines the vortex flow in a torque converter?

Fill in the Blanks

1. The combined weight of the _____, _____ _____, and _____ serve as the flywheel for the engine on cars equipped with an automatic transmission.

2. The _____ the engine speed, the _____ the impeller turns, and _____ force is transferred from the impeller to the turbine by the fluid.

3. A simple fluid coupling is composed of three basic members: a _____ , an _____ , and a _____.

4. In a dual-stage positive-displacement oil pump, the primary and secondary driven gears supply pressurized fluid flow at _____ speeds. At _____ speeds only the primary gear supplies line pressure.

5. Variable vane-type oil pumps have the ability to _____ the displacement of the pump without changing its rotational speed.

6. A torque converter has reached its coupling phase when turbine speed is about _____ percent of the speed of the impeller. At this point, there is no _____ _____.

7. Most lockup converters consist of the three basic elements: _____, _____, and _____ , plus a _____ and _____ _____ assembly, special _____ _____, and _____ _____.

8. Three types of pumps are commonly used: the _____ type, _____ type, and _____ type. The _____ type is the most common.

9. A viscous converter clutch assembly consists of a _____ , _____ , _____ _____, and _____ _____.

10. It is _____ flow that allows for torque multiplication. Torque multiplication occurs when there is _____ impeller speed and _____ turbine speed.

MULTIPLE CHOICE

1. *Technician A* says all RWD and most FWD vehicles have the torque converter mounted in line with the transmission's input shaft.
   Technician B says some transaxles use a drivechain to connect the converter's output shaft with the transmission's input shaft.
   Who is correct?
   A. A only
   B. B only
   C. Both A and B
   D. Neither A nor B

2. While discussing typical electronic controlled converter clutch systems:
   Technician A says most use an electrical solenoid mounted on the valve body to engage and disengage the converter clutch.
   Technician B says the operation of the clutch is controlled by a relay, which is activated by the coolant temperature sensor.
   Who is correct?
   A. A only
   B. B only
   C. Both A and B
   D. Neither A nor B

3. *Technician A* says that in a vane-type pump, the vane ring contains several sliding vanes that seal against a slide mounted in the pump housing.
   Technician B says vane-type pumps are variable-displacement pumps whose output can be reduced when high fluid volumes are not necessary.
   Who is correct?
   A. A only
   B. B only
   C. Both A and B
   D. Neither A nor B

4. Which of the following statements is *not* true?
   A. A stator is sometimes called a reactor.
   B. The vanes of the turbine, impeller, and stator are sometimes called blades or fins.
   C. The amount of torque multiplication a converter will produce is sometimes referred to as the converter ratio.
   D. A viscous converter clutch system is sometimes called a centrifugal clutch system.

5. While discussing clutch engagement solenoids:
   Technician A says some systems use a PM generator-type solenoid, which allows for full-time, but controlled, slip.
   Technician B says some converter control systems use a pulse width–modulated solenoid.
   Who is correct?
   A. A only
   B. B only
   C. Both A and B
   D. Neither A nor B

6. Pulse width is:
   A. The length of time something is energized.
   B. The amount of "off time."
   C. The number of cycles per second.
   D. A characteristic of an analog signal.

7. *Technician A* says when the speed of the turbine nearly equals the speed of the impeller, fluid flows against the stator vanes in the same direction as the fluid from the impeller.
   Technician B says when there is little vortex flow and the engine's torque is carried through the converter by the rotary flow of the fluid, the coupling phase of the torque converter occurs and no torque multiplication takes place.
   Who is correct?
   A. A only
   B. B only
   C. Both A and B
   D. Neither A nor B

8. While discussing the construction of typical converter clutch assemblies:
   Technician A says the clutch plate is splined to the turbine shaft, and when the piston is applied, the plate connects the turbine to the converter.
   Technician B says the piston and clutch plate assembly has frictional material on the outside of the plate with a spring-cushioned damper assembly in the center.
   Who is correct?
   A. A only
   B. B only
   C. Both A and B
   D. Neither A nor B

9. *Technician A* says the rear hub of the torque converter is bolted to the flexplate.

Technician B says the flexplate is designed to be flexible enough to allow the front of the converter to move forward or backward if it expands or contracts because of heat or pressure.

Who is correct?

A. A only
B. B only
C. Both A and B
D. Neither A nor B

10. While discussing the operation of the stator:

Technician A says the stator redirects the fluid leaving the turbine back to the impeller.

Technician B says only a portion of the fluid returning from the turbine to the impeller passes through the stator.

Who is correct?

A. A only
B. B only
C. Both A and B
D. Neither A nor B

Chapter 7

HYDRAULIC CIRCUITS AND CONTROLS

UPON COMPLETION AND REVIEW OF THIS CHAPTER, YOU SHOULD BE ABLE TO:

- Describe the design and operation of the hydraulic controls used in modern transmissions.
- Describe the basic types of valves used in automatic transmissions.
- Describe the various configurations of a spool valve and explain how it can be used to open and close various hydraulic circuits.
- Explain the role and operation of the pressure regulation valve, throttle valve, governor assembly, manual valve, shift valves, and kickdown assembly.
- Describe how fluid flow and pressure can be controlled by a solenoid.

- Describe the different designs of a governor and explain the operation of each type.
- Explain why load-sensing devices are necessary for automatic transmission efficiency.
- Identify the various pressures in the transmission and explain their purpose.
- Describe the purpose of a transmission's valve body.
- Trace through the oil circuit for a transmission and describe where the fluid flows in each transmission range.

INTRODUCTION

An automatic transmission receives engine power through a torque converter, which is indirectly attached to the engine's crankshaft. Fluid in the converter allows power to flow from the engine to the transmission's input shaft. The input shaft drives a planetary gearset, which provides the different forward gears, a neutral position, and one reverse gear. Power flow through the gears is controlled by multiple-friction disc packs, one-way clutches, or friction bands. These hold or drive a member of the gearset when hydraulic pressure (Figure 7-1) is applied to them. Hydraulic pressure is routed to the correct apply device by the transmission's valve body, which controls the pressure and direction of the hydraulic fluid.

The various gear ratios are selected by the transmission according to engine and vehicle speeds, engine load, and other operating conditions. Both upshifts and downshifts occur automatically. The transmission can also be manually shifted into a lower forward gear, reverse, neutral, or park. In order to accomplish all of this, a transmission contains many hydraulic circuits, valves, and devices.

VALVE BODIES

For efficient operation of the transmission, the brakes and clutches must be released and applied at the proper time. It is the responsibility of the valve body to control the hydraulic pressure being sent to the different reaction members in response to engine and vehicle load, as well as to meet the needs and desires of the driver.

A BIT OF HISTORY

The Chevrolet Powerglide was a popular transmission in the 1950s. During its time it had a very complex hydraulic system. The valve body and accumulator assembly contained a total of seven valves, two of which were simple check valves.

Shop Manual

Chapter 7, page 333

FIGURE 7-1 An automatic transmission fitted with a transfer case for all-wheel-drive.

The valve body is machined from aluminum or iron castings and has many precisely machined bores and fluid passages (Figure 7-2). Fluid flow to and from the valve body is routed through passages and bores. Valves control fluid pressure and open, close, or modulate the size of the fluid passages in the valve body. The hydraulic circuits extend to the transmission and pump housing and are connected either by direct mounting or through oil tube passages.

Valve bodies are normally fitted with three different types of valves: disc (poppet), check ball, and spool valves (Figure 7-3). The purpose of these valves is start, stop, or regulate and direct the flow of fluid throughout the transmission. Fluid flow through the passages and bores is controlled either by a single valve or a series of valves. Most of the control valves operate automatically to direct the fluid as needed to perform a certain function.

The valve body assembly may be comprised of two or more main parts. One of these parts may be a separator plate (Figure 7-4) and/or **transfer plate**. The separator and transfer plates are designed to seal off some of the passages and they contain some openings that help to control and direct fluid flow through specific passages. The three parts are typically bolted together and mounted as a single unit to the transmission housing.

Other transmission designs may have separate valve body assemblies, each with a purpose. For example, the pressure regulator assembly shown in Figure 7-5 is bolted to the main

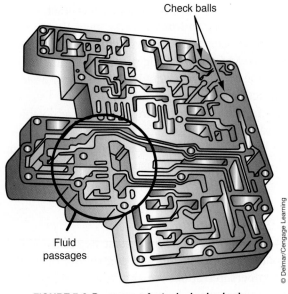

Check balls

Fluid passages

FIGURE 7-2 Passages of a typical valve body.

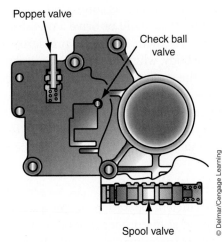

Poppet valve

Check ball valve

Spool valve

FIGURE 7-3 Various types of valves make up a typical valve body.

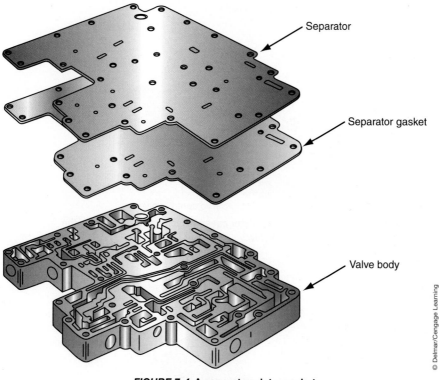

FIGURE 7-4 A separator plate, gasket, and valve body.

Separator

Separator gasket

Valve body

© Delmar/Cengage Learning

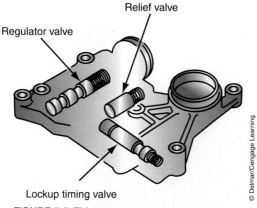

Relief valve

Regulator valve

Lockup timing valve

© Delmar/Cengage Learning

FIGURE 7-5 This pressure regulator valve body assembly is part of the main valve body assembly.

valve body in Figure 7-6. In other designs, the valve body is comprised on three or more major units (Figure 7-7); each unit is separated from the other by a gasket and a separator plate. Often, the shift solenoids are mounted to one of the valve body assemblies.

A multiple part valve body is found in many Chrysler transmissions and transaxles. The valve body assembly is comprised of the main valve body, regulator valve body, servo body, top accumulator body, and accumulator body, all assembled together as a single assembly. The main valve body's primary purpose is to control fluid pressure and to control where the pressure will go in the rest of the hydraulic control system. The regulator valve body maintains constant hydraulic pressure from the oil pump to the hydraulic control system and supplies fluid to the torque converter. The servo and accumulator valve bodies work to control shift quality at the various bands and multiple-disc packs and are the mounting point for the shift solenoid valves.

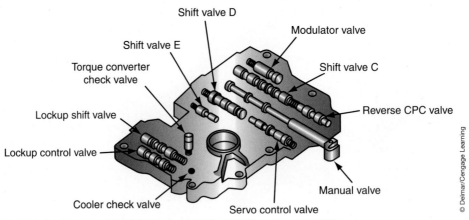

FIGURE 7-6 The pressure regulator assembly in Figure 7-5 bolts to the top of this valve body assembly.

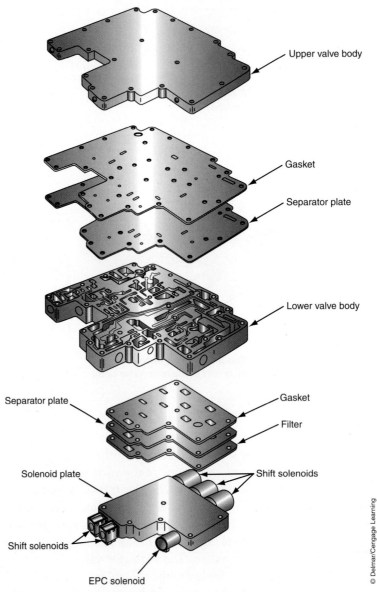

FIGURE 7-7 This valve body assembly is made up of three separate units, each mounted with gaskets and a separator plate.

HYDRAULIC CONTROLS

Although the pump is the source of all fluid flow through the transmission, the valve body regulates and directs the fluid flow to provide for the gear changes. Many different valves and hydraulic passages are used to regulate, direct, and control the movement of the fluid after it leaves the pump. These control valves either open or close a circuit and do so without changing the pressure of the hydraulic fluid. The valves react to mechanical and hydraulic forces. The basic types of hydraulic control used in automatic transmissions are orifices, check valves, and spool valves.

Orifices

The hydraulic system in a transmission has many restrictions, such as connecting lines, small bores, and orifices. An **orifice** is a bore in a passage that is smaller than the connecting passages to cause a restriction to fluid flow and slow down fluid flow (Figure 7-8). An orifice is used to dampen a fluid flow to prevent sudden or undue movement of a control valve.

Often, orifices are found in the valve body's spacer or separator plate and/or the plate to valve body gasket (Figure 7-9). Orifices may also be cast as hydraulic passages in the transmission's case or valve body, or they can be line plugs with precisely drilled holes.

The pressure of a fluid is affected by the fluid's movement from one point to another through a restriction. An orifice provides a resistance to fluid flow and therefore causes a pressure drop as long as fluid is flowing through it. The restriction causes the fluid's pressure to increase at the inlet side of the restriction and decrease at the outlet side. The amount of pressure drop is relative to the size of the orifice and flow volume. The smaller the orifice, the greater the resistance to flow and the larger the pressure drop. Pressure drops only when fluid flows through the orifice. As soon as the flow stops, the orifice no longer has an affect on the flow, and pressure on both sides becomes equal. Orifice sizes are specifically selected to meet the needs of the transmission (Figure 7-10).

Two or more orifices can be placed in series within a single fluid passage or line. These restrictions reduce the shock caused by the immediate application of a device caused by immediate hydraulic force to a component. Normally, a series of orifices are used to gradually engage an apply device that improves shift quality.

> In the separator plate of a valve body, the smaller holes are orifices. The larger rectangular holes do not act as orifices.

Check Valves

Check valves are used to hold fluid in cylinders and to prevent fluid from returning to the reservoir. A check valve opens when fluid is flowing and closes when the flow stops. The valve also closes when fluid pressure is applied to the outlet side of the valve. The direction of the fluid flow controls and operates the check valve (Figure 7-11). Ball-type valves are the most commonly used check valves, but some transmissions are fitted with disc or poppet check valves.

Most check valves are spring loaded and will open whenever fluid pressure on the valve exceeds the spring's tension. The valve will stay open until the fluid's pressure drops below

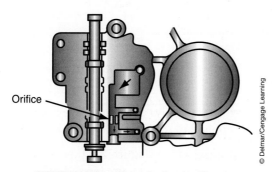

FIGURE 7-8 Location of a fixed orifice to restrict fluid flow.

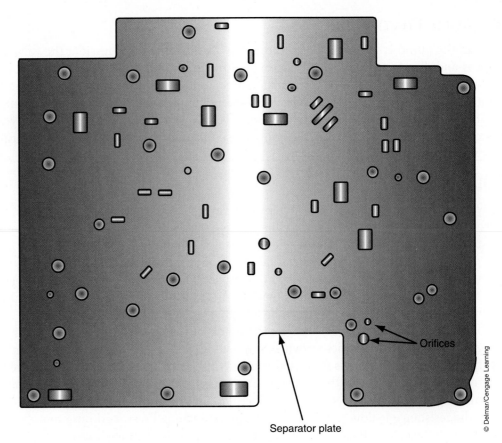

FIGURE 7-9 A spacer plate with orifices and other bores for a valve body.

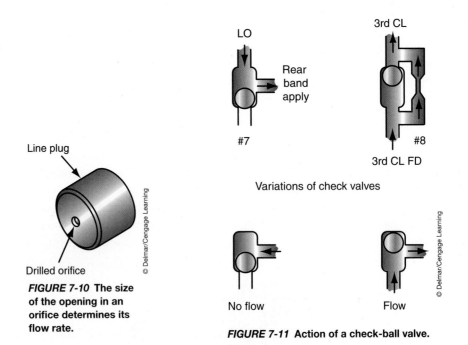

FIGURE 7-10 The size of the opening in an orifice determines its flow rate.

Variations of check valves

FIGURE 7-11 Action of a check-ball valve.

the tension of the spring. If the check valve does not have a return spring, the valve will not totally close until pressure is applied in the opposite direction.

Poppet (Disc) Valves. A **poppet valve** (Figure 7-12) is normally a flat disc with a stem to guide the valve's operation. The stem normally fits into a hole acting as a guide to the valve's opening and closing. Poppet valves tend to pop open and close, hence their name. This type

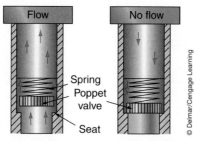

FIGURE 7-12 Typical poppet valve operation.

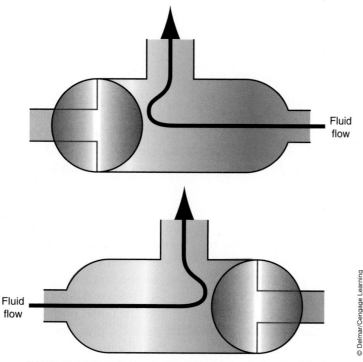

FIGURE 7-13 Fluid is directed by the movement of a check ball.

of check valve is often used to prevent back flow. A return spring forces the valve closed as soon as the pressure of the spring is greater than the pressure of the fluid.

Check Balls. Ball-type check valves are commonly used to redirect fluid flow (Figure 7-13) or to stop it from back flowing to the reservoir. Ball check valves either open or close a fluid passage. They are basically a round steel or plastic ball positioned over a hole in the separator plate of the valve body (Figure 7-14) or in the valve body itself. Check balls may also be used in various apply devices. Needle-type check valves are also used to prevent back flow. However, they are not frequently used although they function in the same way as a ball-type valve.

When fluid pressure forces the ball against its seat, fluid flow is blocked (Figure 7-15). When fluid pressure is applied to the opposite of the ball, the ball is unseated and fluid can flow. Check ball valves may also be fitted with a return spring to accomplish the same task as a poppet valve. Check balls and poppet valves can be normally open, which allows free flow of fluid pressure, or normally closed, which blocks fluid pressure flow.

The operation of a simple lockup torque converter clutch can serve as an example of how a check ball works. When the conditions are such that the TCC should be applied, the PCM energizes the clutch relay and lockup solenoid. This moves the solenoid's check ball

Shop Manual
Chapter 7, page 343

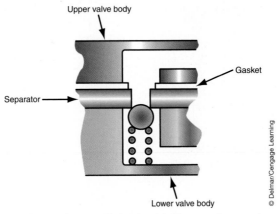

FIGURE 7-14 The seat for a check ball-type valve may be located in the valve body's separator plate.

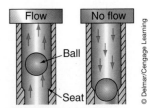

FIGURE 7-15 The operation of a check-ball valve.

onto its seat. This stops the lockup solenoid from exhausting line pressure. This allows pressure to build and the increasing line pressure forces the lockup switch valve to move against spring tension. The fluid is then directed to the torque converter. Fluid flows from the impeller and turbine to fill the space behind the torque converter clutch piston and forces clutch engagement.

Two-Way Check Valves. A ball-type check valve without a return spring can be used as a **two-way check valve** (Figure 7-16) also called a shuttle valve. The ball is positioned between two side-by-side holes. Each hole leads to a different hydraulic passage. Fluid flow through one of the holes causes the ball to move over and seat against the other hole. Flow through the second hole causes the ball to move over and seal the first hole.

A two-way valve can be used at points where fluid from two different sources can be sent to the same outlet port. The ball normally rests over the outlet port, closing it. As the inlet fluid enters the valve, the ball will move according to the pressure from one of the inlets. When pressure on one side of the ball is stronger than the pressure on the other side, the ball will move to the weaker side and close that inlet port. The ball will toggle between the inlet ports in response to differing pressures. When the pressure from the inlets is equal, the ball will be centered and will block the outlet port.

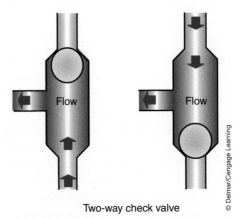

Two-way check valve

FIGURE 7-16 A check valve using a ball to direct fluid flow.

Spool valves

The most commonly used valve in a valve body is the **spool valve**, which is used as a flow-directing valve. It is called a spool valve because it looks like a spool for sewing thread (Figure 7-17). The valve's movement is controlled by hydraulic pressure, spring pressure (Figure 7-18), mechanical linkage, or a combination of hydraulic and spring pressure. As a result, fluid flow can be directed into hydraulic passages according to the operating conditions.

The large circular parts of the valve are called the spools. Every spool valve has at least two spools. The outside circumference of the spools is a precisely machined surface called the **land**. The lands are sized to fit into a bore in the valve body.

The land rides on a very thin film of fluid in the bore. It is machined so that it does not allow fluid to escape while being able to move easily within its bore. The outer edges of the lands have sharp corners to help prevent dirt from wedging between the land and

A spool valve will always move in the direction of the greater force.

Shop Manual
Chapter 7, page 344

A land is the sealing area of a spool valve.

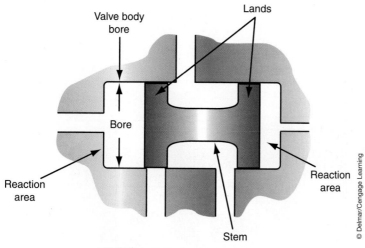

FIGURE 7-17 Typical spool valve.

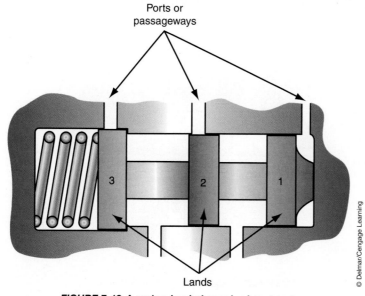

FIGURE 7-18 A spring-loaded spool valve at rest.

the bore. The edges also give the valve a self-cleaning action as it moves back and forth in the bore.

The lands on a spool valve are together by the valve's stem. The stem has a smaller diameter than the lands and is not a precisely machined part of the valve. The face of each spool is flat and serves as a pressure surface, called a **reaction area**, to produce valve movement. The area or space between the lands is called the **valley**. The valley of the valve is smaller than the spools to allow fluid to pass between the lands and through to the adjoining open ports (Figure 7-19).

Some valves may have a stem or extension on one end to mount a return spring or to limit valve travel. Some spool valves have a series of lands and valleys to control fluid flow through two or more passages at the same time. The lands of these valves can have different diameters to provide different-sized reaction surfaces or areas. A larger diameter provides for more surface area for the hydraulic pressure to work against (Figure 7-20). If pressure is applied to the valley between two differently sized lands, the valve will move to the direction of the larger land, simply because there is greater pressure there.

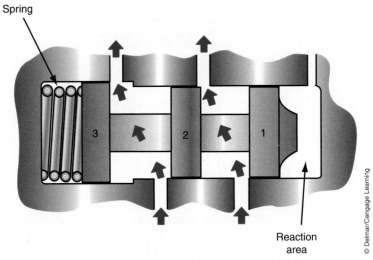

FIGURE 7-19 A spring-loaded spool valve with sufficient pressure on the reaction area to overcome the tension of the valve's spring.

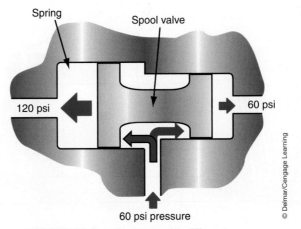

FIGURE 7-20 By using lands with different surface areas, spool valves will have more force on one end of the valve.

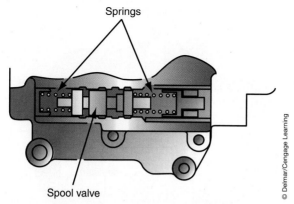

Springs

Spool valve

© Delmar/Cengage Learning

FIGURE 7-21 Some spool valves are set in place by springs that must be overcome by hydraulic pressure before the valve can move in the other direction.

A few spool valves ride on a sleeve or bushing that allows the auto manufacturer to use the same valve body in different applications. By using a different sized sleeve, a spool valve with different sized spools can be used.

As a spool valve moves in its bore, the lands open (uncover) or close (cover) fluid passages. This allows the valve to allow or block fluid flow from one port to another. The movement of the valve is primarily based on the action of the transmission's solenoid valves or by mainline, throttle, and governor pressures.

When a spool valve is controlled by a spring and hydraulic pressure, the tension of the spring keeps the valve at one end of the bore. On the other end of the valve (opposite of the spring), fluid enters through a port. When the fluid pressure on this end of the valve is greater than the tension of the spring, the valve moves against the spring. When a spool valve with different sized lands is fitted with a spring (Figure 7-21), less hydraulic pressure is needed to overcome the tension of the spring. If the spring is behind the smaller land, hydraulic pressure on the larger land effectively increases the spring tension on the valve. The pressure on the smaller land must be now greater before it can overcome the tension of the spring and move the spool valve.

Relay Valves. A **relay valve** is a spool valve that can have several lands and reaction areas. It is used to control the direction of fluid flow and does not control pressure. A constant supply of fluid is fed to the valve and the valve's position directs the fluid to the appropriate hydraulic passageway. If a relay valve is fitted into a bore with one inlet and three separate outlets, the position of the valve will determine where the fluid will be directed.

A relay valve is held in position in a bore by spring tension, auxiliary fluid pressure, or a mechanical linkage. Forces from hydraulic pressure or mechanical linkages oppose spring tension and move the relay valve to a different position in its bore. In each position, the valve's valley may or may not be aligned outlet ports. Fluid flows from an inlet port across the valve's valley to an outlet port, if the port is not blocked by a valve land (Figure 7-22). If the land is blocking the port, fluid flow to that circuit is stopped.

A relay valve is used in many torque converter clutch circuits. Lockup is controlled by the Transmission Control Module (TCM) based on inputs from many sensors. The TCM determines the status of the clutch by comparing the engine rpm to the speed of the turbine. The TCM further calculates the actual operating gear by comparing the turbine speed to the speed of the counter gear. When conditions are right, the TCM applies control voltage to a solenoid. When the solenoid is turned on, it sends pressure to the lockup relay valve and to engage the clutch.

Spool valves that rely on spring tension and hydraulic force for movement are sometimes called **hybrid valves**.

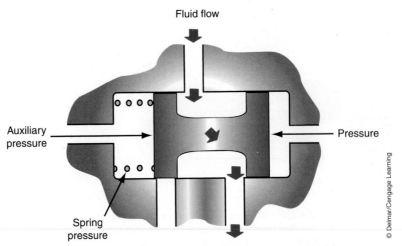

FIGURE 7-22 A relay valve responds to opposing forces on each end of the valve. A high pressure always moves to a lower pressure.

Shift Valves

Shop Manual
Chapter 7, page 344

Transmissions have several **shift valves** that provide for control of all possible gear changes. Shift valves are spool valves (Figure 7-23) that control the upshifting and downshifting of the transmission by controlling the flow of fluid to the apply devices that engage the different gears. Most shift valves are controlled by two different hydraulic pressure sources that oppose each other. As one pressure gains strength over the other, the valve moves in the direction of the lower pressure.

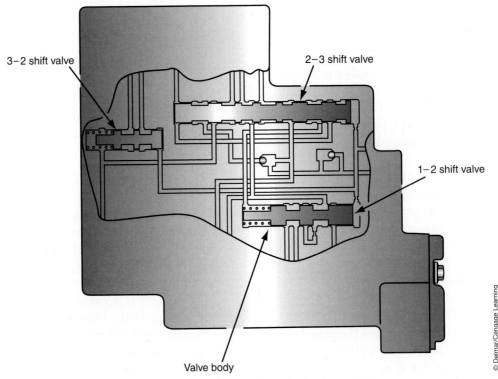

FIGURE 7-23 The shift valves in a valve body assembly.

For example, when the governor sends a signal that would normally force an upshift, throttle pressure may delay the shift, which allows for continued operation in a lower gear when the vehicle still needs the gear reduction. While operating under a load, governor pressure must overcome the throttle pressure plus the spring tension on the shift valve before it can force an upshift.

Many shift valves are held at the end of their bore by spring tension. Therefore, fluid pressure must overcome the tension of the spring before the valve will move.

Manual Valve. The **manual valve** is a spool valve operated manually within the valve body by the gear selector linkage. The valve directs fluid to components when a drive range is selected. If the driver selects the Drive range, the manual valve is moved into a position that allows fluid pressure to flow across the manual valve valley through its outlet port to activate the forward circuit. If the gear selector is moved to the reverse position, the manual valve is moved to open the reverse inlet and outlet ports to activate the reverse circuit.

Kickdown Valve. A valve body may be also fitted with a kickdown circuit to provide a downshift when additional torque is needed. When the throttle is quickly opened, throttle pressure rapidly increases and is directed to the kickdown valve. This moves the kickdown valve to allow mainline pressure to flow against the shift valve. The spring tension on the shift valve, kickdown pressure, and throttle pressure push the end of the shift valve moving it the downshift position, forcing a downshift.

A transmission can also automatically downshift under a load, such as climbing a hill. During this time, throttle pressure exceeds governor pressure and forces a downshift.

Some transmissions have a kickdown solenoid operated by a switch located on the linkage of the throttle pedal. When the throttle is fully depressed, the solenoid allows pressure to move against the kickdown valve, which forces an immediate downshift.

Kickdown is a forced downshift accomplished by increasing throttle pressure using a kickdown valve.

PRESSURES

Hydraulic pressure is used to apply the clutches, brakes, and bands that hold or connect members of the planetary gearsets. The fluid must be pressurized in order to flow through the transmission and lubricate and cool components. The required amount of pressure varies with the hydraulic circuit within the transmission. High pressure is required during demands for high torque. The high pressure is needed to reduce the chance of clutch and band slippage by applying them firmly. Lower pressure is required when operating at cruising speeds.

All automatic transmissions use three basic pressures to control their operation: mainline pressure, throttle pressure, and governor (most do not have a governor) pressure (Figure 7-24).

Mainline Pressure

Mainline pressure is a regulated pump flow that is the source of all other circuits in the transmission. Mainline pressure may fill the torque converter and typically lubricates the transmission, supplies fluid to the valve body, and is used to apply the brakes and clutches.

The pump provides a flow of fluid to the mainline circuit. Engine speed and the pressure regulator valve limit this flow. The pressure regulator valve develops mainline pressure when it blocks or resists the flow of fluid. Mainline pressure is the source of all other pressures used by the transmission. It is the pressure used to engage or apply the clutches and bands within the transmission.

The pressure regulator valve controls line pressure to meet the needs of the transmission regardless of operating speeds. As engine speed increases, the pump works faster and delivers a greater flow than when the engine is running at low speeds. The pressure regulator keeps line

Shop Manual
Chapter 7, page 329

Mainline pressure is sometimes referred to as baseline or **line pressure**.

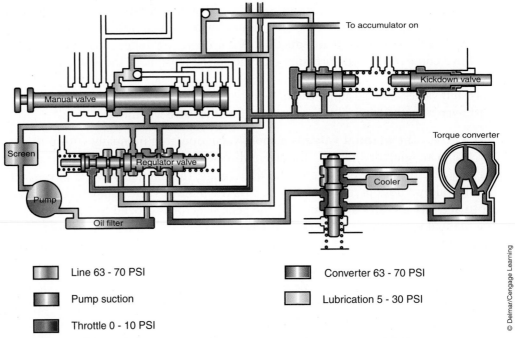

To accumulator on

Kickdown valve

Torque converter

Manual valve

Screen

Regulator valve

Cooler

Pump

Oil filter

Line 63 - 70 PSI

Pump suction

Throttle 0 - 10 PSI

Converter 63 - 70 PSI

Lubrication 5 - 30 PSI

© Delmar/Cengage Learning

FIGURE 7-24 The various pressures in a typical automatic transmission.

pressure from building to a point where it could damage the transmission. Line pressure also increases as the pump fills all of the circuits in the hydraulic system. Again, the pressure regulator prevents excessive pressure buildup by exhausting some of the fluid flow from the pump when the pressure reaches a predetermined limit.

To describe how the line pressure is used, a look at a simplified hydraulic circuit is presented. The oil pump, driven by the torque converter, rotates with the engine's crankshaft. Therefore, when the engine is running, the pump is operating. ATF is pulled from the fluid reservoir through a filter and the pump pressurizes the fluid. This pressurized fluid becomes line pressure and is controlled by the regulator valve. This regulated line pressure moves through the regulator valve to the main valve body and then through the manual valve. At the same time, line pressure enters the torque converter. From the manual shift valve, the fluid flows to the appropriate circuits for the various reaction members at the gears.

When the PCM commands the appropriate shift solenoids to turn on or off, line pressure is directed to the shift valve. This action causes the shift valve to move and pressurized fluid flows to the corresponding reaction member. Line pressure is also present at the lockup clutch solenoid and the fluid will engage the clutch when the solenoid is commanded to open by the PCM. When a gear change is necessary, the PCM will stop fluid flow to the lockup clutch and control the shift solenoids to send fluid to the next reaction member and exhaust the fluid at the previous one. Again, once conditions are met, the PCM will open the lockup clutch solenoid to apply line pressure to the clutch.

When one pressure is designed to work against another, we are utilizing the balanced valve principle, which is commonly used in pump pressure regulator valves, governor valves, and throttle valves.

Throttle Pressure

Throttle pressure is a regulated fluid flow that varies with engine load or throttle position. Throttle pressure interacts with governor pressure to control shifting.

Throttle pressure is a signal pressure. The amount of throttle pressure for many transmissions depends directly on the engine's load. Engine load is sensed by mechanical linkages relaying the amount of throttle opening, the effect of engine vacuum on a modulator, a combination of these two, or various sensors tied to a PCM. Many late-model transmissions use a pulse-modulated solenoid, controlled by the computer, to regulate throttle pressure in response to load inputs to the computer. Most new transmissions rely on signals from a MAP sensor to determine when to shift.

Governor Pressure

Governor pressure is a regulated fluid flow that varies with vehicle speed. When governor pressure is higher than throttle pressure, an upshift occurs. Governor pressure is also a boosted line pressure. Early transmission designs used a mechanical governor made of springs, weights, and a spool valve to create this signal. Late-model transmissions use a computer-controlled solenoid to control this pressure based on vehicle speed.

A transmission's change of gears is actually caused by the movement of a shift valve. When the shift valve moves, line pressure is directed to the appropriate apply device. Since throttle and governor pressures only control the movement of the shift valve, they are only signal pressures. Signal pressures control the direction of the line pressure by their action on shift valves.

Most electronically controlled transmissions do not use governors; rather, a speed sensor is used to help the PCM or TCM decide when to activate a shift solenoid.

> EATs do not have a governor; therefore, there is no governor pressure rather the EPC responds to the inputs from speed sensors.

Boost Pressures

During certain operating conditions, fluid pressure must be increased above its mainline pressure. Increased pressure is needed to hold brakes and clutches more tightly. Increasing the pressure above normal mainline pressures allows the vehicle to overcome heavy loads, such as pulling a trailer or climbing a steep hill.

When the vehicle is placed under heavy load, throttle pressure is applied to a booster valve at the pressure regulator. This pressure acting on the booster valve assists the pressure regulator valve's spring in pushing the regulator valve up against the line pressure. Line pressure is able to continue to increase until the pressure on the regulator valve overcomes the spring in the pressure regulator and throttle pressure. The pressure regulator valve now opens its exhaust port with a boosted line pressure, which is used to hold the reaction members tightly to resist slippage. Some transmissions are equipped with two boost valves. The second boost valve is used for improved shifting into reverse gear and for greater holding power when operating in reverse.

Newer transmissions use an EPC solenoid to increase the pressure to provide tight clamping of the brakes and clutches.

PRESSURE CONTROL VALVES

Pressure control valves are widely used to control line pressure, as well as the pressures used to control or signal shift points. Hydraulic force can move the regulating spool valves against spring pressure or move the valve back and forth in its bore. Regulator valves have a spring that holds the valve in a position until the fluid pressure is great enough to compress the spring. At that time, the valve moves and allows some fluid to exhaust into the fluid reservoir. Once the fluid begins to exhaust, the pressure at the valve decreases.

The valve remains in the exhausted position until the fluid's pressure cannot work against the tension of the spring. At this point, the valve closes and pressure is able to build again. When the pressure again reaches a specified level, the port will open to allow the pressure to drop. The toggling back and forth controls the fluid pressure at the valve.

Several different pressure control valves are found in transmissions. Each model may have different numbers and types of these valves, but the most common control valves include pressure regulator, pressure relief, and governor and/or throttle valves.

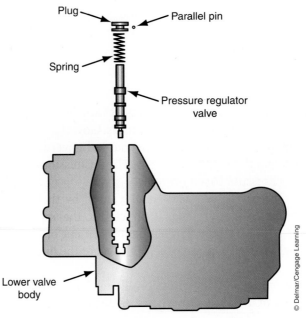

Plug — Parallel pin

Spring — Pressure regulator valve

Lower valve body

© Delmar/Cengage Learning

FIGURE 7-25 A pressure regulator valve.

Shop Manual

Chapter 7, page 332

The fluid pressure from the regulator is sometimes referred to as the main control pressure.

Pressure Regulator Valves

Most automatic transmissions have a positive displacement oil pump. This means that as the pump delivers fluid to the transmission, the pressure increases with an increase in operating speed. Enough pressure can be generated to stall the pump. To prevent this stalling, a **pressure regulator valve** is normally located in the valve body (Figure 7-25).

Also, since engine power is used to create fluid pressure, creating high pressure when it is not needed wastes engine power, which affects fuel economy and vehicle performance. High pressure also means the fluid will have a higher temperature. The high temperature can break down ATF causing it to lose some of its ability to lubricate and clean the internals of the transmission. High pressure, when not required, will also cause harsh shifting as the brakes and clutches apply too quickly and with great force. The pressure regulator valve allows the pump to produce enough pressure for the transmission to function but will not allow excessive pressure (Figure 7-26).

The resulting pressure from the regulator valve is called line pressure, which is the highest oil pressure in a transmission. Line pressure is used to apply most clutches and brakes. The valve controls line pressure before it is sent through the transmission. The position of the valve in its bore determines the amount of pressure at its outlet. Its position is determined by throttle pressure, line pressure, and spring tension. Until the line pressure reaches a specified limit, the valve's spring keeps it in position to prevent fluid from exhausting, thereby allowing pressure to build. When the pressure exceeds that limit, the spring tension is overcome and some of the fluid is exhausted and the pressure decreases.

Nearly all pressure regulating valves are spool valves with several lands. The different lands allow fluid to flow into other circuits of the transmission, such as the torque converter. The valve maintains the desired pressure for the circuits by opening and closing an exhaust port. This toggling back and forth is an ongoing process while the transmission is operating. The valve's movement results, primarily, from the difference between the fluid's pressure and the tension of the spring. When both forces are equal, the valve is in its balanced position.

When higher than normal pressure is needed, such as when operating under load, throttle pressure is applied to the end of the valve. This pressure along with the spring

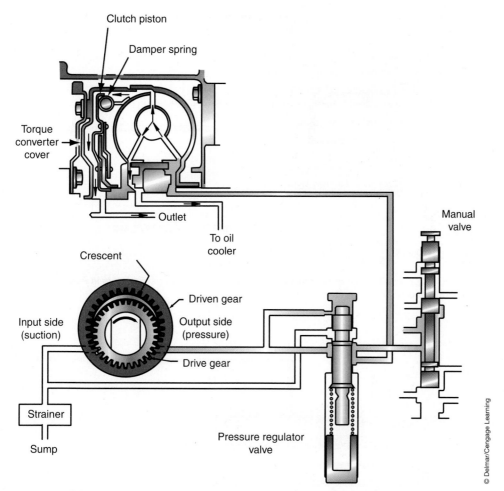

FIGURE 7-26 Oil pump output is regulated by the pressure regulator valve.

Labels in figure:
- Clutch piston
- Damper spring
- Torque converter cover
- Outlet
- To oil cooler
- Crescent
- Input side (suction)
- Driven gear
- Output side (pressure)
- Drive gear
- Strainer
- Sump
- Pressure regulator valve
- Manual valve

© Delmar/Cengage Learning

tension keeps the valve in position to block the exhaust and pressure increases. As throttle pressure decreases, the valve is held in position by the spring and normal regulating valve operation continues. Typically, the more the throttle is open, the more line pressure will increase.

When the manual shift valve is in the Park position (Figure 7-27), fluid is flowing through the transmission but has very low pressure. Once the manual shift valve is moved, fluid from the pump is sent to the pressure regulator valve. As the pressure on the valve increases beyond a predetermined level, normally about 60 psi, the regulator valve is moved against the tension of the springs. This movement opens an additional outlet port that sends fluid out to the torque converter and to the appropriate apply device (Figure 7-28). Fluid flows into the converter circuit and the converter will become pressurized.

Line pressure may also increase in response to torque demands. This is accomplished by monitoring the stator's torque reaction. The torque converter's stator is splined to the stator shaft. At the end of the stator shaft is a pressure regulator spring cap. When the vehicle has a high torque demand, such as during acceleration or going up a hill, the stator's reaction to the torque demand moves the stator shaft. The shaft in turn pushes the regulator spring cap in proportion to the reaction. This causes the stator reaction spring to compress and the regulator valve moves to increase the line pressure. The line pressure reaches its maximum value when the torque reaction by the stator is at its maximum.

> The pressure regulator valve may block off fluid flow to the converter if there is a large decrease in pressure.

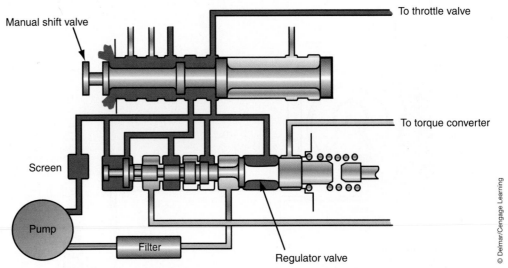

FIGURE 7-27 The path of fluid in a transmission when the transmission is in Park and the engine speed is low.

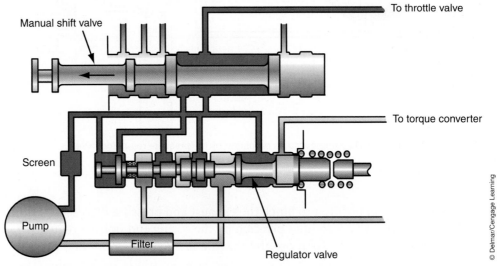

FIGURE 7-28 The path of fluid in a transmission when the transmission is not in Park and power can be transmitted through the torque converter into the transmission.

Relief Valves

A check valve fitted with a calibrated spring designed to prevent the valve from opening until a specific pressure is reached is called a pressure relief valve (Figure 7-29). This valve is used to protect the system from damage due to excessive pressure. When the pressure builds beyond the rating of the spring, the valve will open and reduce the pressure by exhausting some fluid back to the reservoir. After the pressure decreases, the valve again closes allowing normal fluid flow.

Electronic Pressure Controls

The hydraulic pressure in many late-model transmissions is controlled by a pressure control solenoid (Figure 7-30), which is controlled by the TCM. An EPC solenoid regulates pressure according to the current that flows through its windings. The resultant magnetic field

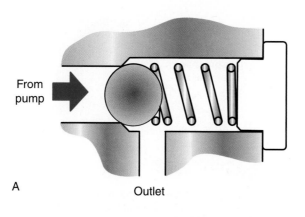

A

From
pump →

Outlet

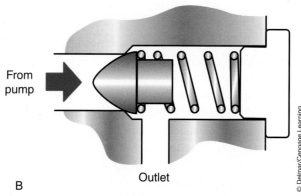

From
pump →

Outlet

B

© Delmar/Cengage Learning

FIGURE 7-29 A relief valve is typically a **(A)** ball
check valve or a **(B)** poppet valve with a spring
holding it in position.

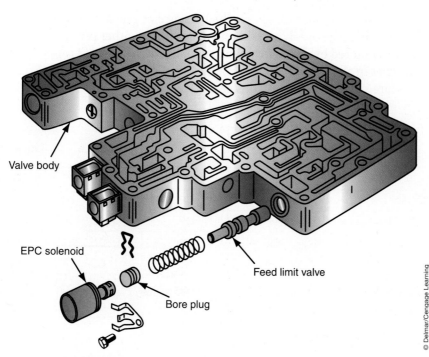

Valve body

EPC solenoid

Bore plug

Feed limit valve

© Delmar/Cengage Learning

FIGURE 7-30 Location of an Electronic Pressure Control (EPC)
solenoid in a valve body.

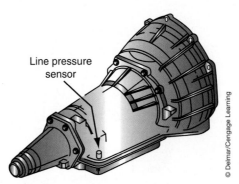

FIGURE 7-31 A typical location for a line pressure sensor in a transmission.

moves the solenoid's valve, which in turn regulates the pressure. The current is provided by the TCM. The PCM changes the amperage to the solenoid by altering the duty cycle of the solenoid. High amperage (long duty cycle) from the PCM causes minimum line pressure and low amperage (short duty cycle) results in high pressure.

The line pressure is monitored by a line pressure sensor (Figure 7-31). The TCM calculates the desired line pressure based on inputs from the transmission and engine. It also calculates torque input to the transmission and uses this value to determine the desired line pressure. The TCM compares the desired line pressure with that measured by the line pressure sensor and adjusts the duty cycle of the solenoid accordingly. When the solenoid if turned off, there is maximum line pressure and the pressure is minimized when the solenoid is turned on.

Pressure Switches

Pressure switches are used to let the TCM know if there is hydraulic pressure at particular band or clutch circuit. These switches do not measure the pressure, rather they merely indicate whether pressure is present or not. When the switch is exposed to hydraulic pressure, the switch is turned on. This supplies the ground for the switch's circuit and this action informs the TCM that pressure is present. Based on the presence of pressure, the TCM assumes the shift solenoids are working properly and that no major leak or other problem is present in the hydraulic circuit monitored by the switch. Since not all switches should report pressure from all pressure switches in all gears, the TCM's memory has the expected normal readings from the various pressure switches in the various gears (Figure 7-32).

Gear position	L/R switch	2C switch	4C switch	UD switch	OD switch
R	OP	OP	OP	OP	OP
P/N	CL	OP	OP	OP	OP
1st	CL*	OP	OP	CL	OP
2nd	OP	CL	OP	CL	OP
2 Prime	OP	OP	CL	CL	OP
3rd	OP	OP	OP	CL	CL
4th	OP	OP	CL	OP	CL

CL = Switch is closed (pressure indicated)

OP = Switch is open (no pressure indicated)

*L/R closed below 100 output rpm in Drive and Manual 2. L/R on in Manual L.

FIGURE 7-32 Examples of the normal state of the pressure switches in various gears and at various reaction members.

GOVERNORS

A **governor** is a control device typically driven by the transmission's output shaft. It senses road speed and sends a fluid pressure signal to the valve body to either upshift or downshift. Basically the governor valve controls transmission shift points according to vehicle speed.

Shop Manual
Chapter 7, page 347

Fluid flows from the pump throughout the transmission and eventually into the governor circuit. When the flow of fluid reaches the shift valves, the forces behind the valves offer a resistance. This resistance to fluid flow causes a build up of pressure in the governor circuit. Depending on the type of governor, it will exhaust fluid flow or restrict fluid flow from the circuit, thereby regulating pressure. The governor pressure increases with road speed.

Many different designs of governors have been used in automatic transmissions. All governors, regardless of the design, provide a variable pressure in relation to road speed. The governors rotate with or are turned by the transmission's output shaft. Governors are mounted on the transmission case and driven by a worm gear on the output shaft or they are mounted directly on the output shaft. On many FWD vehicles, the governor is driven by a gear on the final drive unit.

A governor assembly is normally a small separate valve body with fluid passages for line pressure, governor pressure, and an exhaust port to send fluid back to the reservoir. When the vehicle is stopped, fluid to the governor is blocked. As the vehicle begins to move, mainline fluid flows into the governor circuit, which sends fluid to the shift valves.

Types of Governors

Governors respond to vehicle speed through the action of movable weights that respond to centrifugal force. As the speed of the output shaft increases, centrifugal force moves the weights farther and farther from their rotating axis. The movement of the weights causes the governor valve to direct more fluid flow to the shift valve. As the amount of fluid increases so does the fluid's pressure. Eventually the spring tension on the shift valve is overcome and the valve moves to the upshift position. This causes an upshift as pressurized fluid flows out of an outlet port at the valve body and through the connecting worm tracks to engage the apply device for the next higher gear. A decrease in output shaft speed will result in a decrease in pressure and will cause a downshift.

Most governors have one or two valves controlled by primary and secondary weights and springs. The heavier primary weights move out at low speeds and the lighter secondary springs move out at higher speeds.

There are three common designs of governors: shaft-mounted, gear-driven with check balls, and gear-driven with a spool valve (Figure 7-33).

Shaft-Mounted Governor. The shaft-mounted governor has a spring-loaded spool valve attached to the output shaft. The governor's weights are mounted so that the primary weight moves against spring tension while the secondary weight moves against mainline pressure. The combined force of the weights and springs on one side of the spool valve is much greater than the force on the other side. Therefore as the output shaft rotates faster, the valve is pulled inward. As speed increases, fluid is exhausted through a port that feeds pressure to the shift valves. With this increased speed, the governor gradually closes the exhaust circuit to allow governor pressure to build, until the governor pressure equals mainline pressure. At this point, fluid is exhausted and the pressure decreases. After the fluid is exhausted, the valve once again is pulled inward by the rotating weights, allowing mainline pressure into the governor passage and the cycle starts again. When the vehicle is stopped, the governor spring forces the valve closed which blocks mainline pressure from the governor circuit.

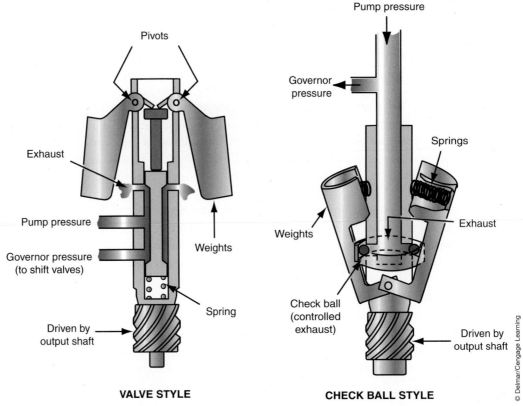

VALVE STYLE   **CHECK BALL STYLE**

FIGURE 7-33 Gear-driven governors with a spool valve and with check balls.

Gear-Driven Governor with Check Balls. Gear-driven governors with check balls mount to the transmission at a right angle to the output shaft. The centrifugal weights work directly on the check balls that open or close two fluid passages. When the vehicle is stopped, the check balls unseat and fluid flow to the governor is allowed to bleed off. As vehicle speed increases, the weights apply force to seat the check balls and restrict the flow of escaping fluid. This increases governor pressure, up to the point where no fluid escapes and governor pressure equals mainline pressure.

Gear-Driven Governor with a Spool Valve. The gear-driven governor with a spool valve (Figure 7-34) operates much like the shaft-mounted governor; however, its mechanical operation is different. The weights move by centrifugal force and indirectly control the spool valve through levers attached to the weights. The levers act against a spool valve located in the governor shaft. As vehicle speed increases, centrifugal force moves the weights out causing the spool valve to move. This closes the exhaust port to increase the pressure and directs the fluid to the shift valves.

Vehicle Speed Sensors

AUTHOR'S NOTE: Don't forget, there is the possibility that many other speed sensors are attached to the transmission, such as the input shaft and output shaft speed sensors. Most of these operate in the same way as a VSS, but have a different, but related, function in the grand scheme of things.

Electronically controlled transmissions do not rely on hydraulic signals from a governor to determine when to shift, nor are they fitted with a speedometer cable. Instead, they use permanent magnet (PM) generators to sense vehicle speeds. This sensor is the VSS (Figure 7-35).

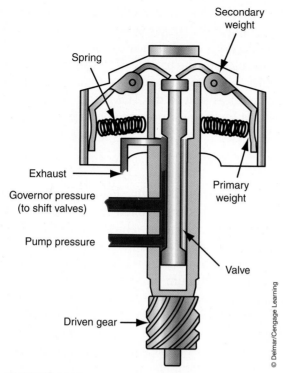

FIGURE 7-34 Components of a gear-driven governor.

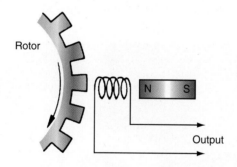

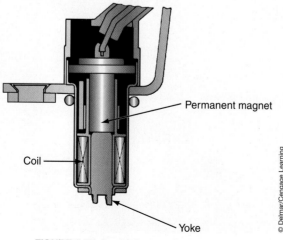

FIGURE 7-35 A vehicle speed sensor.

PM generators are often referred to as pulse generators.

A PM generator is an electronic device that utilizes a magnetic pickup sensor located on the transmission/transaxle housing and a trigger wheel mounted on the output shaft. The trigger wheel has a number of projections evenly spaced around it. As the wheel rotates, the projections move by the magnetic pickup. The passing of the projections produces a voltage in a coil inside the pickup assembly. This voltage is actually AC voltage, but is changed to a digital DC voltage by a converter before it is sent to the TCM. The frequency of the pulsations and the amount of voltage generated by the pick-up assembly is translated into vehicle speed by the computer.

Load Sensors

Shift timing and quality should vary with engine load, as well as with vehicle speed. Throttle pressure is used to increase the pressure applied to the apply devices to hold them tightly and reduce the chance of slipping while the vehicle is operating under heavy load. Throttle pressure interacts with line pressure to coordinate shift points and shift quality with vehicle and engine load. Most current transmissions rely on a MAP or similar sensor (Figure 7-36).

On some transmissions, a vacuum modulator is used to control throttle pressure (Figure 7-37). Transmissions that are not equipped with a vacuum modulator use a throttle cable (Figure 7-38) or electronic devices, such as a MAP sensor, to sense engine load and change fluid pressures. It is not safe to assume that all electronic transmissions are not equipped with a vacuum modulator, some are. For example, GM's 4T60-E uses a vacuum modulator, in addition to electronic inputs and shift solenoids, to control shift timing and quality.

The terms *modulator* and *throttle* are used to describe the same function and pressure in a transmission. Although a vacuum modulator is very different from an engine load sensor, they both generate an engine torque signal based on engine load. As a result of this signal, the transmission can have many different possible shift points. Each of those shift points is dictated by engine speed and load. This feature allows for a delay in shifting when load demands are high.

FIGURE 7-36 A MAP sensor.

© Delmar/Cengage Learning

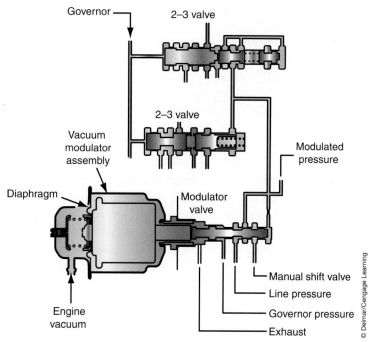

Governor

2–3 valve

2–3 valve

Vacuum
modulator
assembly

Modulated
pressure

Diaphragm

Modulator
valve

Engine
vacuum

Manual shift valve

Line pressure

Governor pressure

Exhaust

© Delmar/Cengage Learning

FIGURE 7-37 Action of a vacuum modulator and the
modulated pressure it generates.

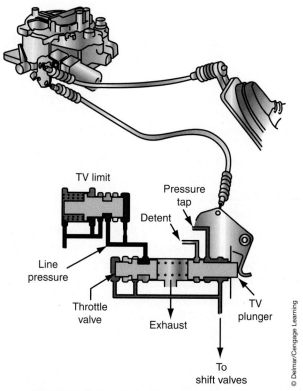

TV limit

Pressure
tap

Detent

Line
pressure

Throttle
valve

Exhaust

TV
plunger

To
shift valves

© Delmar/Cengage Learning

FIGURE 7-38 A typical cable-type throttle
pressure control system.

OIL CIRCUITS

The valve body is the master flow control for an automatic transmission. It contains the passages and numerous bores fitted with many spool and other valves. Each passage, bore, and valve forms a specific **oil circuit** with a specific function. These oil circuits can be traced and followed through the use of flow diagrams. These diagrams can also be used to clearly see how each valve functions to allow the transmission to operate in any particular gear range.

By tracing through an oil circuit, the fluid flow through passages, bores, and valves to the components that make the actual shift can be traced.

To fully understand how a particular transmission works, locate a flow diagram for that transmission and trace the oil circuit for each gear. Automatic transmission hydraulic systems vary with each design and model, as well as model year. By carefully analyzing the oil circuits, you can gain a better understanding of the operation of an automatic transmission.

The following are some guidelines to follow that will make tracing through an oil circuit's flow diagram easier:

1. Trace through one circuit at a time. (Flow diagrams are available for each of the possible gear ranges plus the following circuits: supply, main control pressure, converter and cooler, governor, throttle valve, boosted throttle, modulated throttle, accumulator, and converter clutch.)
2. Begin at the applied member or devices for a particular gear and trace backwards.
3. Determine what type and the amount of pressure the manual shift valve circuit and the modulator valve circuit is receiving.
4. Trace the main path of the circuit first, then trace the effect of the alternate circuits.
5. Since all diagrams do not show the actual position of the valves, follow the flow based on your knowledge.
6. Flow diagrams do not show the exact location of the components so don't be misled by them. They are simply a summary of the flow.
7. Trace through the fluid passages as they are drawn.
8. Pay particular attention to the direction of the fluid flow and the placement of the orifices and check valves.

SUMMARY

- An automatic transmission receives engine power through a torque converter, which is indirectly attached to the engine's crankshaft.
- The various gear ratios are selected by the transmission according to engine and vehicle speeds, engine load, and other operating conditions.
- The valve body regulates and directs the fluid flow to provide for the gear changes.
- The basic types of valves used in automatic transmissions are ball, poppet, or needle check valves, and relief valves, orifices, and spool valves.
- A simple check valve is used to block a hydraulic passageway. It normally stops fluid flow in one direction, while allowing it in the opposite direction.
- A relief valve prevents or allows fluid flow until a particular pressure is reached. Then it either opens or closes the passageway. Relief valves are used to control maximum pressures in a hydraulic circuit.
- Orifices are used to regulate and control fluid volumes.
- Spool valves are normally used as flow-directing valves. They are the most commonly used type of valve in an automatic transmission.

- Fluid passages in the bore are closed and opened as the lands of the spool cover and uncover the passage openings.
- The movement of a spool valve is controlled by either mechanical or hydraulic forces.
- A relay valve is a spool valve with several spools, lands, and reaction areas. It is used to control the direction of fluid flow. A relay valve does not control pressure.
- The valve body is machined from aluminum or iron castings and has many precisely machined bores and fluid passages. Fluid flow to and from the valve body is routed through passages and bores.
- A valve body can be composed of two or three main parts: a valve body, separator plate, and transfer plate. The separator and transfer plates are designed to seal off some of the passages and they contain some openings that help control and direct fluid flow through specific passages. The parts are typically bolted together and mounted as a single unit to the transmission housing.
- All nonelectronically shifted automatic transmissions use three basic pressures to control their operation: mainline pressure, throttle pressure, and governor pressure.
- Mainline pressure is the source of all other pressures in the transmission.
- Governor pressure is a regulated hydraulic pressure that varies with vehicle speed.
- Throttle pressure is a regulated pressure that varies with engine load or throttle position.
- Pressure regulator valves use principles of both a pressure relief valve and a spool valve.
- The pressure regulator valve has three primary purposes and modes of operation: filling the torque converter, exhausting fluid pressure, and establishing a balanced condition by maintaining line pressure.
- A governor senses road speed and sends a fluid pressure signal to the valve body to either upshift or downshift.
- There are three common designs of governors: shaft-mounted, gear-driven with check balls, and gear-driven with a spool valve.
- A PM generator is an electronic device that utilizes a magnetic pickup sensor located on the transmission/transaxle housing and a trigger wheel mounted on the output shaft.
- Throttle pressure causes an increase in the fluid pressure applied to the apply devices of the planetary units to hold them tightly to reduce the chance of slipping while the vehicle is operating under heavy load.
- Transmissions that are not equipped with a vacuum modulator use a throttle cable or electronic devices to sense engine load and change fluid pressures.
- The throttle valve converts mainline pressure into a variable throttle pressure based on the position of the throttle plate.
- Engine load can be monitored electronically through the use of various electronic sensors that send information to an electronic control unit, which in turn controls the pressure at the valve body. The most commonly used sensor is the manifold absolute pressure (MAP) sensor. The MAP sensor senses air pressure in the intake manifold.
- Shift valves control the upshifting and downshifting of the transmission by controlling the flow of fluid to the apply devices that engage the different gears.
- The manual valve is a spool valve operated manually by the gear selector linkage.
- The valve body is the master flow control for an automatic transmission. It contains the passages and numerous bores fitted with many spool-type and other type valves. Each passage, bore, and valve forms a specific oil circuit with a specific function.
- All transmissions are designed to change gears at the correct time, according to engine speed, load, and driver intent.

TERMS TO KNOW

Ball-type check valve

Governor

Hybrid valves

Land

Line pressure

Manual valve

Oil circuit

Orifice

Poppet valve

Pressure regulator valve

Reaction area

Relay valve

Shift valve

Spool valve

Transfer plate

Two-way check valve

Valley

REVIEW QUESTIONS

Short-Answer Essays

1. Describe the path of line pressure when a GM 4T60-E is operating in first gear with the gear selector in overdrive.

2. What are the basic types of valves used in an automatic transmission's valve body?

3. What is the purpose of a check valve?

4. What supplies mainline pressure and what regulates it?

5. What is the purpose of the transfer and separator plates in a valve body assembly?

6. Explain each of the different operating pressures in a typical transmission.

7. Name the three common types of governors found in transmissions.

8. What is the purpose of the shift valves?

9. How does current flow to an EPC solenoid affect line pressure?

10. How can a ball-type check valve be used as a two-way check valve?

Fill in the Blanks

1. _____ are commonly used to redirect fluid flow or to stop it from backflowing to the reservoir.

2. _____ _____ valves change the pressure of the oil to control the shift points of the transmission. _____ _____ valves direct the pressurized oil to the appropriate reaction members, which cause a change in gear ratios.

3. During certain operating conditions, fluid pressure must be _____ _____ pressure is needed to hold brakes and clutches more tightly. This allows the vehicle to overcome _____ _____, such as pulling a trailer or climbing a steep hill.

4. _____ valves are valves that change the pressure of the fluid. The pressure change occurs as the position of the valve allows some of the fluid to _____.

5. Changes in amperage to the EPC solenoid affect the pressure in a transmission. Normally, _____ amperage from the PCM causes minimum line pressure and _____ amperage results in high pressure.

6. The _____ valve is a spool valve operated manually by the gear selector linkage.

7. The _____ of a spool valve ride on a very thin film of fluid in the valve's bore. As the valve moves in its bore, the lands cover and uncover ports in the valve's bore, which opens and closes _____. The valley of the valve forms a _____ that is smaller to allow fluid to pass between the lands and through the valve. The movement of the valve allows fluid to flow from various inlets to various outlets. Fluid passages in the bore are closed and opened as the _____ cover and uncover the passage openings.

8. The valve body is composed of two or three main parts: a _____ _____, _____ _____, and _____ _____.

9. All automatic transmissions use three basic pressures to control their operation: _____ pressure, _____ pressure, and _____ pressure.

10. There are three common designs of governors: _____-mounted, _____-driven with _____, and _____-driven with a _____.

MULTIPLE CHOICE

1. Which of the following is NOT a true statement about simple check valves?
   A. A simple check valve can be used to block a hydraulic passageway.
   B. A check valve stops fluid flow in one direction, while allowing it in the opposite direction.
   C. A check valve can be used to divert pressure to other hydraulic passageways.
   D. A check valve allows fluid to flow in one direction only.

2. *Technician A* says the pressure regulator valve usually has the responsibility for filling the torque converter.
   Technician B says the pressure regulator valve usually has the responsibility of keeping the fluid flowing by making sure the pressure is never balanced.
   Who is correct?
   A. A only
   B. B only
   C. Both A and B
   D. Neither A nor B

3. Which of the following statements is *not* true?
   A. A simple check valve is used to block a hydraulic passageway. It normally prevents fluid from flowing in any direction.
   B. Relief valves are used to control maximum pressures in a hydraulic circuit.
   C. Orifices are used to regulate and/or control fluid pressures.
   D. Spool valves are normally used as flow-directing valves and are the most commonly used type of valve in an automatic transmission.

4. Which of the following has the primary purpose of supplying fluid to the main valve body?
   A. Governor
   B. Manual shift valve
   C. Shift solenoid assembly
   D. Pressure regulator valve

5. Which of the following valves does not regulate pressure?
   A. Manual valve
   B. Throttle valve
   C. Pressure regulator valve
   D. Governor valve

6. *Technician A* says spool valves are the most commonly used type of valve in an automatic transmission.
   Technician B says spool valves are normally used as relief valves.
   Who is correct?
   A. A only
   B. B only
   C. Both A and B
   D. Neither A nor B

7. *Technician A* says engine load can be monitored electronically through the use of various electronic sensors that send information to an electronic control unit.
   Technician B says the most commonly used load sensor is a vehicle speed sensor.
   Who is correct?
   A. A only
   B. B only
   C. Both A and B
   D. Neither A nor B

8. *Technician A* says check valves are used to allow fluid to back flow into the fluid reservoir.
   Technician B says check valves are used to hold fluid in a cylinder or redirect fluid flow.
   Who is correct?
   A. A only
   B. B only
   C. Both A and B
   D. Neither A nor B

9. *Technician A* says mainline pressure is the source of all other pressures used by the transmission.
   Technician B says mainline pressure is the pressure used to apply clutches and brakes.
   Who is correct?
   A. A only
   B. B only
   C. Both A and B
   D. Neither A nor B

10. *Technician A* says the movement of a spool valve can be controlled by mechanical forces.
    Technician B says the movement of a spool valve can be controlled by hydraulic forces.
    Who is correct?
    A. A only
    B. B only
    C. Both A and B
    D. Neither A nor B

Chapter 8

GEARS AND SHAFTS

UPON COMPLETION AND REVIEW OF THIS CHAPTER, YOU SHOULD BE ABLE TO:

- Identify the components in a basic planetary gearset and describe their operation.

- Describe how different gear ratios are obtained from a single planetary gearset.

- Describe the construction and operation of typical Simpson gear-based transmissions.

- Describe the construction and operation of Ravigneaux gear-based transmissions.

- Describe the construction and operation of transmissions that use planetary gearsets in tandem.

- Describe the purpose of a differential.

- Identify the major components of a differential and explain their purpose.

- Describe the various gears in a differential assembly and state their purpose.

- Explain the operation of a FWD differential and its drive axles.

- Describe the different designs of four-wheel-drive systems and their applications.

- Discuss the purpose of the various shafts found in today's automatic transmissions and transaxles.

INTRODUCTION

An automatic transmission is based on the principle of levers, and automatically allows the engine to move heavy loads with little effort. As the heavy load decreases or the vehicle begins to move, less leverage or lower gear ratios are required to keep the vehicle moving. By providing different gear ratios, a transmission provides for performance and economy over the entire driving range.

Most automatic transmissions use planetary gearsets to provide for the different gear ratios. The gear ratios are selected manually by the driver or automatically by the hydraulic control system that engages and disengages the clutches and brakes used to shift gears. Planetary gears (Figure 8-1) are always in constant mesh; therefore, they allow quick, smooth, and precise gear changes without the worry of clashing or partial engagement.

SIMPLE PLANETARY GEARSETS

Planetary gears provide for the different gear ratios needed to move a vehicle in the desired direction at the correct speed. They also provide for a neutral position that allows for power input but no power output. Planetary gears are used in automatic transmissions, as well as final drive units and transfer cases, because:

- Planetary gears are compact, yet strong enough to handle great amounts of torque.
- All members of a planetary gearset share a common axis, which provides for a very compact unit.

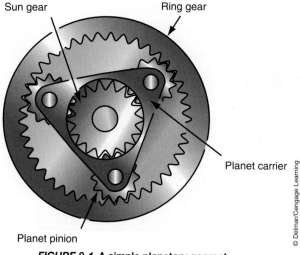

Sun gear

Ring gear

Planet carrier

Planet pinion

© Delmar/Cengage Learning

FIGURE 8-1 A simple planetary gearset.

- The gears are typically helically cut gears, which offer quiet operation.
- Planetary gears are always in constant mesh. This reduces the risk of gear damage and allows for smooth and quick gear ratio changes.
- Holding a gearset member or locking them together is easy because of the location of the different members of the gearset.
- The gears within the gearset can function independently of one another and can turn on their own centers while revolving around the sun gear.

One of the keys to understanding the power flows in a transmission is understanding the action of a planetary gearset. At the center of the planetary gearset is the sun gear. Planet **pinion gears** surround the sun gear and are mounted to and supported by the planet carrier. A planetary gearset normally has three or four planet pinion gears and each pinion gear spins on its own separate shaft. For heavy loads, the number of planet gears is increased to spread the workload over more gear teeth. The planet pinion gears are in constant mesh with the sun and ring gears. The ring gear is the outer gear and has internal teeth. The ring gear surrounds the rest of the gearset and its teeth are in constant mesh with the planet gears. Each member of a planetary gearset can act as an input or output gear, or it can be held stationary.

When the ring gear and the pinion gears are able to rotate at the same time, the pinions always rotate in the same direction as the ring gear. The sun gear, however, always rotates in the opposite direction as the pinion gears. This action can be explained by looking at the gears. The ring gear has internal gear teeth and the planet pinions and sun gears have external teeth. When an external gear drives another external gear, the output gear will always rotate in the direction opposite that of the input gear. But, when an external gear is in mesh with an internal gear, the two gears will rotate in the same direction.

When the planet carrier is the output member, it always goes in the same direction as the input gear. Likewise, when the planet carrier is in the input, the output gear member will always rotate in the same direction as the carrier.

Basic gear theory must also be considered when studying planetary gearsets. Remember, whenever a small gear drives a larger gear, output torque is increased and output speed is decreased. Also, whenever a large gear drives a smaller gear, output torque is decreased and output speed is increased (Table 8-1).

The planet carrier is the bracket in a planetary gearset on which the planet pinion gears are mounted on pins and are free to rotate.

The planet gears can be referred to as the **pinion gears**.

The ring gear is also known as the annulus or internal gear.

TABLE 8-1 The basic laws of simple planetary gear action.

Sun	Carrier	Ring	Speed	Torque	Direction
Input	Output	Held	Maximum reduction	Maximum increase	Same as input
Held	Output	Input	Minimum reduction	Minimum increase	Same as input
Output	Input	Held	Maximum increase	Maximum reduction	Same as input
Held	Input	Output	Minimum increase	Minimum reduction	Same as input
Input	Held	Output	Reduction	Increase	Opposite of input
Output	Held	Input	Increase	Reduction	Opposite of input

Increasing the engine's torque is generally known as *reduction* because there is always a decrease in the output speed, which is proportional to the increase in the output torque.

Holding a member of the gearset prevents it from turning.

Freewheeling means something is able to spin on its own shaft without transferring any useable power. Input to a member that is freewheeling is ineffective.

Neutral

When there is an input to the gearset but no reaction member, the gearset is in neutral. A reaction member is a stationary member of the gearset. Gearset members become stationary through brakes, clutches, and one-way clutches. These hold or stop a member from turning with the other members of the gearset. When no member of the planetary gearset is held stationary, there will be no output regardless of which member receives the input. This is the normal neutral position of a planetary gearset.

Consider the planetary gearset shown in Figure 8-2; the input gear is the sun gear and it rotates clockwise. The output is the planet carrier, which is held stationary by the weight of the vehicle on the drive wheels. This allows the pinions to rotate on their own shafts and drive the ring gear in a counter clockwise direction. The planet pinions and the ring gear simply rotate or freewheel around the sun gear.

In some transmissions, the input member is disconnected or de-clutched from the input shaft. The result of doing this is that there is no input to the gearset. With no input, there will be no output and the gearset is in neutral.

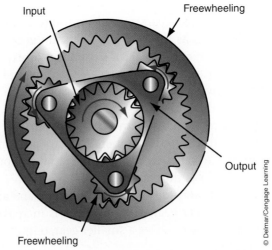

FIGURE 8-2 A planetary gearset in its neutral position.

© Delmar/Cengage Learning

Gear Reduction

When the output member turns at a slower speed than the input member, gear reduction has occurred and this results in increased torque on the output member. This can be accomplished in a planetary gearset by holding the sun gear. If power is applied to the ring gear, the planet gears will spin on their shafts in the planet carrier (Figure 8-3). Because the sun gear is being held, the spinning planet gears will walk around the sun gear and carry the planet carrier with them. This causes a gear reduction because one complete revolution of the ring gear will not cause one complete revolution of the planet carrier. The planet carrier is turning at a slower speed and therefore is increasing the torque output. By holding the sun gear, the planet carrier moves in the same direction as the ring gear.

To calculate the amount of gear reduction or the gear ratio of the gearset in Figure 8-3, count the number of teeth on the sun gear and the ring gear. The ring gear is the driving gear and has 42 teeth. The sun gear is held and is not the driven gear, but since the carrier rotates around the sun, the number of teeth on the sun is used to calculate the ratio. In this case, the sun gear has 18 teeth. The formula for determining the gear ratio when the planet carrier is the output is:

Gear reduction is a condition that exists in a gearset when the input driving gear turns faster than the output driven gear.

When the ring gear and carrier pinions are free to rotate at the same time, the pinions will always follow the same direction as the ring gear.

$$\frac{\text{Number of sun gear teeth} + \text{Number of ring gear teeth}}{\text{Number of teeth on the driving member}}$$

or

$$\frac{18 + 42}{42}$$

or

$$\frac{60}{42}$$

or

$$1.429$$

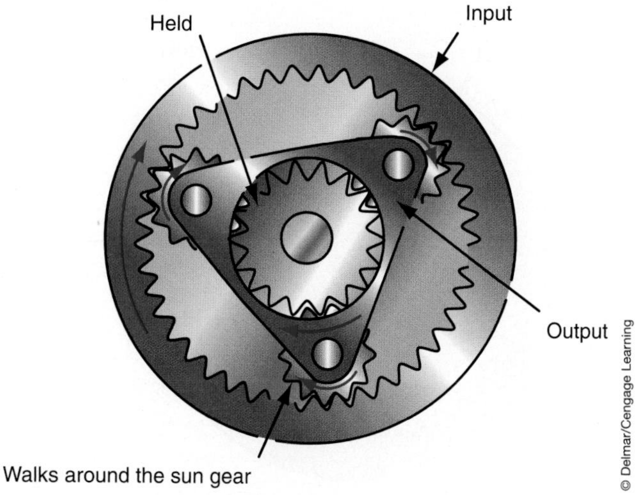

FIGURE 8-3 Gear reduction with the ring gear as the input.

Held

Input

Output

Walks around the sun gear

© Delmar/Cengage Learning

The result of this calculation identifies the gear ratio as 1.429:1. This means that if 75 ft.-lbs. were input through the ring gear, the carrier would rotate at a little more than 107 ft.-lbs. But with this torque increase came a decrease in rotating speed. If the ring gear were rotating at 1000 rpm, the carrier would rotate at approximately 700 rpm.

Gear reduction can also occur if the ring gear is held and the sun gear is the input gear (Figure 8-4). The planet pinions will rotate on their own shafts and walk around the inside of the ring gear, but in the opposite direction as the sun gear. This causes the carrier to rotate in the same direction as the sun gear. The planet carrier will rotate more slowly than it did when the ring gear drove the planet carrier. This results in greater speed reduction and therefore greater torque.

To calculate the amount of gear reduction or the gear ratio of the gearset in Figure 8-4, count the number of teeth on the sun gear and the ring gear. The sun gear is the driving gear and has 18 teeth. The ring gear is held and is not the driven gear, but since the carrier rotates around the inside of the ring gear, the number of teeth on the ring gear is used to calculate the ratio. In this case, the ring gear has 42 teeth. The formula for determining the gear ratio when the planet carrier is the output is:

$$\frac{\text{Number of sun gear teeth} + \text{Number of ring gear teeth}}{\text{Number of teeth on the driving member}}$$

or

$$\frac{18 + 42}{18}$$

or

$$\frac{60}{18}$$

or

$$3.333$$

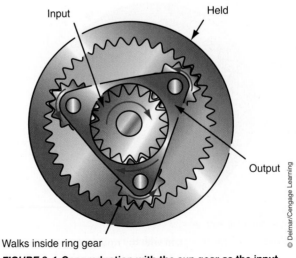

FIGURE 8-4 Gear reduction with the sun gear as the input.

Input

Held

Output

Walks inside ring gear

© Delmar/Cengage Learning

The result of this calculation identifies the gear ratio as 3.333:1. This means that if 75 ft.-lbs. were input through the ring gear, the carrier would rotate at about 250 ft.-lbs. Again, with this torque increase comes a decrease in rotating speed. If the sun gear were rotating at 1000 rpm, the carrier would rotate at approximately 300 rpm.

To have gear reduction, a planetary member must be held stationary. To do this a brake band, multiple-disc clutch, or one-way clutch is used. A brake band is used to clamp around a rotating drum and stop it from rotating. Either the sun or ring gear is attached to the drum. When the band is applied, the member in the drum becomes the reaction member of the gearset.

Multiple-friction disc packs are also used to hold a planetary member stationary. The outer edges of the steel clutch discs are splined and fit into internal splines in the transmission housing. Either the sun or ring gear is attached to the friction discs. When the assembly is applied, the friction discs squeeze against the steel discs and the sun or ring gear is locked to the case.

One-way clutches are also used to hold members. If the sun gear is connected to a one-way clutch, it will only be able to rotate freely in one direction. If the input member is the ring gear, the planet carrier will rotate with the ring gear. This would cause the sun gear, if not held by another brake, to attempt to rotate in the opposite direction. This tendency will cause the one-way clutch to lock and stop the sun gear from turning.

Holding a member is sometimes called grounding it, because often the member gets locked to the transmission housing.

Overdrive

When the planet carrier is the driving member of the gearset and the sun gear is held, overdrive occurs (Figure 8-5). When the planet carrier is the input member of the gearset and the sun gear is held, overdrive occurs. As the planet carrier rotates, the pinion gears are forced to walk around the held sun gear, which drives the ring gear faster. One complete rotation of the planet carrier causes the ring gear to rotate more than one complete revolution in the same direction. This provides more output speed but less torque or overdrive.

Overdrive is a condition in which the output member of the gearset rotates at a greater speed than the input gear. The output speed is increased but the torque is decreased.

To calculate the amount of gear ratio for this overdrive condition, again count the number of teeth on the sun gear and the ring gear. The sun gear is held and has 18 teeth. The ring gear is the driven member and has 42 teeth. The formula for determining the gear ratio when the planet carrier is the input is:

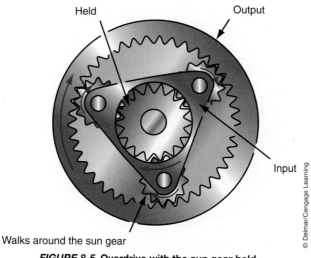

Held   Output

Input

Walks around the sun gear

© Delmar/Cengage Learning

FIGURE 8-5 Overdrive with the sun gear held.

$$\frac{\text{Number of teeth on the driven member}}{\text{Number of sun gear teeth} + \text{Number of ring gear teeth}}$$

or

$$\frac{42}{18 + 42}$$

or

$$\frac{42}{60}$$

or

$$0.700$$

The result of this calculation identifies the gear ratio as 0.7:1. This means if 75 ft.-lbs. were input through the carrier, the ring gear would rotate at about 53 ft.-lbs. With this torque decrease comes an increase in rotating speed. If the carrier were rotating at 1000 rpm, the ring gear would rotate at approximately 1429 rpm.

A higher speed overdrive is possible by holding the ring gear stationary. With input on the planet carrier, the pinion gears are forced to walk around the inside of the ring gear, driving the sun gear clockwise. The planetary carrier rotates much less than one turn to rotate the sun gear one complete revolution. The result is a great reduction in torque output and a maximum increase in output speed.

Direct Drive

> Direct drive is a condition in which the speed and torque of the output is the same as that of the input.

If any two of the planetary gearset members receive power in the same direction and at the same speed, the third member is forced to move with the other two (Figure 8-6). If the ring gear and the sun gear are the input members, the internal teeth of the ring gear will try to rotate the planetary pinions in one direction, while the external teeth of the sun gear will try to drive them in the opposite direction. This action locks the planetary pinions between the other members and the entire planetary gearset rotates as a single unit. The input is now

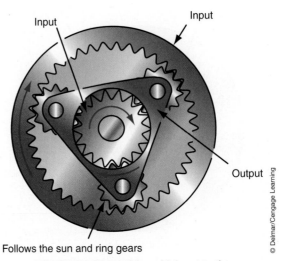

Input   Input

Output

Follows the sun and ring gears

© Delmar/Cengage Learning

FIGURE 8-6 Direct drive with input to the ring and sun gears.

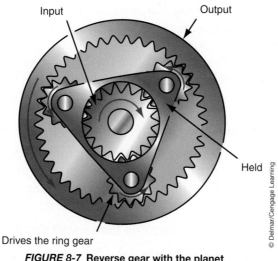

Input   Output

Held

Drives the ring gear

FIGURE 8-7 Reverse gear with the planet
carrier held and the ring gear as the output.

locked to the output, which results in direct drive. One input revolution equals one output revolution for a gear ratio of 1:1.

Reverse

If the planet carrier is held and the sun gear is rotated in a clockwise rotation (Figure 8-7), the ring gear will rotate in a counterclockwise direction. The planet gears cannot travel around the teeth of an enmeshed gear; rather, the carrier is held in place and the planet gears spin on their shafts. The sun gear spins the planet gears, which drive the ring gear in the opposite direction but at a slower speed. Therefore, the gearset is providing reverse gear with reduction.

To calculate the gear ratio of this combination, divide the number of teeth on the driven gear by the number of gear teeth on the driving gear. The sun gear has 18 teeth and is the driving member. The ring gear is the driven member and has 42 teeth. Therefore:

$$\frac{driven}{driving}$$

or

$$\frac{42}{18}$$

or

$$2.333$$

The gear ratio for this reverse with gear reduction is 2.333:1.

Reverse with overdrive is possible by applying the input to the ring gear and using the sun gear as the output member. A reversal of direction is obtained whenever the planet carrier is stopped from turning and power is applied to either the sun gear or the ring gear. This causes the pinion gears to act as idler gears, thus driving the output member in the opposite direction as the input, with a great amount of overdrive. The gear ratio, using the previous gearset, would be approximately 0.429:1.

Multiple Planetary Gearsets

To provide additional gear ratios, transmissions are equipped with more than one planetary gearset. The additional gearsets are arranged in a number of different ways, which vary among transmission models.

An idler gear is a gear that does not contribute to the input or output torque; it merely transmits torque from one gear to another, usually in the opposite direction.

From a single planetary gearset, five forward gear combinations are possible. These are maximum speed reduction, minimum speed reduction, direct drive, minimum overdrive, and maximum overdrive. Keep in mind that the actual gear ratios from the gearset depend on the size or number of teeth on each member of the set. From two planetary units working together, about 25 forward gears are possible. For example, here is a summary of 10 of those combinations. (To better understand what is happening, refer to Table 8-1.)

Speed number 1   Maximum torque multiplication

Gearset number 1: The sun gear is the input and the ring gear is held. The carrier is the output member and drives the sun gear of gearset number 2.

Gearset number 2: The sun gear is the input and the ring gear is held. The carrier is the output member.

Speed number 2   Less torque multiplication than speed number 1

Gearset number 1: The sun gear is the input and the ring gear is held. The carrier is the output member and drives the ring gear of gearset number 2.

Gearset number 2: The ring gear is the input and the sun gear is held. The carrier is the output member.

Speed number 3   Less torque multiplication than speed number 2

Gearset number 1: The sun gear is the input and the ring gear is held. The carrier is the output member and drives two members of gearset number 2.

Gearset number 2: Two members are serving as the input member and receive the output from gearset number 1. The other is the output member.

Speed number 4   Less torque multiplication than speed number 3

Gearset number 1: The ring gear is the input and the sun gear is held. The carrier is the output member and drives the ring gear of gearset number 2.

Gearset number 2: The ring gear is the input and the sun gear is held. The carrier is the output member.

Speed number 5   Less torque multiplication than speed number 4

Gearset number 1: Two members are serving as the input member and the other is the output, which drives the ring gear of gearset number 2.

Gearset number 2: The ring gear is the input and the sun gear is held. The carrier is the output member.

Speed number 6   Direct drive

Gearset number 1: Two members are serving as the input member and the other is the output, which drives two members of gearset number 2.

Gearset number 2: Two members are serving as the input member and receive the output from gearset number 1. The other is the output member.

Speed number 7   Overdrive with minimum speed reduction

Gearset number 1: The sun and ring gears are input and the carrier is the output member and drives the carrier gear of the gearset number 2.

Gearset number 2: The carrier gear is the input and the sun gear is held. The ring is the output member.

Speed number 8   Overdrive with less speed reduction than number 7

Gearset number 1: The sun is held and the carrier is the input gear. The ring gear is the output and drives the carrier gear of gearset number 2.

Gearset number 2: The carrier gear is the input and the sun gear is held. The ring is the output member.

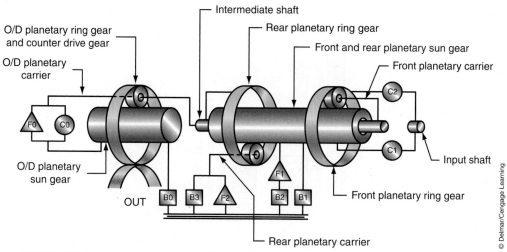

O/D planetary ring gear
and counter drive gear

O/D planetary
carrier

Intermediate shaft

Rear planetary ring gear

Front and rear planetary sun gear

Front planetary carrier

O/D planetary
sun gear

OUT

Input shaft

Front planetary ring gear

Rear planetary carrier

© Delmar/Cengage Learning

FIGURE 8-8 A compound planetary gearset. Note the different reaction and holding devices: those labeled with a "B" are brakes, "C" are clutches, and "F" are one-way clutches.

Speed number 9   Overdrive with less speed reduction than number 8

Gearset number 1: The sun is held and the carrier is the input gear. The ring gear is the output and drives the carrier gear of gearset number 2.

Gearset number 2: The carrier gear is the input and the ring gear is held. The sun is the output member.

Speed number 10   Overdrive with minimum speed reduction

Gearset number 1: The ring is held and the carrier is the input gear. The sun gear is the output and drives the carrier gear of gearset number 2.

Gearset number 2: The carrier gear is the input and the ring gear is held. The sun is the output member.

From these two planetary gearsets, many different reverse gear ratios are also available. There are many more possibilities for forward gears when a third planetary gearset is added. The limiting factor in being able to use all of the possible gear ratios is the space required to house the reaction elements. This is why most transmissions use compound gearsets. These units are comprised of two gearsets that share a common member. This allows one reaction element to control one member of both gearsets (Figure 8-8).

The two common compound arrangements are the Simpson gearset, which has two planetary gearsets sharing a common sun gear, and the Ravingeaux gearset, which has two sun gears, two sets of planet gears, and a common ring gear. Some transmissions add a simple planetary gearset to a compound gearset to gain additional gear ratios. Others use multiple gearsets connected in series to obtain the various forward speeds.

SIMPSON GEARSET

The Simpson geartrain is an arrangement of two separate planetary gearsets with a common sun gear, two ring gears, and two planetary pinion carriers (Figure 8-9). A Simpson geartrain is the most commonly used compound planetary gearset and is used to provide three forward reduction gears. The two planetary units do not need to be the same size or have the same number of teeth on their gears. The size and number of gear teeth determine the actual gear ratios obtained by the compound planetary gear assembly.

Gear ratios and direction of rotation are the result of applying torque to one member of either planetary unit, holding at least one member of the gearset, and using another member as the output (Figure 8-10). For the most part, each automobile manufacturer uses different

Shop Manual
Chapter 8, page 361

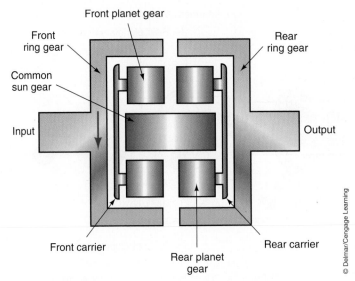

FIGURE 8-9 Components of a Simpson gearset.

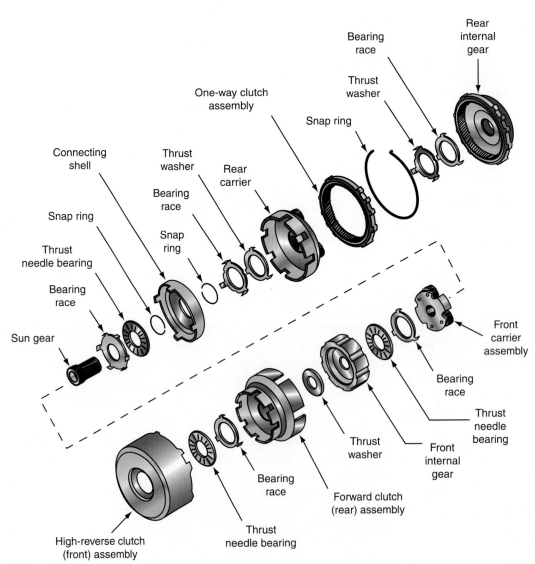

FIGURE 8-10 A Simpson planetary gearset.

parts of the planetary assemblies as input, output, and reaction members; this also varies with the different transmission models from the same manufacturer. There are also many different apply devices used in the various transmission designs.

A Simpson gearset can provide the following gear ranges: neutral, first reduction gear, second reduction gear, direct drive, and reverse. The typical power flow through a Simpson geartrain when it is in neutral has engine torque being delivered to the transmission's input shaft by the torque converter's turbine. No planetary gearset member is locked to the shaft; therefore, engine torque enters the transmission but goes nowhere else.

When the transmission is shifted into first gear (Figure 8-11), engine torque is again delivered into the transmission by the input shaft. The input shaft is now locked to the front planetary ring gears that turns clockwise with the shaft. The front ring gear drives the front planet gears, also in a clockwise direction. The front planet gears drive the sun gear in a counterclockwise direction. The rear planet carrier is locked; therefore, the sun gear spins the rear planet gears in a clockwise direction. These planet gears drive the rear ring gear that is locked to the output shaft in a clockwise direction. The result of this power flow is a forward gear reduction, normally with a ratio of 2.5:1 to 3.0:1.

When the transmission is operating in second gear (Figure 8-12), engine torque is again delivered into the transmission by the input shaft. The input shaft is locked to the front planetary ring gear that turns clockwise with the shaft. The front ring gear drives the front planet gears, also in a clockwise direction. The front planet gears walk around the sun gear because it is held. The walking of the planets forces the planet carrier to turn clockwise. Since the carrier is locked to the output shaft, it causes the shaft to rotate in a forward direction with some gear reduction. A typical second gear ratio is 1.5:1.

When operating in third gear (Figure 8-13), the input is received by the front ring gear as in the other forward positions. However, the sun gear also receives the input. Since the sun and ring gears are rotating at the same speed and in the same direction, the front planet carrier is locked between the two and is forced to move with them. Since the front carrier is locked to the output shaft, direct drive results.

To obtain a suitable reverse gear in a Simpson geartrain, there must be a gear reduction, but in the direction opposite that of the input torque (Figure 8-14). The input is received by the sun gear, as in the third gear position, and rotates in a clockwise direction. The sun gear then drives the rear planet gears in a clockwise direction. The rear planet carrier is held;

The sun gear always rotates opposite the rotation of the pinion gears.

When the planet carrier is the output, it always follows the direction of the input member.

Forward gear reduction always occurs when the planet carrier is the output member of the gearset.

When the planetary carrier is held, the output of the gearset will be in the opposite direction as the input and reverse gear will be produced.

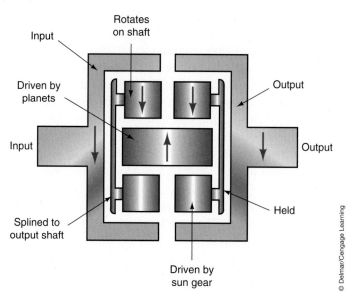

FIGURE 8-11 **Power flow in a Simpson gearset during first gear.**

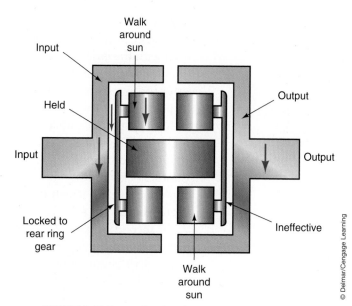

FIGURE 8-12 **Power flow in a Simpson gearset during second gear.**

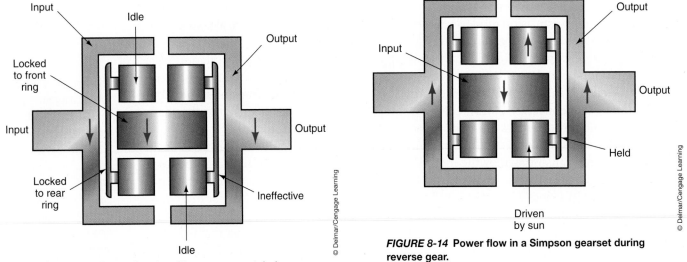

FIGURE 8-13 Power flow in a Simpson gearset during third gear or direct drive.

FIGURE 8-14 Power flow in a Simpson gearset during reverse gear.

© Delmar/Cengage Learning

therefore, the planet gears drive the rear ring gear in a counterclockwise direction. The ring gear is locked to the output shaft, which turns at the same speed and in the same direction as the rear ring gear. The result is a reverse gear with a ratio of 2.5:1 to 2.0:1.

Typically, when the transmission is in neutral or park, no apply devices are engaged, allowing only the input shaft and the transmission's oil pump to turn with the engine. In park, a pawl is mechanically engaged to a parking gear that is splined to the transmission's output shaft, locking the drive wheels to the transmission's case.

RAVIGNEAUX GEARSET

The Ravigneaux gearset is designed to use two sun gears, one small and one large. It also has two sets of planetary pinion gears—three long pinions and three short pinions. The planetary pinion gears rotate on their own shafts that are fastened to a common planetary carrier. A single ring gear surrounds the complete assembly (Figure 8-15).

The small sun gear is meshed with the short planetary pinion gears. These short pinions act as idler gears to drive the long planetary pinion gears. The long planetary pinion gears mesh with the large sun gear and the ring gear.

Typically, when the gear selector is in neutral position, engine torque, through the converter turbine shaft, drives the forward clutch drum. Since the forward clutch is not applied, the power is not transmitted through the geartrain and there is no power output.

When the transmission is operating in first gear, engine torque drives the small sun gear clockwise. The planetary carrier is prevented from rotating counterclockwise by a one-way clutch or another brake; therefore, the small sun gear drives the short planetary pinion gears counterclockwise. The direction of rotation is reversed as the short pinion gears drive the long pinion gears, which drive the ring gear and output shaft in a clockwise direction with greater torque but at a lower speed than the input (Figure 8-16).

In second-gear operation (Figure 8-17), the large sun gear is held. The small sun gear receives the input and rotates in a clockwise direction. The small sun gear drives the short pinion gears counterclockwise. The direction of rotation is reversed as the short pinion gears drive the long pinion gears that walk around the stationary large sun gear. This walking drives the ring gear and output shaft in a clockwise direction and at a gear reduction. This reduction is less than when the carrier was held.

Shop Manual
Chapter 8, page 361

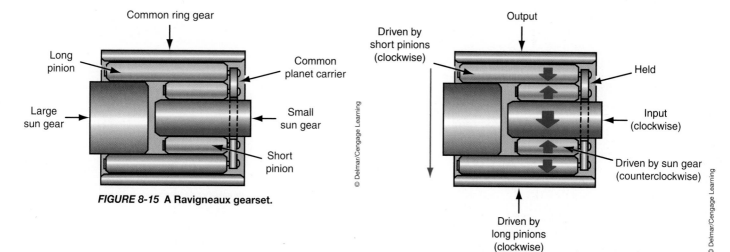

FIGURE 8-15 A Ravigneaux gearset.

FIGURE 8-16 A Ravigneaux gearset in low gear (typically first gear).

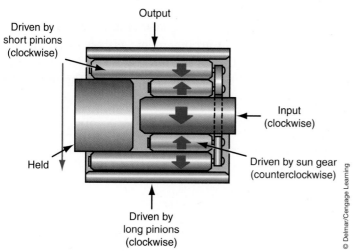

FIGURE 8-17 A Ravigneaux gearset in gear reduction (typically second gear).

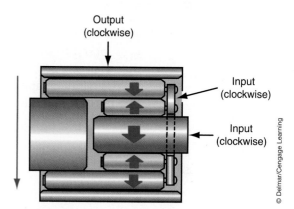

FIGURE 8-18 A Ravigneaux gearset with two inputs and in direct drive.

During third-gear operation (Figure 8-18), there are two inputs into the planetary geartrain. As in other forward gears, the turbine shaft of the torque converter drives the small sun gear in a clockwise direction. Input is also received at the planetary gear carrier. Since two members of the geartrain are being driven at the same time, the planetary gear carrier and the small sun gear rotate as a unit. The long pinion gears transfer the torque in a clockwise direction through the gearset to the ring gear and output shaft. This results in direct drive.

To operate in overdrive (Figure 8-19) or fourth gear, input drives the planetary carrier in a clockwise direction. The long pinion gears walk around the stationary large sun gear in a clockwise direction and drive the ring gear and output shaft. This results in an overdrive condition with an approximate ratio of 0.75:1.

During reverse gear operation, input is received at the large sun gear (Figure 8-20). The planetary gear carrier is held. The clockwise rotation of the large sun gear drives the long pinion gears in a counterclockwise direction. The long pinions then drive the ring gear and output shaft in a counterclockwise direction with a gear reduction.

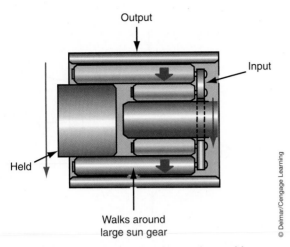

Output

Input

Held

Walks around
large sun gear

FIGURE 8-19 A Ravigneaux gearset in overdrive.

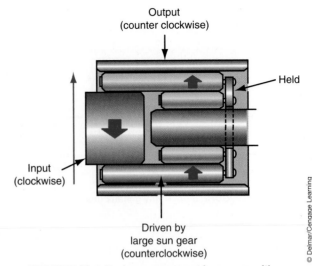

Output
(counter clockwise)

Held

Input
(clockwise)

Driven by
large sun gear
(counterclockwise)

FIGURE 8-20 A Ravigneaux gearset in reverse with
gear reduction.

© Delmar/Cengage Learning

PLANETARY GEARSETS IN TANDEM

Rather than relying on a compound gearset, some automatic transmissions use two simple planetary units in series (Figure 8-21). In this type of arrangement, gearset members are not shared; rather, certain members are locked together or are integral with each other. For example, the front planetary carrier may be locked to the rear ring gear and the front ring gear locked to the rear planetary carrier (Figure 8-22). Through such an arrangement, the output of one gearset can become the input for the next.

Using the simple planetary gearsets already discussed, we can see the effect of putting two of them in tandem or series. If one planetary unit is in gear reduction and the resulting

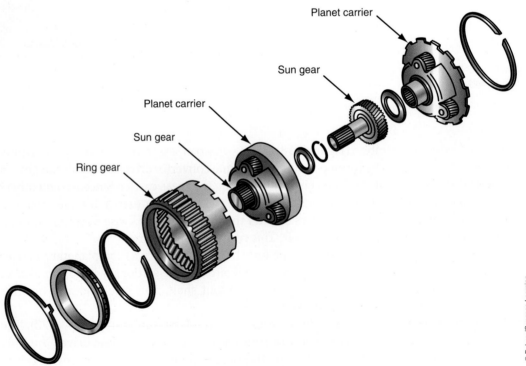

Planet carrier

Sun gear

Planet carrier

Sun gear

Ring gear

© Delmar/Cengage Learning

FIGURE 8-21 Two planetary units in series.

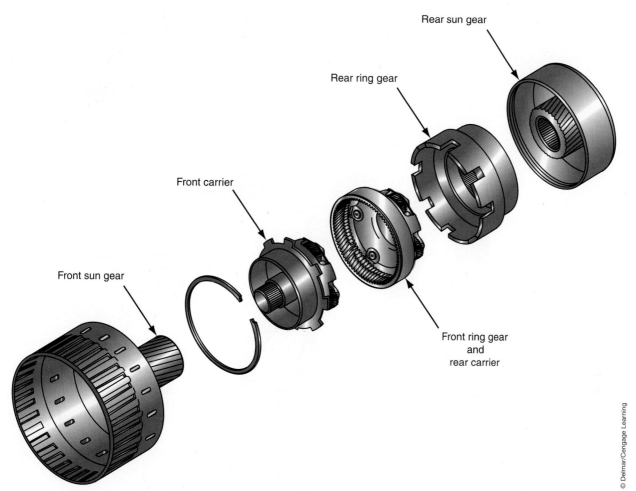

FIGURE 8-22 Two planetary units with the ring gear of one gearset connected to the planet carrier of the other.

Rear sun gear

Rear ring gear

Front carrier

Front sun gear

Front ring gear
and
rear carrier

gear ratio is 3.33:1 or 250 lbs.-ft. at 300 rpm, this output would be sent as the input to the second gearset. If the second planetary unit were in position to provide an overdrive with a ratio of 0.7:1, the gear ratio at the output shaft would be 2.33:1, which means about 175 lbs.-ft. at 430 rpm. This gear ratio falls between the ratios available during gear reduction of a simple planetary unit, thus providing an additional forward gear with reduction.

AUTHOR'S NOTE: To get a good picture of the possible gear ratios available by having planetary units in series, I suggest you play with the different possible ratios from each single unit, then put them together using the output of one as the input of the next. After you have done this for two sets, add a third. You may think this would be a boring thing to do, but you would be wrong. This can be a fun and interesting exercise.

The placing of planetary gearsets in series is one way to achieve many different forward gear ratios without the complexity of adding compound units.

Wilson Gearset. A few transmissions use the Wilson gearset. The most common application of this type of gearset is in BMW five-speed transmissions (models A5S440Z and A5S560Z). The gearset is made up of three planetary gearsets connected in tandem. The unit has three planetary carriers, three ring gears, and two sun gears. The sun gear for the second

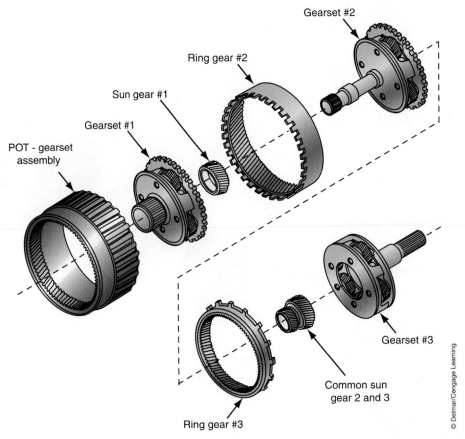

FIGURE 8-23 A Wilson gearset.

and third planetaries is a single unit sometimes called a "double sun gear." The ring gear of the first gearset, the planetary carrier of the second gearset, and the ring gear of the third planetary gearset are connected together by a "pot" (Figure 8-23). The pot is an internally and externally splined cylinder that slides over the gearsets to form a single assembly.

Lepelletier Gear Designs

Some late-model five-, six-, seven-, and eight-speed transmissions use the Lepelletier system. By using this design, the transmissions are more compact and lighter than nearly all four or five speed transmissions. The **Lepelletier** design connects a simple planetary gearset to a Ravigneaux gearset (Figure 8-24). This gearset positions the Ravigneaux gearset on the same mainshaft as the single planetary gearset, which allows the shaft to be short and durable. Although this design has been around for many years, electronics have now made it practical.

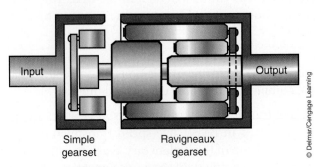

FIGURE 8-24 A Lepelletier gearset.

The transmission's input shaft is always connected to the ring gear of the single planetary unit. Input can also be simultaneously directed to the carrier and/or large sun gear in the compound gearset. The carrier of the single planetary gear is connected, by clutches, to the small or large sun gear of the compound gear. The sun gear of the single planetary gearset is connected to the housing and therefore cannot rotate. In some transmissions, the sun gear is connected by a one-way clutch to the housing and so it can freewheel or be braked. The ring gear of the compound gearset is the output member of the transmission.

As a result of this arrangement, the sun gears and the carrier of the compound planetary gearset can be driven at different speeds. The sun gears are driven by the output of the single planetary gearset and the carriers of the compound gearset are driven by the input into the transmission. As a result, the sun gears and the carrier have different speeds. This results in basically two inputs to the compound gearset and the combinations of those provide for the various forward speed ratios. Shifting between gears is accomplished by engaging a clutch at the same time the clutch for the previous gear is disengaged. This process is commonly called "clutch-to-clutch actuation."

In the six-speed model, the various speeds are provided by five multiple-disc packs and, at times, a one-way clutch (Figure 8-25). A key feature to the operation of this transmission design is that two of the clutches are applied at the same time. As in other transmissions, the clutches connect or hold specific members of the planetary gearsets. The transmission's input shaft drives the ring gear of the single planetary gear and, at the same time, can be connected to the carrier of the Ravigneaux gear. The sun gear of the planetary gear is held by a one-way clutch. The carrier of the planetary gear is connected, by clutches, to the large and small sun gears of the Ravigneaux gear. The ring gear of the Ravigneaux gearset is the output member for the transmission.

The seven-speed model is based on the six-speed except it relies on six different multiple-disc packs. It also requires that three multiple-disc packs be applied at the same time. The sun gear of the single planetary gearset is connected to the housing by a clutch to the housing, which allows it to be held. The carrier of the planetary gear is connected, by another clutch, to the small sun gear of the Ravigneaux gearset. The input shaft is always connected to the ring of the planetary gear and, in addition, can be connected to the carrier and large sun gear of the Ravigneaux gear using separate clutches. The ring gear of the Ravigneaux gearset serves as the output.

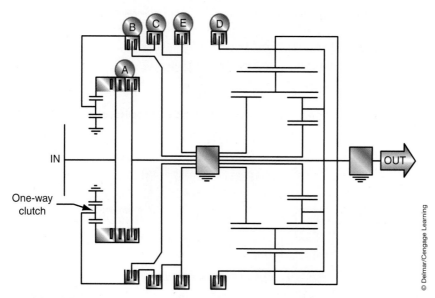

FIGURE 8-25 A six-speed Lepelletier transmission based on a planetary gear and a Ravigneaux gear and five multiple-disc packs (marked A through E).

Powerflow. The power flow through a Lepelletier gearset is based on the concept of splitting the input into two ratios: the ratio of the single set and the ratio of the second. The overall ratio is the combination of the two. Here is an example of the powerflow for a typical six-speed unit.

Drive: First Gear—For first gear, the rear carrier is held. This transfers torque from the sun gear in the rear planetary assembly to the ring gear (Figure 8-26). The transmission is now in first gear.

Drive: Second Gear—When shifting into second gear, the large sun gear of the rear planet is held. The sun gear now transfers torque through the short and long planetary pinions to the ring gear, which is the output.

Drive: Third Gear—Third gear is achieved by locking the input shaft and both of the sun gears in the compound gearset together. This forces both of the rear planetary assemblies to lock and drive the output ring gear.

Drive: Fourth Gear—For fourth gear, the input is transferred from the single gearset's carrier to the large sun gear and the planetary carriers in the compound gearset (Figure 8-27). The ring gear is then driven with a slight reduction.

Drive: Fifth Gear—In fifth gear, the input moves from the single planetary unit's carrier to the small sun gear and the planetary carrier in the compound gearset. The ring gear is then driven at an overdrive.

Drive: Sixth Gear—Sixth speed made by holding the front sun gear in the compound gearset. The input shaft is locked to the carrier of the compound gear. The carrier then drives the ring gear to provide for an overdrive output (Figure 8-28).

Reverse—For reverse, the ring and planet gears of simple planetary rotate together. The single sun gear is held and rotates the carrier and the sun gear of the compound planetary gear (rear) assembly. The rear planet carrier is also held. The long planets in the compound gear rotate with the rear sun gear. This causes the ring gear to rotate backward and drives the output shaft in reverse.

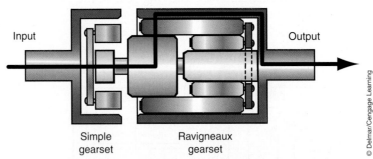

FIGURE 8-26 Power flow in first gear for a typical Lepelletier gearset.

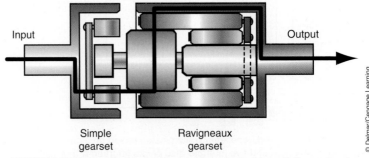

FIGURE 8-27 Power flow in fourth gear for a typical Lepelletier gearset.

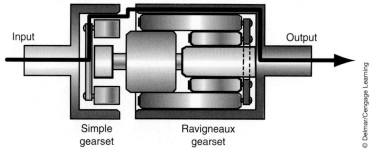

FIGURE 8-28 Power flow in sixth gear for a typical Lepelletier gearset.

Various Configurations. In a Ford 6R60 transmission, the single planterary's sun gear is mounted to the stator support. GM's 6L80 is constructed in a similar way except the sun gear is splined to the stator support. In both cases, the planetary carrier connects the sun to the ring gear.

Jatco's RE5R05A mounts the sun gear on the stator support through a one-way clutch to allow it to freewheel in one direction. It also has a brake that prevents the sun gear from freewheeling when necessary. This transmission also has the sun gear splined to the carrier. This feature allows for a direct drive ratio, most other designs do not offer a direct drive.

NONPLANETARY-BASED TRANSMISSION

Honda and Saturn transaxles are unique in that they do not use a planetary gearset to provide for the different gear ranges. Constant-mesh helical and square-cut gears (Figure 8-29) are used in a manner similar to that of a manual transmission.

These transaxles have a mainshaft and countershaft on which the gears ride. The two shafts run parallel to each other and some later models have a third and/or fourth parallel shaft. To provide the forward gears and one reverse gear, different pairs of gears are locked to the shafts by hydraulically controlled clutches (Figure 8-30). Reverse gear is obtained on some models through the use of a shift fork that slides the reverse gear into position (Figure 8-31). The power flow through these transaxles is also similar to that of a manual transaxle.

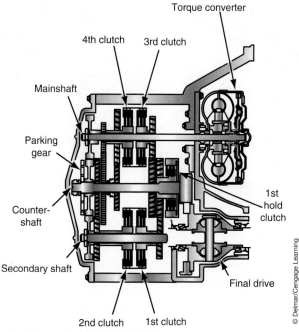

FIGURE 8-29 Honda automatic transaxles use constant-mesh helical gears instead of planetary gearsets.

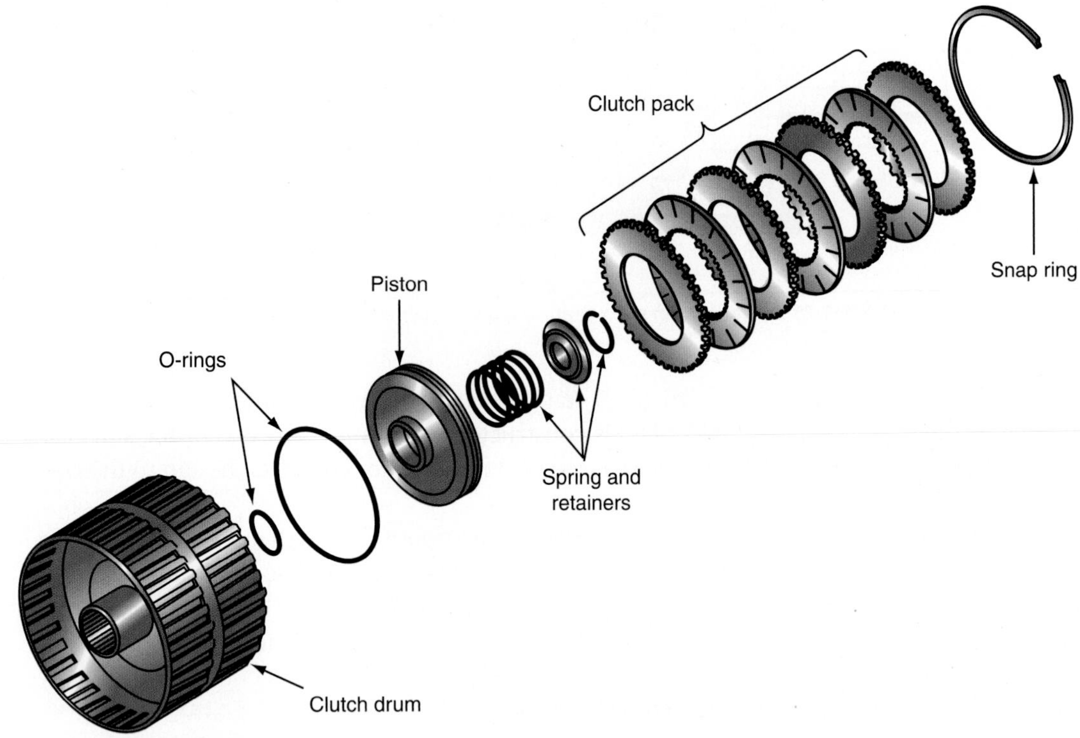

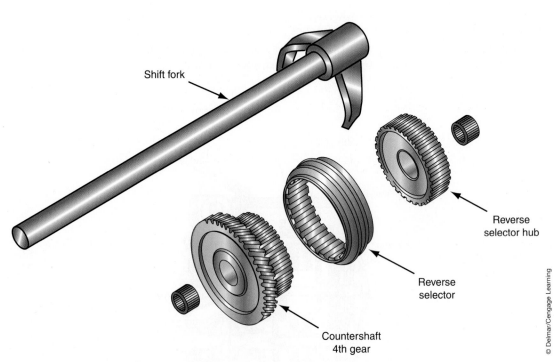

FIGURE 8-30 Different pairs of gears are locked to the shafts by hydraulically controlled multiple-disc clutches.

Clutch pack

Snap ring

Piston

O-rings

Spring and retainers

Clutch drum

© Delmar/Cengage Learning

Shift fork

Reverse selector hub

Reverse selector

Countershaft 4th gear

FIGURE 8-31 Reverse gear is obtained by sliding a gear into position.

© Delmar/Cengage Learning

FINAL DRIVES AND DIFFERENTIALS

The last set of gears in the drivetrain is the final drive. In most RWD cars, the final drive is located in the rear axle housing. On most FWD cars, the final drive is located within the transaxle. Some FWD cars with longitudinally mounted engines locate the differential and final drive in a separate case that bolts to the transmission.

Shop Manual

Chapter 8, page 371

Differentials

Although the differential can be part of the rear drive axle and is typically covered in detail in a manual transmission course, the differential may be in the transaxle and may be the cause of a problem in a transaxle. The basics of how a differential works are important for both automatic and manual transmission specialists. The differential (Figure 8-32) is a geared mechanism located between two drive axles. It rotates the drive axles at different speeds when the vehicle is turning, and at the same speed when the vehicle is traveling in a straight line. The drive axle assembly directs driveline torque to its driving wheels. The gear ratio between the pinion and ring gear is used to increase torque and improve driveability. The gears within the differential serve to establish a state of balance between the forces or torques between the drive wheels, which allows the drive wheels to turn at different speeds as the vehicle turns a corner.

The term *differential* means relating to or exhibiting a difference or differences.

Power is delivered to the differential assembly on the pinion gear. The pinion teeth engage the ring gear, which is mounted upright at a 90-degree angle to the pinion. Therefore, as the driveshaft turns, so do the pinion and ring gears.

The ring gear is fastened to the differential case. Holes machined through the center of the differential housing support the differential pinion shaft. The pinion shaft is retained in the housing case by clips or a specially designed bolt. Two beveled differential pinion gears and thrust washers are mounted on the differential pinion shaft. In mesh with the differential pinion gears are two axle side gears splined internally to mesh with the external splines on the left and right axle shafts. Thrust washers are placed between the differential pinions, axle side gears, and differential case to prevent wear on the inner surfaces of the differential case.

The ring gear in a transaxle is sometimes referred to as the differential drive gear.

The differential case is located between the side gears and is mounted on bearings so that it is able to rotate independently of the drive axles. Pinion shafts, with small pinion gears, are fitted inside the differential case. The pinion gears mesh with the side gears.

Engine torque is delivered to the drive pinion gear, which is in mesh with the ring gear and causes it to turn. Power flows from the pinion gear to the ring gear. The ring gear drives the differential case that is bolted to the ring gear. The differential case extends from the side of the ring gear and normally houses the pinion gears and the side gears. The side gears are mounted so they can slip over splines on the ends of the axle shafts.

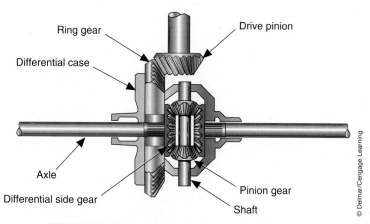

FIGURE 8-32 Major components in a basic differential.

269

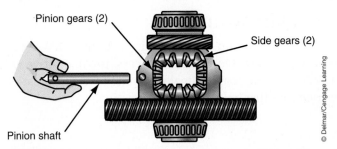

FIGURE 8-33 The small pinion gears are held in the differential case by pinion shafts.

The pinion gears are located between and meshed with the side gears, thereby forming a square of gears inside the differential case. The differential pinion gears are free to rotate on their own centers and can travel in a circle as the differential case and pinion shaft rotate. The side gears are meshed with the pinion gears and are also free to rotate on their own centers. However, since the side gears are mounted at the centerline of the differential case, they do not travel in a circle with the differential case, as do the pinion gears.

The small pinion gears are mounted on a pinion shaft (Figure 8-33). The pinion gears are in mesh with the axle side gears that are splined to the axle shafts. In operation, the rotating differential case causes the pinion shaft and pinion gears to rotate end-over-end with the case. Since the pinion gears are in mesh with the side gears, the side gears and axle shafts are also forced to rotate.

When a car is moving straight ahead, both drive wheels are able to rotate at the same speed. Engine power comes in on the pinion gear and rotates the ring gear. The differential case is rotated with the ring gear. The ring gear carries around the pinion shaft and pinion gears and all of the gears rotate as a single unit. Each side gear rotates at the same speed and in the same plane as does the case and they transfer their motion to the axles. Each axle rotates at the same speed and the vehicle moves in a straight course.

As the vehicle goes around a corner, the inside wheel travels a shorter distance than the outside wheel. The inside wheel must therefore rotate more slowly than the outside wheel. In this situation, the differential pinion gears will walk forward on the slower-turning or inside side gear. As the pinion gears walk around the slower side gear, they drive the other side gear at a greater speed. An equal percentage of speed is removed from one axle and given to the other; however, an equal amount of torque is applied to each wheel.

When one of the driving wheels has little or no traction, the torque required to turn the wheel without traction is very low. The wheel with good traction is, in effect, holding the axle gear on that side stationary. This causes the pinions to walk around the stationary side gear and drive the other wheel at twice the normal speed but without any vehicle movement. With one wheel stationary, the other wheel turns at twice the speed shown on the speedometer.

Final Drive Assemblies

Shop Manual
Chapter 8, page 371

The pinion and ring gears and differential assembly are normally located within the transaxle housing. The differential section of the transaxle has the same components as the differential gears in a RWD axle and basically operates in the same way. The drive pinion gear is connected to the transmission's output shaft and the ring gear is attached to the differential case. The pinion and ring gearset provides for a multiplication of torque.

Some transaxles route power from the transmission's output shaft through two helical-cut gears to a transfer shaft (Figure 8-34). A helical-cut pinion gear attached to the opposite end of the transfer shaft drives the differential ring gear and carrier (Figure 8-35). The differential assembly then drives the axles and wheels.

Rather than using helical-cut or spur gears in the final drive assembly, some transaxles use a simple planetary gearset for its final drive (Figure 8-36). The sun gear of this planetary unit

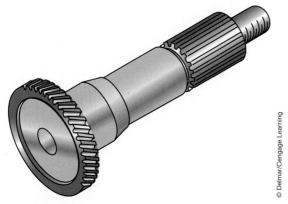

FIGURE 8-34 A transfer shaft for a final drive unit.

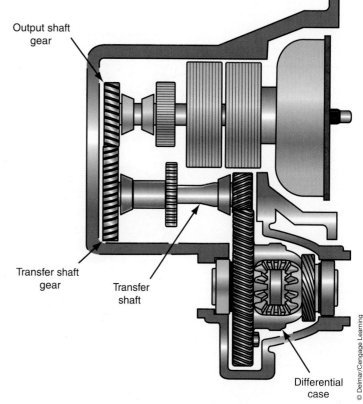

FIGURE 8-35 The entire final drive system for a transaxle with a transfer gear and shaft.

is driven by the output of the transaxle's gearsets. Typically a shaft is used to connect both the carrier of one planetary unit and the ring gear of the other to the final drive's sun gear. Doing this allows for input to the final drive unit when a forward and reverse gear are operating.

The final drive sun gear meshes with the final drive planetary pinion gears that rotate on their shafts in the planetary carrier. The pinion gears mesh with the ring gear that is held in the transaxle case. The ring gear of a planetary final drive assembly has lugs around its outside diameter that fit into grooves machined inside the transaxle housing. These lugs and grooves hold the ring gear stationary. The pinion gears walk around the inside of the stationary ring gear. The rotating planetary pinion gears drive the planetary carrier and differential case. The planetary carrier is part of the differential case, which contains typical differential gearing: two pinion gears and two side gears.

As you recall, when the ring gear is held and input is sent to the sun gear, forward gear reduction takes place. This gear reduction is the final drive gear ratio.

Chain-drive final drive assemblies use a multiple-link chain to connect a drive sprocket, connected to the transmission's output shaft, to a driven sprocket (Figure 8-37), which is connected to the differential case. This design allows for remote positioning of the differential within the transaxle housing. Final drive gear ratios are determined by the size of the driven sprocket compared to the drive sprocket. The driven sprocket is attached to the differential case, which provides differential action for the drive wheels.

FOUR-WHEEL-DRIVE DESIGN VARIATIONS

Four-wheel-drive (4WD) vehicles normally have some sort of transfer case to distribute the transmission's output to two or four drive wheels. Some transfer cases are mounted directly onto the transmission housing, while others are separate and mounted to the frame between the transmission and the axles.

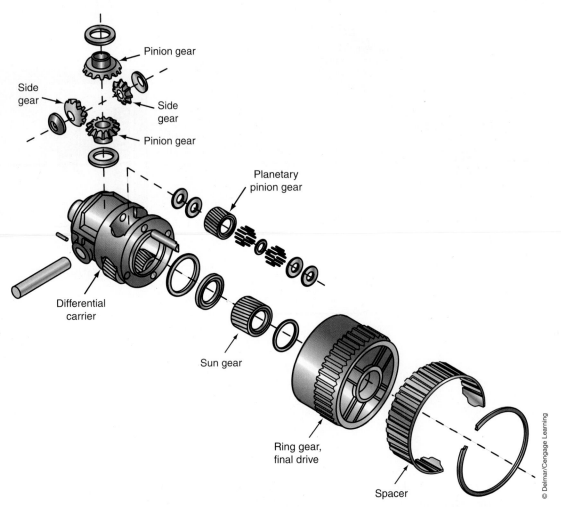

FIGURE 8-36 Some transaxles use a planetary gearset as the final drive and differential unit.

Pinion gear

Side gear

Side gear

Pinion gear

Planetary pinion gear

Differential carrier

Sun gear

Ring gear, final drive

Spacer

© Delmar/Cengage Learning

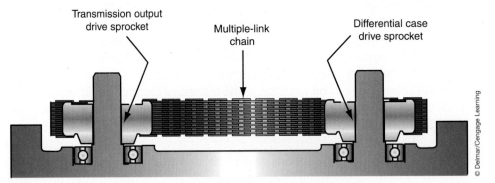

Transmission output drive sprocket

Multiple-link chain

Differential case drive sprocket

© Delmar/Cengage Learning

FIGURE 8-37 The chain setup for a final drive unit.

A BIT OF HISTORY

The first known gasoline-powered four-wheel-drive automobile was the Spyker, built in the Netherlands in 1902.

The center differential is commonly referred to as the interaxle differential.

   With four-wheel-drive, engine power can flow to all four wheels. This action can greatly increase a vehicle's traction when traveling in adverse conditions and can also improve handling because side forces generated by the turning of a vehicle or by wind gusts will have less of an effect on a vehicle that has power applied to the road on four wheels.

Transfer Cases

Four-wheel-drive is most useful when a vehicle is traveling off the road or in deep mud or snow. However, some high-performance cars are equipped with four-wheel-drive to improve the handling characteristics of the car. Most of these cars are front-wheel-drive models

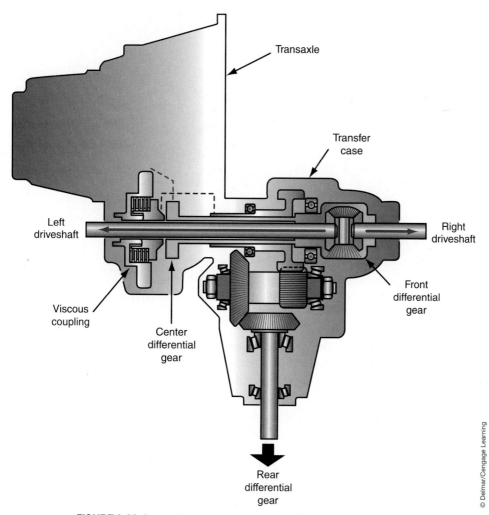

FIGURE 8-38 A transfer case mounted to a transaxle to provide 4WD.

converted to four-wheel-drive. Normally, FWD cars are modified by adding a transfer case (Figure 8-38), a rear driveshaft, and a rear axle with a differential. Although this is the typical modification, some cars are equipped with a center differential or clutch assembly in place of the transfer case. These allow the rear and front wheels to turn at different speeds.

The typical 4WD system on trucks and SUVs consists of a front-mounted, longitudinally positioned engine, either an automatic or manual transmission, front and rear driveshafts, front and rear drive axle assemblies, and a transfer case.

The transfer case (Figure 8-39) is usually mounted to the side or rear of the transmission. When a driveshaft is not used to connect the transmission to the transfer case, a chain or gear drive within the transfer case receives the transmission's output and transfers it to the driveshafts leading to the front and rear drive axles.

The driveshafts from the transfer case shafts connect to differentials at the front and rear drive axles. As on 2WD vehicles, these differentials are used to compensate for road and operating conditions by altering the speed of the wheels connected to the axles.

An electric switch or shift lever, located in the passenger compartment, controls the transfer case so that power is directed to the axles selected by the driver. Power can typically be directed to all four wheels, two wheels, or none of the wheels. On many vehicles, the driver can also select a low-speed range for extra torque while traveling in very adverse conditions.

While most 4WD trucks and utility vehicles are design variations of basic RWD vehicles, most passenger cars equipped with 4WD are based on FWD designs. These modified FWD systems consist of a transaxle and differential to drive the front wheels, plus some type of

Full-time 4WD systems use a center differential, which accommodates speed differences between the two axles, necessary for on-highway operation.

Part-time 4WD systems can be shifted in and out of 4WD.

Integrated full-time 4WD systems use computer controls to enhance full-time operation, adjusting the torque split depending on which wheels have traction.

FIGURE 8-39 A typical transfer case for an AWD vehicle.

mechanism for connecting the transaxle to a rear driveline. In many cases this mechanism is a simple clutch or differential. Some vehicles are fitted with a compact transfer case bolted to the front-drive transaxle. A driveshaft assembly carries the power to the rear differential. The driver can switch from 2WD to 4WD by pressing a dashboard switch. This switch activates a solenoid vacuum valve that applies vacuum to a diaphragm unit in the transfer case. The linkage of the diaphragm unit locks the output of the transaxle to the input shaft of the transfer case.

The driver cannot select between 2WD or 4WD in an all-wheel-drive (AWD) system. These systems always drive four wheels. AWD vehicles are not designed for off-road operation; rather, they are designed to increase vehicle performance in poor traction situations, such as icy or snowy roads. AWD allows for maximum control by transferring a large portion of the engine's power to the axle with the most traction. Most AWD designs use a center differential to split the power between the front and rear axles. On some designs, the center differential locks automatically or the driver can manually lock it with a switch. AWD systems may also use a viscous coupling to allow variations in axle speeds.

On-demand 4WD systems power a second axle only after the first begins to slip.

Shop Manual

Chapter 8, page 364

SHAFTS

The shafts inside a transmission have an important role: they transfer torque from one component to another. All transmissions have an input shaft and an output shaft. The input shaft (Figure 8-40) connects the output of the torque converter to the driving members inside the transmission. Each end of the input shaft is externally splined to fit into the internal splines of the torque converter's turbine and the driving member in the transmission. Normally, the front clutch pack's hub is attached to the driving member of a planetary gearset.

The output shaft (Figure 8-41) connects the driven member of the gearsets to the final drive gearset. The rotational torque and speed of the output shaft varies with input speed and the gear ratio of the operating gear. The output shaft may be splined to any member of the planetary gearset, but only one. The output member will be the planet carrier or the ring gear, depending on the operating gear. On some transaxles with a planetary gear final drive unit, the output shaft is connected to two different members of the gearset through two different apply devices. This allows the output to reach the final drive gear when the transmission is operating in any forward range and in reverse gear.

An additional shaft is also found in some transaxles. These shafts are typically referred to as transfer shafts and transfer the torque on the output shaft to the final drive unit.

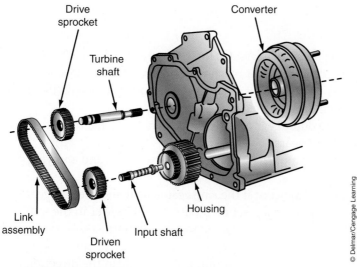

FIGURE 8-40 Engine torque is delivered from the torque converter to the input (turbine) shaft and through a chain and sprockets to the transaxle's input shaft.

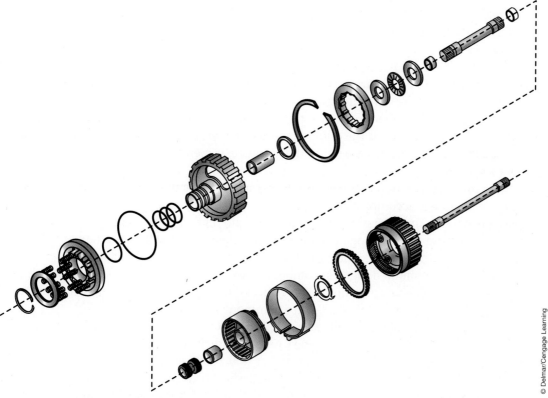

FIGURE 8-41 Examples of some of the small shafts used to connect components inside a transmission or transaxle.

Some transaxles have additional shafts. These shafts are actually a continuation of the input and output shafts. They are placed in parallel where the rotating torque can be easily transferred from one shaft to another. The shafts are divided to keep the transaxle unit compact.

On RWD vehicles, the output shaft is connected to the final drive unit in the rear drive axle housing by the driveshaft.

There are also small shafts (Figure 8-41) throughout transmissions and transaxles that serve to connect the planetary units to each other or to apply devices.

SUMMARY

- The sun gear in a planetary gearset is the central gear, which other gears revolve around.
- The planet gears are in constant mesh with the sun and ring gears.
- The planet carrier is the bracket in a planetary gearset on which the planet pinion gears are mounted on pins and are free to rotate.
- The ring gear is the outer gear of the gearset. It has internal teeth that are in constant mesh with the planet gears.
- Gear reduction is accomplished by holding the sun gear and applying power to the ring gear.
- Gear reduction can also occur if the ring gear is held and the sun gear is the input gear.
- If any two of the planetary gearset members receive power in the same direction at the same speed, the third member is forced to move with the other two and direct drive results.
- When the planet carrier is held, the output will rotate in the direction opposite that of the input.
- When a large gear drives a smaller gear, output torque is decreased and output speed is increased.
- When an external gear drives another external gear, the output gear will rotate in the direction opposite that of the input gear.
- When an external gear is in mesh with an internal gear, the two gears will rotate in the same direction.
- When there is input into a planetary gearset, but a member is not held, neutral results.
- When the ring gear and carrier pinions are free to rotate at the same time, the pinions will always follow the same direction as the ring gear.
- The sun gear always rotates opposite the rotation of the pinion gears.
- When the planet carrier is the output, it always follows the direction of the input member.
- When the planet carrier is the input, the output gear member always follows the direction of the carrier.
- When there is a reaction member and the planet carrier is the input, overdrive results.
- Gear reduction is accomplished in a planetary gearset by holding the sun gear. If power is applied to the ring gear, the planet gears will spin on their shafts in the planet carrier. Because the sun gear is being held, the spinning planet gears will walk around the sun gear and carry the planet carrier with them. This causes a gear reduction.
- There are two common designs of compound gearsets: the Simpson gearset in which two planetary gearsets share a common sun gear, and the Ravigneaux gearset, which has two sun gears, two sets of planet gears, and a common ring gear.
- The Lepelletier gear design connects a simple planetary gearset to a Ravigneaux gearset.
- Honda transaxles do not use a planetary gearset; rather, constant-mesh helical and square-cut gears are used in a manner similar to that of a manual transmission.
- As the vehicle goes around a corner, the inside wheel travels a shorter distance than the outside wheel. The inside wheel must therefore rotate more slowly than the outside wheel. An equal percentage of speed is removed from one axle and given to the other; however, an equal amount of torque is applied to each wheel.
- There are four common configurations used as the final drives on FWD vehicles: helical gear, planetary gear, hypoid gear, and chain drive.
- The transfer case is usually mounted to the side or rear of the transmission.
- All transmissions have at least two shafts: an input shaft and an output shaft.
- The input shaft of a transmission connects the output of the torque converter to the driving members inside the transmission. The output shaft connects the driven member of the gearsets to the final drive gearset.

TERMS TO KNOW

Pinion gears

REVIEW QUESTIONS

Short-Answer Essays

1. How does direct drive result in a planetary gearset?

2. What do the input shaft and the output shaft connect to?

3. What are the two common designs of compound planetary gearsets? How do they differ?

4. How is reverse gear accomplished in a planetary gearset?

5. How can a gear reduction be obtained from a planetary gearset?

6. What is the most important function of a differential?

7. What is the purpose of a transfer case?

8. In basic terms, explain how a helical-gear, constant mesh automatic transmission works.

9. What happens in a planetary gearset when no member is held? Explain your answer.

10. When a transmission is described as having two planetary gearsets in tandem, what does that mean?

Fill in the Blanks

1. When an _____ gear is in mesh with an _____ gear, the two gears will rotate in the same direction.

2. Gear reduction can occur if the _____ gear is held and the _____ gear is the input gear.

3. If any two of the planetary gearset members receive power in the same _____ at the same _____, the third member is forced to move with the other two and _____ drive results.

4. The major components of a planetary gearset are the: _____ _____, _____ _____, and _____ _____.

5. The ring gear is also known as the _____ or _____ gear.

6. When a small gear drives a larger gear, torque is _____ while speed is _____.

7. Forward gear reduction always occurs when the _____ _____ is the output member of the gearset.

8. The center differential in an AWD system is commonly refferred to as the _____ _____.

9. A Simpson gearset has two sets of _____ gears, two _____ gears, and a common _____ gear.

10. A Ravigneaux gearset has two _____ gears, two sets of _____ gears, and a common _____ gear.

MULTIPLE CHOICE

1. *Technician A* says that when the planetary carrier is the input member, the gearset produces a gear reduction.
   Technician B says that when the planetary carrier is held, the output of the gearset will be in the opposite direction as the input, and reverse gear will be produced.
   Who is correct?
   A. A only   C. Both A and B
   B. B only   D. Neither A nor B

2. Which of the following statements is *not* true?
   A. Gear reduction is accomplished by holding the ring gear and applying power to the sun gear.
   B. When the planet carrier is the driving member of the gearset and the sun gear is held, a low-reduction reverse gear results.
   C. When the planet carrier is held, the output will rotate in the opposite direction as the input.
   D. When the planet carrier is the input, the output gear member always follows the direction of the carrier.

3. In a simple planetary gearset, which of the following will result in reverse gear?
   A. The ring gear is held, the input is sent to the sun gear, and the output is at the ring gear.
   B. The sun gear is held, the input is sent to the carrier, and the output is at the ring gear.
   C. No element is held in the gearset.
   D. The carrier is held, the sun gear is the input, and the output is at the ring gear.

4. While discussing gear ratios available from a planetary gearset:
   Technician A says a single planetary gearset can provide only one reverse gear.
   Technician B says a single planetary gearset can provide only one gear reduction forward gear.
   Who is correct?
   A. A only
   B. B only
   C. Both A and B
   D. Neither A nor B

5. *Technician A* says that when a small gear drives a larger gear, output torque is increased and output speed is decreased.
   Technician B says that when a large gear drives a smaller gear, output torque is decreased and output speed is increased.
   Who is correct?
   A. A only
   B. B only
   C. Both A and B
   D. Neither A nor B

6. *Technician A* says that when the planet carrier is the output, it always follows the direction of the input member.
   Technician B says that when the planet carrier is the input, the output gear member always moves in the direction opposite that of the carrier.
   Who is correct?
   A. A only
   B. B only
   C. Both A and B
   D. Neither A nor B

7. While discussing differentials:
   Technician A says that as a vehicle goes around a corner, the differential side gears walk on the slower-turning or inside pinion gear. As the side gears walk around the slower pinion gear, they drive the other pinion gear at a greater speed.
   Technician B says an equal percentage of speed is removed from one axle and given to the other; however, an equal amount of torque is applied to each wheel.
   Who is correct?
   A. A only
   B. B only
   C. Both A and B
   D. Neither A nor B

8. *Technician A* says that when an external gear drives another external gear, the output gear will rotate in the direction opposite that of the input gear.
   Technician B says that when an external gear is in mesh with an internal gear, the two gears will rotate in the same direction.
   Who is correct?
   A. A only
   B. B only
   C. Both A and B
   D. Neither A nor B

9. *Technician A* says that when the ring gear and carrier pinions are free to rotate at the same time, the pinions will always follow the same direction as the ring gear.
   Technician B says the sun gear always rotates opposite the rotation of the pinion gears.
   Who is correct?
   A. A only
   B. B only
   C. Both A and B
   D. Neither A nor B

10. *Technician A* says a Simpson gearset is two planetary gearsets that share a common sun gear.
    Technician B says a Ravigneaux gearset has two sun gears, two sets of planet gears, and a common ring gear.
    Who is correct?
    A. A only
    B. B only
    C. Both A and B
    D. Neither A nor B

Chapter 9

REACTION AND FRICTION UNITS

UPON COMPLETION AND REVIEW OF THIS CHAPTER, YOU SHOULD BE ABLE TO:

- Describe the purpose and operation of the common reaction members.

- Explain how a brake band works and what its purpose is.

- Identify the basic components in a hydraulic servo and describe their function.

- Describe the different types of one-way clutches used in automatic transmissions.

- Explain how a roller-type or sprag-type one-way clutch works.

- Identify the components in a hydraulic multiple-disc clutch and describe their function.

- Describe the conditions required for automatic shifting.

- List the conditions that may have an effect on shift feel.

- Explain the purpose and operation of an accumulator and modulator valve.

INTRODUCTION

The reaction and friction devices in an automatic transmission are the devices that make everything else work. The hydraulic system with its complexity does no more than activate these devices. The planetary gearsets respond to the action of these devices. The devices covered in this chapter are the various brakes and clutches used in transmissions (Figure 9-1).

A discussion on how multiple-disc clutches, brakes, and one-way clutches affect gear changes is also presented.

REACTION MEMBERS

Reaction members are those parts of a planetary gearset that are held in order to produce an output motion. Other members of the planetary gearset react against the stationary or held member. Devices such as multiple-friction disc packs, brakes, and one-way overrunning clutches are used in automatic transmissions to hold or drive members of the planetary gearset in order to provide for the various gear ratios and directions. One-way overrunning clutches are purely mechanical devices, whereas clutches and brakes are hydraulically controlled mechanical devices. Most automatic transmissions use more than one type of these devices; some use all three.

The bulk of a transmission's interior comprises these devices. Each member of the compound planetary gear unit is attached to some sort of holding or driving device. This results in at least five apply or brake devices attached to a simple compound planetary unit. There can be many more than five because each member of the gearset can be used as the input, output, or reaction member.

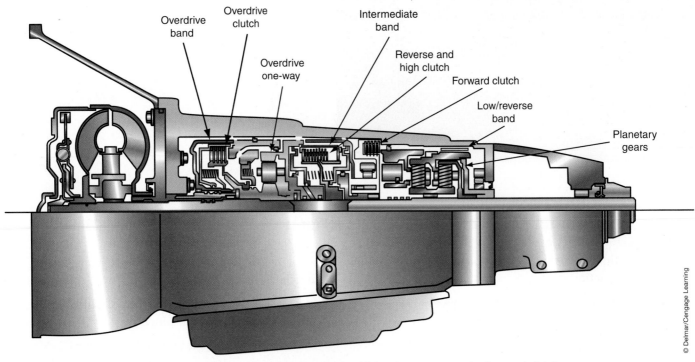

FIGURE 9-1 An automatic transmission shown with its planetary gears, brakes, and clutches.

A reaction unit is a brake. Brake bands work by holding a drum that is attached to a planetary member stationary. Multiple-friction discs can ground or lock a planetary member to the transmission housing. One-way clutches prevent a planetary member from rotating in one direction.

To understand how these devices affect gear ratios within a transmission, a quick review of planetary gear action is necessary (Figure 9-2).

- When the sun gear is the input, the carrier is the output, and the ring gear is the reaction member, the output will have maximum reduction in the same direction as the input.
- When the sun gear is the reaction member, the carrier is the output, and the ring gear is the input, the output will have minimum reduction in the same direction as the input.

SUN	CARRIER	RING	SPEED	TORQUE	DIRECTION
Input	Output	Held	Maximum reduction	Maximum increase	Same as input
Held	Output	Input	Minimum reduction	Minimum increase	Same as input
Output	Input	Held	Maximum increase	Maximum reduction	Same as input
Held	Input	Output	Minimum increase	Minimum reduction	Same as input
Input	Held	Output	Reduction	Increase	Opposite of input
Output	Held	Input	Increase	Reduction	Opposite of input

FIGURE 9-2 The basic laws of simple planetary gear action.

- When the sun gear is the output, the carrier is the input, and the ring gear is the reaction member, the output will have maximum overdrive in the same direction as the input.
- When the sun gear is the reaction member, the carrier is the input, and the ring gear is the output, the output will have minimum overdrive in the same direction as the input.
- When the sun gear is the input, the carrier is the reaction member, and the ring gear is the output, there will be gear reduction and the output will be in the direction opposite that of the input.
- When the sun gear is the output, the carrier is the reaction member, and the ring gear is the input, there will be an overdrive condition and the output will be in the direction opposite that of the input.
- When any two members are held together, direct drive results and the output speed and direction are the same as the input.

> **AUTHOR'S NOTE:** To help understand these laws, examine the drawings given in this chapter for each gear combination.

Reviewing the above, you will notice that the planet carrier is the key member. When the carrier is the output, gear reduction always takes place. When it is the input, overdrive always takes place. And reverse gear always results from using the carrier as the reaction member.

BRAKE BANDS

A band is an externally contracting brake assembly that is positioned around the outside of a drum (Figure 9-3). The drum is connected to a member of the planetary gearset. In most cases, a band is used to hold the sun or ring gear. Bands are simply flexible metal strips lined with either a semi-metallic or organic friction material. Semi-metallic linings are very durable and can work well under high pressures. This type of lining is typically used in reverse gear because it tends to wear the drum surface when it is tightened around a drum that is rotating with high torque. In reverse, the band doesn't need to stop the drum while it is rotating; rather, it prevents it from rotating.

Shop Manual
Chapter 9, page 407

© Delmar/Cengage Learning

FIGURE 9-3 A band from a late-model transaxle.

Organic friction lining materials are commonly used on bands that stop a rotating drum with high torque. The lining material is typically a soft paper pulp-based or cellulose-based material. These materials are preferred because they don't readily wear away the surface of the drum and provide for a good clamp on the drum. The heat produced by the friction of the band during braking is very high (up to 800°F). The fluid present on the drum or in the band's lining quickly evaporates in the presence of this heat, thus creating more frictional heat. The soft lining, however, is saturated with fluid and as the band tightens, the fluid in the lining is squeezed out. The fluid replaces the fluid lost by evaporation and cools the band. To aid in this cooling process, bands are made with grooves cut into the friction surface.

When a band is applied, it wraps around a drum to hold a drum attached to a member of a planetary gearset and keeps it from rotating (Figure 9-4). The construction of the drum is critical to band action. If the drum is made of soft iron, the squeezing action of the band can distort the drum. However, the rough surface of this type of drum aids in the clamping effort of the band. The rough surface also holds fluid in its pores and this fluid can cause glazing on the drum and band after some use. Drums made with hard iron will not distort and will withstand high clamping forces. This type of drum can be finished with a smooth finish that does not readily wear the band's surface.

A band is applied hydraulically by a **servo** assembly. Hydraulic pressure moves the servo piston, which compresses the servo spring and applies the band through a mechanical linkage (Figure 9-5). To release the band, hydraulic pressure to the servo is diverted and the spring moves the piston back into its bore.

A band and its servo are always used as holding devices and are never used to drive a member of the gearset. A band stops and holds the reaction member of the gearset. A band is anchored at one end and force is applied against the other end. As this force is applied, the band contracts around the rotating drum and squeezes it to a stop. The amount of pressure that stops the drum from rotating is determined by the length and width of the band, and the amount of force applied against the band's unanchored end. Rods or struts are used to apply force against the band. They may be placed between the anchor and the band or be at both sides of the band.

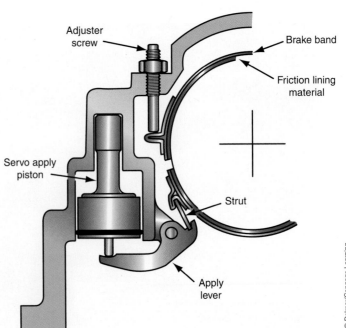

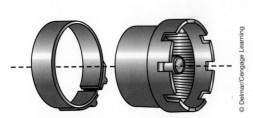

FIGURE 9-4 A clutch drum and band assembly.

FIGURE 9-5 Band in place with its servo and apply lever.

A band can be positioned to allow the applying pressure to move against the direction of drum rotation or with it. If the band is mounted so that the force is applied in the same direction as drum rotation, the movement of the drum adds to the applying force and less hydraulic pressure is needed. When the band moves in the direction opposite that of drum rotation, the drum opposes the band and more pressure is needed to stop the drum.

Although all transmission bands are made of flexible steel and their inside surfaces are lined with frictional material, they differ in size and construction depending on the amount of work they are required to do. A band that is split with overlapping ends is called a *double-wrap band*. A one-piece band that is not split is called a *single-wrap band*.

Two types of single-wrap bands (Figure 9-6) are commonly used in transmissions today. One type is made of a light and flexible steel and the other type is made of heavy and more rigid cast iron. The heavy bands are typically made with a metallic lining material that can withstand large gripping pressures. Light bands are lined with a less abrasive material that helps limit drum wear.

Double-wrap bands (Figure 9-7) have a smoother and more uniform grip, and lend themselves more to self-energizing. A double-wrap band readily conforms to the circular shape of a drum; therefore, it can provide greater holding power for a given application force. A double-wrap band also requires less hydraulic pressure than a single-wrap band to produce the same amount of holding power. All of these features allow a double-wrap band to provide smooth gear changes.

To prevent harsh changing of gears, which results from quickly stopping the movement of a gearset member, bands are designed to slip a little as they are being applied. The amount that a band slips increases as its lining wears. This wear increases the clearance between the band and the drum, and reduces the holding force of the band. Because of this wear, the bands in older automatic transmissions need to be adjusted periodically. However, as band designs have improved, periodic band adjustment is not needed on most new transmissions. On units that require periodic adjustment of the bands, the clearance between the band and the drum is set with an adjustment screw that also serves as the anchor for the band. Excessive slippage will cause a band to burn or become glazed, as well as poor power transfer through the transmission.

Several factors contribute to the effectiveness of a band:

- The type of frictional material used for the lining.
- The composition of the drum and its surface finish.
- Sufficient fluid flow to cool the band and drum.

Shop Manual
Chapter 9, page 408

Rods are round metal bars, whereas struts are flat metal plates. Both are often referred to as operating links.

Double-wrap bands are also called split bands or dual bands.

Shop Manual
Chapter 9, page 409

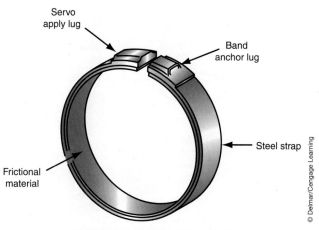

FIGURE 9-6 A single-wrap band.

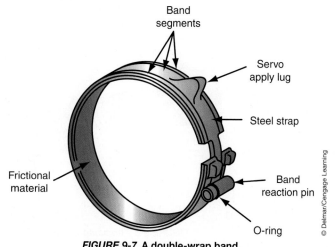

FIGURE 9-7 A double-wrap band.

- Condition and type of transmission fluid.
- Proper grooving of the lining material to aid in the cooling process.
- Proper adjustment of the band.
- The force used to apply the band.

SERVOS

A servo assembly is a hydraulic piston and cylinder assembly that controls the contraction or application and release of a band.

Shop Manual

Chapter 9, page 411

A band is applied hydraulically by a servo unit (Figure 9-8). The servo contracts the band when hydraulic pressure pushes against the servo's piston and overcomes the tension of the servo's return spring. This action moves an operating rod toward the band that squeezes the band around the drum. When hydraulic pressure to the servo apply port is stopped and exhausted, the band and servo are released and the return spring on the opposite side of the piston returns the piston to its original position. On some transmissions, the return of the piston is aided by hydraulic pressure sent to the release side of the servo (Figure 9-9). Once the piston has returned, this release pressure keeps the band from applying until apply pressure is again sent to the servo.

A servo unit provides for the application and deenergizing of the transmission's bands (Figure 9-10). A band is energized by the action of the servo's piston on one end of the band while its other end is anchored to the transmission case. A servo unit consists of a piston in a cylinder and a piston return spring. The cylinder may be part of the transmission housing's casting or it may be a separate unit bolted to the case (Figure 9-11). The force from the servo's piston acts directly on the end of the band by an apply pin or through a lever arrangement that provides mechanical advantage or a multiplying force. The band squeezes a drum attached to a gearset member.

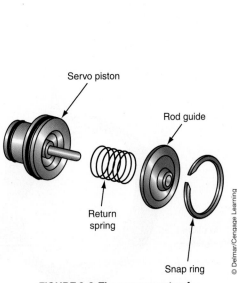

FIGURE 9-8 The components of a piston-type servo.

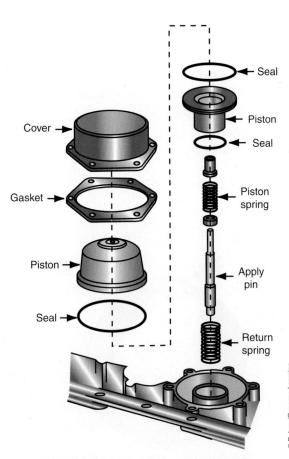

FIGURE 9-9 A typical dual-piston servo unit.

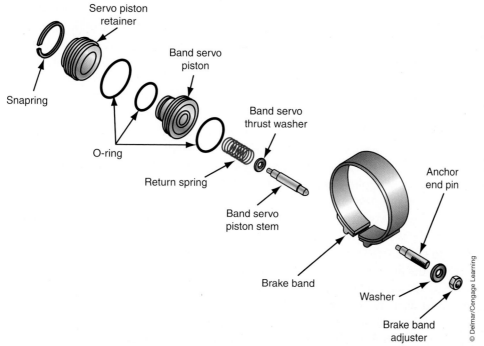

FIGURE 9-10 A typical band and servo assembly.

FIGURE 9-11 This servo assembly fits into a bore in the transmission housing.

The servo unit must apply the band securely to rigidly hold a member of the planetary gearset in order to cause forward or reverse gear reduction. To assist the hydraulic and mechanical apply forces, the servo and band anchor are positioned to take advantage of drum rotation. When the band is applied, it becomes self-energized and wraps itself around the drum in the same direction as drum rotation. This self-energizing effect reduces the force that the servo must produce to hold the band.

To release the servo apply action on the band, the servo apply oil is exhausted from the circuit or a servo release oil is introduced on the servo piston that opposes the apply oil. When servo release oil is introduced on a shift, the hydraulic mainline pressure acting on the top of the piston plus the servo return spring will overcome the servo apply oil and move the piston down.

The application force exerted on it by the servo determines the clamping pressure of a band. Because the area of the servo's piston is relatively large, the application force of the servo is considerably greater than the line pressure delivered to the servo. This increase in force is necessary to clamp the band tightly enough around the spinning drum to bring it to a halt.

The servos in most transmissions are designed to react to varying pressures and apply different amounts of application forces to meet the needs of the transmission. The total force of a piston is equal to the pressure applied to it multiplied by the surface area of the piston. Therefore, if the servo piston has a surface area of 2.5 square inches and the normal pressure sent to the servo is 50 psi, the application force of the servo on the band will be 125 psi. When the pressure sent to the servo increases to 100 psi, such as during heavy loads, the application force on the operating rod would increase to 250 psi (Figure 9-12). This increase in force allows for increased clamping pressures during heavy load conditions and times when slipping will most likely take place.

The operating rod of a servo tightens a band through one of three basic linkage designs: straight, lever, or cantilever. The straight linkage design uses a straight rod or strut to transfer the piston's force to the free end of the band (Figure 9-13). This type of linkage is only used when the servo is placed where it can act directly on a band and when the servo's piston is large enough to hold the band when maximum torque is applied to the drum. Some transmissions that use a straight linkage have specially designed rods that are graduated. These rods are designed to minimize the need for periodic band adjustment. Band adjustment takes place during assembly by selecting the proper rod length. When the correct rod is installed in the servo, the piston will move a specific distance and apply a specific amount of force against a band.

The lever-type linkage uses a lever to move the rod or strut that actually applies the band. A lever-type linkage normally increases the application force of the piston because the **fulcrum** of the lever is closer to the band (Figure 9-14).

A **fulcrum** is the support and pivot point of a lever.

Shop Manual

Chapter 9, page 412

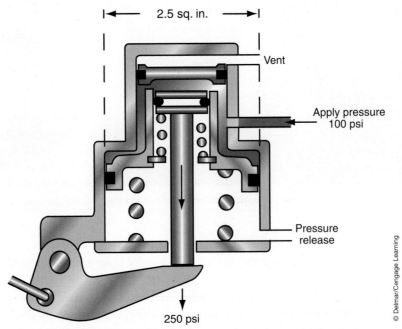

FIGURE 9-12 The surface area of a servo piston increases fluid pressure so more force is on the operating rod.

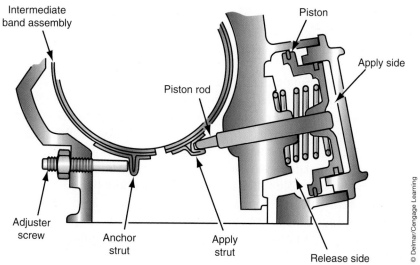

FIGURE 9-13 A servo assembly with an external adjustment screw at the end of the activation rod.

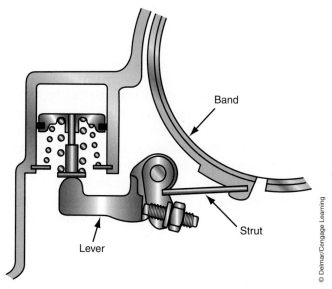

FIGURE 9-14 A lever-type servo linkage.

A **cantilever**-type linkage uses a lever and a cantilever to act on both ends of an unanchored band. As the servo piston applies force to the operating rod, the rod moves the lever and applies force to one end of the band. The movement of the piston also pulls the cantilever toward its pivot pin, thereby clamping the ends of the band together and tightening the band around the drum. A cantilever linkage increases band application force and allows for smooth band application because the band self-centers and contracts evenly around the drum.

ONE-WAY CLUTCHES AND BRAKES

One-way or overrunning clutches are holding or braking devices. These clutches operate mechanically, not hydraulically, and are considered apply devices. The main difference between a multiple-disc pack or band and an overrunning clutch is that the one-way clutch allows rotation in only one direction and operates at all times, whereas a disc pack or a band allows or stops rotation in either direction and operates only when hydraulic pressure is applied to them.

A cantilever linkage acts on both ends of a band to tighten it around a drum. A **cantilever** is a lever that is anchored and supported at one end by its fulcrum or pivot pin and provides an opposing force at its opposite end.

Shop Manual
Chapter 9, page 434

One-way clutches can freewheel. This means they can turn without affecting the input or output of the planetary gearset. When one-way clutches are freewheeling, they are off or ineffective. Freewheeling normally takes place when the clutch is rotating in a counterclockwise direction.

Most one-way overrunning clutches can be either roller or sprag type. A *roller clutch* utilizes roller bearings held in place by springs to separate the inner and outer race of the clutch assembly (Figure 9-15). Around the inside of the outer race are several cam-shaped indentations. The rollers and springs are located in these pockets. Rotation of one race in one direction locks the rollers between the two races, causing both to rotate together. When a race is rotated in the opposite direction, the roller bearings move into the pockets and are not locked between the races. This allows the races to turn independently.

A one-way *sprag clutch* consists of a hub and a drum separated by figure-eight-shaped metal pieces called sprags. The sprags are shaped this way so they can lock between the races when a race is turned in one direction only. Between the inner and outer races of the clutch are the sprags, cages, and springs. The sprags are longer than the distance between the two races. The cages keep the sprags equally spaced around the diameter of the races. When a race turns in one direction, the sprags tilt and allow the races to move independently. When a race is moved in the opposite direction, the sprags straighten and lock the two races together.

All types of one-way clutches apply and release quickly and evenly in response to the rotational direction of the races. This allows for smooth gear changes. Either type can be used to hold a member of the planetary gearset by locking the inner race to the outer race that is held by the transmission housing (Figure 9-16). Both types also are effective as long as the engine powers the transmission. While the transmission is in a low gear and is coasting, the drive wheels rotate the transmission's output shaft with more power than is received on the input shaft. This allows the sprags or rollers to unwedge and begin freewheeling. This, in turn, puts the planetary gearset into neutral, disallowing engine compression to aid slowing down or braking. Engine braking is provided in many different ways by the various transmission designs. It is normally provided by a band, which holds regardless of the power source.

Some aftermarket and a few manufactured transmission one-way clutches are ratchet types (Figure 9-17). These work in the same way as a ratchet wrench. The 4R70W is equipped with this type of one-way clutch.

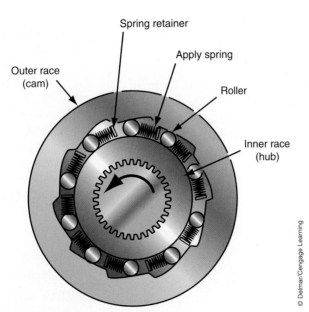

FIGURE 9-15 A roller-type one-way clutch.

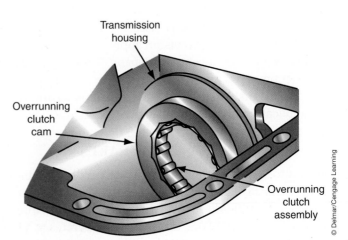

FIGURE 9-16 An overrunning (one-way) clutch secured in a transmission housing.

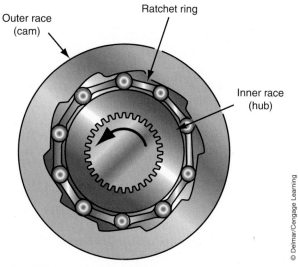

Outer race (cam)

Ratchet ring

Inner race (hub)

FIGURE 9-17 A ratchet-type one-way clutch assembly.

MULTIPLE-FRICTION DISC CLUTCH AND BRAKE ASSEMBLIES

The multiple-disc assembly (Figure 9-18) can be used to drive or hold a member of the planetary gearset. These units can serve as a brake by locking a gearset member to the transmission housing to prevent it from rotating. They can also be used to lock one race of a one-way clutch. As a brake, a multiple-friction disc pack serves the same purpose as a band; however, because there is much more surface area on the friction discs, it has greater holding capabilities. The multiple-friction disc pack can also be used to connect and hold two planetary members together.

Shop Manual
Chapter 9, page 417

Multiple-disc assemblies are often referred to as clutch packs.

The multiple-friction disc clutch assembly used in automatic transmissions is a "wet clutch" assembly.

FIGURE 9-18 A multiple-disc assembly.

Construction

When used as a clutch to connect and hold two members together, the assembly (Figure 9-19) is typically made up of the following components: friction discs, steel plates, clutch drum and hub, apply piston, and return spring(s). Typically, when a multiple-friction disc pack is used as a brake, it has the same basic parts but uses the transmission housing instead of a clutch drum.

The multiple-disc pack contains of several plates lined with friction material and several **steel** separator **discs** that are placed alternately inside a clutch drum (Figure 9-20). The friction plates are lined with rough frictional material on their faces, whereas the steel discs have smooth faces without friction material. The friction plate has friction material bonded to both sides of a steel plate (Figure 9-21). Metallic, semi-metallic, and paper-based materials are used as this frictional lining. Paper cellulose is the most commonly used friction material because it offers good holding power without the high frictional wear of metallic materials. The friction plates often have grooves cut in them to help keep them cool, thereby increasing their effectiveness and durability. Friction discs are always mounted between two steel plates.

Steel discs are sometimes also called apply plates or steels.

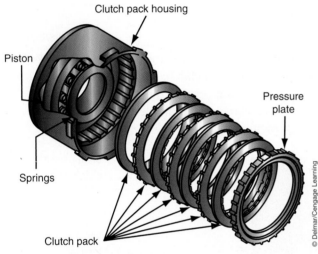

FIGURE 9-19 The main components of a multiple-disc assembly.

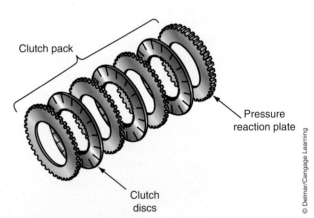

FIGURE 9-20 The frictional and steel discs are placed alternately in the assembly.

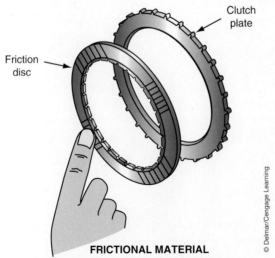

FRICTIONAL MATERIAL

FIGURE 9-21 The frictional discs have a friction facing on both sides and the steel plates have no facing.

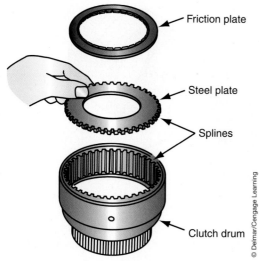

FIGURE 9-22 The clutch discs have internal splines, and the clutch plates have external splines.

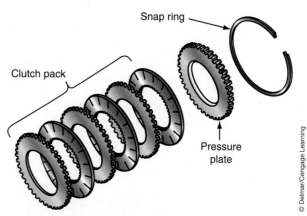

FIGURE 9-23 The location of the pressure plate in a multiple-friction disc assembly.

The steel plates provide a smooth surface for the friction discs to engage. Plates can be installed steel-to-steel to create a specific clearance for the clutch pack. The set of friction or steel plates has splines on its inner edges, while the other set is splined on its outer edges (Figure 9-22). The splines of each set fit into matching splines on a shaft, drum, member of the gearset, or the transmission case.

The discs are mounted in the clutch drum or in the housing of the transmission. The drum or housing also contains the apply piston, seals, and return springs. Hydraulic pressure moves the apply piston against return spring pressure and clamps the plates against the pressure plate. The friction between the plates locks them together, causing them to turn as a unit. The **pressure plate** (Figure 9-23) is a heavy metal plate that provides the clamping surface for the plates and is installed at one, or both, ends of the pack. The seals hold in the hydraulic pressure when the clutch pack is applied. In a typical pack, the apply piston at the rear of the drum is held in place by the return springs and a spring retainer secured by a snap ring.

The apply piston in a multiple-friction disc pack is retracted by one large coil spring (Figure 9-24), several small springs (Figure 9-25), or a single Belleville spring (Figure 9-26). The type and number of return springs used in the pack is determined by the pressure needed to release the piston quickly enough to prevent dragging. However, the amount of spring tension is limited to minimize the resistance to moving the piston. Pistons fitted with multiple springs have spring pockets machined into the assembly. Often there are fewer springs used than there are pockets. This is an indication that the manufacturer uses the same assembly for different applications and springs are added or subtracted to meet the needs of those applications. The manufacturer may also make other changes to the basic assembly to change the load-carrying capacity of the pack. One of these changes may be to use a different number of plates in the assembly. This is done by using a different thickness of the piston or retainer plate, or by changing the location of the snaping groove in the drum.

A **Belleville spring** acts to improve the clamping force of the assembly, and as a piston return spring. The spring is locked into a groove inside the drum by a snap ring. As the piston moves to apply the pack, it moves the inner ends of the Belleville spring fingers into contact with the pressure plate to apply the assembly. The spring's fingers act as levers against the pressure plate and increase the application force of the pack. When hydraulic pressure to the piston is stopped, the spring relaxes and returns to its original shape. The piston is forced back and the pack is released.

Splines are used as a method for attaching a gear, hub, drum, or other component to a housing or shaft by aligning interior slots of one component with matching exterior slots on the other so that the two parts are locked together.

A Belleville spring is also called a diaphragm or **over-center spring**.

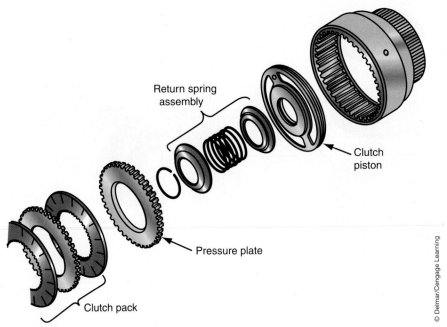

FIGURE 9-24 A clutch assembly that uses a single coil return spring for release.

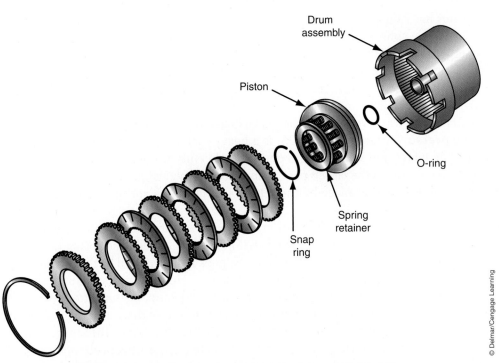

FIGURE 9-25 A multiple-disc assembly with its piston, multiple small return springs, retainer, and snap ring.

Some clutch drums or pistons have a check ball and vent port. This is necessary because the clutch drum rotates while the vehicle is moving; it is difficult to move fluid from the clutch when it is released. The vent port and check ball are used to relieve any residual pressure when the pack is released (Figure 9-27). The check ball is forced against its seat when full hydraulic pressure is applied to the pack. This holds the pressure inside the drum. When the pressure supply is stopped, only residual pressure remains on the check ball. Centrifugal force pulls the ball from its seat and allows the fluid to escape from the drum through the open vent

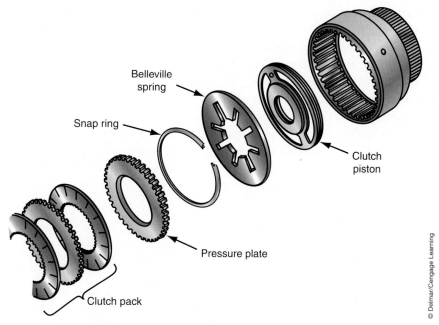

FIGURE 9-26 A multiple-friction disc pack using a single Belleville-type return spring.

- Belleville spring
- Snap ring
- Clutch piston
- Pressure plate
- Clutch pack

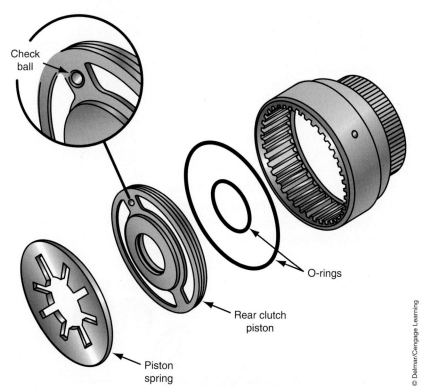

FIGURE 9-27 The typical location of the check ball in the piston for a clutch assembly.

- Check ball
- O-rings
- Rear clutch piston
- Piston spring

© Delmar/Cengage Learning

port. If the residual pressure is not relieved, some pressure may remain on the piston causing partial engagement of the pack. Not only would this cause the discs to wear prematurely it would also adversely affect shift quality.

Even when the clutch is not applied, some fluid remains behind the piston. As the drum rotates, centrifugal force moves this fluid to the outer diameter of the drum where it can create pressure on the piston. The pressure may not be enough to fully engage the clutch, but it may reduce the clearance between the friction discs and the steel plates. The pressure can cause the clutch to disengage too slowly and/or not release completely, leading to erratic shifting, heat build-up, and excessive clutch plate wear.

When a multiple-disc pack is used as a brake, there is no need for the check ball. Since the unit does not rotate, centrifugal force will have no affect on the fluid behind the piston.

Newer transmissions do not have a vent port or a check ball. Rather they have a centrifugal fluid pressure canceling system. This system has opposing fluid chambers that apply equal amounts of centrifugal force to both sides of the piston. As the clutch drum rotates, fluid in the canceling fluid pressure chamber counters the pressure built up inside the drum pressure chamber. This cancels out the effects of centrifugal force on piston movement and leads to smoother shifts.

Operation

Multiple-disc packs can either be holding or driving devices; this depends on what they are splined to. When multiple-disc packs lock two members of a planetary gearset together, it is a driving device (Figure 9-28). To apply the clutch, hydraulic pressure is routed to the side of the piston opposite the return springs. When the pressure overcomes the tension of the return spring(s), the springs compress and the piston squeezes the friction discs and steel plates tightly together so they rotate as a unit.

One set of discs is splined to one member of the planetary gearset and the other set of discs is splined to another member of the gearset. One set of plates (discs) are driven by the input at a gearset member, they are called the drive discs. The other set of plates is connected to another member of the gearset and these are the driven plates. When the drive discs engage with the driven discs, the clutch drum and the attached gearset member rotate at the same speed. In some cases, the driven plates are connected directly to the output shaft. Also, the drive plates may be splined to a clutch drum that is splined directly to the input shaft.

A multiple-disc pack may also be used to hold one member of the planetary gearset. The friction discs are splined on their inner edges and are fit into matching splines on the outside of a drum. The steel discs are splined on their outer edges and fit into matching splines machined into the transmission case (Figure 9-29). When the pack is applied, the gearset member cannot rotate as it is locked to the transmission case.

When hydraulic pressure to the piston is stopped and exhausted, the return spring(s) move the piston and allow for a clearance between the sets of plates. This allows the friction discs and steel plates to rotate independently of each other.

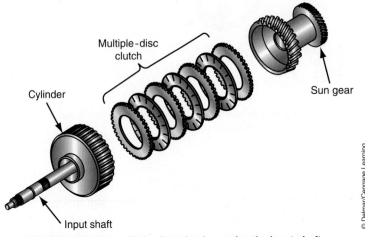

FIGURE 9-28 This multiple-disc clutch couples the input shaft, through the clutch cylinder, to the sun gear.

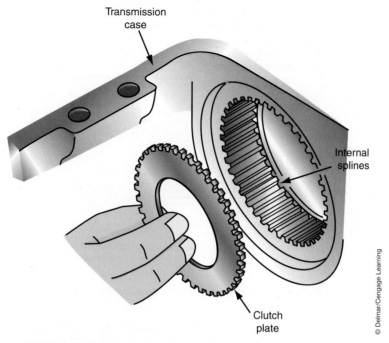

Transmission
case

Internal
splines

Clutch
plate

© Delmar/Cengage Learning

FIGURE 9-29 Multiple-friction disc pack used to lock a planetary gearset member to the transmission case.

Holding Capabilities

The following factors contribute to the effectiveness of a clutch pack:

- The type of frictional material used on the friction discs.
- The composition of the steel plates and their surface finish.
- Sufficient fluid flow to cool the clutch pack.
- Condition and type of transmission fluid.
- Proper grooving of the lining on the friction plates to aid in the cooling process.
- Proper clutch plate clearances.
- The force used to apply the clutch.

GEAR CHANGES

An automatic transmission will change gear ratios and direction automatically or at the command of the driver. Automatic shifting allows for forward gear ratio changes in response to engine speed and load, and, in the case of electronically controlled units, in response to the PCM's interpretation of operating conditions. The transmission will not automatically move into park, reverse, or neutral. The driver makes these gear selections. The driver can also select certain forward gears. The forward gears available for manual selection vary with each transmission model.

A discussion on how gear changes are made is a summary of the discussions on all of the transmission's components. Keep in mind that the way a particular transmission operates depends on its construction and components. This discussion is based on a common four-speed design using a Simpson gearset and an add-on overdrive planetary unit (Figure 9-30). There are no electronic controls mentioned in this discussion because it is important for you to understand how the gearset responds to the various hydraulic components in a transmission before you can understand how electronics improve the performance of a transmission. The clutch, brake, and band applications for this typical transmission are shown in Figure 9-31.

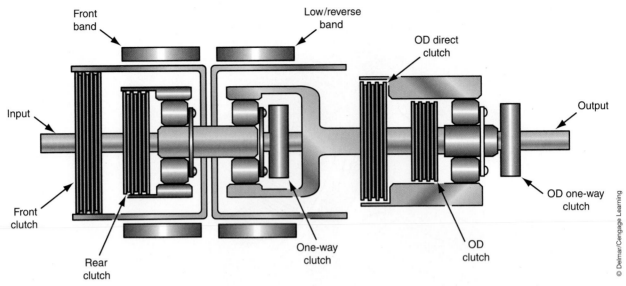

FIGURE 9-30 The planetary gears and friction and reaction units referred to in the discussion of gear changes.

	Transmission clutches and bands					Overdrive clutches		
Lever position	Front clutch	Rear clutch	One-way clutch	Front band	Low/rev band	OD clutch	Direct clutch	One-way clutch
P—Park								
R—Reverse	X				X		X	
N—Neutral								
D—Drive								
First		X	X				X	X
Second		X		X			X	X
Third	X	X					X	X
Fourth	X	X				X		
2—Range second		X	X	X			X	X
1—Range first		X	X		X		X	X

FIGURE 9-31 The clutch, brake, and band application chart for the transmission in Figure 9-30.

> The front clutch is often called the direct clutch in Simpson gearsets.

Park/Neutral

When park is selected, the shift linkage moves the park pawl or lever into the park gear, located on the output shaft or final driveshaft (Figure 9-32). The shift linkage also moves the manual shift valve to block the fluid passages to the clutches and bands. No member of the gearset receives input torque and no member is held; therefore, there is no output. Fluid flows to feed the torque converter and to lubricate and cool the transmission. In the neutral position, all of the conditions of the park mode exist, except the park lever is not engaged to the park gear.

Reverse

When reverse is selected, the manual shift valve is moved and fluid is sent to the front clutch and the low/reverse band. The pressure of the fluid is greater than mainline because boost pressure is added to the circuit at the pressure regulator valve. This added pressure is necessary to provide good clamping and to prevent clutch and band slippage.

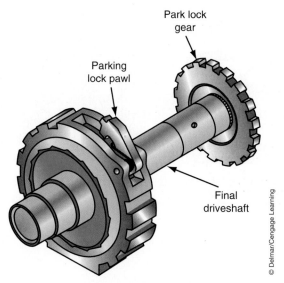

FIGURE 9-32 A typical setup for a parking pawl.

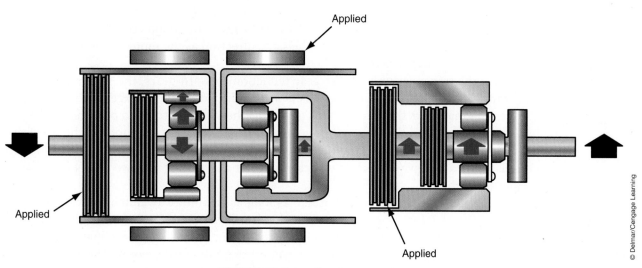

FIGURE 9-33 Power flow in reverse gear.

The front clutch locks into the drive shell and the sun gear becomes the input. The low/reverse band holds the rear planetary carrier (Figure 9-33). The sun gear rotates in a clockwise direction and drives the rear planet pinions in a counterclockwise direction. The pinions then cause the rear ring gear to rotate in a counterclockwise direction. Since the rear ring gear is splined to the output shaft, the output shaft rotates in the direction opposite that of the input shaft.

While the transmission is in reverse and in the gear reduction forward gears, the separate overdrive planetary gearset is in direct drive. This is accomplished by the engagement of the overdrive direct clutch that locks the sun and ring gears together. This clutch also provides for engine braking in reverse and the forward reduction gears.

Drive Range

When the shift linkage is moved to the D or O position, the manual shift valve allows fluid flow to engage the rear clutch. Fluid pressure is also directed to the shift valve. When the vehicle is stopped and in first gear, the various gear shift valves are closed by spring tension.

Input is transmitted through the rear clutch to the front ring gear in a clockwise direction. The front carrier is splined to the output shaft and is held by the weight of the vehicle at

The rear clutch is also referred to as the forward clutch in Simpson gearsets since it is the input clutch for all forward gears.

The shaft that connects the compound planetary unit to the overdrive planetary gearset is called the intermediate shaft.

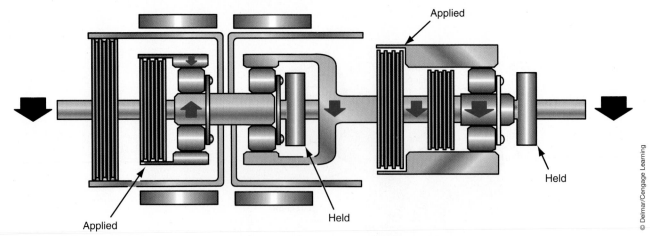

Applied

Applied

Held

Held

FIGURE 9-34 Power flow in first gear.

the drive wheels. The front ring gear drives the front planet pinions in the same direction. The planets, in turn, drive the common sun gear in the opposite or counterclockwise direction. The rear carrier is held by the one-way clutch; as the planet pinions attempt to rotate the rear ring gear, they walk the rear carrier in a counterclockwise direction. The ring gear is splined to the output shaft and the output is in the same direction as the input but at greater torque (Figure 9-34). This is first gear.

While the transmission is in the gear reduction forward gears, the separate overdrive planetary gearset is in direct drive. This is accomplished by the engagement of the overdrive clutch that locks the sun and ring gears together. This clutch also provides for engine braking in reverse and the forward reduction gears. The direct drive mode of the overdrive planetary unit keeps transmission output connected to the input of the transmission. This unit also uses a one-way clutch that locks the shaft connecting the two planetary units to the output shaft.

This mode of operation will continue as long as the one-way clutch is locked or until the 1–2 shift valve forces an upshift. The clutch will remain locked until engine torque is released from the planetary unit or when the inertia of the vehicle is driving the planetary unit through the output shaft. The latter condition exists during **coast**. During coast, the rear carrier rotates in a clockwise direction with the output shaft. This releases the one-way clutch. Since there is no longer a reactionary member, the transmission is effectively in neutral.

As the vehicle accelerates in first gear, the automatic movement of the shift valve controls progressive shifts through the other forward gears. As the vehicle moves forward, throttle pressure builds and helps the shift valve spring keep the valve in position. As output shaft and vehicle speed increase, governor pressure builds. Once governor pressure is great enough to overcome the combined pressure of the spring and throttle pressure, the 1–2 shift valve is pushed open. Fluid is directed to the front band, which holds the driving shell and the sun gear.

The rear clutch is still engaged and drives the front ring gear and front planet pinions in a clockwise direction. The front planets drive the front carrier and the output shaft with some reduction. This is second gear. The rear planetary unit is ineffective because the rear carrier is able to freewheel in the one-way clutch (Figure 9-35).

As output shaft and vehicle speed increase, governor pressure continues to build. Much more pressure is needed to move the 3–2 shift valve than was required to move the 1–2 shift valve. This is because it is held by a heavier spring tension and throttle pressure. Once governor pressure is great enough to overcome the pressure holding the 2–3 shift valve, the shift valve moves and opens the fluid circuit to the front clutch. At this

Coast is an operational condition that results from using engine braking during deceleration. Releasing the throttle while moving typically causes it to happen.

Engine flare-up is best described as a sudden increase in engine speed that does not affect power output.

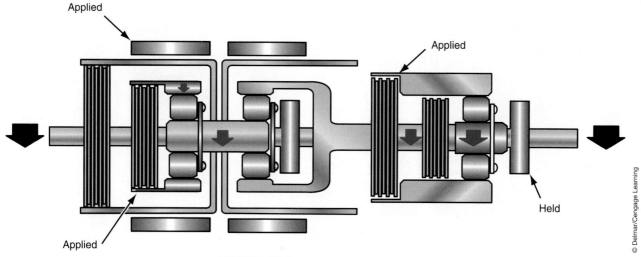

Applied

Applied

Applied

Applied

Held

© Delmar/Cengage Learning

FIGURE 9-35 Power flow in second gear.

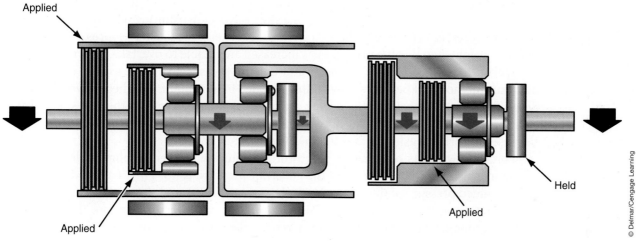

Applied

Applied

Applied

Applied

Held

© Delmar/Cengage Learning

FIGURE 9-36 Power flow in third gear.

time, fluid flow to the front band is stopped and the sun gear is released. The rear clutch remains applied and input is applied to the front ring gear through the rear clutch and to the sun gear through the front clutch. The front carrier is locked between these two input members and drives the output shaft at the same speed as the input (Figure 9-36). This is third gear.

To move the 3–4 shift valve requires more pressure than the 1–2 and 2–3 shift valves need. Again this is due to high spring tensions. Once vehicle and output shaft speed is great enough to increase governor pressure enough to move the 3–4 shift valve, fluid is sent to the overdrive clutch. When this happens, the 2–3 shift valve remains open and direct drive output is available to the overdrive planetary.

When pressure is first applied to the overdrive clutch, the overdrive clutch piston compresses the spring for the direct clutch and releases it. Then the pressure engages the overdrive clutch and locks the sun gear to the transmission housing. In the time between the release of the direct clutch and the full engagement of the overdrive clutch, the one-way clutch is used to carry power to the output shaft (Figure 9-37). This eliminates the possibility of engine flare-up and provides for a smooth shift into fourth gear.

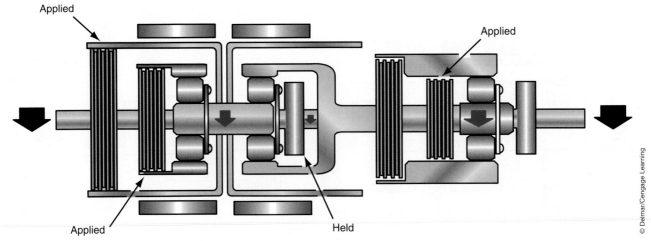

FIGURE 9-37 Power flow in fourth gear.

Automatic Downshifting

The key to understanding how automatic downshifts are made is knowing how the upshifts were made. Upshifts occur because governor pressure builds up enough to overcome the spring tension and the throttle pressure on a shift valve. Downshifting occurs when governor pressure decreases to the point where it can no longer overcome those pressures. Downshifting will also occur when throttle pressure has increased and is able to overtake governor pressure.

During a coast condition, governor and throttle pressure decrease as the vehicle slows. The transmission will begin its downshift sequence by responding to these lower pressures. Since the 3–4 shift valve has the highest spring tension on it, it will be the first shift valve to move with the decreased governor pressure. The last shift valve to move is the 1–2 shift valve because the spring tension on it is the lowest of all shift valves.

Forced downshifts occur when the throttle is quickly opened during acceleration. This action increases throttle pressure. If the governor pressure is not great enough to overcome the combined pressure of the shift valve spring and the increased throttle pressure, the open and operating shift valve will close. This action drops the transmission into the lower gear until governor pressure builds up enough to overtake the pressure on the shift valve.

Manual Low

When the driver selects a gear other than D or O, the manual shift valve controls the action of the various shift valves. Basically, by selecting a gear, the driver tells the transmission to stay in that gear or, in some cases, start off in that gear. The latter is often a possibility when initiating movement of a vehicle when it is on ice or other slippery surfaces. In this case, the driver can select Drive 2, which tells the transmission to start out in second gear rather than first. Manual downshifting can also occur if the conditions for doing so are right.

The movement of the gear selector into a manual gear inhibits a shift valve. When the driver selects Drive 2 while the transmission is in third gear, the transmission may downshift regardless of vehicle and engine speed. The manual shift valve cuts off line pressure to the 2–3 shift valve and the transmission drops to second gear.

When Drive 2 is selected for startup, the manual valve closes the pressure feed to the 2–3 shift valve and allows line pressure to flow through the second gear circuit at the 1–2 shift valve.

SHIFT FEEL

All transmissions are designed to change gears at the correct time, according to engine speed and load and driver intent. However, transmissions are also designed to provide for positive change of gear ratios without jarring the driver or passengers. If a brake or clutch is applied too quickly, a harsh shift will occur. "Shift feel" is controlled by the pressure at which each

Shift feel is sometimes referred to as shift quality.

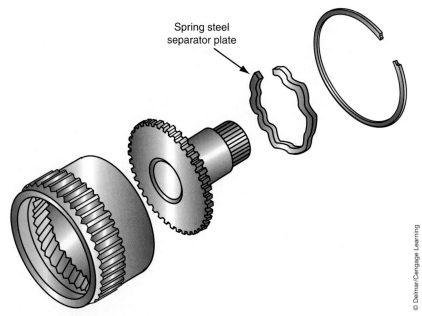

Spring steel
separator plate

© Delmar/Cengage Learning

FIGURE 9-38 Some multiple-friction disc assemblies and planetary gearsets are fitted with a wavy spring.

reaction member is applied or released, the rate at which each is pressurized or exhausted, and the relative timing of the apply and release of the members.

Shift feel is also affected by fluid type, the momentary engagement of a component in a different circuit, pulsed pressures, the clearance of the apply devices, and many more design features of the various transmission models.

To improve shift feel during gear changes, a band is often released while a multiple-disc pack is being applied. The timing of these two actions must be just right or both components will be released or applied at the same time, which would cause engine flare-up or driveline shudder. Several other methods are used to smoothen gear changes and improve shift feel.

Multiple-disc packs sometimes contain a wavy spring-steel separator plate that helps smooth the application of the clutch (Figure 9-38). Shift feel can also be smoothed out by using a restricting orifice or an accumulator piston in the brake or clutch apply circuit. A restricting orifice in the passage to the apply piston restricts fluid flow and slows the pressure increase at the piston by limiting the quantity of fluid that can pass in a given time. An accumulator piston slows pressure buildup at the apply piston by diverting a portion of the pressure to a second piston in the same hydraulic circuit. This delays and smooths the application of a clutch or brake.

The use of hydraulic power to apply brakes and clutches is confined to a single device, the hydraulic piston. The pistons are housed in cylinder units known as servo and clutch assemblies. The function of the piston is to convert the force in the fluid into a mechanical force capable of handling large loads. Hydraulic pressure applied to the piston strokes the piston in the cylinder and applies its load. During the power stroke, a mechanical spring or springs are compressed to provide a means of returning the piston to its original position. The springs also determine when the apply pressure buildup will stroke the piston. This is critical to clutch/brake life and shift quality.

Accumulators

Shift quality with a brake or a clutch is dependent upon how quickly the apply device is engaged by hydraulic pressure and the amount of pressure exerted on the piston. Some apply circuits use an **accumulator** to slow down application rates without decreasing the holding force of the apply device.

Shop Manual
Chapter 9, page 423

Shop Manual
Chapter 9, page 411

An accumulator (Figure 9-39) works like a shock absorber and cushions the application of servos and disc packs. An accumulator cushions sudden increases in hydraulic pressure by temporarily diverting some of the apply fluid into a parallel circuit or chamber. This allows the pressure to gradually increase and provides for smooth engagement of a brake or clutch.

An accumulator is basically a large diameter piston located in a bore and held in position by a heavy, calibrated spring that acts against hydraulic pressure. Accumulators are placed in the hydraulic circuit between the shift valve and the holding device.

A piston-type accumulator is similar to a servo in that it consists of a piston and cylinder (Figure 9-40). However, its purpose is to cushion the application of an apply device. The pressure required to activate the piston in an accumulator is controlled by a spring or by fluid pressure. As the spring compresses, pressure at the servo or disc pack increases. This increase in pressure activates the accumulator, which causes a delay in the servo or pack receiving the high pressure (Figure 9-41). As a result, the shift takes slightly longer but is less harsh.

When the driver or the valve body orders a change of gears, the shift valve provides a rapid flow of pressurized fluid to the servo or disc pack. As the hydraulic pressure works to compress the return springs of the holding device return springs, pressure in the circuit builds. The increasing pressure must also move the piston against the spring in the accumulator. The movement of the piston controls the amount of pressure applied to the servo

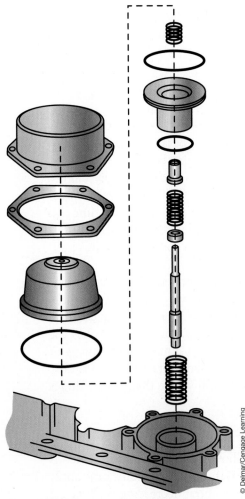

FIGURE 9-39 An accumulator assembly.

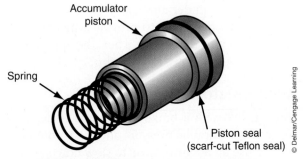

FIGURE 9-40 A typical accumulator assembly.

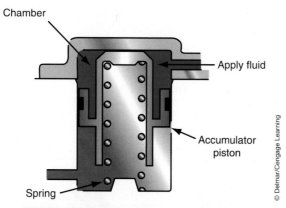

Figure labels: Chamber, Apply fluid, Accumulator piston, Spring

© Delmar/Cengage Learning

FIGURE 9-41 An accumulator works like a shock absorber to cushion the application of servos and clutches.

or clutch assembly. Pressure cannot reach its highest level until the accumulator spring is compressed and the accumulator piston is fully seated. This means the pressure in the holding device builds slowly and the application of the holding device application is softened.

The tension of the spring in an accumulator controls its action. The pressure on the spring side of the accumulator is controlled by an accumulator control valve, which is a pressure modulating valve. The control valve adjusts line pressure to the accumulators according to engine load. Since the engine torque produced is low when the throttle is partially open, accumulator back pressure is reduced. This prevents shift shock when the brakes and clutches are applied.

Many accumulators also rely on throttle pressure and mainline pressure to control the action of the accumulator. During heavy loads, line pressure is increased and throttle pressure is acting on the base of the accumulator control valve. This increases back pressure at the accumulators and allows for quick, firm shifts when the throttle opening is large.

Valve-type accumulators are not as commonly used as the piston-type. When a shift is ordered, these units allow just enough fluid pressure to allow the brake or clutch to apply. After initial engagement of the clutch or brake, the accumulator valve allows more pressure to the apply device and eventually allows full pressure to the brake or clutch. This delay is accomplished through an interaction of springs, valves, and an orifice. The orifice allows some pressure to reach the apply device initially. As the fluid passes through the orifice, pressure builds and begins to overcome the tension of the spring. At this time, more fluid pressure is sent to the brake or clutch. In a short period of time, pressure has built up enough to totally overcome the tension on the spring and full pressure is applied to the brake or clutch.

Some transmission designs do not rely on an accumulator for shift quality; rather, they have a restrictive orifice in the line to the servo or multiple-disc pack's piston. This restriction decreases the amount of initial apply pressure but will eventually allow for full pressure to act on the piston.

Servo/Accumulator Units. Several transmissions use servo units that also work as an accumulator (Figure 9-42). These units are typically used with the intermediate band and for upshifts out of second or third gear. This servo/accumulator unit actually keeps the band applied during the initial engagement of the clutch pack. As the clutch becomes more engaged, the band is released. This action prevents the harsh engagement of the upshift.

Other designs have multiple accumulators built into the servo assembly (Figure 9-43). Doing this allows the accumulators to respond directly to the action of the servo.

A BIT OF HISTORY

The idea for using a band as a brake or holding device was first applied to the vehicle's brakes. Before cars featured hydraulic internal brakes, a band was fitted around the outside of a brake drum and was tightened mechanically through a series of levers.

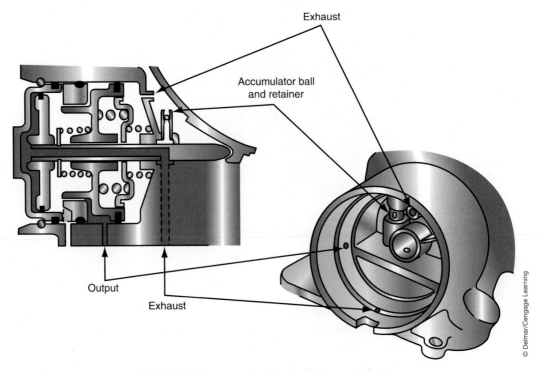

FIGURE 9-42 A servo unit with a built-in accumulator.

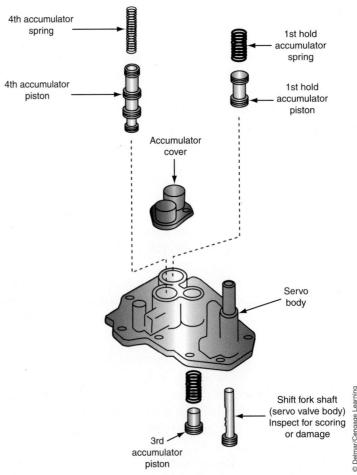

FIGURE 9-43 A servo assembly with multiple accumulators built into it.

SUMMARY

- A band is an externally contracting brake assembly that is positioned around the outside of a drum. The drum is connected to a member of the planetary gearset.
- A servo is a hydraulically operated piston assembly used to apply a band.
- A servo contracts the band when hydraulic pressure pushes against the servo's piston and overcomes the tension of the servo's return spring. This action moves an operating rod toward the band that squeezes the band around the drum.
- The operating rod of a servo tightens a band through one of three basic linkage designs: straight, lever, or cantilever.
- One-way overrunning clutches are purely mechanical devices.
- The main difference between a multiple-disc pack or band and a one-way clutch is that the one-way clutch allows rotation in only one direction at all times, whereas the others allow or stop rotation in either direction and operate only when hydraulic pressure is applied to them.
- The purpose of a multiple-friction disc pack can be the same as for a band; however, it can also be used to drive a planetary member.
- The multiple-friction disc pack can be used to drive or hold a member of the planetary gearset by connecting it to the transmission's case or to a drum.
- When multiple-disc packs lock two members of a planetary gearset together, the pack is a driving device.
- A multiple-disc pack is also used to hold one member of the planetary gearset.
- Disc pack apply pistons are retracted by one large coil spring, several small springs, or a single Belleville spring.
- One-way overrunning clutches can be either roller or sprag type.
- Shift feel is controlled by the pressure and relative timing at which each reaction member is applied or released and the rate at which each is pressurized or exhausted.
- Shift feel is also affected by fluid type, the momentary engagement of a component in a different circuit, pulsed pressures, the clearance of the apply devices, and many more design features of the various transmission models.
- Some apply circuits use an accumulator to slow down application rates without decreasing the holding force of the apply device.

TERMS TO KNOW

Accumulator
Band
Belleville spring
Cantilever
Coast
Fulcrum
Over-center spring
Pressure plate
Servo
Steel discs

REVIEW QUESTIONS

Short-Answer Essays

1. What is a reaction member and why is it called that?

2. What allows for a band to be self-energized?

3. Briefly describe how a multiple-friction disc pack works.

4. How can a multiple-friction disc assembly be used to drive or hold a member of the planetary gearset?

5. What are the different types of single-wrap bands found on today's transmissions and when is each used?

6. Why do some multiple-disc packs have a Belleville spring?

7. What is the purpose of an accumulator?

8. What determines shift feel or quality?

9. What is the difference between a roller-type and a sprag-type one-way clutch?

10. How is a band applied?

Fill in the Blanks

1. In most cases, a band is used to hold the _____ or _____ gear.

2. A band that is split with overlapping ends is called a _____ - _____ band. A one-piece band that is not split is called a _____ - _____ band.

3. Upshifts occur because _____ pressure built up enough to overcome the spring tension and the _____ pressure on a shift valve. Downshifts occur when _____ pressure decreases to the point where it can no longer overcome those pressures. Downshifting will also occur when _____ pressure has increased and is able to overtake governor pressure.

4. In rotating friction disc units, a problem usually arises when the pack is not engaged. With the pack off, the drum still spins. The high-speed rotation could create sufficient centrifugal force in the remaining fluid in the apply cylinder to partially engage the pack. This creates an unwanted drag between the plates. To prevent this problem, a _____ _____ _____ valve is incorporated in the drum or piston.

5. Multiple-friction disc pack apply pistons are retracted by one large _____ spring, several _____ _____ springs, or a single _____ spring.

6. The operating rod of a servo tightens a band through one of three basic linkage designs: _____ , _____ or _____ .

7. When multiple-friction disc packs lock two members of a planetary gearset together, the pack is a _____ device.

8. The total force of a servo piston is equal to the pressure applied to it multiplied by the _____ _____ of the piston.

9. When multiple-disc packs lock two members of a planetary gearset together, the pack is a _____ _____ . One set of discs is _____ to one member of the planetary gearset and the other set of discs is _____ to another member of the gearset. The two members of the gearset will rotate _____ at _____ _____ speed, when the pack is applied.

10. _____ _____ are those parts of a planetary gearset that are held in order to produce an output motion.

MULTIPLE CHOICE

1. Which of the following statements about the operation of a multiple-disc assembly is *not* true?
   A. The seals of the pack hold in the hydraulic pressure during application of the pack.
   B. The pressure plate provides the clamping surface for the plates and is installed at one or both ends of the pack.
   C. The purpose of a multiple-friction disc pack can be the same as a band.
   D. The multiple-disc assembly can be used to hold a member of the planetary gearset by connecting it to a clutch drum.

2. While discussing brake band operation:
   Technician A says a servo is a hydraulically operated piston assembly used to apply the band.
   Technician B says an accumulator is a hydraulic piston assembly that helps a servo to quickly apply the band.
   Who is correct?
   A. A only   C. Both A and B
   B. B only   D. Neither A nor B

3. *Technician A* says one-way overrunning clutches can be either roller or sprag type.
   Technician B says a roller clutch utilizes roller bearings held in place by sprags that separate the inner and outer race of the clutch assembly.
   Who is correct?
   A. A only   C. Both A and B
   B. B only   D. Neither A nor B

4. *Technician A* says a multiple-friction disc pack can be used to hold one member of the planetary gearset.
   Technician B says a multiple-friction disc pack can be used to drive one member of the planetary gearset.
   Who is correct?
   A. A only   C. Both A and B
   B. B only   D. Neither A nor B

5. *Technician A* says one-way clutches are effective as long as the engine powers the transmission. While the transmission is in a low gear and is coasting, the drive wheels rotate the transmission's output shaft with more power than is received on the input shaft. This allows the sprags or rollers to unwedge and begin freewheeling.
Technician B says a one-way clutch is often used to allow engine compression to aid in slowing the vehicle down.
Who is correct?
A. A only
B. B only
C. Both A and B
D. Neither A nor B

6. *Technician A* says driving clutches can have a set of friction plates splined to the transmission's input shaft.
Technician B says driving clutches can also have a drum splined directly to the output shaft.
Who is correct?
A. A only
B. B only
C. Both A and B
D. Neither A nor B

7. *Technician A* says an accumulator relies on the action of a piston or a valve to delay the delivery of high pressure to a clutch or band.
Technician B says an accumulator cushions sudden increases in hydraulic pressure by temporarily diverting some of the apply fluid into a parallel circuit or chamber.
Who is correct?
A. A only
B. B only
C. Both A and B
D. Neither A nor B

8. Which of the following statements about "shift feel" is NOT true?
A. It is controlled by the pressure at which each reaction member is applied or released, the rate at which each is pressurized or exhausted, and the relative timing of the apply and release of the members.
B. It is affected by fluid type, the momentary engagement of a component in a different circuit, pulsed pressures, and the clearance of the apply devices.
C. To improve shift feel, multiple-disc packs may contain an additional friction plate between each of the steel plates.
D. It can be improved by using a restricting orifice or an accumulator piston in the brake or clutch apply circuit.

9. *Technician A* says when a band is applied, it always wraps itself around the drum in the same direction as drum rotation.
Technician B says the self-energizing effect of a band reduces the force that the servo must produce to hold the band.
Who is correct?
A. A only
B. B only
C. Both A and B
D. Neither A nor B

10. Which of the following factors does *not* contribute to the effectiveness of a band?
A. The composition of the drum and its surface finish.
B. Condition and type of transmission fluid.
C. The direction of rotation of the drum.
D. The force used to apply the band.

Chapter 10

COMMON AUTOMATIC TRANSMISSIONS

UPON COMPLETION AND REVIEW OF THIS CHAPTER, YOU SHOULD BE ABLE TO:

- Describe the construction and operation of common Chrysler Corporation transmissions.

- Describe the construction and operation of common Ford Motor Company transmissions.

- Describe the construction and operation of common General Motors Corporation transmissions.

- Describe the construction and operation of Honda's nonplanetary gear automatic transmissions.

- Describe the construction and operation of common Nissan's transmissions.

- Describe the construction and operation of common Toyota's transmissions.

- Describe the construction and operation of other typical Simpson–gear based transmissions.

- Describe the construction and operation of other typical Ravigneaux–gear based transmissions.

- Describe the construction and operation of constantly variable transmissions (CVTs).

INTRODUCTION

Most automatic transmissions use planetary gearsets to provide for the different gear ratios. The gear ratios are selected manually by the driver or automatically by the hydraulic and electronic control system, which engages and disengages the clutches and brakes used to shift gears. To provide for these gear changes, controls are used to provide an input and output to the gearset. Controls are also used to hold a member of the gearset.

Automatic transmissions (Figure 10-1) vary by overall design, size, work capacity, number of forward gears, type of hydraulic controls and apply devices, and their electronic control system. Transmission models also vary by the type of gears or gearsets they use. This variable is actually the one that dictates the basic design of the controls. Since there are specific results from combining members of a gearset, a particular combination of gears is always required from a type of gearset to achieve the required gear ratios.

The Simpson-type planetary gearset is the most commonly used, although some transmissions use a Ravigneaux gearset, tandem planetary gearsets, or helical gearsets. Since engineers have worked to increase the number of forward gears to gain fuel economy and performance, many late-model transmissions use the Lepelletier system. Remember, gear ratios and the direction of rotation are the result of applying torque to one member of a planetary unit, holding at least one member of the gearset, and using another member as the output.

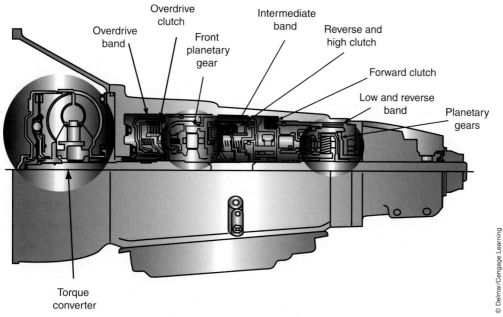

Overdrive band
Overdrive clutch
Overdrive band
Front planetary gear
Intermediate band
Reverse and high clutch
Forward clutch
Low and reverse band
Planetary gears
Torque converter

© Delmar/Cengage Learning

FIGURE 10-1 A cutaway view of a typical four-speed automatic transmission.

Simpson Geartrain

The Simpson geartrain is an arrangement of two separate planetary gearsets with a common sun gear, two ring gears, and two planetary pinion carriers. One half of the compound set or one planetary unit is referred to as the front planetary and the other planetary unit is the rear planetary. The two planetary units do not need to be the same size or have the same number of teeth on their gears. The size and number of gear teeth determine the actual gear ratios obtained by the compound planetary gear assembly.

Ravigneaux Geartrain

The Ravigneaux geartrain uses two sun gears, one small and one large, and two sets of planetary pinion gears, three long pinions and three short pinions. The planetary pinion gears rotate on their own shafts, which are fastened to a common planetary carrier. The small sun gear is meshed with the short planetary pinion gears. These short pinions act as idler gears to drive the long planetary pinion gears. The long planetary pinion gears mesh with the large sun gear and the ring gear. A single ring gear surrounds the complete assembly.

Several transmissions/transaxles use a Ravigneaux gearset, primarily because of the increased tooth contact area, which means it can withstand heavy torque loads, and compact size.

Gears-in-Tandem

Rather than relying on the use of a compound gearset, some transmissions use two or three simple planetary units in series. Gearset members are not shared, instead, the holding devices are used to lock different members of the planetary units together. The output of one gearset is applied to the next and the output is the result of combining the gearsets.

Lepellietier Geartrain

A Lepellietier gearset is also based on gearsets connected in tandem. However, one of the gearsets is a compound gear. The output ratio is a combination of the output of the Ravigneaux gearset combined with the output of a simple planetary unit. The operation of the gearset depends on an elaborate system of clutches and brakes. These transmissions are often referred to as clutch-to-clutch units because more than one clutch is applied in all gears. For example, in a seven-speed unit three separate clutches are applied to provide each forward gear.

CHRYSLER TRANSMISSIONS

Chrysler Corporation introduced the Torqueflite transmission in 1956. This transmission was the first modern three-speed automatic transmission with a torque converter and the first to use the Simpson geartrain. All Torqueflite-based transmissions and transaxles use a Simpson geartrain (Figure 10-2).

Since then, Chrysler has used many different models and designs of transmissions. Those based on a Simpson gearset operate similarly. Three-speed models relied on one compound gearset and the four- and five-speed models use that same planetary unit with an additional planetary unit for fourth and fifth gears. In recent years, Chrysler has used transmissions and transaxles with planetary gears in tandem built by Mitsubishi and Mercedes.

RWD transmissions vary with the intended work load and the engine powering the vehicle. Chrysler transaxles are based on the RWD designs; however, their basic construction depends on the position of the engine in the vehicle. Chrysler has placed the engine in FWD

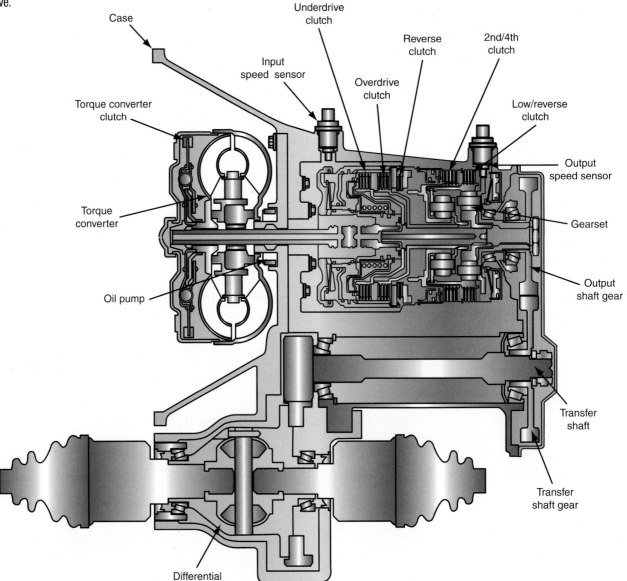

FIGURE 10-2 A Chrysler Simpson-based transaxle.

vehicles, either transversely or longitudinally. The placement of the engine affects the overall construction of the transmission.

Through the years, Chryslers' transmissions have been lightened and/or strengthened, fitted with converter clutches, received hydraulic control enhancements, and, of course, advanced electronic controls were added.

Chrysler designates their transmissions according to the following pattern:

- First digit = number of forward speeds
- Second digit = Torque rating (1–9, with 9 being the highest)
- Third digit = The orientation of the unit—(R)ear, (L)ongitudinal, (T)ransverse
- Fourth and fifth digits = Method of control—(E)lectronic, (F)ull (E)lectronic, (H)ydraulic

Therefore, the 42LE is a longitudinally mounted four-speed unit with a relatively low torque rating and electronic controls. The 68RFE is a six-speed unit with a high torque rating for RWD vehicles and is fully electronically controlled. Transmissions built by other manufacturers use the designation assigned by that manufacturer.

Simpson-Based Units

Chrysler's Simpson-based FWD and RWD transmissions have been available in both standard and wide-ratio gearsets. The original Torqueflite transmission was called the A488. It was big and heavy but served as the basis for many different transmissions and transaxles. In 1960, this transmission was modified for light-duty and was called the A904, which was designated later as the 30RH. The A998/A999 (later called the 31RH and 32RH) was a heavier duty, wide-ratio version of the A904 used behind large V-6s and medium-power V8 engines. The heavy-duty A488 was replaced in 1962 with the A727 (later 36RH and 37RH). This design had an aluminum case and incorporated a parking pawl. Later 36RH and 37RH units had electronic T/C clutch control and were used until a few years ago.

Early transaxles were actually modifications of the A904 Torqueflite. The modifications included the incorporation of a final drive unit. Most used helical gears and an intermediate/idler gear or shaft to transfer power from the output of the transmission to the differential unit. The helical gears provided the necessary speed reduction.

In the 1970s, Chrysler released the A404 three-speed transaxle for their new subcompact cars. Through the years, this basic unit would be modified for additional duties and loads. The A404 was strengthened to become the A413 (later called the 31TH) for use with slightly larger engines. Through the life of this transaxle, it was with and without a T/C lockup clutch. The basic transaxle was again for use with a more powerful engine and was called the A470. The basic A404 and its derivatives were used for more than 10 years. It was eventually replaced by the four-speed Ultradrive transaxle, the A604.

All Torqueflite transmissions have at least two multiple-disc clutches, an overrunning clutch, two servos and bands, and a compound gearset to provide three or more forward gear ratios and a reverse ratio. The two multiple-disc clutches are called the front clutch and rear clutch packs. The servos and bands are also referred to by their location, front and rear, or by their function, kickdown, and low/reverse.

The front and rear clutches serve as the input devices. The transmission's input shaft rotates clockwise. Since the front clutch hub and rear clutch drum are splined together at one end of the input shaft, they also rotate clockwise (Figure 10-3). In the transaxles, the rear ring gear carries the output of the transmission to the final drive unit.

The inner edges of the front clutch friction discs are splined to the outer edges of the front clutch hub and therefore turn with the hub. The outer edges of the front clutch steel plates are splined to the inner edges of the front clutch drum. The drum rotates with the input shaft whenever the front clutch is applied.

An input shell is splined to the front clutch drum and the common sun gear, which rotates on the output shaft but is not splined to it. Since the front and rear pinion gears mesh with

Torqueflite transmissions are also referred to as Loadflite transmissions.

On some transmissions, the front clutch hub and rear clutch drum are built as a single assembly.

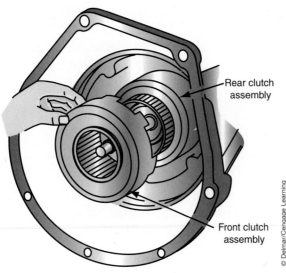

FIGURE 10-3 The front clutch hub and rear clutch drum are splined together.

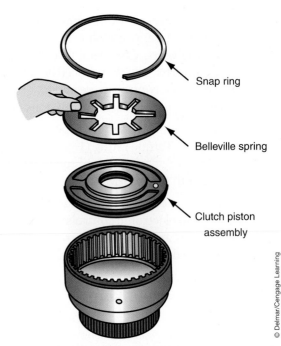

FIGURE 10-4 The rear clutch uses a Belleville return spring.

the sun gear, the drum, the input shell, and the sun gear rotate with the input shaft when the clutch is applied.

The front clutch is applied in third and reverse gears to drive the sun gear. The front clutch is released by either one large coil spring or several small coil springs when hydraulic pressure at the clutch is released.

The rear clutch is applied in all forward gears. It uses a Belleville spring to multiply the applying force of the piston and to help the piston retract into its bore (Figure 10-4).

The front and rear bands and the one-way overrunning clutch are the holding devices for these transmissions. The front or **kickdown band** is used only in second gear and holds the input shell and the sun gear stationary.

The rear or low/reverse band is applied in reverse and manual low and holds the rear planetary carrier. In manual low, the band is ineffective because the one-way overrunning clutch is holding. The rear band is effective during coasting or deceleration when the one-way clutch begins to freewheel; the rear band holds the drum to allow for engine braking. If the rear band slips or cannot hold the drum, the one-way clutch will hold but will not provide engine braking.

Most Torqueflite transmissions have a **controlled load** front **servo** (Figure 10-5). This servo has two pistons and allows for the quick release of the band during shifts from second to third gear and from second to first gear. The servo also allows for smooth engagement of the band during first to second and third to second gearshifts. Some transmissions do not have a controlled-load front servo.

Power Flow. Power flow through these transmissions occurs by the engagement and disengagement of the clutches and bands. Refer to the clutch and band application chart (Figure 10-6) while reading through the power flow.

Power Flow in Park or Neutral—When the gear selector is in Park and Neutral, input power ends at the front and rear clutches because neither is applied. When the selector is placed in Park, a parking pawl is moved by the linkage and locks the parking gear to the transmission case. The parking gear is on the outer circumference of the governor support.

Power Flow in Reverse—When the gear selector is placed into Reverse, the front clutch is applied and engages the input shell and the sun gear. The rear band is also applied and

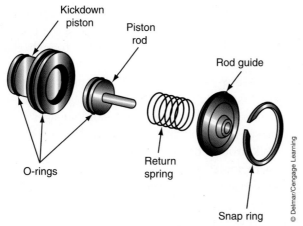

FIGURE 10-5 An exploded view of a Chrysler controlled-load servo unit.

Gear Selector Position	Operating Gear	Front Clutch	Rear Clutch	Overrunning Clutch	Kickdown Band	Low/Reverse Band	Overdrive Clutch	One-way Clutch	Direct Clutch
D	1st gear		x	x				x	x
	2nd gear		x		x			x	x
	3rd gear	x	x					x	x
	Overdrive	x	x				x		
2	1st gear		x	x				x	x
	2nd gear		x		x			x	x
1	1st gear		x	x		x		x	x
R	Reverse	x				x			x

FIGURE 10-6 Clutch and band application chart for most Chrysler transmissions with an add-on fourth gear.

holds the rear planetary carrier. The front clutch drives the sun gear in a clockwise direction. Because the rear planet carrier is held, the sun gear drives the rear planet gears in a counterclockwise direction. The rotation of the planet gears drives the rear ring gear and the output shaft in a counterclockwise direction, resulting in reverse gear with gear reduction.

Power Flow in First Gear—When the gear selector is moved to the Drive position, fully automatic shifting is available. When the transmission is operating in first gear, the rear clutch is applied and serves as the input member for the gearset. Because the rear clutch is applied, input flows through the input shaft to the rear clutch, which drives the front ring gear in a clockwise direction.

The front planetary carrier is splined to the output shaft and is therefore held by the weight of the vehicle and the drive wheels. This causes the ring gear to drive the front planet gears in a clockwise direction. The pinion gears, which are in mesh with the sun gear, rotate it counterclockwise. The rotation of the common sun gear causes the rear planet pinion gears to rotate in a clockwise direction. The planet gears cause the rear ring gear and the output shaft to turn in a clockwise direction.

The rear planet carrier is held by the overrunning clutch whenever the engine's torque is driving the planetary unit. During coast or deceleration, the weight of the vehicle and its momentum drive the planetary units through the output shaft. This causes the rear carrier to rotate clockwise and release the one-way clutch. This results in a neutral condition and doesn't allow for engine braking.

When the gear selector is placed in Manual Low (1), the rear band and clutch are applied. The overrunning clutch holds the rear planet carrier, as does the rear band. The band holds the carrier during deceleration when the one-way clutch freewheels, this does allow for engine braking during coasting.

Power Flow in Second Gear—While in Drive, the transmission will shift from first to second gears when the vehicle's speed has reached a particular point and some load is overcome. The rear clutch remains applied and the front band engages to hold the sun gear stationary. The input shaft causes the front ring gear to rotate clockwise. Because the sun gear is held, the planet gears walk around the sun gear and drive the front planetary carrier and output shaft in clockwise with some torque multiplication. The rear planet carrier is rotating clockwise and allows the one-way clutch to freewheel, thereby providing a neutral condition in the rear planetary gearset.

Power Flow in Third Gear—Third gear is a direct drive gear. When the transmission shifts into third, both the front and rear clutches become inputs for the gearsets. Whenever two members of a gearset are locked together, the pinion gears are unable to rotate on their individual shafts and the planetary gearset is locked and direct drive results. In this transmission, the two members that are locked together are the front ring gear and the sun gear.

Fourth and Fifth Gears. Chrysler introduced a four-speed automatic transmission in 1989. The power flow for the first three forward gear ratios is the same as earlier models. Fourth gear is provided by a separate planetary gearset and controlled by an overdrive clutch, direct clutch, and overrunning clutch. On fifth gear models, the additional gearset allows for an additional overdrive ratio or a direct drive in fourth plus an overdrive in fifth gear.

For RWD vehicles, the A727 was modified and became the A518 (later called the 46RH or 46RE). The overdrive gear is at the rear of the unit. In the mid-1990s, a heavier duty version of the A518 was released. This was the A618 (later called the 47RH or 47RE). It was used in trucks and vans with diesel or V-10 engines. In 2004, another heavy duty version was released, the 48RE which is an electronically controlled transmission. The basic A904 was also modified to include an additional gearset mounted in the tailshaft of the transmission. All of the resulting versions are considered light to medium duty transmissions. The A500 (later called the 40RH/42RH or 40RE/42RE), was used in trucks and vans with V-6s and small V-8s. By 2004, these transmissions were replaced by the 42RLE.

To control the operation of the additional overdrive planetary, two multiple-disc clutches (direct and overdrive clutches) and an overrunning clutch was used. The intermediate shaft is locked to the output shaft whenever the overrunning clutch is locked. This locking results in bypassing the overdrive planetary and provides for direct drive.

The direct clutch locks the sun and ring gears together to prevent freewheeling of the overrunning clutch during coasting and deceleration. This provides for engine braking. The spring used in the direct clutch assembly is a heavy tension single-coil spring that applies great holding force onto the clutch discs.

The intermediate shaft also drives the planetary carrier of the overdrive gearset. When the transmission shifts from third to fourth gear, the overdrive clutch piston moves the clutch's hub to relieve the spring tension on the direct clutch assembly. It also applies pressure to the overdrive clutch which locks the sun gear to the transmission case. With the sun gear held, the planet carrier forces the ring gear and output shaft to rotate in an overdrive condition.

Units with Gears-in-Tandem

In 1989, Chrysler began their departure from Simpson gearsets to gearsets in tandem. These units had two or three gearsets connected in series and are commonly called Ultradrive transmissions. The first Ultradrive was called the A604 and this basic design has evolved into the transmissions and transaxles found on today's Chrysler vehicles. The most commonly found Chrysler transmissions/transaxles are the 41TE/41AE, 42LE/42RLE, 62TE, 45RFE, 545RFE,

68RFE, and units made by Mitsubishi and Mercedes. The introduction of the different transmissions was largely due to advances made in electronic control and the required load capabilities.

AUTHOR'S NOTE: The original Ultradrives were plagued with problems. They had poor shifting and would suddenly drop into the limp hime mode, meaning they became locked in second gear. Chrysler made nine design changes in an attempt to fix these problems, in addition to clutch failure. Also, four modifications were made to stop excessive shifting when driving under load. The problem was so prevalent that Chrysler contacted every owner of a vehicle equipped with an A604 to let them know if the dealer could not fix their transmission, Chrysler would buy their vehicle. Chrysler's ultimate fix for the A604 was the 41TE.

Chrysler 41/42 Series. The 41TE is a four-speed transaxle with two gears in tandem. It was used from 1989 to 2008 in a variety of vehicles. It had a low load rating and was designed as a transaxle with an integral final drive assembly. Many different variations in the final drive units are found in the 41 series. They may have transfer gears, a transfer shaft, helical cut gears or sprockets, and/or a chain. This transaxle uses no bands or one-way clutches. A special version of the 41TE was made for all-wheel-drive vehicles and was called the 41AE. Versions of the 41TE, such as the 40TES and 41TES were made for specific applications. The last digit, the "S," meant it had a compact T/C and a shallower bell-housing.

Shop Manual
Chapter 10, page 456

The 42LE is a four-speed transaxle designed for longitudinally placed engines and has a high load rating. The construction is similar to a 41TE except it had a final drive housing mounted to the transmission case. A chain and sprocket were used to transfer the output from the transmission to a transfer shaft. This shaft also served as the pinion gear for the hypoid final drive unit.

The 41/42 series is controlled through the use of adaptive electronic controls and basic computer logic. Optimum shift scheduling is accomplished through continuous real-time sensor feedback provided to the TCM. The TCM receives information from various inputs and the modulated bi-directional bus system is called Chrysler Collision Detection (CCD) bus (Figure 10-7) and controls a solenoid assembly.

Shop Manual
Chapter 10, page 461

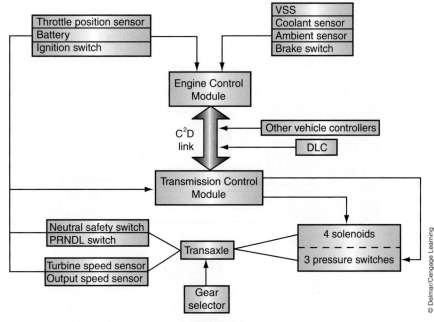

FIGURE 10-7 Schematic of the electronic control system used in Chrysler transaxles. Note the CCD (C^{2}D) bus link.

Engine speed, throttle position, temperature, engine load, and other engine-related inputs are used by the TCM to determine the best shift points. In addition, there are three pressure switches that send feedback to the TCM. These switches are located in the solenoid assembly. There are also speed sensors at the input and output shafts. Once the TCM determines it is time to shift, it sends·commands to the solenoid assembly. The solenoid assembly contains four solenoids that control hydraulic pressure to four of the five clutches in the transaxle and to the clutch of the torque converter (Figure 10-8).

All gear ratios are achieved by the action of the clutches (Figure 10-9). During a shift, one clutch is released and another clutch is applied. The three input clutches (UD, underdrive; OD, overdrive; and Reverse) are houses in an input clutch assembly. The two holding clutches are the (L/R) Low and Reverse clutch and the (2/4) second and fourth clutch (Figure 10-10). The input clutch assembly also contains the UD hub, which connects the UD clutch to the rear sun gear and the OD hub that connects the OD clutch to the front carrier and rear ring gear.

The solenoids are controlled by the TCM, which completes the ground of the solenoids when a particular solenoid should be activated. The 2–4 and UD solenoids are normally

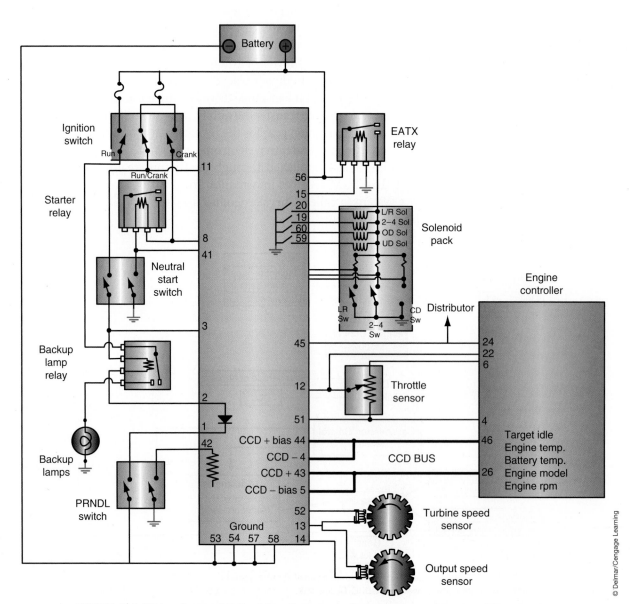

FIGURE 10-8 Wiring diagram for the electronic transmission controls on a Chrysler 41TE transaxle.

© Delmar/Cengage Learning

Gear Selector Position	Operating Gear	Underdrive Clutch	Overdrive Clutch	2/4 Clutch	Low/Reverse Clutch	Reverse Clutch
OD	1st gear	x			x	
	2nd gear	x		x		
	3rd gear	x	x			
	4th gear		x	x		
D	1st gear	x			x	
	2nd gear	x		x		
	3rd gear	x	x			
L	1st gear	x			x	
	2nd gear	x		x		
R	Reverse				x	x

FIGURE 10-9 Clutch and band application chart for late-model Chrysler transaxles.

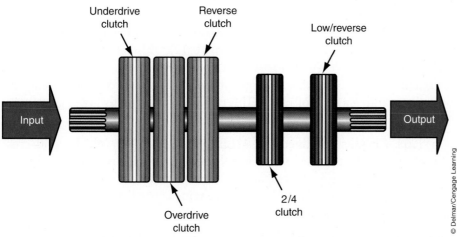

FIGURE 10-10 The placement of the various clutches in Chrysler's 41/42 series transmission.

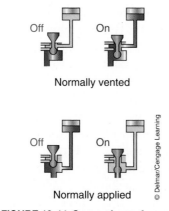

FIGURE 10-11 Comparison of normally vented and normally applied solenoids.

applied and allow fluid to move to their associated clutches when they are not energized. The other two solenoids are normally vented and only allow fluid to pass when they are energized (Figure 10-11). When a solenoid is energized, the solenoid valve opens or closes to either apply or release a clutch. If there is a major control or hydraulic problem with the transmission, the TCM will cause the transmission to enter into the "limp-home" mode. During

Gear	Original gear	Original ratio	Ratio of compounder gear	Final ratio
1st	First	2.84:1	1.45:1	4.11:1
2nd	First	2.84:1	1.00:1	2.84:1
3rd	Second	1.57:1	1.45:1	2.27:1
4th	Third	1.00:1	1.45:1	1.45:1
5th	Third	1.00:1	1.00:1	1.00:1
6th	Fourth	0.68:1	1.00:1	0.68:1
Reverse	Reverse	2.21:1	1.45:1	3.20:1

© Delmar/Cengage Learning

FIGURE 10-12 This is how the compounder gear in a 62TE provides for six forward speeds.

this time, the normally applied 2–4 and UD solenoids only allow fluid flow to provide Park, Reverse, Neutral, and second gears.

62TE. In 2007, the basic 41TE, which had two planetary gearsets and five clutch packs, was modified to provide six forward speeds. The six-speed, the 62TE, is currently used in many FWD Chryslers. It has an additional planetary unit, referred to as a compounder gearset. This gearset is mounted on the output shaft, with two multiple-disc clutch packs and an overrun clutch. The gearset is used to modify the output from the original two gearsets. The compounder planetary gearset provides either a 1.45:1 or 1.00:1 ratio to the output shaft of the transaxle.

The ratios of the forward speeds are a combination of the ratios from the original gearsets and the ratio of the compounder gearset (Figure 10-12). This setup actually provides for a seventh gear, that being a different ratio for fourth gear when the transmission is kicked down while traveling in sixth gear. The kickdown gear ratio is based on second gear being applied to the compounder gear that a ratio of 1.00:1. This provides for a slightly lower gear (numerically higher) ratio of 1.57:1 for fourth gear during kickdown.

42RLE. In 2003, Chrysler modified the 42LE for use in RWD vehicles and created the 42RLE. The modification included removing the integral final drive unit and the transfer chain and sprockets. Obviously, the transmission case had to be modified and an extension housing added, but the internal transmission is the same as the original. The 42RLE is used in many late-model Chrysler products.

RFE Series. The transmissions in the RFE series are for RWD vehicles and they are very similar to the 41/42 series. These series use basically the same input assemblies and share electronic features. The most commonly used model in the series is the 45RFE. The 45RFE is a fully adaptive four-speed automatic transmission designed for light to medium duty applications.

The transmission has several features that distinguish it from other Chrysler RWD units. 45RFE is a four-speed with three simple gearsets rather than the two normally used in a four-speed automatic. There are seven control devices which include three multiple-disc input clutches, three multiple-disc holding clutches, and one overrun clutch. It uses no brake bands, has a dual internal filter system (one filter for the transmission sump and one for the fluid cooler return system), and has a dual-stage oil pump, and the main regulator and torque converter valves are in the oil pump rather than the valve body. Two control devices are applied in every gear range and the transmission is controlled by seven solenoids. The transmission also has two second gears: one for normal operation and one used during kickdown from fourth to second. This additional second gear allows for smoother kickdowns at high engine speeds and provides increased passing power. The 45RFE was introduced in 1999 and was used until 2003. The gear ratios of this transmission are as follows:

First gear—3.0 0:1
Second gear—1.67:1
Second prime gear—1.50:1
Third gear—1.00:1
Fourth gear—0.75:1
Reverse gear—3.00:1

Shop Manual

Chapter 10, page 461

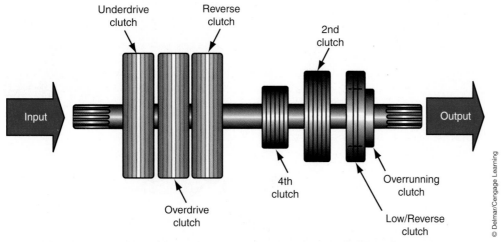

Underdrive clutch

Reverse clutch

2nd clutch

Input

Output

4th clutch

Overrunning clutch

Overdrive clutch

Low/Reverse clutch

© Delmar/Cengage Learning

FIGURE 10-13 The placement of the various clutches in Chrysler's RFE series transmission.

One major change from the 41/42 series is the number and location of the holding clutches (Figure 10-13). In the 41/42 series, the holding clutches are the L/R clutch and the 2/4 clutch. In the RFE, the holding clutches are the L/R clutch, 2C (second) clutch, 4C (fourth) clutch, and an overrunning clutch. The Underdrive (UD), Overdrive (OD), and Reverse (REV) clutches are part of the input clutch assembly.

Electronic control is accomplished by the TCM, which can be a stand-alone module or integrated with the PCM. The solenoid and switch assembly is mounted directly to the valve body. The assembly houses seven solenoids and five pressure switches. Refer to Figure 10-14 for the status of each solenoid during the different drive ranges. The Overdrive (OD), fourth clutch (4C), second clutch (2C), and Low/Reverse (L/R) solenoids are normally vented. The Underdrive (UD) solenoid is normally applied as is the Multi–Select (MS) solenoid. The PC solenoid is used to control line pressure. The system allows the vehicle to be driven (in "limp-in" mode) in the event of an electronic control system failure, or a situation that the TCM recognizes as potentially damaging to the transmission.

		Solenoids					
Manual Valve Lever Position	Gear	NA UD	NV OD	NV 4th Clutch	NV 2nd Clutch	NV L/R	NA MS
R	R						∧X∧
P/N	P/N					{x}	
OD	D1					{x}	x
OD	D2				x		x
OD	*D2*			x			x
OD	D3		x				x
OD	*D3*						
OD	D4	x	x	x			x
OD	*D4*	x		x			
OD L/I	3rd						
M2 L/I	2nd						

{x} = effective   *x* = operating in

∧ ∧ = On only if shift to reverse is above 8 mph.

© Delmar/Cengage Learning

FIGURE 10-14 Solenoid application chart for 45RFE.

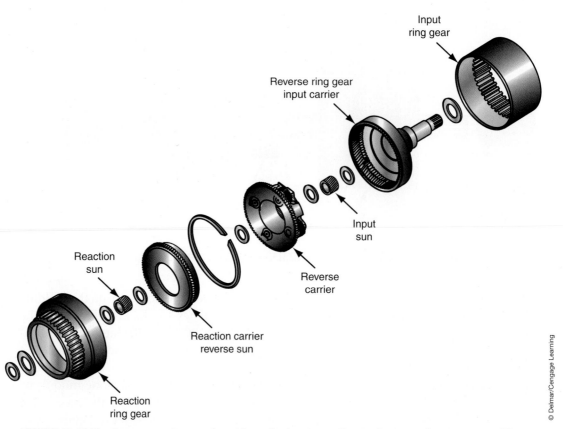

FIGURE 10-15 The three gearsets are referred to as the input, reaction, and reverse planetary assemblies.

The three gearsets are referred to as the reaction set at the front, a reverse set in the middle, and the input at the rear (Figure 10-15). Here is a summary of the action of the planetary gearset members and their associated clutches:

- **First Gear**—Input from the torque converter enters through the UD hub, which is connected to the input sun gear. The L/R clutch is applied and holds the input ring gear. The input sun gear drives the input planet pinion gears around the stationary input ring gear. This causes the input carrier, which is the output of the gearset, to rotate.
- **Second Gear**—Input from the torque converter enters through the UD hub, which is connected to the input sun gear. The second clutch is applied to hold the reverse sun gear. The input sun gear turns the input planet pinion gears and they drive the input ring gear and the reverse carrier. The pinions of the reverse carrier walk around the reverse sun gear, which is being held. The output from this combination becomes the reverse ring and input carrier assembly.
- **Third Gear**—Input from the torque converter enters through the UD hub, which is connected to the input sun gear. Input is also sent to the reverse ring and input carrier assembly through the application of the OD clutch. This results in direct drive.
- **Fourth Gear**—Input from the torque converter enters through the UD hub, which is connected to the input sun gear. The reaction ring gear is held by 4C and the reaction sun gear drives the reaction planet pinion gears. The pinion gears walk around the ring gear and drive the reaction carrier and reverse sun gear assembly. These gears, in turn, drive the reverse ring and input carrier assembly to provide an overdrive output.
- **Reverse Gear**—Input is at the reverse hub, which drives the reaction ring gear. The L/R clutch is applied to hold the reaction sun gear and reverse carrier assembly. The reaction ring gear causes the reaction carrier and reverse sun gear assembly to walk around the reaction sun gear. The reverse sun gear drives the reverse pinion gears, which drive the reverse ring and input carrier assembly to provide an output in reverse.

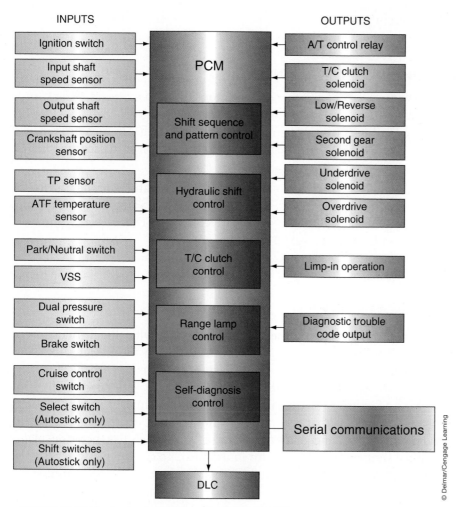

FIGURE 10-16 The basic electronic system for a 545RFE transmission with Autostick.

545RFE. The 45RFE was replaced by the five-speed 545RFE on some models beginning in 2001 and is still in use today. The 545RFE offers a second overdrive gear with a 0.67:1 ratio. Like the 45RFE, this transmission has an alternative second gear for smoother fourth to second kickdowns at high speeds. This transmission uses an Electronically Modulated Converter Clutch (EMCC) that provides for partial engagement in third, fourth, or fifth gears. Naturally, it also provides for full torque converter clutch engagement when certain operating conditions are met. The transmission also features Electronic Range Select (ERS), which is another name for Autostick (Figure 10-16). The 545RFE has two ATF filters, one in the oil sump and the other, an external canister-type filter, positioned in the oil return circuit for the oil pump.

68RFE. This transmission was introduced in 2007 for the Dodge Ram 2500 and 3500 Pickups with the 6.7L Cummins ISB Diesel engine and is still in use. The 68RFE is a six-speed unit designed for use with high output engines. Its basic design and operation is the same or similar to the 45RFE and 545RFE transmissions, except it has a larger bell-housing, stronger internal

AUTHOR'S NOTE: Current Dodge Ram 3500/4500/5500s use an Aisin AS68RC, which is a heavy duty transmission with a PTO (Power Take Off). The 68RFE does not provide for a PTO.

components, different gear ratios, and no second prime gear, and the TCM is programmed to work with high output engines, such as diesels. The limp-in mode uses fourth gear.

FORD MOTOR COMPANY TRANSMISSIONS

Ford has used a wide variety of gearset configurations through the years. In fact, the transmissions found in current vehicles can be Simpson, Ravigneaux, gears-in-tandem, or Lepellietier gear based units. Ford uses transmissions they build in North America or France, as well as units built by ZF and JATCO. To define the transmission types, Ford uses the following codes:

First digit = the number of forward gears
Second and sometimes the fourth or fifth digit = the features of the transmission
(E = electronically controlled; F = FWD; HD = heavy duty; L = lockup clutch;
N = nonsynchronous; O [OD] = overdrive; R = RWD; S = synchronous;
W = wide ratio; X = transaxle)
Third and fourth digits = the maximum input torque capacity (e.g., 44 = 440 ft.-lbs.,
55 = 550 ft.-lbs., and 70 = 700 ft.-lbs.)

Simpson-Based Units

Ford Motor Company began to use the Simpson geartrain with the introduction of the C-4 transmission in 1964. Previous transmissions were based on the Ravigneaux design, as are some of their current models. Ford has used many transmissions that use the Simpson geartrain and all of these units use a gear and crescent-type oil pump. The power flow through Ford transmissions with a Simpson geartrain is similar to that of a Chrysler Torqueflite transmission.

All transmission models that rely on a Simpson gearset operate similarly. Three-speed models relied on one compound gearset and four-speed models use that same planetary unit with an additional simple planetary unit for fourth gear. To control planetary gear action, the units typically use two multiple-disc clutches, two bands, and one overrunning clutch. The different transmission models vary in their control of the planetary units; therefore, on some models a multiple-disc pack may be used instead of a band or vice versa.

Through the years, the transmissions were lightened or strengthened depending on their application, fitted with converter clutches, received hydraulic control enhancements, and, of course, electronic controls were added. In appearance, the transmissions are similar and have a separate and removable extension housing. The major external difference between the models is the size and the reinforcements of the case.

Input Devices. The front planetary carrier and rear ring gear are splined to the output shaft; therefore, either of these will always serve as the output member. Both the front carrier and the rear ring gear can either drive the output shaft or be driven by it, depending on which bands and clutches are applied.

The input shaft is splined to the **forward clutch** drum and high/reverse clutch hub. The high/reverse clutch friction discs are splined to the outer edge of the **high/reverse clutch** hub and rotate with it. The outer edges are splined to the inner edge of the high/reverse clutch drum. When the high/reverse clutch is applied, the input shaft rotates the drum clockwise. The sun gear is splined to the input shell and becomes the input member when the clutch is applied.

The high/reverse clutch is applied in third and reverse gears to hold or drive the sun gear. The forward clutch friction discs are splined to the forward clutch drum, which is also splined attached to the input shaft. The forward clutch steel discs are splined to the forward clutch hub. When the forward clutch is applied, the front ring gear becomes the input member of the gearset. The outside of the front ring gear is either part of the forward clutch hub or splined to it. The forward clutch is applied in all forward gears.

Holding Devices. As input is received on the front ring gear, the front pinion gears rotate in the front carrier and drive the sun gear. Torque is sent to the rear planetary unit through the common sun gear. The rear planetary carrier is splined to the inside of the low/reverse hub and a one-way roller clutch is attached to the rear of the hub. The one-way clutch locks and holds the rear carrier when the low/reverse hub tries to rotate counterclockwise. This happens when the gear selector is placed in Drive and the transmission is operating in Low.

The rear carrier can also be held by the low/reverse band or clutch. In most models, a low/reverse band on the low/reverse drum can hold the rear carrier. In other models, the friction discs of a low/reverse clutch are splined to the outside of the low/reverse drum and the steel discs are splined to the transmission case. The low/reverse band or clutch is applied in reverse and manual low.

The intermediate band is wrapped around the high/reverse clutch drum. When the transmission is operating in second gear, the intermediate band is applied and the clutch drum, the input shell, and the sun gear are held stationary.

When the gear selector is placed in Park, a mechanical lever, rod, or pawl engages with a large gear on the transmission's output shaft and locks it to the transmission case.

A4LD. The A4LD transmission (Figure 10-17) is a four-speed with a torque converter clutch. It was the first Ford automatic to use an electronically controlled T/C lockup clutch. To provide for fourth gear, an additional planetary gearset is placed in front of the previously used (the C3 transmission) three-speed Simpson geartrain. This transmission was introduced in 1985 for four- and six-cylinder engines only and was last used in the 2002 Mazda Series B pickup. Electronic shift control for the 3–4 shift was added in 1987, the other gears remained hydraulically controlled. It began to be replaced by the 4R44E and 4R55E in 1995.

Gear shifting is controlled by three multiple-disc clutches, three bands, and a one-way clutch. Input passes through the overdrive planetary gearset (Figure 10-18) before it reaches the Simpson gearset. To control planetary action of the overdrive gearset, the A4LD uses an **overdrive band**, an overdrive clutch, and an **overdrive one-way clutch**. The overdrive band holds

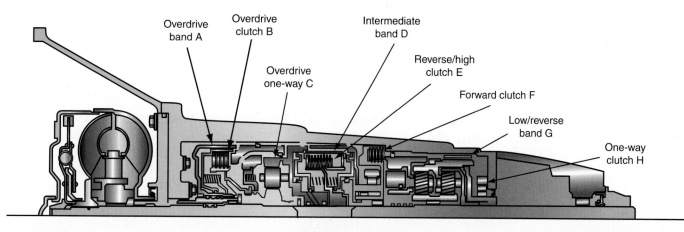

Gear	Overdrive band A	Overdrive clutch B	Overdrive one-way clutch C	Intermediate band D	Reverse and high clutch E	Forward clutch F	Low and reverse band G	One-way clutch H	Gear ratio
1- Manual first gear (low)		Applied	Holding			Applied	Applied	Holding	2.47:1
2- Manual second gear		Applied	Holding	Applied		Applied			1.47:1
D- Drive auto-1st gear		Applied	Holding			Applied		Holding	2.47:1
D- O/D auto-1st gear			Holding			Applied		Holding	2.47:1
D- Drive auto-2nd gear		Applied	Holding	Applied		Applied			1.47:1
D- O/D auto-2nd gear			Holding	Applied		Applied			1.47:1
D- Drive auto-3rd gear		Applied	Holding		Applied	Applied			1.0:1
D- O/D auto-3rd gear			Holding		Applied	Applied			1.0:1
D- Overdrive automatic fourth gear	Applied				Applied	Applied			0.75:1
R- Reverse		Applied	Holding		Applied		Applied		2.1:1

FIGURE 10-17 A Ford A4LD transmission.

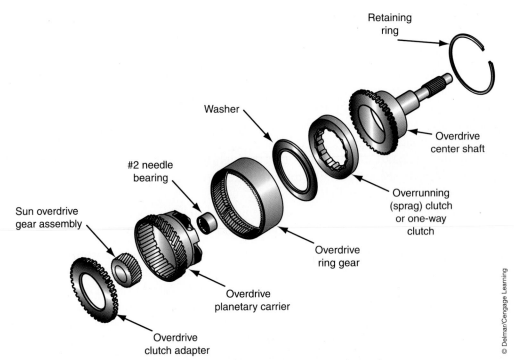

FIGURE 10-18 Add-on planetary gearset that provides fourth gear.

the overdrive sun gear to provide for fourth gear overdrive. The overdrive clutch locks the overdrive sun gear to the overdrive planetary carrier to provide for engine braking while the transmission is operating in Drive. The overdrive one-way clutch locks the input shaft directly to the input of the Simpson geartrain by locking the carrier to the ring gear of the single planetary gearset. Direct drive through the overdrive planetary unit is present whenever the transmission is operating in first, second, third, and reverse gears. During these conditions, the overdrive clutch is applied to provide for engine braking. When overdrive is selected, the overdrive band is applied and holds the sun gear. The overdrive clutch is released and the planet carrier becomes the input member driving the ring gear at an overdrive ratio of approximately 0.75:1.

4R100/E4OD. The C6 transmission was a heavy-duty automatic transmission used from 1966 to 1996. It had three forward speeds and was used primarily in trucks, although it was also used in some cars with larger engines. The C6 began to be replaced by the E4OD in 1989. This transmission was modified to include a fourth forward gear and electronic shift controls. It was the first electronically controlled transmission for trucks. To provide for a fourth gear, an overdrive planetary gearset, band, one-way clutch, and multiple-disc clutch were added.

The action of the E4OD is controlled by the PCM, as is the operation of the converter clutch. Five solenoids are used. One of the solenoids controls converter clutch operation and an EPC solenoid is used to control line pressure. There are also two shift solenoids and a coast clutch solenoid. Figure 10-19 shows the operation of these three solenoids according to gear selector position.

Gear Selector Position	Operating Gear	Shift Solenoid #1	Shift Solenoid #2	Coast Clutch Solenoid
P, N, R	Park, Neutral, Reverse	ON	OFF	OFF
OD	1st	ON	OFF	OFF
OD	2nd	ON	ON	OFF
OD	3rd	OFF	ON	OFF
OD	4th	OFF	OFF	OFF
2	2nd	OFF	OFF	ON
1	1st	ON	OFF	ON

FIGURE 10-19 Shift solenoid activity for an E4OD/4R100.

The E4OD was updated in 1998 and renamed the 4R100. It was basically the same as the E40D, but was built with some reinforcements to improve reliability. This model was last used in the 2004 E-series cargo vans.

4R44E, 4R55E, and 5R55E. For the 1995 model year, the 4R44E appeared as an updated version of the A4LD. The 4R44E is a four-speed unit that has a simple gearset in front of a Simpson unit. It also had full electronic control. Two versions, the 4R44E and 4R55E, were used. The transmissions are similar in design and vary only in their torque capacity. The 4R44E was used until 2001 when it was replaced by the 5R44E. The 4R55E was replaced by the 5R55E after the 1997 model year.

The 5R55E is a five-speed version of the 4R55E (Figure 10-20). An additional gearset was not added to achieve the fifth gear, rather control devices were added and the TCM's software was upgraded. This resulted in new combinations of apply and holding devices. This was the first five-speed automatic transmission used by an American automaker. These transmissions are equipped with five or six solenoids mounted to the valve body. These solenoids control shift quality and timing and torque converter clutch operation. The transmission offered more precise computer controls, clutch-to-clutch shifting, and introduced adaptive shift control. A version of the 5R55E was made for lighter duty applications: the 5R44E. Both versions are currently used.

A few other versions of this transmission have also been used. Each is similar in design but have different controls, gear ratios, or programming designed to match the transmission to the vehicle it will be used in. The 5R55N was built for the Jaguar S-Type and Lincoln LS luxury sedans. It provided smoother and quieter operation. The 5R55S was used in late-model Thunderbirds. It featured Ford's SelectShift, which allows for true manual shifting of the automatic transmission. The 5R55S, without SelectShift, is currently used in the Mustang. The 5R55W is used in late-model Explorers, Mariners, and Rangers.

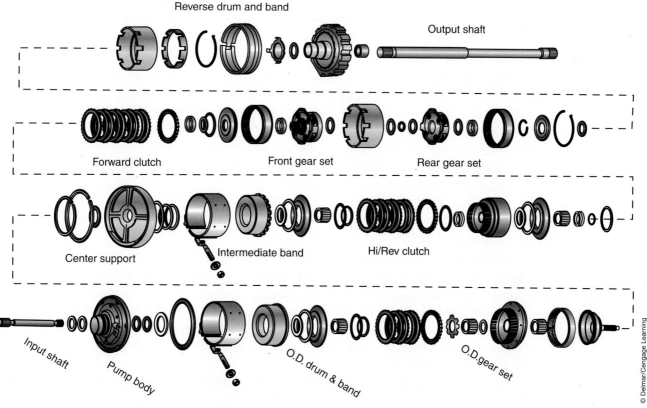

FIGURE 10-20 The main components of a 5R55E transmission.

© Delmar/Cengage Learning

5R110W. The 5R110W, referred to by Ford as the "TorqShift" transmission, is an upgrade of the 4R100 transmission. It was introduced in 2003 in F-Series trucks and Excursions equipped with a 6.0L diesel engine.

The 5R110W is a five-speed, high-capacity, rear-wheel-drive transmission. A unique feature in this transmission is that it actually has six forward speeds but only five are used. When the transmission is operating in its cold mode, it shifts from first to second to third to fourth to sixth. When it is the hot mode, it shifts from first to second to third to fifth to sixth. The change in shift strategy according to temperature improves the efficiency and durability of the transmission.

This transmission offers the Tow/Haul Mode. During this, upshifts occur at a higher speed. Downshifts also occur at a higher speed to provide some additional engine braking.

Ravigneaux Gear-Based Units

Through the years, Ford has probably used the Ravigneaux gearset more than other manufacturers. The AOD was Ford's first four-speed automatic overdrive transmission and was based on a Ravigneaux gearset. In 1980, the AOD replaced a majority of Ford's older transmissions, including the Cruise-O-Matic and C5. In 1991, the AOD was redesigned and equipped with electronic controls; it was then called the AODE. The transmission was replaced by the 4R70 series with the 1993 model vehicles. This transmission was used continuously, with some modifications, until 2008 on some vehicles.

The torque converter on early models of the AOD and AODE split the power flow from the engine. The converter drove two input shafts to provide both a hydraulic and mechanical input into the transmission. A typical input shaft from the converter's turbine hydraulically drove the forward clutch drum. An additional input shaft mechanically linked the converter and the output of the engine to the direct clutch. The amount of input from either or both sources varied according to the operating gear. When the transmission was operating in first, second, and reverse gears, 100 percent of the input was delivered hydraulically by the turbine. When operating in third gear, the geartrain received approximately 40 percent of its input from the turbine and 60 percent was delivered through the mechanical linkage. In fourth gear, all of the input to the geartrain was through the mechanical linkage. Later models of this transmission had a conventional input shaft and torque converter with a clutch.

4R7xx Series. Many improvements have been made to the 4R70W since 1993, most of this were advanced electronic controls and/or improved strength. In 2003, the 4R75W was introduced in some cars and trucks with a V-8 engine. This transmission has greater torque capacity ($75 \times 10 = 750$ lbs.-ft.) than the previous model. Subsequent variances are designated as the 4R70E and 4R75E. Members of the gearset are held by bands or clutches and only driven by multiple-disc clutches. The speed ratios are as follows:

First—2.84:1
Second—1.55:1
Third—1.00:1
Fourth—0.70:1
Reverse—2.32:1

The Ford F-150 used the 4R70E or 4R75E, depending on application. These transmissions have wide-ratio gears, electronic shifting, torque converter clutch, and transmission pressure controls. They have two bands, two overrunning clutches, and four friction clutches. See the clutch and band application chart (Figure 10-21). The action of the planetary gearset is controlled by solenoids, called SSA and SSB. Both shift solenoids are normally open. There is also a solenoid that controls the operation of the torque converter clutch (refer to Figure 10-21 for the action of the solenoids).

This series uses the common ring gear as the output member (Figure 10-22). The output shaft is connected to the ring gear. The forward clutch hub and input shaft transfer torque from the T/C to the transmission. This shaft is splined to the T/C's turbine at one end and to

Solenoid Operation Chart

Gear Lever Position	PCM Commanded Gear	Solenoids		
		SSA	SSB	TCC
P/R/N	1	On	Off	HD
(D)	1	On	Off	UD
(D)	2	Off	Off	EC
(D)	3	Off	On	EC
(D)	4	On	On	EC
w/OD OFF				
1	1	On	Off	HD
2	2	Off	Off	EC
3	3	Off	On	EC
Manual 2	2	Off	Off	EC
Manual 1	1	On	Off	HD
1[a]	2	Off	Off	EC

[a] When a manual pull-in occurs above a calibrated speed, the transmission will not downshift from the higher gear until the vehicle speed drops below this calibrated speed.
EC = Electronically controlled.
HD = Hydraulically disabled.

FIGURE 10-21 Action of the solenoids in a Ford 4R70E transmission.

© Delmar/Cengage Learning

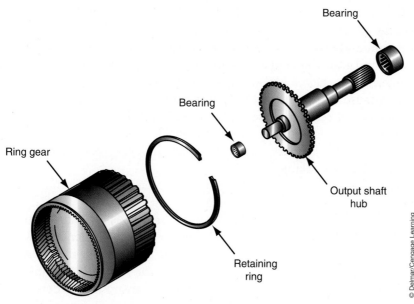

FIGURE 10-22 The 4R70W uses the common ring gear as the output member.

© Delmar/Cengage Learning

the forward clutch sun gear and stub shaft on the other end. The stub shaft transfers torque from the input shaft to the planetary carrier (through the direct clutch) during third and fourth gear operation.

The input shaft drives the front sun gear and the forward clutch is applied in first, second, and third gears (Figure 10-23). The clutch is applied hydraulically and is released by the tension of a single large coil spring when hydraulic pressure at the clutch is exhausted. The clutch is not applied in fourth gear.

The planetary carrier is to be held by the **low one-way clutch** to prevent it from turning counterclockwise while the transmission is in drive and accelerating in first gear. The long pinion gears are in constant mesh with the ring gear, the reverse sun gear, and the short

Gear Selector Position	Operating Gear	Intermediate Clutch	Intermediate One-way Clutch	Overdrive Band	Reverse Clutch	Forward Clutch	Planetary One-way	Low/Reverse Band	Direct Clutch
OD	1st gear					x	x		
	2nd gear	x	x			x			
	3rd gear	x				x			x
	4th gear	x		x					x
3	1st gear					x	x		
	2nd gear	x	x			x			
	3rd gear	x				x			x
1	1st gear					x	x	x	
	2nd gear	x	x	x		x			
R	Reverse				x			x	

FIGURE 10-23 Clutch and band application chart for a Ford 4R70W transmission.

pinion gears. The short pinion gears are in constant mesh with the forward sun gear and the long pinions. The short pinions do not mesh with the ring gear but can drive it through the long pinion gears. The ring gear is splined to a flange on the output shaft.

The direct clutch is applied in third and fourth gears. This clutch connects the input shaft to the carrier through the stub shaft. The direct clutch is released by several small coil springs when hydraulic pressure to the clutch is relieved.

The reverse clutch couples the input shaft to the reverse sun gear and is applied only in reverse. The reverse clutch is released by a Belleville spring, which also increases the clamping force of the clutch.

The **intermediate clutch** along with the **intermediate one-way clutch** holds the reverse sun gear in second gear. This clutch prevents the reverse sun gear from rotating counterclockwise. The intermediate clutch is also applied in third and fourth gears, but is only effective in second gear because the intermediate one-way clutch freewheels in third and fourth gear. The intermediate clutch is released by several small coil springs.

The low/reverse band is applied during manual low and reverse gear operation. The low/reverse band holds the planetary carrier stationary. The reverse band is also applied in manual 1 to provide engine braking.

The overdrive band holds the reverse clutch drum in fourth gear and manual 2. This holds the reverse sun gear. This causes the reverse sun gear to be held in these ranges.

This transmission uses a gerotor front pump. Pressure is regulated by the main regulator valve. The pump has an internal boost circuit which is more efficient at lower engine speeds.

The operation of these transmissions is controlled by the PCM. The PCM uses logic to control shift scheduling, shift feel, and converter lockup. The PCM relies on information from the engine control system, as well as information from the transmission to determine the optimum shift timing. Information about the operation of the transmission is received by the PCM through signals from an Output Shaft Sensor (OSS), a Vehicle Speed Sensor (VSS), a **Transmission Oil Temperature (TOT) sensor**, and a Transmission Range sensor (TR sensor).

Besides the shift solenoids, these transmissions have an EPC solenoid and a modulated converter clutch (MCCC) solenoid. The EPC solenoid controls line pressure at all times. Likewise, the operation of the converter clutch is also totally controlled by the PCM through the MCCC. The only exception to this is during first and reverse gear operation when the clutch is disabled to prevent lockup regardless of the commands by the computer.

The EPC solenoid is a **variable force solenoid (VFS)** and contains a spool valve and spring. To control fluid pressure, the PCM sends a varying signal to the solenoid. This varies the amount the solenoid will cause its spool valve to move. When the solenoid is off, the spring tension keeps the valve in place to maintain maximum pressure. As more current is

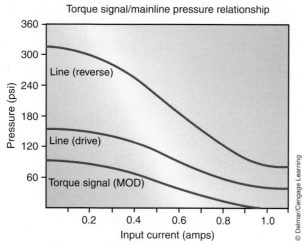

FIGURE 10-24 The EPC solenoid controls hydraulic pressures throughout the transmission by controlling the current going to the solenoid.

applied to the solenoid, the solenoid moves the spool valve more, which moves to uncover the exhaust port, thereby causing a decrease in pressure (Figure 10-24).

F4E. F4E is a modern version of the 4EAT which was a redesign of Mazda's G4A-EL. This unit was Mazda's first four-speed transmission and was introduced in 1987. It had three shift solenoids, a lockup solenoid, and a vane-type pump. It was redesigned for 1988 and used by Ford and called the 4EAT-G. The F-4EAT was released in 1990 and was a fully electronically controlled unit. Originally it was manufactured by Mazda and JATCO, but Ford took over production from 1991 to 2003. The transmission was redesigned again in 1993 to include seven solenoids and a rotor-type pump.

Gears in Tandem

The first of this design was the AXOD which was introduced in the 1986. It was a four-speed transaxle for transversely mounted engines. The torque converter did not directly drive the gears inside the transaxle. Rather, the converter drove a sprocket, which is connected, by a chain, to a sprocket on the input shaft. The transaxle used a planetary gearset mounted on the output shaft for the final drive gears. The construction and operation of this transmission is similar to GM's 4T60. The AXOD was updated with electronic controls in 1991 and is referred to as the AXOD-E. Released in 1994, the AX4S was a refined version of the AXOD-E. It had updated electronics. The AX4N is a modified version of the AX4S. Changes included the elimination of a brake band used in the AXOD and AX4S. The transaxle was then classified as a nonsynchronous unit. The AXOD family was used until the end of 2007.

AX4S Transaxle. The AX4S uses four multiple-disc clutches, two band assemblies, and two one-way clutches. Output from the planetary units is through the rear planetary carrier and front ring gear. All of the multiple-disc clutches are released by several small coil springs when hydraulic pressure is diverted from the clutch's piston. For more insight on the operation of the AX4S, refer to the clutch and band application chart (Figure 10-25).

Input to the geartrain is received by the front sun gear, the front planetary carrier, or the rear ring gear. The low one-way clutch is locked by the action of the forward clutch in low, second, third, and reverse gears. The one-way clutch is only effective in reverse and low, because it freewheels in second and third gear.

The direct one-way clutch is locked by the action of the direct clutch in third and fourth gears and when the transmission is operating in manual low. The one-way clutch is only effective in third gear and manual low, it freewheels in fourth gear. In third gear, the direct one-way clutch transmits input torque from the direct clutch to the front sun gear. Also in third

Shop Manual
Chapter 10, page 477

Gear Selector Position	Operating Gear	Low/Intermediate Band	Overdrive Band	Forward Clutch	Intermediate Clutch	Direct Clutch	Reverse Clutch	Low One-way Clutch	Direct One-way Clutch
OD	1st gear	x		x				x	
	2nd gear	x		x	x				
	3rd gear			x	x	x			x
	Overdrive		x		x	x			x
D	1st gear	x		x				x	
	2nd gear	x		x	x			x	
	3rd gear			x	x	x			
Low	1st gear	x		x		x		x	x
R	Reverse			x			x	x	

FIGURE 10-25 Clutch and band application chart for a Ford AX4S transaxle.

gear, the direct clutch drives the front sun gear through the direct one-way clutch and locks the geartrain to provide for engine braking.

The intermediate clutch is applied in second, third, and fourth gears and drives the front planetary carrier and rear ring gear in second and third gears.

The low/intermediate band holds the rear sun gear stationary in first and second gears. The overdrive band is applied in fourth gear and holds the front sun gear. The reverse clutch holds the front planetary carrier and rear ring gear for reverse gear.

Five solenoids are mounted on the AX4S valve body. One solenoid is used for modulated converter clutch control. Another, the EPC solenoid, is used to control line pressure. The remaining three solenoids are shift solenoids.

These solenoids are on/off solenoids that are normally off and in the open position. Being open, the solenoid valves allow line pressure to the bore of the shift valve and keeps the shift valve closed. When the shift solenoids are energized, they block the line pressure and exhaust line pressure from the valve. This allows the shift valve to open. A summary of the shift solenoid activity is shown in Figure 10-26.

AX4N/4F50N. The AX4N was renamed 4F50N in 2001. It is a four-speed unit with two bands, five multiple-disc clutches, and three one-way clutches. This transaxle is nonsynchronous. Nonsynchronous transmissions tend to have smoother shifts. In the AX4N, to eliminate the need for the synchronized release of the intermediate band with the application of the direct clutch, the intermediate band was eliminated. In its place, there is a low-intermediate clutch and an intermediate one-way clutch. Actually, the intermediate band was not eliminated; rather, it is now called the coast brake. With the new name came a new job, to allow for engine braking in first and second gears when the gear selector is in the manual 1 position (Figure 10-27).

Gear Selector Position	Operating Gear	Shift #1 Solenoid	Shift #2 Solenoid	Shift #3 Solenoid
P or N	Park or Neutral	OFF	ON	OFF
R	Reverse	OFF	ON	OFF
OD	First	OFF	ON	OFF
OD	Second	ON	ON	OFF
OD	Third	OFF	OFF	ON
OD	Fourth	ON	OFF	ON
D	First	OFF	ON	OFF
D	Second	ON	ON	OFF
D	Third	OFF	OFF	OFF
1	First	OFF	ON	OFF
1	Second	OFF	ON	OFF

FIGURE 10-26 Shift solenoid activity for an AX4S.

						Shift Solenoid Operation Chart			

Transaxle range selector lever position	Powertrain control module Gear commanded	AX4N solenoids			
		Engine braking	SS 1	SS 2	SS 3
P/N[a]	P/N	No	Off[b]	On[b]	Off
R (Reverse)	R	Yes	Off	On	Off
Overdrive	1	No	Off	On	Off
	2	No	Off	Off	Off
	3	No	On	Off	On
	4	Yes	On	On	On
D (Drive)	1	No	Off	On	Off
	2	No	Off	Off	Off
	3	Yes	On	Off	Off
Manual 1	2[c]	Yes	Off	On	Off
	3[c]	Yes	Off	Off	Off
		Yes	On	Off	Off

[a] When transmission fluid temperature is below 50°F then SS 1=off, SS 2=on, SS 3=on to prevent cold creep.

[b] Not contributing to powerflow.

[c] When a manual pull-in occurs above calibrated speed, the transaxle will downshift from the higher gear until the vehicle speed drops below this calibrated speed.

FIGURE 10-27 Shift solenoid operation for an AX4N transaxle.

CD4E Transaxle. The CD4E was used in cars from 1994 to 2007. The transaxle is similar in design to the AX4S. It is a four-speed unit with electronic control. It uses two simple planetary gearsets and a planetary final drive gearset. A chain drive is used to connect the output of the planetary gears to the final drive.

The transaxle uses five multiple-disc clutches, two one-way clutches, and a band (Figure 10-28). They may also be equipped with a Transmission Control Switch or a Manual Hold Switch. The type of switch also determines the availability of certain gears during manual shift operations, the function of holding members during coast, and the electronic control system. This transaxle is fitted with a lockup torque converter that is controlled by modulated pressure.

Gear Selector Position	Operating Gear	Intermediate Overdrive Band	Reverse Clutch	Direct Clutch	Forward Clutch	Forward One-way Clutch		Coast Clutch	Low/ Reverse Clutch	Low One-way Clutch	
						Drive	coast			Drive	Coast
D	1st gear		x		x		Overrunning			x	
	2nd gear	x			x		Overrunning			Overrunning	Overrunning
	3rd gear			x	x		Overrunning			Overrunning	Overrunning
	4th gear	x		x	x	Overrunning	Overrunning			Overrunning	Overrunning
S	1st gear				x				x	x	Overrunning
	2nd gear	x			x				x	Overrunning	Overrunning
	3rd gear			x	x				x	Overrunning	Overrunning
L	1st gear				x				x	x	x
R	Reverse								x		

FIGURE 10-28 Band and clutch application chart for a Ford CD4E transaxle.

The main valve body has no check balls. However, it does contain all of the transaxles' valves, accumulators, and solenoids. The solenoid body contains a built-in filtering screen. There are five solenoids and one sensor connected to the main valve body. The Electronic Pressure Control (EPC) solenoid varies hydraulic pressure in response to directions from the PCM. There are two shift solenoids and a 3-2T/CCS (3-2 Timing/Coast Clutch solenoid). This solenoid varies pressure to two valves. One valve controls the apply of the coast clutch, while the other regulates the synchronous release of the direct clutch and the apply of the Intermediate/Overdrive band during 3–2 downshifts. The TCC solenoid is also mounted to the main valve body assembly. The engagement of the TCC is controlled by the PCM, which varies the pulse-width of the solenoid.

4F27. The 4F27E is similar to the CD4E and is also used by Mazda as the FN4A-EL. The 4F27E has a transfer-shaft gear final drive. The control system has six electronically controlled solenoids for shift feel (through line pressure control), shift scheduling (through shift valve position control), and TCC apply. This transaxle is found in the late-model Ford Focus.

Lepellietier Gear-Based Units

Ford has two basic families of Lepellietier gear-based units. The transaxle built by Ford is the result of a cooperative engineering effort with General Motors. Other designs are built by Aisin or ZF. At the current time, Ford uses this gear arrangement to provide six forward speeds for RWD and FWD vehicles.

6R Transmissions. The 6R is a six-speed built primarily for SUVs and pickups. It was introduced in the 2006 models of the Ford Explorer and Mercury Mountaineer. The designations of these transmissions are 6R60 and 6R80, according to torque capacity.

Like most Lepellietier-based transmissions, the input shaft is always connected to the ring of the planetary gear and can be simultaneously connected to the carrier of the Ravigneaux gear using a separate clutch. The sun of the planetary gear is connected to the housing and cannot rotate, while the carrier of the planetary gear is connected by clutches to the large and small sun gears of the Ravigneaux gear. The output shaft is connected to the ring of the Ravigneaux gear.

The transmission does not use a band and all shifting is done by computer-controlled clutches. The T/C clutch can be locked in any of the forward gears.

Heavy-duty units have a new remote transmission oil cooler mounted to the engine block via a three-way coolant thermostat. This system allows engine coolant to help warm-up the transmission fluid during start-up and provides improved fluid cooling when the vehicle is operating under heavy loads.

The electronics also provide for many additional features such as grade control and adaptive learning. When the vehicle is going down hill with the brakes applied, grade control logic will cause the transmission to downshift to increase the amount of engine braking. The adaptive learning or "Driver Recognition" matches transmission operation to the current driving style by monitoring acceleration and deceleration rates, brake and throttle applications, and cornering speed.

6F Transaxles. The 6F transaxle is constructed like the 6R transmission. Currently, this series is available as three versions: 6F35, 6F50, and 6F55. It was first used in 2007 in the Ford Edge and Lincoln MKX crossovers. This transaxle was codeveloped by GM and Ford; however, each company makes their own units. This series is currently being used in many FWD vehicles.

The geartrain consists of the planetary gearsets, apply devices, holding devices, planetary final drive gearset, and differential and provides the following ratios:

First—4.148:1
Second—2.370:1
Third—1.556:1
Fourth—1.155:1
Fifth—0.859:1

Gear	Clutch #1	Clutch #2	Clutch #3	Brake #1	Brake #2	One-way Clutch
1st	X					X
2nd	X			X		
3rd	X		X			
4th	X	X				
5th		X	X			
6th		X		X		
Reverse			X		X	

FIGURE 10-29 Clutch and brake application chart for a Ford 6F50 transaxle.

Sixth—0.686:1
Reverse—3.394:1
Final drive—3.46:1

To provide the necessary gear changes, the transaxle uses three multiple-disc clutches, two brakes, and an overrunning clutch (Figure 10-29).

- The C1 clutch connects the front planetary carrier to the rear sun gear of the rear planetary gearset and is applied in first, second, third, and fourth gears.
- The C2 clutch connects the intermediate shaft to the rear planetary carrier and is applied in fourth, fifth, and sixth gears.
- The C3 clutch connects the front planetary carrier to the sun gear of the rear planetary gearset and is applied in third and fifth gears. It is also applied in reverse, if the vehicle's speed is 4 mph (7 km/h) or less.
- B1 brake holds the sun gear of the rear planetary gearset and is applied in second and sixth gears.
- B2 brake holds the rear planetary carrier while in reverse, when the vehicle speed is 4 mph (7 km/h) or less.
- F1 clutch (overrunning clutch) locks the counterclockwise rotation of the rear planetary carrier during first gear operation.

Four linear-controlled solenoids (Figure 10-30) are used to control shift changes: these are called SSC, SSD, SSE, and SSF. The solenoids control hydraulic pressure to the three clutches (C1, C2, and C3) and one brake (B1). The pressure control solenoid (PCA) responds to signals from the TCM regarding throttle opening, engine load, brake pedal position, vehicle speed, gear selector position, and other operating conditions. T/C clutch is regulated according to output speed, operating temperature, and other inputs. The clutch can be on or off, depending on operating conditions, or can be modulated to allow some slip. The PCA is regulated according to throttle opening, engine load, and values programmed into the TCM. The pressure sent to the clutches and bands varies to allow for smooth shifting in all operating conditions.

The prefix THM stands for Turbo-Hydramatic, which is the division of GM responsible for automatic transmission development and production.

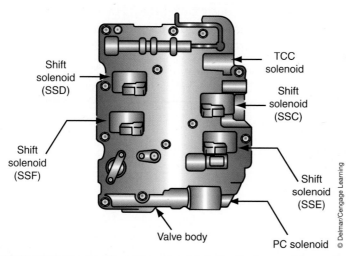

FIGURE 10-30 The location of the solenoids in Ford's six-speed transaxle.

GVWR stands for gross vehicle weight rating.

General Motors Transmissions

GM has a long history of providing automatic transmissions. The original Hydra-matic series of transmissions was used until they were replaced by the Turbo-Hydramatic series in the 1960s. The Turbo-Hydramatic was used by all GM divisions until the mid-1990s and served as the basic design for the company's current Hydra-matic line. Although all early transmissions were for RWD vehicles, GM modified them to develop transversely and longitudinally mounted transaxles. The basic design was further enhanced with the addition of electronics.

GM uses a simple naming scheme for their transmissions, with the *Hydra-Matic* name used on most automatics across all divisions. Most late-model transmissions do not have the E designation, because nonelectronic units are not available.

> First digit = number of forward gears
> Second digit = Basic design (L = Longitudinal, T = Transverse)
> Third and Fourth digits = GVWR
> Fifth digit = Features (E= electronic, HD = Heavy duty)

GM transmissions have been used by many different manufacturers, such as Rolls Royce, Jaguar, Isuzu, Daimler, and BMW. The transmission model found in a vehicle depends entirely on its application. Nearly all GM's early transmissions were based on the Simpson geartrain. Very few used a Ravigneaux gearset and those will not be discussed in this section. As GM modified their transmissions and transaxles, they began to use gearsets in tandem. This trend, along with the need for additional forward gears, led to the use of Lepellietier gear-based units.

Simpson Gear-Based Units

All Simpson-based units operate similarly, regardless of the number of speeds and their work capacity. The power flow through GM's Simpson-based units is similar to that of the power flow through a Chrysler Torqueflite transmission. There are some slight variations within the different model types, but these are dependent upon the operating characteristics of each model.

Most GM's transmissions and transaxles use a variable displacement vane-type oil pump. Older models used a gear-type pump. Although there are differences between the various models of GM transmissions with a Simpson geartrain, most use three or four multiple-disc clutches, one band, and a single one-way roller clutch to provide the various gear ratios. Each clutch is applied hydraulically and some are released by several small coil springs and others use waved spring for clutch release in place of small coil springs.

When GM first introduced a four-speed automatic transmission, they added a planetary unit to the existing transmissions. By adding a planetary unit in front of the Simpson gearset, a 3T40 became a 4T40 and a 3L80 became the 4L80, and so on.

The addition of the planetary unit brought with it additional clutch and brake devices. For example, when the 3L30 (previously called the THM 200) was modified to become the 4L30, the transmission used the same multiple-disc clutches as the 3L30 plus three additional clutches: the drive clutch, the fourth clutch, and an overdrive one-way clutch (Figure 10-31).

4L80-E. The 4L80 is a four-speed version of the Turbo-Hydramatic 400 that was released in 1965. It was used in a variety of GM pickups, vans, and SUVs until recently. There are two versions of this transmission, the 4L80-E and the 4L85-E that is able to handle more torque. Both of these transmissions offer the same gear ratios:

> First—2.48:1
> Second—1.48:1
> Third—1.0:1
> Fourth—0.75:1
> Reverse—2.07:1

All automatic upshifts and downshifts are electronically controlled by the PCM on gasoline engines or the TCM with diesel engines (Figure 10-32). Once the PCM has processed the

Gear Selector Position	Operating Gear	Overrun Clutch	Intermediate Band	Overdrive Roller Clutch	Direct Clutch	Low Roller Clutch	4th Clutch	Forward Clutch	Low/Reverse Clutch
OD	1st gear			x		x		x	
	2nd gear		x	x				x	
	3rd gear			x	x			x	
	Overdrive			x			x	x	
3	1st gear	x		x		x		x	
	2nd gear	x	x	x				x	
	3rd gear	x		x	x			x	
L2	1st gear	x		x		x		x	
	2nd gear	x	x	x				x	
L1	1st gear	x		x		x		x	x
R	Reverse			x	x				x
N/P	Neutral			x					x

FIGURE 10-31 Clutch and band application chart for a THM 4L30 transmission.

© Delmar/Cengage Learning

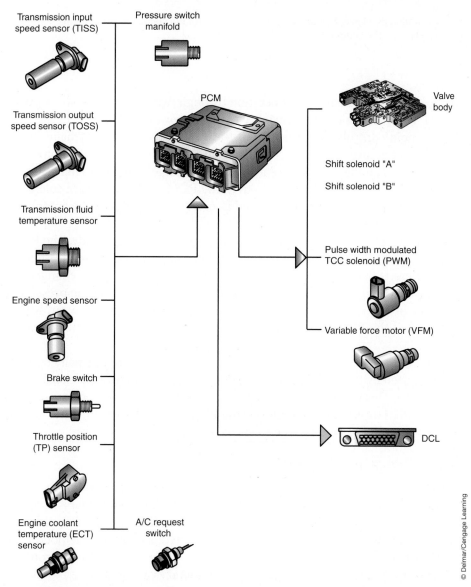

Transmission input speed sensor (TISS)

Pressure switch manifold

PCM

Valve body

Transmission output speed sensor (TOSS)

Shift solenoid "A"

Shift solenoid "B"

Transmission fluid temperature sensor

Pulse width modulated TCC solenoid (PWM)

Engine speed sensor

Variable force motor (VFM)

Brake switch

Throttle position (TP) sensor

DCL

Engine coolant temperature (ECT) sensor

A/C request switch

© Delmar/Cengage Learning

FIGURE 10-32 Layout of components for a typical GM electronically controlled transmission.

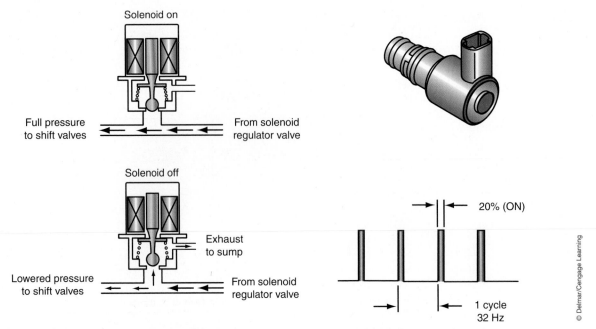

FIGURE 10-33 The action of a typical pressure control solenoid and the signal representing the control or ordered duty cycle from the computer.

input signals, it controls transmission operation through two on/off solenoids, a **pulse width modulated (PWM)** solenoid, and a **variable force motor (VFM)** located in the valve body. The PWM solenoid controls the operation of the T/C clutch. It is a normally closed valve. When the solenoid is turned off, the clutch's signal fluid exhausts and the converter clutch remains released. When the solenoid is energized, the plunger moves a metering ball to allow the fluid to move to the clutch's regulator valve. The PCM cycles the PWM solenoid on and off and varies the length of time it is energized in each cycle to control the amount of clutch lockup.

The VFM is controlled by the PCM and is located on the valve body. The PCM cycles the solenoid on and off 292.5 times per second, but varies its duty cycle. When the duty cycle is zero, there is maximum line pressure (Figure 10-33). The VFM changes line pressure in response to engine speed and vehicle load. It also adjusts the line pressure according to the wear of the transmission. As the parts of the transmission wear, shift overlap times increase and the PCM adjusts line pressure to maintain proper shift timing in spite of the wear.

Two shift solenoids are also attached to the valve body and are normally open. When a solenoid is energized by the PCM, the plunger of the solenoid moves a check ball to block the fluid pressure feed. This closes the exhaust passage and causes the fluid pressure to increase. When the ground to the solenoid is cutoff by the PCM, fluid pressure unseats the check ball. This allows fluid to exhaust through the solenoid to decrease the pressure.

Units with Gears-in-Tandem

The increased use of gears-in-tandem came about with advancements made in electronic control. Although the geartrain is based on two simple planetary gearsets operating in tandem, the combination of the two planetary units does function much like a compound unit (Figures 10-34 and 10-35). The two units do not share a common member; rather, certain members are locked together or are integral to each other. Normally the front planetary carrier is locked to the rear ring gear and the rear planetary carrier is locked to the front ring

According to SAE J1930, a VFM should be called a pressure control solenoid (PCS). Most literature still refers to it as a VFM.

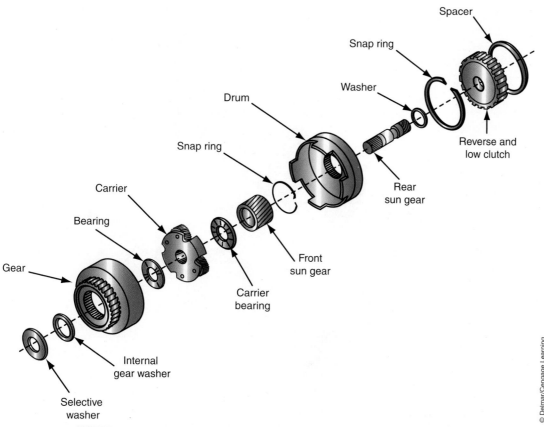

FIGURE 10-34 The front planetary gearset, low and reverse clutch, and the rear sun gear for a GM gears-in-tandem transaxle.

gear. Most transaxles house a third planetary unit, which is used only as the final drive unit and not for overdrive.

4T60 Transaxle. The 4T60 was the first domestic four-speed automatic transaxle built for FWD vehicles. It was introduced in 1984 and was the transitional transaxle from the old THM325 to the new electronically controlled transverse family. Its electronics were improved and it became the 4T60-E in 1991. By the mid-1990s, the 4T60-E was used in most GM FWD vehicles. A heavy-duty 4T60-E HD was introduced in 1991 and replaced by the 4T65 in 1997.

The 4T65-E is a modified version of the 4T60-E transaxle. It was designed to provide smoother shifting and make it more durable. With the inclusion of improved electronics, the vacuum modulator found on the 4T60 was replaced with a PCS. The 4T65-E is designed to handle vehicles up to 6500 lbs (2948 kg) GVWR and up to 280 ft.-lbs. (380 Nm) of torque.

These transaxles were mounted to the side of the engine and used a pair of sprockets and a drive chain to transfer power from the T/C turbine shaft to the transmission's input shaft (Figure 10-36). This allowed the geartrain to be positioned away from the crankshaft's centerline. The final drive is a hypoid gearset in a separate housing attached to the transmission.

They had a variable-displacement vane-type oil pump, four multiple-disc clutches, two bands, and two one-way clutches to provide the various gear ranges. One of the one-way clutches is a roller clutch, the other is a sprag. The four multiple-disc clutches are released by several small coil springs when hydraulic pressure is diverted from the clutch's piston.

The 4T60 transaxle used to be referred to as the 440-T4.

Shop Manual
Chapter 10, page 501

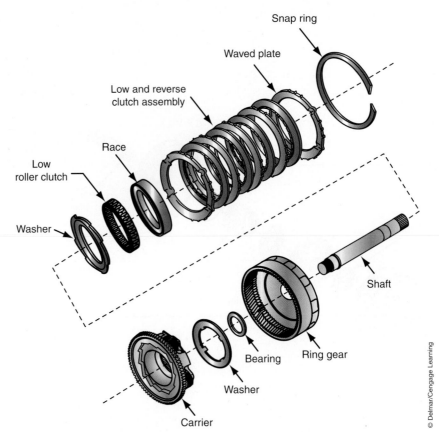

FIGURE 10-35 The low roller clutch, low and reverse clutch, and rear planetary gearset for a GM gears-in-tandem transaxle.

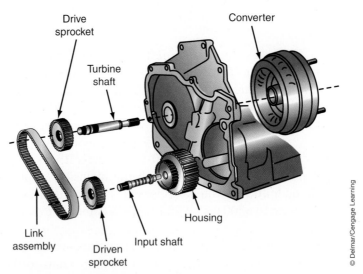

FIGURE 10-36 Engine torque is delivered to a 4T60 transaxle through a sprocket and drive link assembly.

In first gear, input is received through the front sun gear, which is locked to and drives the rear ring gear. The 1–2 band is applied and holds the rear sun gear. This forces the rear carrier to be driven by the rear ring gear at a speed reduction and a ratio of 2.92:1 (Figure 10-37).

The front planetary carrier receives input for second gear operation when the second clutch is applied. Since the rear ring gear is an integral part of the front planetary carrier, the

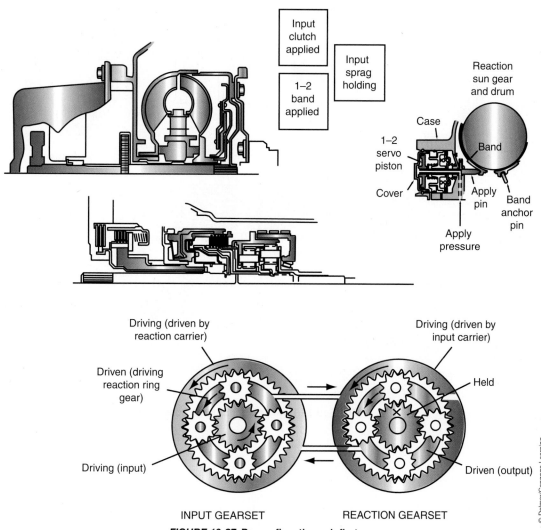

Input clutch applied

Input sprag holding

1–2 band applied

Reaction sun gear and drum

Case

1–2 servo piston

Band

Cover

Apply pin

Band anchor pin

Apply pressure

Driving (driven by reaction carrier)

Driving (driven by input carrier)

Driven (driving reaction ring gear)

Held

Driving (input)

Driven (output)

INPUT GEARSET

REACTION GEARSET

© Delmar/Cengage Learning

FIGURE 10-37 **Power flow through first gear.**

two rotate at the same speed and in the same direction. The rear sun gear is held stationary by the one-way clutch; therefore, the rear planetary carrier rotates around it at a gear reduction ratio of 1.57:1 (Figure 10-38).

Third gear offers direct drive. Input is received by the geartrain through the front sun gear by the application of the third clutch. The second clutch is still applied and directs input through the front planetary carrier. Since two of the three planetary members are rotating at the same speed, the ring gear is forced to rotate with them. The ring gear drives the rear planetary carrier at the same speed and direct drive results (Figure 10-39).

In fourth gear, the second clutch remains applied and input is received through the front planetary carrier. The fourth clutch is also applied and it holds the front sun gear. This causes the front ring gear to rotate around the carrier at a speed greater than input speed. Since the rear planetary carrier is connected to the front ring gear, the output from the gearsets is an overdrive with a ratio of 0.70:1 (Figure 10-40).

Reverse operation is achieved by holding the front planetary carrier with the reverse band. With the carrier held, the input on the front sun gear causes the front ring gear to rotate in the opposite direction. Because the front ring gear is locked to the rear carrier, the output of this gear combination is in the opposite direction as the input and at a ratio of 2.38:1 (Figure 10-41).

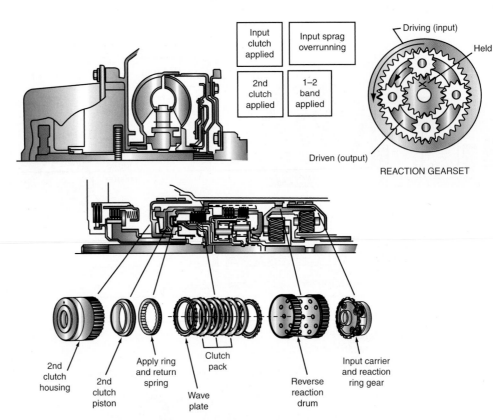

Input clutch applied	Input sprag overrunning
2nd clutch applied	1–2 band applied

Driving (input)

Held

Driven (output)

REACTION GEARSET

2nd clutch housing

2nd clutch piston

Apply ring and return spring

Clutch pack

Wave plate

Reverse reaction drum

Input carrier and reaction ring gear

FIGURE 10-38 Power flow through second gear.

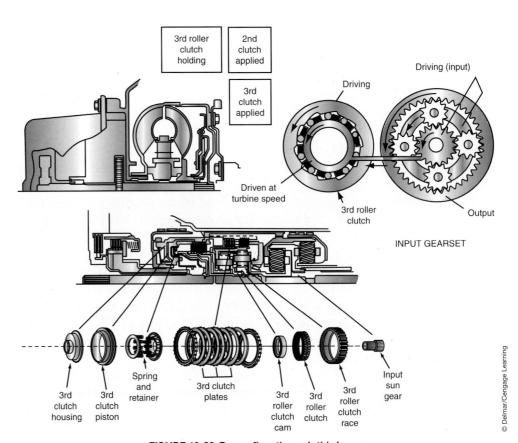

3rd roller clutch holding	2nd clutch applied
3rd clutch applied	

Driving

Driving (input)

Driven at turbine speed

3rd roller clutch

Output

INPUT GEARSET

3rd clutch housing

3rd clutch piston

Spring and retainer

3rd clutch plates

3rd roller clutch cam

3rd roller clutch

3rd roller clutch race

Input sun gear

FIGURE 10-39 Power flow through third gear.

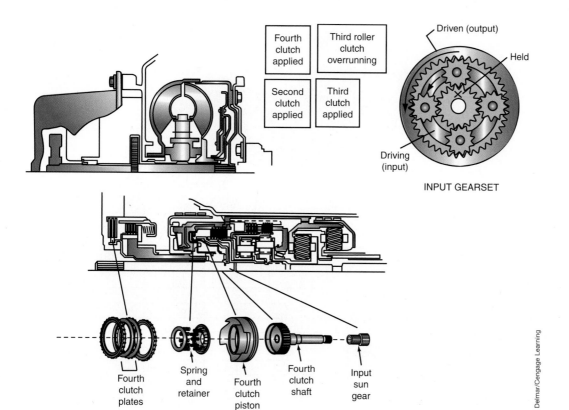

FIGURE 10-40 Power flow through fourth gear.

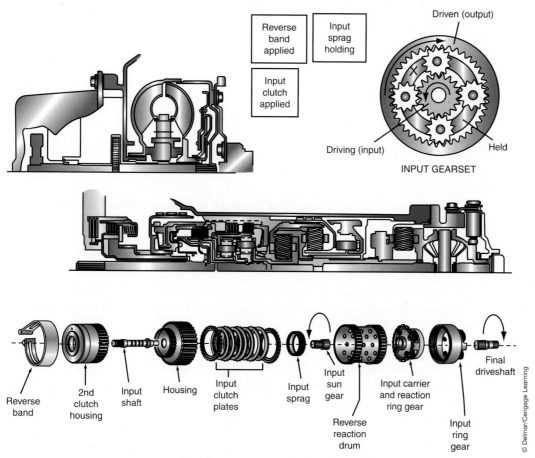

FIGURE 10-41 Power flow through reverse gear.

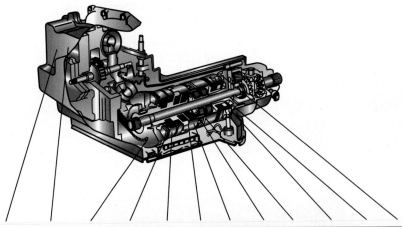

Range	Gear	A solenoid	B solenoid	4th clutch	Reverse band	2nd clutch	3rd clutch	3rd roller clutch	Input clutch	Input sprag	Forward band	1–2 roller clutch	2–1 band
P - N		On	On						*	*			
D	1st	On	On						Appl	Hold	Appl	Hold	
	2nd	Off	On			Appl			*	Orun	Appl	Hold	
	3rd	Off	Off			Appl	Appl	Hold			Appl		
	4th	On	Off	Appl		Appl	*	Orun			Appl		
D	3rd	@ Off	@ Off			Appl	Appl	Hold	Appl	Hold	Appl		
	2nd	@ Off	@ On			Appl			*	Orun	Appl	Hold	
	1st	@ On	@ On						Appl	Hold	Appl	Hold	
2	2nd	@ Off	@ On			Appl			*	Orun	Appl	Hold	Appl
	1st	@ On	@ On						Appl	Hold	Appl	Hold	Appl
1	1st	@ On	@ On				Appl	Hold	Appl	Hold	Appl	Hold	Appl
R	Rev	On	On		Appl				Appl	Hold			

★ Applied by not effective

On = Solenoid energized
Off = Solenoid deenergized
Appl = Applied
Hold = Holding
Orun = Overrunning

@ The solenoid's state follows a shift pattern that depends on vehicle speed and throttle opening, not on the selected gear.

© Delmar/Cengage Learning

FIGURE 10-42 Range reference chart for a 4T60-E transaxle.

The shifting of this four-speed transaxle is controlled by two shift solenoids that are controlled by the PCM. The shift solenoids operate and control the operating gears in the typical manner, as shown in Figure 10-42.

The transaxle has the ability to change line pressure in response to normal clutch, seal, and spring wear. The PCS is duty cycled controlled, which varies the current flow through its windings. As current flow increases, the magnetic field around the windings also increases. This increased strength moves the solenoid's plunger farther away from the fluid exhaust port. Allowing less fluid to exhaust will cause the pressure of the fluid to increase.

The T/C clutch is controlled by two solenoids, one for apply and release and the other to control its feel during the application and release of the clutch. Some 4T60-E transaxles use a PWM solenoid to control converter clutch engagement.

4T80. The 4T80 and 4T80-E were developed specifically for use with V-8 FWD cars. The transaxle is able to handle vehicles up to 8000 lbs (3629 kg) GVWR. The 4T80-E is still used today. The original 4T80 used a viscous clutch for T/C lockup which was replaced with an electronically controlled torque converter clutch in 2005 with the release of the 4T80-E.

The transaxles has three oil pumps: a primary pump that supplies the needs for the transaxle; a second pump that helps meet the demands during shifts; and a scavenger pump that is used to move the fluid from the bottom oil pan to the side pan. All three pumps are in the valve body and are driven by a driveshaft from the T/C.

Gear selector position	Operating gear	2–4 band	Reverse input clutch	Overrun clutch	Forward clutch	Forward Sprag clutch	3–4 clutch	Low roller clutch	Low/reverse clutch
OD	1st gear				x	x		x	
	2nd gear	x			x	x			
	3rd gear				x	x	x		
	Overdrive	x			x		x		
D	1st gear			x	x	x		x	
	2nd gear	x		x	x	x			
	3rd gear			x	x	x	x		
2	1st gear			x	x	x		x	
	2nd gear	x		x	x	x			
R	Reverse		x						x

FIGURE 10-43 Clutch and band application chart for a THM 4L60 transmission.

4Lxx Series. This series was used in most full-size RWD GM cars and trucks. The 4L60 was introduced in 1993. It was a renamed version of the THM700, which had been around for over 10 years. Through time, the 4L60-E was introduced. It was basically the same transmission as the 4L60 but did not have a governor or throttle valve linkage. The 4L60-E was updated in 2002 and became the 4L65-E. A major difference between the 4L60-E and 4L65-E is the number of planetary pinion gears. 4L60s have four pinion gears and there are five pinion gears in the 4L65-E. The 4L70-E is a version of the 4L65-E. It featured a stronger output shaft to handle higher engine torque. Stronger versions, the 4L80 and 4L85, were later introduced. Today, the 4L60 and 4L80 transmissions are found in GM vehicles with medium and large displacement engines. All of these transmissions have adaptive electronic shift control and Electronic Controlled Capacity Clutch (ECCC) technology, which allows for a controlled amount of clutch slip to dampen engine pulses.

These transmissions rely on two planetary gearsets, five multiple-disc clutches, one sprag clutch, one roller clutch, and a band. (Refer to the clutch and band application chart, Figure 10-43 for more details.) The multiple-disc clutches are released by several small coil springs.

The forward clutch, forward sprag clutch, 3–4 clutch, and reverse input clutch are possible input devices for these transmissions. The forward clutch is applied in all forward gears and drives the front sun gear. The forward sprag clutch also drives the front sun gear in first, second, and third gears but freewheels in fourth gear. The 3–4 (high/overdrive) clutch is applied in third and fourth gears and drives the front ring gear and rear planetary carrier. The reverse input clutch is applied only in reverse gear and locks the rear sun gear to the input shaft.

The low/reverse clutch, low roller clutch, and 2–4 (intermediate/overdrive band) clutch are the holding devices for this transmission (Figure 10-44). The low/reverse clutch is applied in reverse and manual low. This clutch locks the rear planetary carrier to the transmission case. The low roller clutch connects the rear carrier and front ring gear in low gear. The drive overrunning clutch is applied in manual third, manual second, and manual low gears. It locks the front sun gear to the input shaft. The overrunning clutch prevents the front sun gear from freewheeling during deceleration in the manual low, manual second, and manual third gear ranges. The 2–4 band is only applied in second and fourth gears. These transmissions use a single servo to apply the 2–4 band.

Lepellietier Gear-Based Units. GM currently has four transmissions and transaxles based on the Lepellietier geartrain. The construction and operation of these units are very similar and the components only vary according to the intended application of the unit. Like other six-speeds, these units have small steps between gears, allowing the transmission to

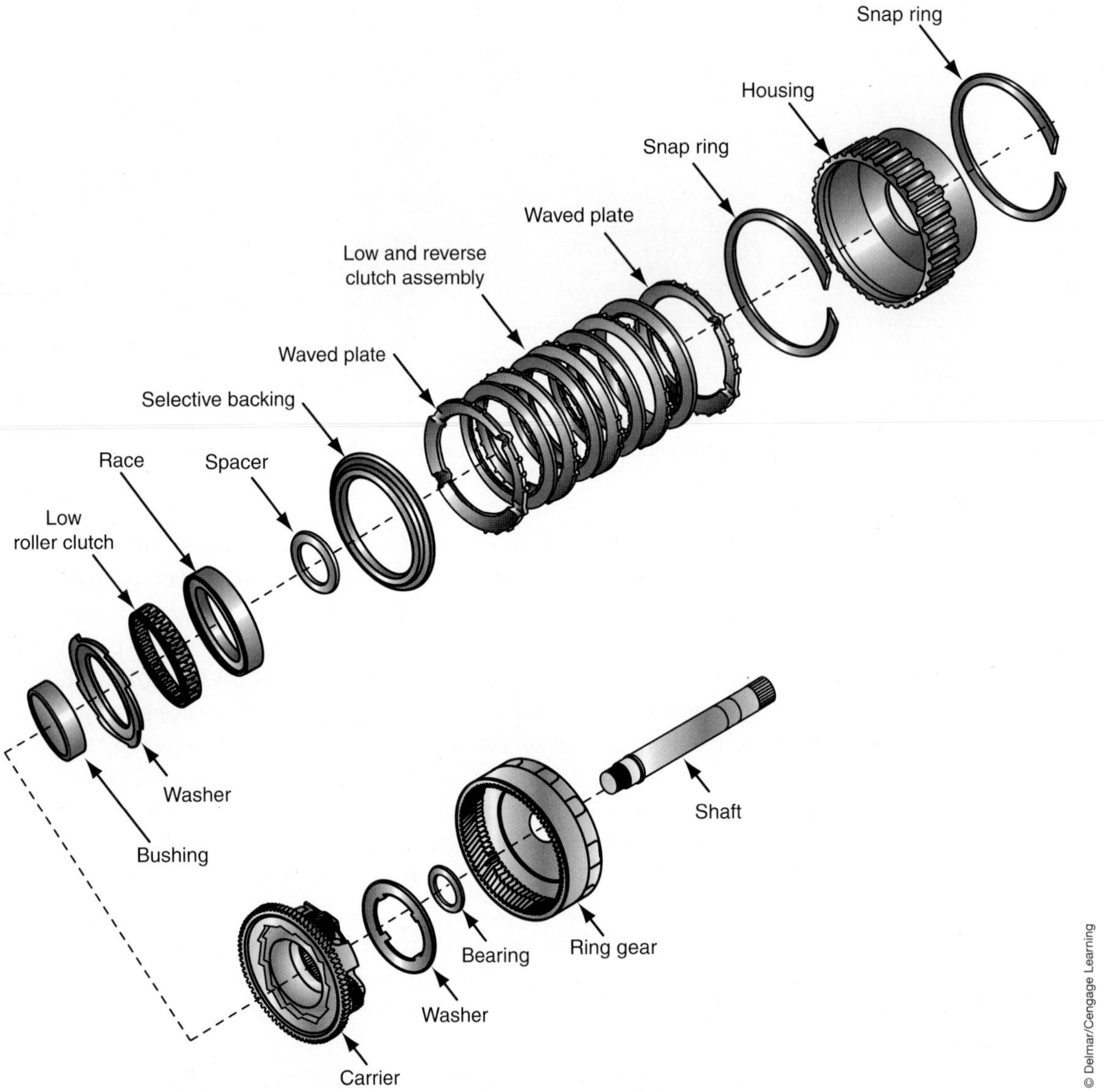

FIGURE 10-44 **The low/reverse clutch and low one-way clutch, in addition to the intermediate band, are used as the holding devices in many THM transmissions, including the 4L series.**

quickly find the ideal operating gear. A numerically high first gear provides strong acceleration from a standing start and the two overdrive gears provide improved fuel economy at cruising speed.

6L80/6L50. The 6L80 six-speed transmission was introduced in 2006 on the Cadillac XLR-V, STS-V, and Chevrolet Corvette. In 2007 it was modified for use in SUVs and trucks. It has an integrated 32-bit electro-hydraulic controller (Figure 10-45). This control module responds to a variety of inputs as it attempts to provide the best shift quality for the operating conditions. It also allows for a tow/haul mode and auto grade braking and manual range selection. The transmission is designed to handle a maximum of 430 ft.-lbs. (583 Nm) engine torque.

A version of the 6L80, the 6L50, was introduced in the 2007 Cadillac STS and the V-8-powered SRX and is currently used in other models. The 6L50 is totally based on the larger 6L80 and uses the same electronic controls. The available gear ratios are nearly identical to those in the 6L80. The transmission is designed to handle up to 332 ft.-lbs. (450 Nm) of engine torque. The transmission relies on five multiple-disc clutches and one sprag-type overrunning clutch. Figure 10-46 shows when each of these is applied.

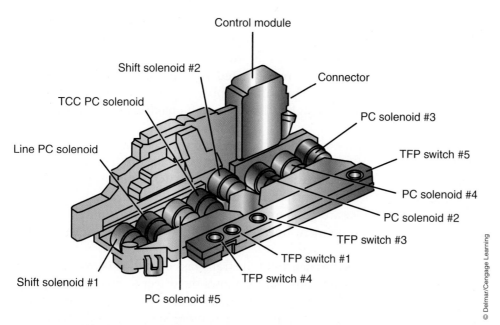

FIGURE 10-45 The control solenoid valve assembly for a 6L80 transmission.

Range	1-2-3-4 Clutch	3-5 Reverse clutch	4-5-6 Clutch	2-6 Clutch	Low/reverse clutch	Low sprag clutch
Park					Applied*	
Reverse		Applied			Applied	
Neutral					Applied*	
1st Braking	Applied				Applied	Holding
1st	Applied					Holding
2nd	Applied			Applied		
3rd	Applied	Applied				
4th			Applied			
5th		Applied	Applied			
6th			Applied	Applied		

FIGURE 10-46 A clutch application chart for a 6L80 transmission. The asterisk (*) after the word *applied* indicates that the device is applied but there is no load.

Gear	Shift sol #1	Shift sol #2	1-2-3-4 Clutch PC sol	2-6 Clutch sol	3-5/rev Clutch sol	low/reverse Clutch PC sol	Resulting gear ratio
Park			OFF	OFF	OFF	ON	--
Reverse	ON	OFF	OFF	OFF	ON	ON	3.06:1
Neutral	ON	ON	OFF	OFF	OFF	ON	--
1st Braking	ON	ON	ON	OFF	OFF	ON	4.03:1
1st	OFF	ON	ON	OFF	OFF	OFF	4.03:1
2nd	OFF	ON	ON		OFF	OFF	2.36:1
3rd	OFF	ON	ON	OFF	ON	OFF	1.53:1
4th	OFF	ON	ON	OFF	OFF	ON	1.15:1
5th	OFF	ON	OFF	OFF	ON	ON	0.85:1
6th	OFF	ON	OFF	ON	OFF	ON	0.67:1

FIGURE 10-47 The activity of the shift and PC solenoids in a 6L80 transmission.

The TCM monitors various inputs and uses this information to command shift solenoids and variable bleed pressure control (PC) solenoids to control shift timing and feel. Figure 10-47 shows the activity of the solenoids during the various gear positions. All of these solenoids are located together in the control solenoid valve assembly.

The basic hydraulic system has a vane-type pump and two control valve body assemblies. The transmissions have a line pressure control system to compensate for the normal wear of the transmission. As the apply devices wear, the amount of time required to apply a clutch increases or decreases. To compensate for these changes, the TCM adjusts the pressure commands to the various PC solenoids in order to maintain the original shift timing. The TCM monitors the signals from the **input speed sensor (ISS)** and **output speed sensor (OSS)** during shifts to determine if a shift is happening too slow or too fast. Adaptive learning is an ongoing process providing a satisfactory shift quality throughout the life of the transmission.

The T/C is an electronically controlled capacity clutch, which was developed to reduce the noise and vibration that may occur when the clutch is applied. In this system, the pressure plate does not always fully lock to the T/C cover. Rather, the pressure plate stays slightly away from it. This allows for some clutch slippage, which can range from 0 to 50 rpm. The clutch can be applied in all forward gears except first. When the clutch is applied and how much slip it will allow is based on many operating conditions, including transmission fluid temperature.

The tow/haul mode reduces the chances of the transmission up- and downshifting while it is towing or hauling a heavy load. This feature allows the driver to block out upper gears and select the desired gears for the driving conditions, such as towing on a steep grade. The manual range selection feature allows the driver to feel more in control of the transmission. When a gear is selected by the driver, the transmission will stay in that gear until the driver changes the gear or when the TCM decides it needs to order a shift to prevent damage to the engine or transmission.

6T70/6T75. In 2007, versions of the Lepellietier-based geartrain were introduced for FWD and AWD vehicles. The 6T70 and 6T75 are six-speeds and operate much in the same way as their RWD predecessors. A planetary gear differential assembly is included in transaxle housing. The final drive gear ratio is different between these units, the 6T70 has a final drive ratio of 2.77:1 and final drive ratio of the 6T75 is 3.16:1. The 6T70 is designed to handle up to 280 ft-lbs (380 Nm) of engine torque, whereas the 6T75 has a slightly higher rating of 301 ft-lbs (406 Nm).

The transaxles were codeveloped with Ford Motor Co. The design and many components are shared by the two manufacturers but each has unique electronic controls and therefore the overall operation of the transaxles differs. GM's transaxles are designed to offer features for the various brands and models of vehicles. These features include manual shifting, auto-grade braking, tow/haul mode, and tapshift range selection.

The TCM, located inside the transaxle, controls all shifting in the same way as the 6L50 and 6L80 transmissions. The gear ratios of the transaxles are as follows:

Shop Manual

Chapter 10, page 502

First—4.48:1
Second—2.87:1
Third—1.88:1
Fourth—1.47:1
Fifth—1.0:1
Sixth—0.74:1
Reverse—2.88:1

Honda and Acura Transmissions

The transmissions used in Honda and Acura vehicles are unique in that they do not use planetary gearsets to provide for the different gear ranges. Constant-mesh helical and square-cut gears are used in a manner similar to that of a manual transmission (Figure 10-48). Many different transmissions have been used by Honda through the years. However, the current five-speed that is commonly found has been used since 2000. In the past, Acura vehicles used

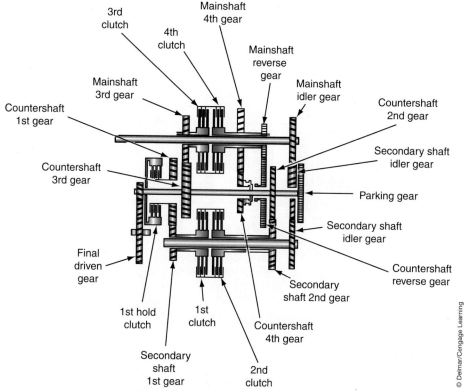

FIGURE 10-48 Current Honda/Acura automatic transmissions have three parallel shafts, helical and straight-cut gears, and hydraulically controlled multiple-disc clutches.

unique transmissions, but today they use the same transmissions found in Honda. Older Acura had longitudinally placed transaxles with an attached hypoid final drive.

Honda has used many different classifications for the transmissions and some are hard to decipher. However, all of the versions can be divided into one of two ways: the number of forward speeds or the number of internal shafts used in the transaxles. Honda has used two-, three-, four-, and five-speed transaxles. Two shaft units were used until the introduction of the five-speed. In a two shaft unit, the mainshaft is positioned in line with the crankshaft. The countershaft runs parallel to the mainshaft and is connected to the final drive. First, second, and fourth clutches are mounted to the mainshaft and the third clutch and a one-way clutch are mounted to the countershaft.

Three shaft units, sometimes referred to as full shaft units, have three parallel shafts. The mainshaft is the input shaft (Figure 10-49). The countershaft is the output to the final drive. A secondary shaft is in parallel with the other shaft (Figure 10-50). The third and fourth clutches and third, fourth, and reverse gears are on the mainshaft. The countershaft has gears for first, second, third, fourth, and reverse along with idler gears and park gear. The secondary shaft has first and second clutches and first, second, and idler gears (Figure 10-51). The gears on the mainshaft and the secondary shaft are in constant mesh with their counter gears on the countershaft.

To provide the forward and reverse gears, different pairs of gears are locked to the shafts by clutches. The action of the clutches is much the same as the action of the synchronizer assemblies in a manual transmission. The power flow through these transaxles is also similar to that of a manual transaxle.

The valve body assembly consists of three separate parts: main valve body, regulator valve body, and servo body. The assembly is bolted to the torque converter housing. The main

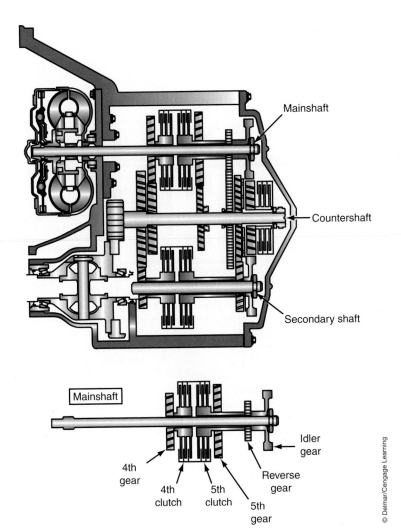

FIGURE 10-49 The mainshaft assembly in a five-speed Honda automatic transaxle.

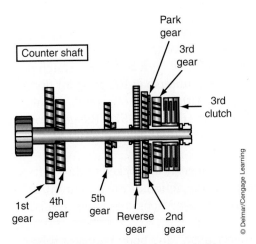

FIGURE 10-50 The countershaft assembly in a five-speed Honda automatic transaxle.

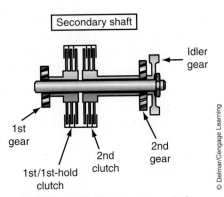

FIGURE 10-51 The secondary shaft assembly in a five-speed Honda automatic transaxle.

valve body contains the manual valve, four shift valves, relief valve, lockup control valve, cooler check valve, servo control valve, and gears for the fluid pump. The regulator valve body contains the regulator valve, torque converter check valve, lockup shift valve, and two accumulators. The servo body contains the servo valve, one shift valve, three accumulators, and shift solenoids.

Five-Speed Units

The Honda/Acura five-speed automatic transaxle uses six multiple-disc clutch packs and a one-way clutch. The transmission is electronically controlled and is equipped with seven solenoids (Figure 10-52). Three of these solenoids are used for gear shifts, three are PC solenoids used to control the pressure applied to the clutches, and one solenoid is used to control the converter's lockup clutch. Like other electronically controlled transmissions, the action of the solenoids is controlled by the PCM in response to various inputs and information stored in the PCM's memory.

To shift gears, the PCM controls the shift solenoids A, B, and C and the associated pressure control solenoids. The shift solenoids move the position of the shift valves, which determines the path for the pressurized fluid. The action of the shift solenoids during the various gear ranges is shown in Figure 10-53. The amount of pressure applied to a particular clutch is controlled by the PC solenoids. These solenoids allow for a slight amount of slippage during engagement and disengagement of gears to provide smooth shifting.

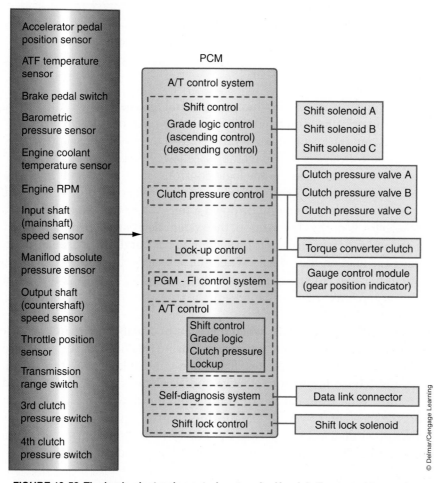

FIGURE 10-52 The basic electronic control system for Honda's five-speed transaxle.

Position	Gear Position	Shift Solenoid Valves		
		A	B	C
D	Shifting from N position	Off	On	Off
	Stays in 1st	On	On	On
	Shifting gears between 1st and 2nd	On	On	On
	Stays in 2nd	On	On	Off
	Shifting gears between 2nd and 3rd	Off	On	Off
	Stays in 3rd	Off	On	On
	Shifting gears between 3rd and 4th	Off	Off	On
	Stays in 4th	Off	Off	Off
	Shifting gears between 4th and 5th	On	Off	Off
	Stays in 5th	On	Off	On
L	Stays in 1st	On	On	On
	Shifting gears between 1st and 2nd	On	On	On
	Stays in 2nd	On	On	Off
R	Shifting from the P and N position	Off	On	On
	Stays in reverse	Off	On	Off
	Reverse inhibit	On	On	On
P	Park	Off	On	Off
N	Neutral	Off	On	Off

© Delmar/Cengage Learning

FIGURE 10-53 Shift solenoid operation for a Honda five-speed transaxle.

The T/C clutch engages in second, third, fourth, and fifth gears when the gear selector is in the D position and only in second and third when the selector is in D3 position. The T/C clutch, PC solenoid, and the lockup control valve control the amount of slippage the T/C will have. The lockup control valve is pulsed by the TCM to regulate the slippage.

These transmissions also have a grade logic control system and a shift hold control system. The PCM compares actual driving conditions with memorized driving conditions, to determine when the vehicle is going up or down a hill. When the vehicle is going up a hill while the gear selector is in the "D" position, the system delays the upshifts to provide more power and to prevent the transmission from switching back and forth through the gears as it attempts to meet the load on the vehicle. When the vehicle is descending, upshifts will take place sooner than normal. However, if the vehicle is operating in fifth gear and the system senses a descent or the brake pedal is depressed, the transmission will downshift into fourth gear to provide engine braking.

The shift-hold control delays shifts when there are frequent changes in throttle opening, such as while driving on a road with many curves. During this time, the throttle is constantly depressed and released, as are the brakes. Shift-hold control keeps the transmission in its current gear as the vehicle moves into a curve and accelerates out of it. Shift-hold control prevents the transmission from constantly shifting up and down. To have this control, the TCM monitors the speed of the vehicle and the throttle opening over a period of time. When the inputs reflect other than normal driving, the TCM will delay all shifting. Once the vehicle is experiencing normal operation, the transmission's shifting characteristics will return to normal.

Operation

With the exception of the one-way clutch, the name of the clutch reflects the gear range it controls. The one-way clutch is between the first clutch hub, which serves as the inner race for the clutch, and first gear on the secondary shaft, which is the clutch's outer race. The one-way clutch locks when torque is transmitted from first gear on the secondary shaft to first gear on the countershaft. The clutch freewheels when the second, third, fourth, and fifth clutches and gears are applied and the gear selector is in the D position.

The input of this transmission is from the T/C to the mainshaft and to the secondary shaft through the third gears on the mainshaft and countershaft and the idler gear on the secondary shaft. A summary of the clutch and gear activity during the various gear selector positions follows:

P Position—No clutches are applied, therefore no torque is transmitted to the countershaft. The park pawl is engaged into the park gear on the countershaft.

N Position—No clutches are applied, therefore no torque is transmitted to the countershaft.

L Position in first gear (Figure 10-54)—The first clutch and first-hold clutch are applied to engage first gear with the secondary shaft. During acceleration, the one-way clutch is locked. First gear on the secondary shaft drives first gear on the countershaft and the final drive gearset. During deceleration, the rotation of the vehicle's wheels unlocks the one-way clutch and drives the secondary shaft idler gear. The torque on the idler gear turns the

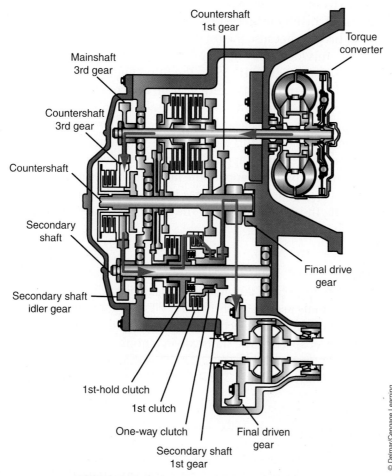

FIGURE 10-54 Power flow when the selector is in the first gear position.

countershaft, which turns the mainshaft through the third gears on both shafts. This allows for engine braking.

L Position in second gear—The second clutch is applied and it engages second gear with the secondary shaft. Power is then transmitted from the secondary shaft to the countershaft and the final drive gears. In this gear range, hydraulic pressure is still applied to the first clutch, but there is no power transmitted at this gear because the rotational speed of the second gear is greater than that of the first gear and this locks the one-way clutch. Pressure to the first-hold clutch is vented and the clutch is not applied.

D Position in first gear—The first clutch is applied to engage first gear with the secondary shaft. The secondary shaft then drives the countershaft and the final drive gears.

D Position in second gear—The second clutch is applied and it engages second gear with the secondary shaft. Power is then transmitted from the secondary shaft to the countershaft and the final drive gears. In this gear range, hydraulic pressure is still applied to the first clutch, but there is no power transmitted at this gear because the rotational speed of the second gear is greater than that of the first gear and this locks the one-way clutch.

D Position in third gear—The third clutch is applied and third gear is engaged to the countershaft. Engine torque rotates the mainshaft and third gear on the shaft drives third gear on the countershaft and output is sent to the final drive unit. Again, hydraulic pressure is still applied to the first clutch, but there is no power transmitted at this gear because the rotational speed of the third gear is greater than that of the first gear and this locks the one-way clutch.

D Position in fourth gear—The fourth clutch is applied to engage the fourth gear to the mainshaft. This gear drives the fourth gear on the countershaft and torque is transmitted to the final drive unit. Again, hydraulic pressure is still applied to the first clutch, but there is no power transmitted at this gear because the rotational speed of the fourth gear is greater than that of the first gear and this locks the one-way clutch.

D Position in fifth gear (Figure 10-55)—The fifth clutch is applied to engage the fifth gear to the mainshaft. This gear drives the fifth gear on the countershaft and torque is transmitted to the final drive unit. Again, hydraulic pressure is still applied to the first clutch, but there is no power transmitted at this gear because the rotational speed of the fifth gear is greater than that of the first gear and this locks the one-way clutch.

R Position—Hydraulic pressure is applied to the servo valve to engage the reverse selector with the countershaft and pressure applies the fifth clutch to engage the reverse gear with the mainshaft. The reverse gear drives the reverse idler gear which in turn drives the reverse gear on the countershaft. The countershaft and the final drive gears rotate in a reverse direction to drive the vehicle in reverse.

Hybrid Models

A modified version of the five-speed automatic transaxle is available in Honda hybrids. The other transmission commonly used in a hybrid is a CVT. The automatic is modified and more compact so that it can fit behind the electric motor mounted at the rear of the engine and occupy the same amount of space as the transaxle in a nonhybrid vehicle. The transaxle has an integrated electric oil pump and different gear ratios that provide for better acceleration, fuel economy, and regenerative braking. These transaxles operate in the same way as other Honda units.

Saturn Transmissions

Although Saturn was a division of GM, some of their vehicles did not use normal GM transaxles. These vehicles had nonplanetary gear transmissions that were electronically controlled four-speeds and referred to the MP6 and MP7.

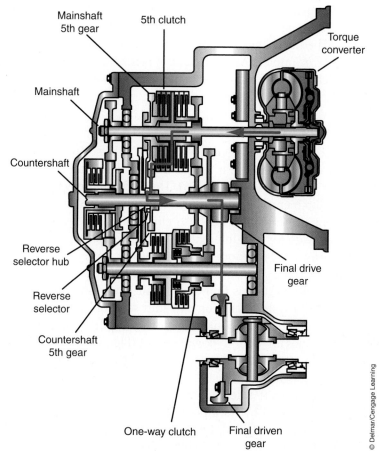

FIGURE 10-55 Power flow in fifth gear when the gear selector is in the D position.

The transaxle relies on five solenoid valves, controlled by the PCM, to regulate shift timing, feel, and TCC application. Two solenoids are used as shift solenoids. They control the delivery of fluid to the manual shift valve. A duty-cycle solenoid is used to control line pressure. Its purpose is to regulate line pressure according to engine running conditions and engine torque. An additional solenoid is used to provide for engine braking during coasting. This solenoid operates when the vehicle is slowing down and the throttle is closed. The other solenoid controls the operation of the lockup converter clutch.

NISSAN TRANSMISSIONS

Nissan and its affiliate, Jatco, have produced a large number of automatic transmissions for the industry. The Japan Automatic Transmission Company started building transmissions in the 1970s. In 1999, Nissan sold its transmission company (TransTechnology), which then combined with the Japan Automatic Transmission Company. This new company officially became JATCO in 2002. Today, vehicles from nearly every manufacturer have used Jatco transmissions, the only exceptions are Honda and Toyota. Honda Motor Company makes all of their transmissions and Toyota Motor Company relies on transmissions built by their subsidiary, Aisin Warner (currently called Aisin AW).

Original Jatco transmissions were designated differently than they are today. For example, look at the L3N71B. The "L" signified "light-duty," the "3" said it was a 3-speed, and the remaining digits defined the series of the transmission. As time passed, prefixes and suffixes were added to further define the unit. The result was a classification that was very difficult

A BIT OF HISTORY

The world's first electronically controlled five-speed automatic transmission, the JR502E, was developed by JATCO in 1989.

to interpret. As most transmission companies changed the way they designated their nits, so did Jatco. The new designations start with "J" for Jatco, "F" for front-wheel-drive, or "R" for rear-wheel-drive. The next digit is the number of gears and the next two digits denote the series number. For example, the JR507E is a Jatco built five-speed transmission for rear-wheel-drive vehicles. It is the seventh version of the series (07) and is electronically controlled.

Simpson Gear-Based Units

Shop Manual
Chapter 10, page 514

Most Nissan RWD vehicles are based on the RE4Rxx series. These transmissions provide four forward gears through the use of a Simpson gearset and an additional planetary unit mounted in front of the Simpson (Figure 10-56). This series has electronic control of shifting and torque converter clutch engagement.

These transmissions use four multiple-disc clutches, three servos and bands, and two one-way clutches to provide for the different ranges of gears. Refer to the clutch and band application chart (Figure 10-57) for more details on power flow.

Nissan uses different transaxles and transmissions in their vehicles. A common transaxle is the RE404A (Ford 4F20E). This four-speed unit has a T/C clutch and electronic shift controls (Figure 10-58). The system relies on two shift solenoids, which are controlled by the PCM (Figure 10-59). In addition to these solenoids, two other solenoids are incorporated. The timing solenoid provides for smooth downshifting (Figure 10-60). The line pressure solenoid provides for smooth upshifting. The system also has a fifth solenoid, which is used to control converter clutch activity.

The PCM activates the shift solenoids according to the shift schedule it selected in response to the signals received by its inputs. The shift solenoids are simple on/off solenoids. When the solenoid is energized, line pressure flows to the appropriate shift valve. One solenoid controls fluid flow to the 2–3 shift valve. The other solenoid activates the 1–2 or 3–4 shift valve through the 2–3 valve. In other words, the position of the first shift valve determines where fluid flow will be directed when the second solenoid is activated. When only the first solenoid is energized, the transaxle will operate in second gear. When both are energized, the transaxle will operate in first gear.

Lepellietier Gear-Based Units

Jatco builds a seven-speed transmission for RWD Nissans. There are two basic versions, the JR710E and JR711E. Each has a different torque capacity and therefore both are used in different vehicles. The JR710E is used in the Infiniti EX37 and the JR711E was built for the Infiniti FX50, which is a SUV. These transmissions offer close ratio gear and two overdrive gears.

The transmissions also have advanced electronic controls which enable the units to provide smoother T/C clutch application in most forward gears, manual shift control, adaptive shift control, and synchronized revolution control.

The adaptive shift control changes shift timing according to the operating conditions, such as normal driving, aggressive driving, and moving up or down a hill. These conditions are determined by a number of inputs to the PCM.

The synchronized revolution control will momentarily increase the engine's speed during downshifting to help synchronize the gears and provide smooth speed changes during downshifting.

TOYOTA TRANSMISSIONS

The synchronized revolution control system accomplishes the same thing an experienced race driver does, only that driver must use a technique called "heel and toe," which involves using one foot to work both the throttle pedal and the brake pedal.

Many different transmissions and transaxles have been used by Toyota, Scion, and Lexus. They are produced by Aisin AW. Aisin AW makes transmissions for light vehicle applications, including hybrids. Aisin AW is owned by Aisin Seiki and Toyota, with Toyota being the

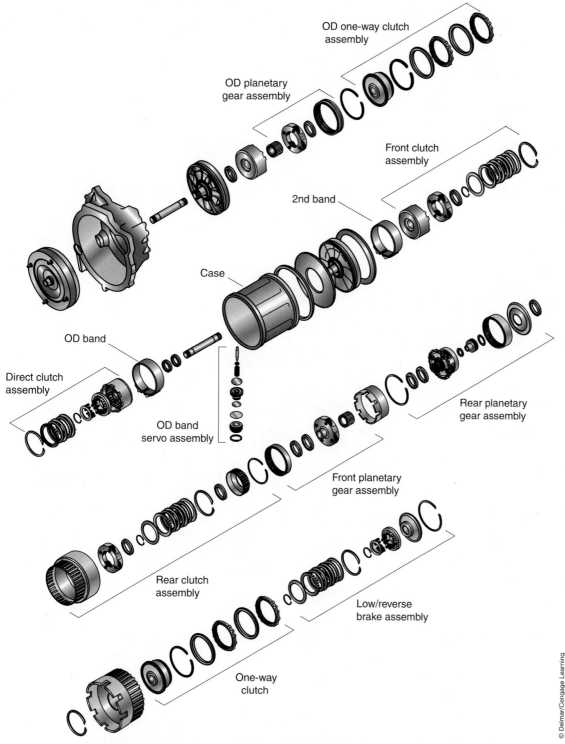

OD one-way clutch
assembly

OD planetary
gear assembly

Front clutch
assembly

2nd band

Case

OD band

Direct clutch
assembly

OD band
servo assembly

Rear planetary
gear assembly

Front planetary
gear assembly

Rear clutch
assembly

Low/reverse
brake assembly

One-way
clutch

© Delmar/Cengage Learning

FIGURE 10-56 Exploded view of a Nissan L4N71B Simpson-based transmission.

majority owner. Aisin Seiki makes transmissions for heavy duty vehicles. The AW after Aisin signifies a previous joint venture between Aisin Seiki and BorgWarner, which was established in 1969 and was known as Aisin-Warner (AW). That joint venture ended in 1987. Although Aisin AW is not a direct part of Toyota, it was set up to be the sole source of RWD automatic transmissions for Toyota, later it started to produce FWD/AWD transmissions. Aisin AW

Gear Selector Position	Operating Gear	Rear Clutch	One-way Clutch	Band	Front Clutch	Low/Reverse Band	Direct Clutch	Overdrive Band	Overdrive One-way Clutch
D	1st gear	x	x				x		x
	2nd gear	x		x			x		x
	3rd gear	x			x		x		x
	4th gear	x			x			x	
2	2nd gear	x		x			x		x
1	1st gear	x	x			x	x		x
R	Reverse				x	x	x		x

© Delmar/Cengage Learning

FIGURE 10-57 Clutch and band application chart for a Nissan L4N71B transmission.

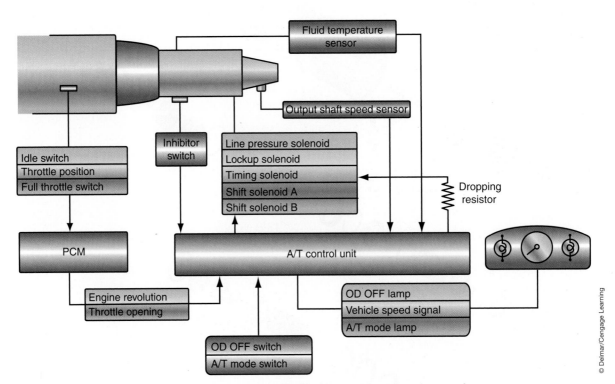

© Delmar/Cengage Learning

FIGURE 10-58 Schematic of a Nissan RE4FO2A transaxle's electronic control system.

A BIT OF HISTORY

Toyota led the way into electronically controlled automatic transmissions. The A140E introduced in 1994 was the first fully electronic transmission.

also shares much of its research and development with Toyota. Aisin AW produces automatic transmissions to 35 different manufacturers around the world. These include Chrysler, GM, Ford, Mitsubishi, Nissan, Porsche, Audi, VW, Volvo, Hyundai, and Isuzu.

There are two basic families of Toyota automatic transmissions, A and U. The A family are those automatic FWD/RWD/AWD transmissions built by Aisin-Warner. In the U family are late-model automatic transaxles built for FWD and AWD vehicles. All Toyota automatic transmission designations begin with an A or a U. This designation is followed by transmission series number, the number of forward gears, and a digit representing the features, such as E for Electronic control, F for 4WD w/mechanical transfer case, H for 4WD w/electronic transfer case, I for Intelligence, and L for Lockup torque converter. When other automakers use Aisin transmissions, they are identified by the prefix AW.

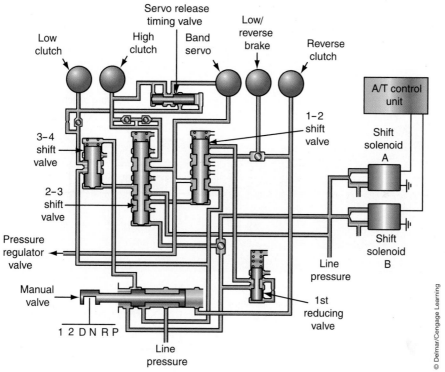

FIGURE 10-59 Two shift solenoids control all forward gear changes.

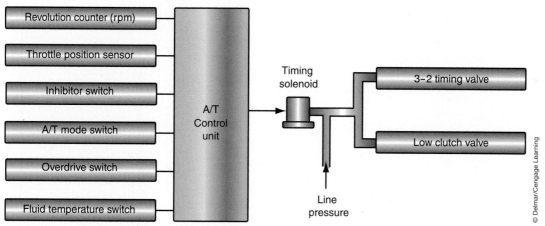

FIGURE 10-60 The timing solenoid controls the pressure to the 3–2 and low valves in order to control the quality of downshifts.

Simpson Gear-Based Units

A common early design is the A140E/A140L (Figure 10-61). This transaxle combines a three-speed transmission with an overdrive assembly. The primary differences between the L and E type transaxles are the main valve body, operating mechanism, and electronic control. The E is referred to an **electronic controlled transaxle (ECT)**. The A140L only uses an electronic overdrive solenoid system.

Although the ECTs are electronically controlled, the basis for operation is a Simpson gear-train in line with a single overdrive planetary gearset. The transaxles use four multiple-disc clutches, two band and servo assemblies, and three one-way clutches to provide the various

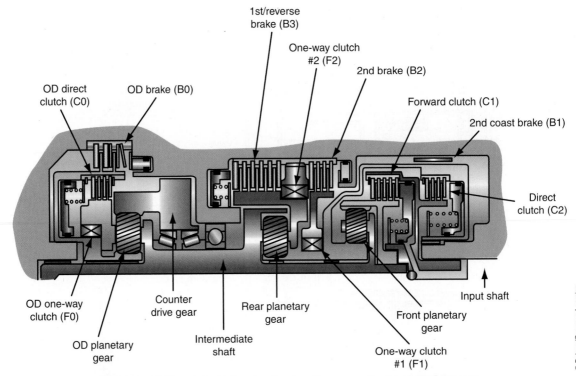

FIGURE 10-61 The apply devices for the planetary gearsets of an A140E transaxle.

Gear selector position	Operating gear	OD clutch	Forward clutch	Direct clutch	2nd Coast band	2nd Coast drum	1st/ Reverse brake	OD One-way clutch	#1 Oneway clutch	#2 Oneway clutch
D-Drive	1st gear	x	x					x		x
	2nd gear	x	x			x		x	x	
	3rd gear	x	x	x		x		x		
	Overdrive		x	x		x				
2-Second	1st gear	x	x					x		x
	2nd gear	x	x		x	x		x	x	
	3rd gear	x	x	x		x		x		
L-Low	1st gear	x	x				x	x		x
	2nd gear	x	x		x	x		x	x	
R	Reverse	x		x			x			
N/P	Neutral	x								

FIGURE 10-62 Clutch and band application chart for a Toyota A140 transmission.

gear ranges. Refer to the clutch and band application chart (Figure 10-62) for details on the power flow through this transmission.

Since the introduction of the A140, Toyota has added these controls to several other transmissions. There has been the A240, A340, and A540, which were modifications of the A140. Variations of the A340 transmission are currently being used. These adapt the basic transmission to a specific application. For example, the A340F is a four-speed transmission with a mechanically-controlled 4WD transfer case. The A340H is also a four-speed transmission but it has an electronically controlled 4WD transfer case.

With the exception of the electronic controls added to the A340H, all A340 transmissions use the same electronic control system (Figure 10-63). This system includes three solenoid valves: two shift solenoids and a TCC solenoid.

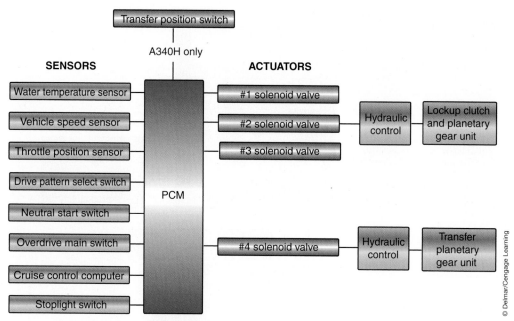

FIGURE 10-63 A simplified look at the control system for a Toyota A541E transaxle.

Gear selector position	Operating gear	#1 Solenoid	#2 Solenoid
Drive	1st	ON	OFF
Drive	2nd	ON	ON
Drive	3rd	OFF	ON
Drive	4th	OFF	OFF
2	1st	ON	OFF
2	2nd	ON	ON
2	3rd	OFF	ON
L	1st	ON	OFF
L	2nd	ON	ON

FIGURE 10-64 Solenoid activity for an A340E transaxle.

The three solenoids are controlled by signals received from the TCM. When the #1 and #2 solenoids are activated, the plunger of the solenoids is moved from their seat. This opens the exhaust port to release line pressure. When either of these solenoids is turned off, the exhaust port is closed. The result of the opening and closing of the exhaust ports is the change of fluid flow to the various apply devices. A summary of the solenoid activity is shown in Figure 10-64.

The TCC solenoid works in the opposite way (Figure 10-65). When it is off, the plunger moves away from its seat thereby opening the exhaust port to release line pressure. Likewise, when the solenoid is on, the exhaust port is closed.

The A540E and A541E transaxles use six multiple-disc clutches, three one-way clutches, and one brake band. The operation of the transaxle and the lockup converter is totally controlled by the PCM. However, a throttle pressure cable is used to mechanically modulate line pressure. The A541E transaxle gained adaptive learning with its new electronics. The band and clutch location, as well as the application of each, is the same as the previous models.

Gears in Tandem

The transaxles in Toyota's U family have two simple planetary gearsets connected in series. The ring and carriers of the two gearsets are connected. The front planetary ring gear is connected to the rear carrier. Both are held to the case to prevent counterclockwise rotation by

Shop Manual
Chapter 10, page 521

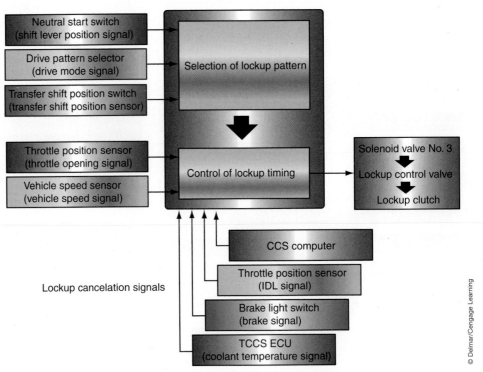

FIGURE 10-65 The basic layout for T/C clutch control for late-model Toyota transmissions.

a one-way clutch and are held in both directions by a band. The rear carrier can be driven by the intermediate shaft through the direct clutch. The front carrier is connected to the rear ring gear. The carrier is also connected to the counter drive gear, which is located in front of the gearsets, to serve as the output for the unit. The rear sun gear is connected to the intermediate shaft through the reverse clutch or to the transmission case through brakes and/or a one-way clutch. The U family also includes the special transmissions used in Toyota hybrids.

Several different U transaxles have been and are being used. The most commonly found are the U140E, U241E, and U151E (Figure 10-66) that were used in Camrys, Scions, and Highlanders. These are four-speed units that were replaced by five- or six-speeds in recent years. The U140E and U241E were transaxles based on two simple planetary gears in tandem and a third set above a counter driveshaft. The counter drive and driven gears are placed in front of the front planetary gearset and the third planetary unit is called the under drive (U/D) gearset. Both of these transaxles are similar in construction and operation, but have different gear ratios. They use three multiple-disc clutches, three bands, and two overrunning clutches (Figure 10-67).

The forward clutch connects the input shaft to the sun gear of the front planetary. The direct clutch connects the input shaft to the sun gear of the rear planetary. And, the U/D direct brake connects the U/D sun gear to the U/D planetary carrier.

The first and reverse brake prevents the carrier of the rear planetary and the ring gear in the front planetary from turning in any direction. The second brake prevents the carrier of the rear planetary from turning either clockwise or counterclockwise. The sun gear of the U/D set is prevented from turning either clockwise or counterclockwise by the U/D brake.

One of the overrunning clutches (called #1) prevents the rear planetary carrier from turning counterclockwise. The U/D one-way clutch prevents the sun gear of the U/D planetary set from turning clockwise.

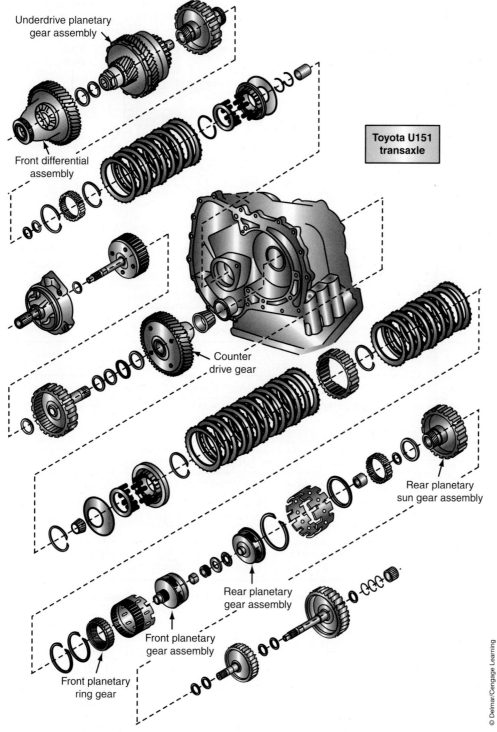

Underdrive planetary
gear assembly

Front differential
assembly

Toyota U151
transaxle

Counter
drive gear

Rear planetary
sun gear assembly

Rear planetary
gear assembly

Front planetary
gear assembly

Front planetary
ring gear

© Delmar/Cengage Learning

FIGURE 10-66 The position of the various planetary gearsets and their controls in an U151E transmission.

The valve body assembly holds seven solenoids: SL1, SL3, SL2, SL4, SR, SLT, and DSL (Figure 10-68). SL1 and SL2 are used to control the flow of fluid to the second brake and the direct clutch. SL4 controls the 3–4 shift valve to allow a shift into fourth gear by changing the fluid pressure applied to the U/D brake band and the U/D brake clutch. SLT is used to control line pressure and the DSL is used to control the first and reverse brake. A summary of the clutch, brake, and solenoid activity is shown in Figure 10-69.

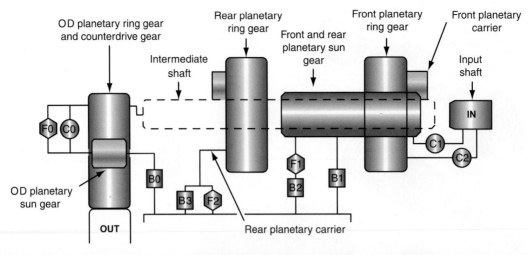

COMPONENT		FUNCTION
Forward clutch	C1	Connects input shaft and front planetary ring gear
Direct clutch	C2	Connects input shaft and front and rear planetary sun gear
2nd coast brake	B1	Prevents front and rear planetary sun gear from turning either clockwise or counterclockwise
2nd brake	B2	Prevents outer race of F1 from turning either clockwise or counterclockwise, thus preventing front and rear planetary sun gear from turning counterclockwise
1st/reverse brake	B3	Prevents rear planetary carrier from turning either clockwise or counterclockwise
#1 one-way clutch	F1	When B2 is operating, prevents front and rear planetary sun gear from turning counterclockwise
#2 one-way clutch	F2	Prevents rear planetary carrier from turning counterclockwise
OD direct clutch	C0	Connects overdrive sun gear and overdrive planetary carrier
OD brake	B0	Prevents overdrive sun gear from turning either clockwise or counterclockwise
OD one-way clutch	F0	When transaxle is being driven by engine, connects overdrive sun gear and overdrive carrier
Planetary gears		These gears change the route through which driving force is transmitted in accordance with the operation of each clutch and brake in order to increase or reduce the input and output speed

FIGURE 10-67 **Clutch and band application chart for a Toyota U151E transmission.**

© Delmar/Cengage Learning

A750x Series. The A750E and A750F transmissions have three planetary gearsets connected in series. The difference between the two versions are based on the A750F being equipped with a new extension housing and output shaft because it is used for four-wheel-drive vehicles. This transmission uses three multiple-disc clutches, four brakes, and three one-way clutches.

The #1 clutch connects the input shaft with the intermediate shaft. It is applied in first through fourth gears. The #2 clutch connects the input shaft to the carrier of the center planetary gearset. It is applied in fourth and fifth gears. The #3 clutch connects the input shaft to the front sun gear and is applied in reverse, third, fourth, and fifth gears. However, it does not contribute to the power flow while the transmission is in fourth gear.

Brake #1 prevents the carrier in the front planetary from turning in either direction. It is applied in reverse and fifth gears. It is also applied when the transmission is manually shifted into "3" and operating in third gear. This allows for some engine braking. Brake #2 prevents the ring gear of the front and center gearsets from turning. It is applied only when the gear selector is in the "2" position and operating in second gear. This allows for engine braking.

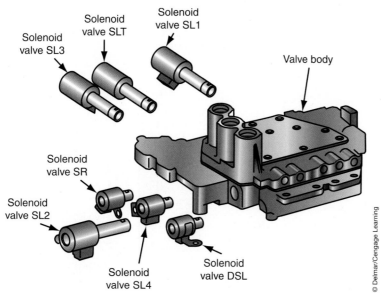

FIGURE 10-68 The U151E transmission is controlled by seven solenoids, all located in the valve body.

GEAR SELECTOR POSITION	GEAR	SL1	SL2	SL4	DSL	FORWARD CLUTCH	DIRECT CLUTCH	U/D BRAKE	2nd BRAKE	1st and REV BRAKE	U/D BRAKE	#1 ONE-WAY	U/D ONE-WAY
P	Park	ON	ON	OFF	OFF	OFF	OFF	OFF	OFF	OFF	ON	OFF	OFF
R	Reverse	ON	OFF	OFF	OFF	OFF	ON	OFF	OFF	ON	ON	OFF	OFF
N	Neutral	ON	ON	OFF	OFF	OFF	OFF	OFF	OFF	OFF	ON	OFF	OFF
D	1st	ON	ON	OFF	OFF	ON	OFF	OFF	OFF	OFF	ON	ON	ON
D	2nd	OFF	ON	OFF	OFF	ON	OFF	OFF	ON	OFF	ON	OFF	ON
D	3rd	OFF	OFF	OFF	ON*	ON	ON	OFF	OFF	OFF	ON	OFF	ON
D	4th	OFF	OFF	ON	ON*	ON	ON	ON	OFF	OFF	OFF	OFF	OFF
2	1st	ON	ON	OFF	OFF	ON	OFF	OFF	OFF	OFF	ON	ON	ON
2	2nd	OFF	ON	OFF	OFF	ON	OFF	OFF	ON	OFF	ON	OFF	ON
L	1st	ON	ON	OFF	ON	ON	OFF	OFF	OFF	ON	ON	ON	ON

*This is ON during T/C lock-up.

FIGURE 10-69 Clutch, band, and solenoid action for U140E and U241E transaxles.

Brake #3 prevents the outer race of F2 from turning. It is applied in second through fifth gears but only contributes to the power flow while the transmission is operating in second gear. Brake #4 prevents the ring gear in the rear planetary set from turning. It is applied in reverse and manual low.

The #1 one-way clutch prevents the carrier of the front planetary set from turning counterclockwise. It is effective in reverse, second, and third gears. However, when the gear selector is manually set in "3," the clutch does not contribute to the power flow. The #2 one-way clutch prevents the sun gear from turning counterclockwise when brake #3 is applied. It is effective only in second gear. The #3 one-way clutch prevents the carrier of the center planetary set and the ring gear of the rear gearset from turning counterclockwise.

A76xx and A96xx Series. The A760 E and H were introduced in the 2006 Lexus. The E series is for RWD and the H is for AWD. The E series is completely electronically controlled (Figure 10-70). Variations of this transmission, such as the A761E, A960E, and A760H are found in many Lexus vehicles.

These are six-speed transmissions based on an A750E. They have four multiple-disc clutches, three planetary gearsets, four one-way clutches, and four brakes (Figure 10-71). The function of each of these is the same as they are in the A750E, with the following

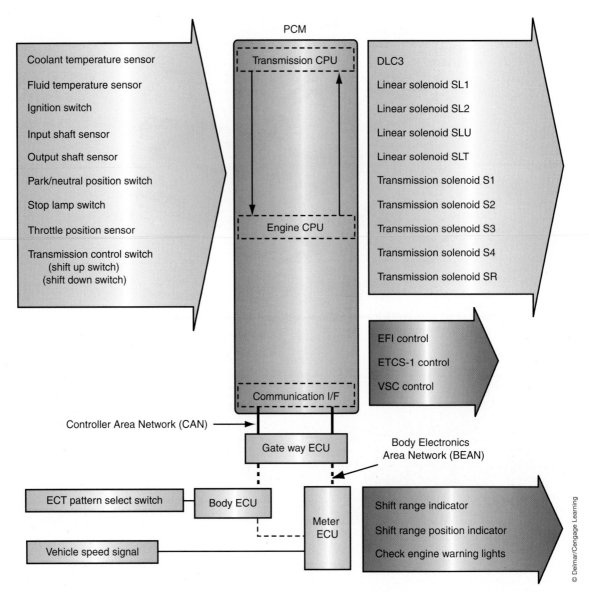

FIGURE 10-70 The basic electrical layout for the A760E transmission.

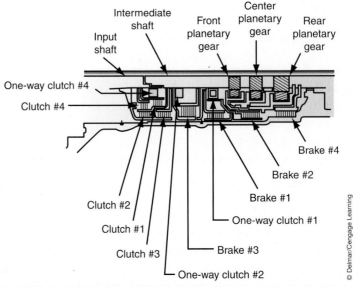

FIGURE 10-71 The location of the clutches, brakes, and gears in an A760E transmission.

exceptions. The new multiple-disc clutch, #4, connects the input shaft to the intermediate shaft. Brake #4 now prevents the carrier in the center planetary set and the ring gear in the rear planetary from turning in either direction. The new #4 one-way clutch prevents the intermediate shaft from turning counterclockwise. Here is a summary of the activity of these devices:

- #1 clutch is applied in all forward gears but does not contribute to the power flow in fifth or sixth gears.
- #2 clutch is applied in fourth, fifth, and sixth gears.
- #3 clutch is applied in reverse, third, fourth, and fifth gears but does not contribute to the power flow in fourth.
- #4 clutch is applied and provides engine braking in fourth, third, second, and first gears.
- Brake #1 is applied in reverse, fifth, and sixth gears but does not affect power flow in sixth.
- Brake #2 is applied in sixth gear and in second gear when manual second gear is selected.
- Brake #3 is applied in second through sixth gears but does not contribute to power flow in third through sixth.
- Brake #4 is applied in reverse and manual low.
- #1 one-way clutch is effective in reverse, second, and third gears.
- #2 one-way clutch is effective in second gear.
- #3 one-way clutch is effective in first gear, except when manual low is selected.
- #4 one-way clutch is effective in first through fourth, except when manual low is selected.

To control the action of the various clutches, the transmission relies on nine solenoids fitted into one of the four assemblies that make up the valve body (Figure 10-72). Solenoids S1, S2, S3, S4, and SR are three-way valves. Solenoids SL1, SL2, SLT, and SLU are linear valves and their operation depends on the current sent to them by the TCM. The purpose of the solenoids follows:

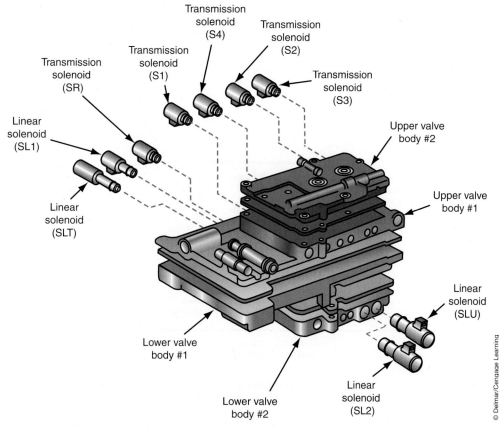

FIGURE 10-72 The placement of the various solenoids used in an A760E transmission.

- S1—Switches the 1–2 shift valve and the SL1 relay valve.
- S2—Switches the 2–3 shift and 5–6 valves.
- S3—Switches the 3–4 shift valve.
- S4—Switches the 4–5 shift valve, the SL1 relay valve, and the reverse sequence valve.
- SR—Switches the clutch apply relay and the B1 relay valves.
- SL1—Controls clutch pressure and accumulator back pressure.
- SL2—Controls brake pressure.
- SLT—Controls line pressure and accumulator back pressure.
- SLU—Controls lockup clutch pressure.

Lepelletier Gear-Based Units

The most widely used Lepelletier gear based Toyota transmission is the AA80E. It was also the world's first mass-produced eight-speed automatic transmission. This transmission was introduced in the 2007 Lexus LS460. The transmission uses four multiple-disc clutches, two multiple-disc brakes, and one overrunning clutch. In this transmission, the Ravigneaux gearset is used as the rear planetary gear unit. This arrangement provides for the following gear ratios:

First	4.596:1
Second	2.724:1
Third	1.863:1
Fourth	1.464:1
Fifth	1.231:1
Sixth	1.000:1
Seventh	0.824:1
Eighth	0.685:1
Reverse	2.176:1

Like many other transmissions, this unit has a centrifugal fluid pressure canceling mechanism (Figure 10-73). It is used in the #1, #2, #3, and #4 clutches that are applied when upshifting from second through eighth gears. The intensity of clutch engagement is affected by the centrifugal fluid pressure that acts on the fluid inside the piston's pressure chamber. To eliminate the effects of centrifugal force, there is a canceling fluid pressure chamber directly

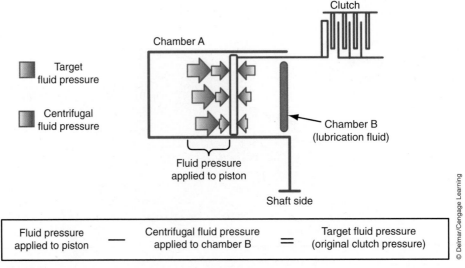

FIGURE 10-73 Most recent transmission models have a centrifugal fluid pressure–canceling mechanism to prevent the generation of pressure, due to centrifugal force, that could keep a clutch applied when the clutch should be released.

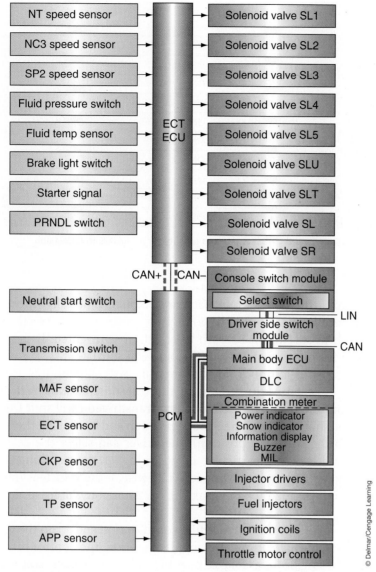

FIGURE 10-74 The communication network for the control of an AA80E transmission.

opposite of the piston's pressure chamber. The pressure developed by centrifugal force moves to the canceling chamber where it is canceled.

The transmission's control unit is isolated from the PCM, but they are always in communication with each other through the CAN (Figure 10-74). All of the solenoids and sensors for transmission control are directly connected to the ECT control unit. The control unit supplies current to six solenoids. The control unit regulates the pressure that is applied directly to the multiple-disc packs by controlling the shift solenoids. Because of these actions and clutch-to-clutch programming, shift control in second and higher gears is done without using the one-way clutch. See Figure 10-75 for the solenoid and clutch and brake activity for an AA80E transmission.

This transmission has many special features, some are unique and others are found on other current transmissions. Optimal line pressure is controlled by a solenoid through the commands of the control module. The line pressure is regulated to meet the precise needs for the engine's and the transmission's current operating conditions, as well as fluid temperature.

The control unit for the transmission determines the road conditions and the intentions of the driver and adjusts shift timing and quality accordingly. This feature is called Artificial

GEAR	SOL #1	SOL #2	SOL #3	SOL #4	SOL #5	SOL L	#1 CL	#2 CL	#3 CL	#4 CL	#1 BR	#2 BR	ONE-WAY
Park/Neutral	ON	OFF	OFF	OFF	OFF	OFF	OFF	OFF	OFF	OFF	OFF	OFF	OFF
Reverse	OFF	OFF	OFF	ON	OFF	ON	OFF	OFF	OFF	ON	OFF	ON	OFF
1st	ON	OFF	OFF	OFF	OFF	OFF	ON	OFF	OFF	OFF	OFF	OFF	ON
2nd	ON	OFF	OFF	OFF	ON	OFF	ON	OFF	OFF	OFF	ON	OFF	OFF
3rd	ON	OFF	ON	OFF	OFF	OFF	ON	OFF	ON	OFF	OFF	OFF	OFF
4th	ON	OFF	OFF	ON	OFF	ON	ON	OFF	OFF	ON	OFF	OFF	OFF
5th	ON	ON	OFF	OFF	OFF	ON	ON	ON	OFF	OFF	OFF	OFF	OFF
6th	OFF	ON	OFF	ON	OFF	ON	OFF	ON	OFF	ON	OFF	OFF	OFF
7th	OFF	ON	ON	OFF	OFF	ON	OFF	ON	ON	OFF	OFF	OFF	OFF
8th	OFF	ON	OFF	OFF	ON	ON	OFF	ON	OFF	OFF	ON	OFF	OFF

FIGURE 10-75 Solenoid and clutch and brake activity for an AA80E transmission.

Intelligence (AI) Shift. The control looks at many inputs, especially throttle opening and vehicle speed to determine if the vehicle is going up or down a hill. When the vehicle is going up, shifts from sixth to seventh or seventh to eighth are prevented. If the vehicle is going downhill, the transmission downshifts to fifth, fourth, or third gear.

Although the transmission does not feature a manual shift option, it does have an S mode. This feature is called Multi-Mode and the transmission will shift only within the selected S-range. For example, in the S6 mode, the transmission will shift from first to sixth. The driver moves the shift lever into the S mode position and by moving the lever to the front ("+" position) or to the rear ("−" position), the driver can select the desired shift range. When the vehicle is driven at a particular speed or higher, downshifting is not permitted as a protection to the engine and transmission. Likewise, if the vehicle is moving at a speed higher than the speed specified for the gear, the transmission will leave the selected range and upshift.

The two overdrive high gears decrease fuel consumption and so do the Coast Downshift Control and Neutral Control features. During deceleration, the control module downshifts and cuts off fuel to the engine for as long as possible. When the vehicle is stopped, the neutral control feature disengages the transmission from the engine to improve fuel economy. It does this by partially disengaging the #1 clutch and Brake #1.

The control of the T/C clutch is also designed to save fuel. In addition to conventional lockup timing control, the transmission has flex lockup clutch control. This feature duty cycles the clutch according to conditions allowing it to be partially or fully engaged. This feature allows the clutch to lock longer than it normally would be. During acceleration, the flex lockup clutch operates when the transmission is in fourth or a higher gear. During deceleration, it operates when the transmission is operating in fifth gear or higher.

OTHER COMMON TRANSMISSIONS

There are many other automatic transmissions found in today's vehicles. Depending on the basic gearsets used, similar designs have similar power flows and major components.

Aisin Transmissions and Transaxles

Many transmissions are built by Aisin and used in other vehicles besides Toyota. A few of these are discussed here.

Aisin AF40/6. The Aisin AF40/6 is used by GM and is based on a Lepelletier gearset. It is a fully electronic-controlled six-speed transaxle that utilizes clutch-to-clutch actuation, where one clutch releases a gear at precisely the same time that the next gear is engaged. It is a wide-ratio unit with overdrive gears in fifth and sixth.

The TCM is located inside the transmission housing along with the shift selector switch. This reduces the wires required for the unit and provides a consistent environment for the TCM. The electronics allow the transaxle to disengage from the engine when it is in neutral and it allows the transmission to stay in a higher gear during deceleration to enhance engine braking. It also allows the T/C to slip slightly to reduce the amount of engine vibration that may transfer into the transaxle.

Aisin AF33. This transaxle is an electronically controlled five-speed unit. It is used by GM, Saturn, Saab, Volvo, and Nissan. The AF33 designation is only used by GM, Nissan calls it the RE5F22A, and the actual designation from Aisin is AW55-50SN (FWD) and AW55-51SN (AWD). The unit can be used behind fairly strong engines, it can handle up to 243 ft.-lbs. (329 Nm) of torque. This unit is unique in that it does not have an overdrive gear. The gear ratios typically are:

First	4.69:1
Second	2.94:1
Third	1.92:1
Fourth	1.30:1
Fifth	1.00:1
Reverse	3.18:1

When fitted for AWD applications, a power take-off unit (PTU) is mounted to the differential housing. The PTU is sealed from the differential housing and does not share fluid with the transaxle, rather the PTU housing is filled with hypoid gear fluid. The differential assembly has an additional ring gear attached to the front differential carrier. This ring gear drives a pinion gear connected to a drive shaft that transfers torque to the rear differential. The rear differential has a single-gerotor pump and clutch pack that automatically engages the rear wheels when one of the front wheels slips.

VW 09D Transmission. The six-speed automatic transmission, 09D, is based on the Lepelletier gearset and was introduced in the 2008 Volkswagen Touareg and is also offered in the Passat. The transmission was developed and is manufactured by Aisin. It is available in two basic configurations: one for a V-6 engine and the other for a V-10. The differences between the two are the number of discs in the brakes and clutches and the size of the torque converter.

All shifting is electronically controlled. Some models have "Tiptronic" controls on the steering wheel. Solenoids regulate the fluid pressure to three multiple-disc clutches or two multiple-disc brakes to cause shifting. There is also one overrunning clutch. The clutches are referred to as K1, K2, and K3 and the brakes are B1 and B2. The overrunning clutch is F, for freewheel (Figure 10-76). Each clutch and brake is fed fluid by an individual solenoid or the manual shift valve. See Figure 10-77 for the activity of the multiple-disc clutches and brakes for this transmission.

The transmission's control unit controls all shifting and monitors all shifts to determine if there are any problems. The transmission has eight solenoids located in the main valve body, their description follows:

Solenoid	Its purpose	Type of solenoid
#1	On for fourth to sixth gears	Switching (On/Off)
#2	On for T/C clutch and B1 in low	Switching (On/Off)
#3	Regulates K1 pressure	Modulating

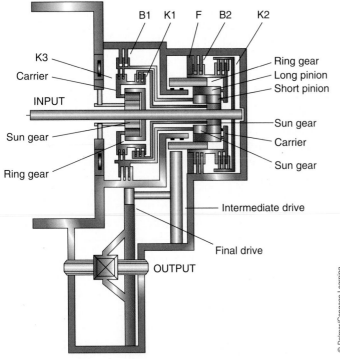

FIGURE 10-76 The major components of a VW six-speed automatic transmission.

GEAR	K1	K2	K3	B1	B2	F
1st	ON	OFF	OFF	OFF	ON*	ON
2nd	ON	OFF	OFF	ON	OFF	OFF
3rd	ON	OFF	ON	OFF	OFF	OFF
4th	ON	ON	OFF	OFF	OFF	OFF
5th	OFF	ON	ON	OFF	OFF	OFF
6th	OFF	ON	OFF	ON	OFF	OFF
R	OFF	OFF	ON	OFF	ON	OFF

*Engine braking in manual first only.

FIGURE 10-77 Activity of the multiple-disc clutches and brakes in a VW six-speed transmission.

#4	Regulates T/C clutch pressure	Modulating
#5	Regulates K3 pressure	Modulating
#6	Regulates mainline	Modulating
#9	Regulates K2 pressure	Modulating
#10	Regulates B1 pressure	Modulating

The operation of all multiple-disc clutches in this transmission is compensated for centrifugal pressure. K1 connects the ring gear of the single planetary gearset to the large sun gear of the dual planetary gearset. It is applied in first to fourth gears and is controlled by the #3 solenoid. K2 connects the input shaft to the carrier of the dual planetary gearset. It is applied in fourth through sixth gear and is controlled by the #9 solenoid. K3 connects the carrier of the single planetary gearset to the small sun gear of the dual planetary gearset. It is applied by the action of solenoid #5 during reverse, third, and fifth gears.

One set of discs in the multiple-disc brakes are splined to the transmission housing. When applied, the member attached to the brake is held by the housing. B1 holds the small sun gear of the dual planetary gearset in second through sixth gear. It is controlled by solenoid #10. B2 holds the carrier of the dual planetary gearset. This brake is not controlled by a solenoid; rather, the manual shift valve sends fluid to apply it. It is applied in reverse and in manual first gear. The latter provides engine braking while the transmission is in the Tiptronic mode.

The overrunning clutch, F, connects the carrier of the dual planetary gearset to the transmission housing. This prevents the carrier from rotating in a clockwise direction. However, when the carrier rotates in the counterclockwise direction, this clutch freewheels. Freewheeling will occur in first gear.

The TCM is connected to the CAN data bus. This allows the TCM to select the ideal gear based on inputs from the various control units including the PCM and ABS. A critical input to the TCM is the input speed sensor located in the housing of the ATF pump. This is a Hall-effect sensor and relies on a ring gear mounted on the input shaft to monitor input speed. This input is used to detect any difference between engine speed and input speed. This difference is used to calculate slip at the T/C. The slip is regulated by a converter bypass coupling that is controlled by the #4 solenoid.

Output speed is also monitored by a Hall-effect unit. This sensor monitors the speed of the transmission output shaft. It relies on teeth on the outside of the rear planetary's ring gear to determine the speed. The calculated speed is used to determine when gears ratios should be changed.

The transmission has a hill-holder feature that prevents the vehicle from rolling back when it is stopped on a hill. When the TCM determines the vehicle is stopped on an incline, it will automatically shift into second gear. The vehicle cannot roll back because the ring gear of the dual planetary gearset is locked to the housing by the one-way clutch. This clutch will allow movement as soon as the engine's torque is greater than the force caused by the descending slope. At that point, the TCM will shift into first gear and the vehicle will accelerate normally.

ZF Transmissions

Many vehicles have transmissions developed and built by ZF Friedrichshafen AG (Zahnradfabrik Friedrichshafen Aktiengesellschaft). This company, commonly known as ZF, is a worldwide supplier of driveline and chassis technology. ZF is major supplier of manual and automatic transmissions for BMW, Audi, Ford, Jaguar, Peugeot, and Alfa Romeo. Perhaps the largest user of ZF transmissions is BMW. BMW automatic transmissions are made by two companies for vehicles sold in North America: ZF and GM. GM's Hydramatic division in Strasbourg, France, supplies automatic transmissions to BMW for four and some six-cylinder vehicles.

ZF uses a simple designation system for their units:

First digit—Number of forward gears
Second and third digits—Type (Hydraulic Planetary = Automatic)
Fourth and fifth digits—ZF internal codes

When BMW does not use the ZF designation, they use the following codes:

First digit—Type (A = automatic or S = Standard)
Second digit—Number of forward gears
Third digit—High gear ratio (D = top gear direct drive or S = top gear overdrive)
Fourth, fifth, and sixth digits—Maximum input torque rating in Nm
Seventh digit—Manufacturer (Z = ZF, R = Hydramatic, G = Getrag)

Through the years, BMW has used Simpson, Ravigneaux, Wilson, and Lepelletier gearsets as the basis of their transmissions. The Simpson gearset is used in 4HP transmissions which are four-speeds. Fourth gear is achieved by an additional single planetary gearset added after the Simpson. The most common Simpson-based transmissions are the 4HP22 and 4HP24.

BMW has used many variations of the Ravigneaux gearset. The A4S310 is a four-speed unit that uses a dual planetary gearset plus a single gearset that provides for fourth gear. The A5S310Z relies on a Ravigneaux gearset and an auxiliary gearset to provide five forward gears. The A5S360 and A5S390R use a Ravigneaux gearset without an additional gearset to provide five forward speeds.

The Wilson gearset is only used in the A5S440Z and A5S560Z. This gearset is made up of three planetary gearsets. The ring gear of the first gearset, the planetary carrier of the second gearset, and the ring gear of the third planetary gearset are directly connected to a cylindrical device that slides over the gears and connects them together. This device is commonly called the "Pot." A Wilson gearset has three planetary carriers, three ring gears, with ring gears one and three connected by the pot, and three sun gears. The sun gears in the second and third gearsets are common and are often called the "Double Sun Gear."

BMW's 6HP26. BMW's 6HP26 (also known as myTronic6) is based on the Lepelletier concept and relies on a control module that combines a hydraulic selector unit and the electronic control module into one unit. This provides for adaptive learning and "Stand-by-Control," which controls the input clutch in the transmission so that the engine is disconnected from the driveline when the vehicle is stationary and the brakes are applied. This six-speed is a high torque capacity transmission with the following gear ratios:

First	4.17:1
Second	2.34:1
Third	1.52:1
Fourth	1.14:1
Fifth	0.87:1
Sixth	0.69:1
Reverse	3.4:1

This basic design is also used in BMW's seven- and eight-speed transmissions. Both of which do not have a torque converter. Instead, they use a dual-mass flywheel with a wet clutch. The 6HP28 used by Jaguar and BMW has a twin-torsional damper in the T/C that reduces driveline vibrations and decreases fuel consumption when used with a diesel engine.

The transmission control system is based on Adaptive Shift Strategy (ASIS). This system constantly monitors conditions and adjusts the operation of the transmission accordingly. In this transmission, the stand-by-control (SBC) allows the torque converter to be disconnected from the engine when the brake pedal is depressed to reduce engine load and fuel consumption.

ZF 8-Speed. BMW and Audi will introduce an eight-speed automatic transmission in their luxury cars in 2010 (Figure 10-78). This transmission, built by ZF, is based on the 6HP series. It is a clutch-to-clutch unit that is capable of extremely quick gear changes. The transmission is designed for RWD and AWD vehicles. Complicated electronics control the operation of the transmission. The unit has many features to improve performance and fuel economy, including engine stop–start.

Mercedes Transmissions

Mercedes-Benz typically develops and produces their own transmissions. Other manufacturers use their units, especially Chrysler. Mercedes transmissions are referred as being in the "NAG 1" series. Commonly found in Chrysler vehicles are five-speed transmissions such as the W5A330, W5A380, W5J400, and W5A580. The acronym "NAG" comes from the German

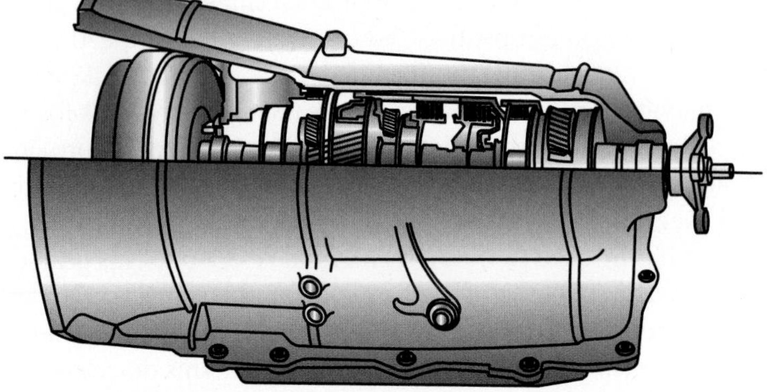

> The 6HP26 was first used in the 2001 BMW 7 Series and Audi used it in their 2003 A8 sedan.

> The electronic control unit for this transmission is called the Mechatronik unit.

FIGURE 10-78 BMW's eight-speed automatic transmission.

© Delmar/Cengage Learning

words for new automatic transmission. The "1" refers to the first iteration of the transmissions. The different model numbers reflect their duty rating, which is expressed as maximum input torque. The W5A380 is found in Sprinters, which is a diesel-powered utility vehicle and the W5A330 is used in the Chrysler Crossfire, which is totally based on a Mercedes-Benz platform. The W5A330 in the Crossfire offers a unique feature; there is a Winter/Standard (W/S) switch at the gear selector. Lower gear ratios are used during first and reverse gear operation when the W/S switch is in the "S" position. When the switch is in the "W" position, these gears operate at a higher ratio.

All transmissions in this series are very similar and operate the same, except for the special features that are the consequence of their application. The W5J400 is an electronically controlled five-speed automatic transmission with a variable lockup torque converter. It uses hydraulically applied clutches and three planetary gearsets, fifth gear is the overdrive gear. The W5A580 was originally introduced in 1995 Mercedes-Benz passenger vehicles. It has a quite a wide spread between the gear ratios, which allows for quicker acceleration in first gear and more fuel efficiency in fifth gear.

The planetary gearsets are controlled (Figure 10-79) by the action of hydraulic multiple-disc packs (B1, K1, K2, B3, K3, and B2), which are controlled by solenoids located in the valve body. These transmissions also have two overrunning clutches (F1 and F2).

7G-Tronic. The 7G-Tronic was introduced in 2003 and was the first seven-speed automatic transmission (Figure 10-80). It replaced the previously used five-speed transmission. The 7G-Tronic is currently used in all eight-cylinder Mercedes-Benz vehicles. It is also available on some six-cylinder models. Some Mercedes-Benz vehicles still have the 5G-Tronic because the 7G cannot handle the torque of a V-12 engine, and it does not work well with AWD.

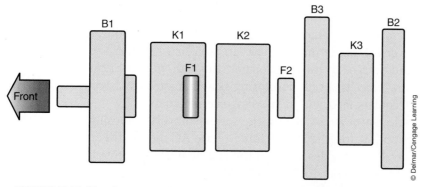

FIGURE 10-79 The placement of the various clutches and brakes in Mercedes' NAG 1 transmissions.

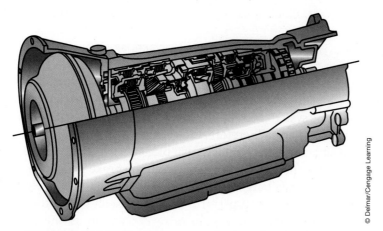

FIGURE 10-80 A 7G-Tronic seven-speed automatic transmission.

This transmission is based on the Lepelletier design with three multiple-disc clutches and four brakes. Another first for this transmission is that its housing is made of magnesium to save weight. It is also a totally electronically controlled transmission that allows for skipping gears during up and downshifts. The unit also has a W/S selector and therefore two reverse gears. While in the Winter mode, Drive starts out in second forward or second reverse gears.

During downshifts, the transmission can skip gears to allow for quicker acceleration. When the driver depresses the throttle to pass or gain speed, the transmission does not always select the next gear in strict order. For example, the transmission may shift from seventh down to fifth and from there go directly into third. It does this by preparing to shift into the next gear at the same time it is preparing to shift into the lower gear before that.

Another feature of the transmission is a lockup clutch in the T/C that can lock in any forward gear. The TCM is programmed to eliminate slip whenever it is possible. This reduces fuel consumption and improves performance.

S400 BlueHybrid. Mercedes-Benz has entered the world of hybrids with its S400 BlueHybrid. This sedan has a 3.5-Liter V6 gasoline engine; an additional electric motor; reconfigured 7G-Tronic transmission; a high-voltage lithium-ion battery; a transformer; and a unique control system. The engine has a maximum output of 279 hp (220kW) and the electric motor is rated at 20 hp (15 kW). Fuel consumption is rated at 30 mpg (8 L/100 km). This vehicle is the first production model to be equipped with a lithium-ion battery specifically developed for automotive use.

The engine has variable valve control and utilizes the Atkinson cycle to improve overall efficiency. The 7G transmission was adapted to meet the needs of the hybrid. Modifications include new software and electric auxiliary oil pump that circulates ATF when the engine is off and the ignition is on.

The lithium-ion battery not only stores energy for the electric motor, but it also powers the 12-volt system in the vehicle. The 120V at the battery is reduced to 12V by a transformer (inverter/converter). Some models have an auxiliary 12V battery that is used for the 12V accessories and charged by the transformer. The high-voltage battery assembly is made up of several lithium-ion cells, a cell monitoring system, battery management system, cooling gel, cooling plate, coolant feed, and high-voltage connectors.

The electric motor is installed in the torque converter housing between the engine and the transmission (Figure 10-81). It is three-phase AC motor. The motor acts as a starter and generator, as well as provides assist to the engine during acceleration and heavy loads. The motor also helps to dampen torsional vibrations in the drivetrain. This reduces the vibrations and noise that enters into the passenger compartment.

The transformer is placed in the right front wheel arch and is accessible. Its location permits jump starting if the battery should lose its charge. The transformer is cooled by a low-temperature circuit. Cooling is necessary to ensure the output from the battery is consistent.

When the vehicle is being stopped, the motor acts as a generator and uses a process known as regeneration to convert kinetic energy into electrical energy. The wheels rotate the generator which causes a drag on the wheels, therefore the vehicle slows down. During deceleration without applying the brakes, the drag of the generator slows the vehicle. When the brakes are applied, the system increases the generator's ability to produce AC voltage and there is more drag on the wheels. Therefore, the vehicle experiences a more rapid slowdown. During both of these situations, the conventional brakes are not initially applied. The brakes only stop the vehicle when there is heavy pressure on the brake pedal. During this time, the regeneration still takes place to charge the battery and to help stop the vehicle.

The system also provides for stop/start. This feature shuts off the engine when the vehicle is near totally stopped. The engine restarts as soon as the driver releases the

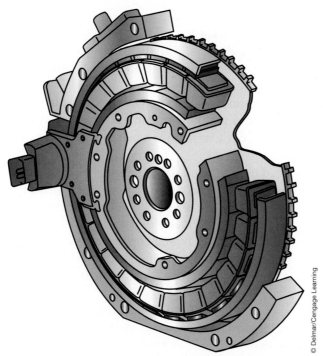

FIGURE 10-81 The hybrid module is a disc-shaped electric motor which is integrated between the engine and the 7G-Tronic and also serves as a starter and an alternator in the Mercedes-Benz S-Class S 400 Hybrid.

© Delmar/Cengage Learning

brake pedal or depresses the throttle. The traction motor serves as the starter motor and quickly cranks the engine. When the engine is off, the A/C compressor and power steering systems remain functional. They are driven by an electric motor rather than an engine-driven belt. The auxiliary oil pump for the transmission also runs when the engine is shut off.

CONSTANTLY VARIABLE TRANSMISSIONS (CVTS)

Basically, a CVT is a transmission with no fixed forward speeds. The gear ratio varies with engine speed and temperature. The simplicity of a CVT makes it an ideal transmission for a variety of devices, not just automobiles. With a CVT, the most ideal drive ratio can be automatically chosen for the existing conditions.

CVTs offer some advantages over conventional, manual, and automatic transmissions. These include the ability to stay in the same power band if/when needed; they allow the engine to remain in its most optimal power range to provide improved fuel economy, and stepless shifting.

Dr. Hub van Doorne invented the first CVT for automotive use in 1958. The transmission was called the Variomatic. The first Variomatic used rubber belts, which were not very durable and could not be used with engines with any sort of power. This led to the development of metal belts that were able to handle more powerful engines. Modern CVTs are made much more durable by using steel link belts.

In the late 1980s and early 1990s, Subaru offered a CVT in their Justy mini-car. In the late 1990s, Nissan designed its own CVT for higher torque engines. This unit, called the Extroid CVT, was based on the toroidal design and is still used in many Nissan models. Honda introduced a CVT in the 1995 Honda Civic VTi. This transmission has been offered in many different vehicles, including their hybrid cars.

A BIT OF HISTORY

It is commonly believed that Leonardo Da Vinci sketched the plans for a CVT in 1490.

A different sort of CVT is the E-CVT that was introduced by Toyota in the 1997 Prius. This basic design is still used in all Toyota hybrids. Technically, these units are not CVTs. Rather, they are torque blending devices. Although they are stepless, the units contain no pulleys or belts. Rather, they operate with a planetary gearset whose members are controlled by electric motors and the engine.

GM designed a CVT for use in small cars and introduced it in the 2002 Saturn Vue. This unit is referred to as the VTi and was an electronically controlled unit with a torque converter. GM stopped production of these in 2005 when the units were replaced by a six-speed automatic transmission.

In 2000, Audi introduced new version of the Variomatic called the Multitronic. This system uses a metal belt (often referred to as a chain). It has been offered as an option on many different models.

In 2005, Ford offered a CVT in the Freestyle, Five Hundred, and Mecury Montego. This unit uses a chain drive and is capable of handling fair amounts of engine torque. This CVT has a torque converter, radial piston pump, planetary gearset for reverse gear, and primary pulley on its primary axis, and a secondary pulley with an integrated parking gear and transfer gear on its secondary axis.

CVTs can now be found in vehicles from Dodge, Fiat, Mitsubishi, Mini-Cooper, Saturn, Nissan, Audi, Honda, Ford, GM, and other manufacturers.

CVTs can be based on many different designs; the chosen design is determined by their application. The basic designs include belt-driven, chain-driven, roller-type (toroidal), hydrostatic, ratcheting-type, and cone-type. Chain and belt drive units and toroidal designs are basically friction units. The others are not currently used in automobiles.

Belt/Chain Drives

Many CVTs use a V-belt to connect two pulleys. The drive ratio is changed by moving the two sections of one pulley closer together and the two sections of the other pulley farther apart. This causes the belt to ride higher on one pulley and lower on the other, which effectively changes diameter of the pulleys and changes the drive ratio. Because the distance between the pulleys does not change, both pulleys must be changed at the same time. If this does not occur, the drive pulley would not be able to turn the driven pulley or the belt would break due to excessive tension. Today's belts are called push belts. This is because as one pulley rotates, it pushes the belt against the other pulley causing it to turn.

Modern push belts consist of many flat, trapezoid-shaped steel plates joined in a loop by steel bands (Figure 10-82). The plates have conical sides and are used to grip the

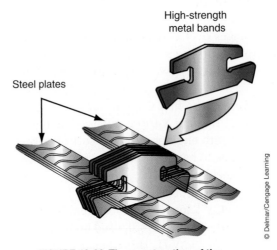

High-strength
metal bands

Steel plates

© Delmar/Cengage Learning

FIGURE 10-82 The construction of the steel belt used in most CVTs.

outermost radius of the pulleys. The longitude steel bands hold the plates in position and give the belt strength. The belt is firmly seated into the pulleys and has a tension of nearly 8000 pounds (3629 kg). A film of thick lubricant is applied to the pulleys to prevent the belt from actually touching the pulleys. Metal belts do not slip and are more durable, quieter, and can handle more engine torque than rubber drive belts. Steel belts are currently used in Saturn Vue, Ford Freestyle and Five Hundred, Honda Civic and Insight, and Nissan Murano CVTs.

Many belt-driven CVTs have a start clutch, which is typically an electromagnetic clutch. The start clutch disengages the engine from the transmission when the vehicle is at a stand-still and engages once the throttle is depressed.

Audi Multitronic CVT. Audi's Multitronic uses a chain drive rather than a belt and is equipped with a torque sensor. The use of a chain enables the transmission to be used behind high torque engines. This unit also uses a multiple-disc clutch rather than a torque converter or electromagnetic clutch. Chain-type CVTs are also called a pull-type or Luk chain CVT. The latter represents the name of the German company that developed the chain. It is also called a pull-type because the drive pulley pulls the chain over the driven one. The chain is pinched by the pulley halves as the pulley's diameter changes. The links of the chain are held together by pins, which also serve as the contact surface on the pulley. The chain is almost as flexible as a V-belt but much stronger.

The torque sensor is used to monitor the contact pressure of the chain against the pulleys. The TCM makes minor adjustments to the pulleys' openings, in response to the torque sensor, to maintain the proper chain tension.

This CVT also offers a manual override that allows the driver to shift sequentially through six forward speeds. Shift controls mounted to the steering wheel are an option.

Toroidal CVTs

Toroidal CVTs do not use a belt or chain; rather, they rely on rollers and discs. Power is transmitted from one disc to another by the rollers (Figure 10-83). This type of CVT can handle great amounts of torque and sets of rollers and discs can be combined to handle even greater amounts of torque. For example, Nissan's Extroid CVT has two pairs of rollers. Toroidal CVTs have been used in FWD and RWD vehicles.

The assembly of rollers and discs is often called the variator. The discs are cone shaped with concave sides. The basic shape is called a torus. The discs are arranged so that their larger diameters face the outside and their narrow ends face each other. The input disc connects to the engine and functions as a driving pulley. The other disc is the output and functions as a driven pulley. In this arrangement, the two discs are close together, but do not make contact.

The space between the discs forms a hollow doughnut shape or toroid. Rollers are positioned in the toroidal space and serve as a link between the two discs. The rollers are attached to hydraulic pistons that can reposition the rollers within that space. The rollers act as a belt or chain and transmit torque from one disc to the other. The rollers ride on a film of lubricant that prevents metal-to-metal contact. Drive ratios are determined by the angle of the rollers. When the rollers are parallel to the discs, direct drive results. As the rollers tilt, the drive ratios change. By tilting, the rollers can contact the discs at varying and distinct diameters, resulting in various drive ratios.

As the rollers tilt, they ride against the discs at different locations. When the roller is riding close to the large diameter of the driven disc, it is riding close to the smallest diameter on the driven disc. This provides a speed increase or torque reduction. When the roller is close to the smallest diameter of the drive disc, the roller is contacting the driven disc at a much larger diameter. This results in a speed reduction and torque increase. A simple tilt of the rollers smoothly and instantaneously changes the drive ratio.

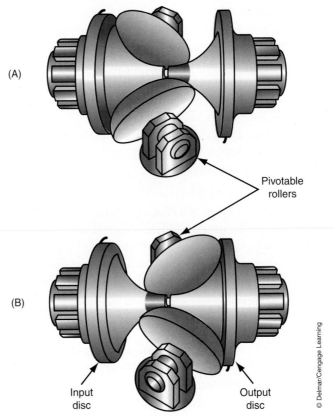

(A)

Pivotable
rollers

(B)

Input
disc

Output
disc

© Delmar/Cengage Learning

FIGURE 10-83 The rollers and discs used in a toroidal CVT. The figure shows the relationship of the rollers and discs when the unit is operating at (A) a high ratio and (B) a low ratio.

Honda's CVTs

Honda's CVT is a steel belt driven unit. The basic unit has been used in many Honda vehicles, including their hybrids. Late-model CVTs are called Multimatics. A Multimatic does not use a torque converter; rather, it uses a dual-mass flywheel and an acceleration clutch. The fly-wheel is designed to dampen engine vibrations and to provide a direct connection between the engine and the transmission. Engine torque is constantly applied to the input shaft.

The acceleration clutch takes the place of a conventional clutch or torque converter and is placed at the output of the transmission. Honda placed it here in order to control torque at the closest possible place to the axle. This placement provides for smooth acceleration as well as provides some creep while the transmission is in gear. The latter feature prevents vehicle movement when it is stopped on a hill. The placement of the acceleration clutch also allows the transmission's discs to be constantly spinning; this means the vehicle can move immediately after the acceleration clutch is engaged.

The torque transferred by the acceleration clutch is controlled by the hydraulic pressure at the clutch's piston. The amount of creep is also controlled by the pressure. When the brakes are applied, the unit will provide very little creep. When the brake pedal is released, the unit will allow more creep.

The operation of the transmission is electronically controlled (Figure 10-84). The drive ratios are based on inputs from various sensors and predetermined patterns set in the unit's software. The TCM looks at the current and ideal engine speeds and adjusts the drive ratios accordingly. The driver can select between two modes of operation: Drive and Sport. In the Drive mode, the unit will change ratios to achieve the best fuel economy. In the Sport mode, the engine's speed can be increased beyond the limits of the Drive mode.

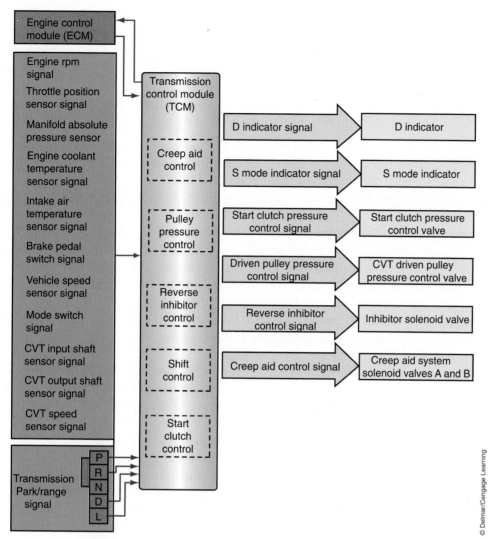

FIGURE 10-84 The electronic control system for a Honda CVT as used in a hybrid vehicle.

The transmission relies on oil-pressure controlled pulleys. The TCM monitors several inputs, including throttle position and vehicle speed, and adjusts the high/low pressure regulator to control the width of the pulleys. For example, when the accelerator is depressed, the width of the drive pulley increases. In the same instance, the width of the driven pulley decreases. This provides for torque multiplication. The pinching force of the pulleys on the belt is also controlled by a pressure regulator.

The gear selector has six positions: Park, Reverse, Neutral, Drive, Second, and Low. When Low is selected, the transaxle shifts into the lowest range of available pulley ratios. This gear is intended for engine braking and power for climbing steep grades or for very heavy loads. Second gear is designed for quicker acceleration while the car is moving at highway speeds and for moderate engine braking. When this range is selected, the transaxle shifts into a lower range of pulley ratios. When this transaxle is used in a hybrid, the selection of second or low increases the amount of regeneration during deceleration. When Drive is selected, the transaxle automatically adjusts the pulley ratios to keep the engine at the best speed for the current driving conditions.

When the selector is in Park, the parking pawl is engaged to the parking gear on the driven pulley shaft. This locks the front wheels. While the transaxle is in Park and Neutral, the acceleration (start) clutch and the forward clutch are released. When the selector is placed into Reverse, the reverse brake clutch is engaged.

In the forward range, the pinion gears do not rotate on their shafts. They revolve with the sun gear. This causes the carrier and forward clutch drum to rotate. The input shaft drives the sun gear. The carrier outputs the power to the drive pulley shaft. In reverse, the reverse clutch locks the ring gear to the housing. The pinion gears revolve around the sun gear causing the carrier to rotate in the opposite direction from the rotation of the sun gear.

Operation. Three solenoids control the flow of hydraulic pressure, which in turn control the action of the transaxle. The transaxle's oil pump is connected to the input shaft by sprockets and a chain. There are two valve body assemblies within the transaxle. The primary valve body, called the lower valve body, is mounted to the lower part of the transaxle case.

The lower valve body includes the main valve body, the secondary valve body, the low pressure (PL) regulator valve body, the acceleration clutch control valve body, and the shift valve body. The main valve body consists of the high pressure (PH) control valve, the lubrication valve, and the pilot regulator valve. The PH control valve supplies high control pressure to the PH regulator valve, which also regulates high pressure. The lubrication valve controls and maintains the pressure of the ATF going to the shafts. The pilot regulator valve controls the acceleration clutch pressure in relationship to engine speed when a fault is detected in the electronic system.

The secondary valve body contains the PH regulator valve, clutch reducing valve, acceleration clutch valve accumulator, and shift inhibitor valve. The PH regulator valve maintains the pressure of the fluid from the oil pump and supplies fluid to the shafts and the rest of the hydraulic control circuit. The clutch reducing valve receives high pressure from the PH regulator valve and controls the clutch reducing pressure. The clutch reducing valve supplies fluid to the manual valve and the start clutch control valve. It also supplies signal pressure to the PH-PL pressure control valve, shift control valve, and inhibitor solenoid valve. The start clutch valve accumulator stabilizes the hydraulic pressure applied to the start clutch. The switching of the hydraulic passages to the start clutch control when there is an electronic failure is the duty of the shift inhibitor valve. This valve also supplies clutch reducing pressure to the pilot regulator valve and the pilot lubrication pipe.

The low pressure valve body contains the PL regulator valve and the PH-PL control valve (Figure 10-85). Two solenoids are mounted to this portion of the valve body: the PH-PL control solenoid and the inhibitor solenoid, both of these solenoids are controlled by the TCM. The PL regulator valve supplies low pressure to the pulleys when necessary.

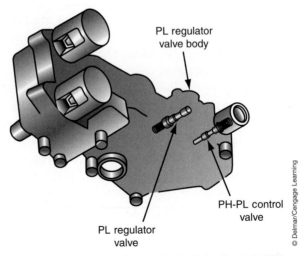

FIGURE 10-85 The regulator valve body for Honda's CVT.

The PH-PL control valve controls the PL regulator valve according to engine torque. The PH-PL control valve supplies control pressure to the PH control valve to regulate PH higher than PL. The action of the PH-PL control valve is controlled by the PH-PL solenoid. The inhibitor solenoid controls the reverse inhibitor valve by cycling on and off. The inhibitor solenoid also controls high control pressure by applying reverse inhibitor pressure to the PH control valve.

The shift valve body contains the shift valve and the shift control valve. The shift valve is controlled by shift valve pressure from the shift control valve. The shift valve distributes high and low pressure to pulleys to change the pulley ratios. The shift control valve controls the shift valve according to vehicle speed and throttle opening. The action of the shift control valve is controlled by the shift control solenoid. When there is an electronic failure, the shift control valve switches the shift inhibitor valve to uncover the port leading to the pilot regulator pressure to the start clutch; this allows the transaxle to work in spite of the electrical problems.

SUMMARY

- Most automatic transmissions use planetary gearsets to provide for the different gear ratios. The gear ratios are selected manually by the driver or automatically by the hydraulic and electronic control system.
- All Chrysler Torqueflite-based transmissions and transaxles use a Simpson geartrain.
- Chrysler designates their transmissions according to the number of forward speeds, torque rating, orientation of the unit, and method of control.
- Four- and five-speed transmissions from Chrysler normally use a Simpson geartrain with a single planetary gearset placed in series.
- In 1989, Chrysler began their departure from Simpson gearsets to using planetary gearsets in tandem. These units had two or three gearsets connected in series and are commonly called Ultradrive transmissions.
- In 2007, the basic 41TE which had a mainshaft with two planetary gearsets and five clutch packs, was modified to provide six forward speeds. It has an additional planetary unit, referred to as a compounder gearset.
- The RFE series is very similar to the 41/42 series and is used in RWD vehicles. These are four- and five-speed units.
- The 68RFE transmission was introduced in 2007 for the Dodge Ram 2500 and 3500 Pickups with the 6.7L Cummins ISB Diesel engine. It is a six-speed unit and similar in operation to the 45RFE and 545RFE transmissions.
- The transmissions found in current Fords can be Simpson, Ravigneaux, gears-in-tandem, or Lepellietier gear based units. Plus they also have CVTs in some models.
- To define the transmission types, Ford's designations are set by the number of forward gears, features of the transmission, and maximum input torque capacity.
- All Ford transmission models that rely on a Simpson gearset operate similarly.
- Four-speed models use the same planetary unit as a three-speed with an additional simple planetary unit for fourth gear.
- Ford's E4OD was updated in 1998 and renamed the 4R100. It was basically the same as the E40D, but was built with some reinforcements to improve reliability. This model was last used in the 2004 E-series cargo vans.
- For the 1995 model year, the 4R44E appeared as an updated version of the A4LD. The 5R55E is a five-speed version of the 4R55E. An additional gearset was not added to the unit to achieve the fifth gear, rather additional control devices were added and the TCM's software was upgraded.

- The AOD was Ford's first four-speed automatic overdrive transmission and was based on a Ravigneaux gearset.
- The AODE was replaced by the 4R70 series with the 1993 model vehicles. This transmission was used continuously, with some modifications, until 2008 on some vehicles.
- Ford's AXOD is based on the gears-in-tandem setup and was a four-speed transaxle for transversely mounted engines. The AXOD-E was renamed AX4S in 1994.
- The AX4N is an improved version of the AXOD and was renamed the 4F50N in 2001.
- The CD4E was similar in design to the AX4S and used in cars from 1994 to 2007.
- Ford has two basic families of Lepelletier-based units. One is a transaxle (6F) built by Ford and the result of a cooperative engineering effort with GM. The other family (6R) is built for RWD vehicles by Aisin or ZF.
- GM's original Hydra-matic series of transmissions was used until they were replaced by versions of the Turbo-Hydramatic series in the 1960s. The Turbo-Hydramatic was used by all GM divisions until the mid-1990s and served as the basic design for the company's current Hydramatic line.
- GM uses a simple naming scheme for their transmissions which include the number of forward gears, basic design, GVWR rating, and features.
- Nearly all of GM's early transmissions were based on the Simpson geartrain.
- When GM first introduced a four-speed automatic transmission, they simply added a planetary unit to the existing transmissions and transaxles.
- The 4L80 is a four-speed version of the Turbo-Hydramatic 400 that was released in 1965. It was used in a variety of GM pickups, vans, and SUVs until very recently.
- GM's 4T60 was the first domestic four-speed automatic built for FWD vehicles. It was based on the gears-in-tandem arrangement.
- The 4T80 and 4T80-E were developed specifically for use with V-8 FWD cars and are used in many large GM cars with FWD.
- The 4Lxx series was used in most full-size RWD GM cars and trucks and in some heavy-duty and high-performance vehicles. The 4L60 was introduced in 1993. The 4L60-E was updated in 2002 and became the 4L65-E. Stronger versions, the 4L80 and 4L85, were later introduced.
- GM currently has four six-speed transmissions and transaxles based on the Lepelletier geartrain, the 6L80, 6L50, 6T70, and 6T75.
- The 6L80 six-speed transmission was introduced in 2006 on the Cadillac XLR-V, STS-V, and Chevrolet Corvette. In 2007, it was modified for use in SUVs and trucks.
- In 2007, six-speed transaxles were introduced, the 6T70 and 6T75. The transaxles were co-developed with Ford Motor Co.
- The transmissions used in Honda and Acura vehicles are unique in that they do not use planetary gearsets to provide for the different gear ranges; rather, they use constant-mesh helical and square-cut gears.
- Honda has used many different classifications for the transmissions. However, all of the versions can be divided into one of two ways: the number of forward speeds or the number of internal shafts used in the transaxles.
- The Honda/Acura five-speed automatic transaxle uses six multiple-disc clutch packs and a one-way clutch and is electronically controlled.
- A modified version of the five-speed automatic transaxle is used in many Honda hybrids. It is fitted with an integrated electric oil pump and different gear ratios that provide for better acceleration, fuel economy, and regenerative braking.
- Some Saturns had nonplanetary gear transmissions that were electronically controlled four-speeds and referred to the MP6 and MP7.

- Nissan and its affiliate, Jatco, have produced a large number of automatic transmissions for the industry.
- Jatco (Nissan) uses the following code to identify their transmissions, "J" for Jatco, "F" for front-wheel-drive or "R" for rear-wheel-drive, the next digit is the number of gears and the next two digits denote the series number.
- Most Nissan RWD vehicles are based on the RE4Rxx series which are four-speeds with Simpson gearset and an additional planetary unit mounted in front of the Simpson.
- A common transaxle from Nissan is the RE404A (Ford 4F20E). This four-speed unit has a converter clutch and electronic shift controls.
- Jatco builds seven-speed transmissions for RWD Nissans. There are two versions of this Lepelletier-based transmission, the JR710E and JR711E.
- Many different transmissions and transaxles have been used by Toyota, Scion, and Lexus and they are produced by Aisin AW.
- Aisin AW produces automatic transmissions for Chrysler, GM, Ford, Mitsubishi, Nissan, Porsche, Audi, VW, Volvo, Hyundai, and Isuzu.
- There are two basic families of Toyota automatic transmissions, A and U. The A family are those automatic FWD/RWD/AWD transmissions built by Aisin-Warner. In the U family are late-model automatic transaxles built for FWD and AWD vehicles.
- All Toyota automatic transmission designations begin with an A or a U. This is followed by the transmission series number, number of forward gears, and a digit representing the features.
- Early four-speed Toyota transaxles used a Simpson plus a single planetary gearset, these are in the A family.
- The transaxles in Toyota's U family have two or three simple planetary gearsets connected in series. The U family also includes the special transmissions used in Toyota hybrids.
- The most widely used Lepelletier gear based Toyota transmission is the AA80E. It was also the world's first mass-produced eight-speed automatic transmission.
- The Aisin AF40/6 is used by GM and is based on a Lepelletier gear arrangement. It is a fully electronic controlled six-speed transaxle that utilizes clutch-to-clutch actuation.
- The Aisin AF33 is an electronically controlled five-speed unit. It is used by GM, Saturn, Saab, Volvo, and Nissan. The AF33 designation is only used by GM, Nissan calls the unit the RE5F22A, and the actual designation from Aisin is AW55-50SN (FWD) and AW55-51SN (AWD).
- ZF is major supplier of manual and automatic transmissions for BMW, Audi, Ford, Jaguar, Peugeot, and Alfa Romeo.
- BMW's A5S310Z relies on a Ravigneaux gearset and an auxiliary gearset to provide five forward gears. The A5S360 and A5S390R use a Ravigneaux gearset without an additional gearset to provide five forward speeds.
- The Wilson gearset is only used in the A5S440Z and A5S560Z.
- BMW's 6HP26 is based on the Lepelletier concept and a six-speed.
- Mercedes-Benz's 7G-Tronic was introduced in 2003 and was the first seven-speed automatic transmission and is currently used in all eight-cylinder Mercedes-Benz vehicles.
- A CVT is an automatic transmission with no fixed forward speeds.
- CVTs can be based on many different designs, including chain and belt drive units and toroidal designs which are basically friction units.
- Toroidal CVTs do not use a belt or chain; rather, they rely on rollers and discs. Power is transmitted from one disc to another by the rollers.
- Honda's CVT is a steel belt driven unit and has been used in many Honda vehicles, including their hybrids.

TERMS TO KNOW

Controlled load servo

Electronic controlled transaxle (ECT)

Forward clutch

High/reverse clutch

Input speed sensor (ISS)

Intermediate clutch

Intermediate one-way clutch

Kickdown band

Low one-way clutch

Output speed sensor (OSS)

Overdrive band

Overdrive one-way clutch

Pulse width modulated (PWM) solenoid

Transmission oil temperature (TOT) sensor

Variable force motor (VFM)

Variable force solenoid (VFS)

REVIEW QUESTIONS

Short Answer Essays

1. Saturn vehicles rely on five different solenoids to control the operation of their transaxle, what do these solenoids do?

2. In GM's transmission designations the fourth and fifth digits represent the GVWR the transmission can handle. What does GVWR represent?

3. How was Chrysler's 41TE modified to become the 42LE and why?

4. What is so unique about a Honda transaxle? How are they controlled?

5. What happens to a typical transmission when the tow/haul mode is selected by the driver?

6. Some transmissions, such as the 4T60, have a unique way of linking the torque converter with the input shaft, what is unique about it?

7. What is the purpose of a VFM?

8. What is the purpose of a centrifugal fluid pressure canceling mechanism?

9. When a Simpson-based transmission is modified to provide for a fourth (overdrive) gear, what is the major modification?

10. Why do some transmissions seldom allow the T/C clutch to fully lock?

Fill in the Blanks

1. When a Simpson gear based transmission is shifted into first gear, the input shaft is locked to the _____ planetary ring gear. The _____ _____ gear drives the front planet gears, which drive the _____ gear. The rear planet carrier is locked; therefore, the sun gear spins the _____ _____ gears. The planet gears drive _____ _____ gear, which is locked to the output shaft, in a clockwise direction. The result is a forward gear reduction.

2. The synchronized revolution control system accomplishes the same thing an experienced race driver does when using a technique called _____ _____ _____, which involves using one foot to work both the _____ _____ and the _____ _____ during downshifting.

3. Most Torqueflite transmissions have a _____ _____ front servo that has two pistons and allows for the quick release of the band during shifts from _____ to _____ gear and _____ to _____ shifts.

4. In some Ford transmissions, an EPC solenoid replaces the conventional _____ setup to provide changes in pressure in response to _____.

5. During reverse gear in a Ravigneaux gearset, input is received at the _____ _____ gear and the planetary gear carrier is held. The clockwise rotation of the _____ _____ gear drives the long pinion gears in a _____ direction. The long pinions then drive the ring gear and output shaft in a _____ direction with a gear reduction.

6. For low pulley ratios in a CVT, an increase of hydraulic pressure at the _____ pulley is proportional to the decrease of pressure at the _____ pulley. The opposite is true for high pulley ratios. _____ hydraulic pressure causes the driven pulley to _____ in size while _____ pressure _____ the size of the drive pulley.

7. All the Torqueflite-based transmissions and transaxles have two multiple-disc clutches, called the _____ and _____ clutch packs. The servos and bands are referred to as _____ and _____, or _____ and _____.

8. Toyota led the way into electronically controlled automatic transmissions. The _____ was the first fully electronic transmission. This transaxle has a _____ compound gearset and a separate simple planetary unit for fourth gear.

9. Common CVTs are based on the following designs: _____ and _____ driven and _____ designs, which are both _____ units.

10. Mercedes-Benz's 7G-Tronic is based on the _____ design with _____ multiple-disc clutches and _____ brakes. Its housing is made of _____ and has a W/S selector which provides for two _____ gear ratios.

MULTIPLE CHOICE

1. *Technician A* says shift solenoids direct fluid flow to and away from the various apply devices in the transmission.

   Technician B says shift solenoids are used to mechanically apply a friction band brake or multiple-disc clutch assembly.

   Who is correct?

   A. A only   C. Both A and B
   B. B only   D. Neither A nor B

2. While discussing the THM 4T60,

   Technician A says the 4T60 was the first domestically produced four-speed automatic transaxle built for FWD vehicles.

   Technician B says the 4T60 uses two tandem planetary units, which do not share a common gearset member.

   Who is correct?

   A. A only   C. Both A and B
   B. B only   D. Neither A nor B

3. *Technician A* says most CVTs that use a steel belt are called push-type units.

   Technician B says CVTs that use a chain to drive the pulleys are called pull-type units.

   Who is correct?

   A. A only   C. Both A and B
   B. B only   D. Neither A nor B

4. While discussing Honda's five-speed transaxle,

   Technician A says it has six multiple-disc clutch packs and a one-way clutch.

   Technician B says it is equipped with seven solenoids.

   Who is correct?

   A. A only   C. Both A and B
   B. B only   D. Neither A nor B

5. While discussing Ford transaxles,

   Technician A says the CD4E is similar in design to the AXOD.

   Technician B says the 4F27 is similar to the CD4E.

   Who is correct?

   A. A only   C. Both A and B
   B. B only   D. Neither A nor B

6. While discussing Jatco transmissions,

   Technician A says these units are used in Nissans.

   Technician B says these units are used in Hondas.

   Who is correct?

   A. A only   C. Both A and B
   B. B only   D. Neither A nor B

7. *Technician A* says Ford's AXOD is a nonsynchronous transmission.

   Technician B says the AX4N is a nonsynchronous unit.

   Who is correct?

   A. A only   C. Both A and B
   B. B only   D. Neither A nor B

8. While discussing GM's 4T80 series transaxles,

   Technician A says the series does not use an electronically controlled torque converter; rather, they use a viscous clutch.

   Technician B says the transaxles have four separate oil pumps.

   Who is correct?

   A. A only   C. Both A and B
   B. B only   D. Neither A nor B

9. *Technician A* says GM's 6L80 was codeveloped with Ford Motor Company.

   Technician B says GM's 6L50 was codeveloped with Ford Motor Company.

   Who is correct?

   A. A only   C. Both A and B
   B. B only   D. Neither A nor B

10. While discussing Toyota's AA80E transmission,

    Technician A says the design is based on three planetary gearsets connected in series.

    Technician B says the unit relies on four multiple-disc clutches, two multiple-disc brakes, and one overrunning clutch.

    Who is correct?

    A. A only   C. Both A and B
    B. B only   D. Neither A nor B

ATRA	Automatic Transmission Rebuilders Association	Oxnard, CA
ATSG	Automatic Transmission Service Group	www.atsg.biz
ETI	Equipment and Tool Institute	Research Triangle Park, NC
MAP	Motorist Assurance Program	Washington, DC
MEMA	Motor & Equipment Manufacturers Association	Research Triangle Park, NC
SEMA	Specialty Equipment Market Association	Diamond Bar, CA
TRNW	Transmission Rebuilders Network Worldwide	www.trnw.net

GLOSSARY
GLOSARIO

Note: **Terms are highlighted in color,** followed by **Spanish translation in bold.**

Abrasion Wearing or rubbing away of a part.

Abrasión El desgaste o consumo por rozamiento de una parte.

Acceleration An increase in velocity or speed.

Aceleración Un incremento en la velocidad.

Accumulator A device used in automatic transmissions to cushion the shock of shifting between gears, providing a smoother feel inside the vehicle.

Acumulador Un dispositivo que se usa en las transmisiones automáticas para suavizar el choque de cambios entre las velocidades, así proporcionando una sensación más uniforme en el interior del vehículo.

Acid A compound that has an excess of H ions and breaks into hydrogen (H^+) ions and another compound when placed in an aqueous (water) solution.

Ácido Un compuesto que tiene un exceso de iones de H y se descompone en iones de hidrógeno (H^+) y otro compuesto cuando se lo coloca en una solución acuosa (agua).

Adaptive learning The ability of a computer to monitor the driver's habits and the operating conditions of its system and make adjustments to its program to correct for them.

Aprender adapante La capacidad de un ordenador de vigilar los hábitos del conductor y las condiciones de funcionamiento de sus sistemas y haga los ajustes a su programa para corregirlos para ellos.

All-wheel-drive (AWD) System for driving up to four wheels on the vehicle based on traction conditions.

Tracción en todas las ruedas (AWD) Sistema para conducir un vehículo con hasta cuatro ruedas recibiendo potencia del motor según las circunstancias de tracción.

Alloy A mixture of different metals such as solder, which is an alloy consisting of lead and tin.

Aleación Una mezcla de diferentes metales como por ejemplo, la soldadura blanda, la cual es una aleación de plomo y estaño.

Amplitude A measurement of a vibration's intensity. The height of a waveform signal strength or the maximum measured value of a signal.

Amplitud Una medición de la intensidad de una vibración. La altura de la fuerza de una señal en forma de onda, o el valor máximo medido de una señal.

Annulus gear Another name for the ring gear of a planetary gearset.

Engranaje annulus Otro nombre para la corona y el engranaje planetario.

Antifriction bearing A bearing designed to reduce friction. This type of bearing normally uses ball or roller inserts to reduce the friction.

Cojinetes de antifricción Un cojinete diseñado con el fin de disminuir la fricción. Este tipo de cojinete suele incorporar una pieza inserta esférica o de rodillos para disminuir la fricción.

Antiseize Thread compound designed to keep threaded connections from damage due to rust or corrosion.

Antiagarrotamiento Un compuesto para filetes diseñado para proteger a las conecciones fileteados de los daños de la oxidación o la corrosión.

Apply devices Devices that hold or drive members of a planetary gearset. They may be hydraulically or mechanically applied.

Dispositivos de aplicación Los dispositivos que sujeten o manejan los miembros de un engranaje planetario. Se pueden aplicar mecánicamente o hidráulicamente.

ATF Automatic transmission fluid.

ATF Fluido de transmisión automática.

Atom The smallest particle of an element in which all the chemical characteristics of the element are present.

Átomo La partícula más pequeña de un elemento, en la cual todas las características químicas del elemento están presentes.

Automatic transmission A transmission in which gear or ratio changes are self-activated, eliminating the necessity of hand-shifting gears.

Transmisión automática Una transmisión en la cual un cambio deengranajes o los cambios en relación son por mando automático, así eliminando la necesidad de cambios de velocidades manual.

Axial Parallel to a shaft or bearing bore.

Axial Paralelo a una flecha o al taladro del cojinete.

Axis The centerline of a rotating part, a symmetrical part, or a circular bore.

Eje La linea de quilla de una parte giratoria, una parte simétrica, o un taladro circular.

Axle The shaft or shafts of a machine on which the wheels are mounted.

Semieje El eje o los ejes de una máquina sobre los cuales se montan las ruedas.

Axle carrier assembly A cast iron framework that can be removed from the rear axle housing for service and adjustment of the parts.

Asamblea del portador del eje Un armazón de acero vaciado que se puede remover del cárter de los ejes traseros para afectuar el mantenimiento o los ajustes de las partes.

Axle housing Designed in the removable carrier or integral carrier types to house the drive pinion, ring gear, differential, and axle shaft assemblies.

Cárter del eje Diseñado en los tipos de portador removible o de portador integral para encajar el piñon de ataque, la corona, la diferencial y las asambleas de las flechas de los ejes.

Axle ratio The ratio between the rotational speed (rpm) of the driveshaft and that of the driven wheel; gear reduction through the differential, determined by dividing the number of teeth on the ring gear by the number of teeth on the drive pinion.

Relación del eje La relación entre la velocidad giratorio (rpm) del árbol propulsor y la de la rueda arrastrada; reducción de los engranajes por medio del diferencial, que se determina por dividir el número de dientes de la corona por el número de los dientes en el piñón de ataque.

Axle shaft A shaft on which the road wheels are mounted.

Flecha del semieje Una flecha en la cual se monta las ruedas.

Backlash The amount of clearance or play between two meshed gears.

Juego La cantidad de holgura o juego entre dos engranajes endentados.

Balance Having equal weight distribution. The term is usually used to describe the weight distribution around the circumference and between the front and back sides of a wheel.

Equilibrio Lo que tiene una distribución igual de peso. El término suele usarse para describir la distribución del peso alrededor de la circunferencia y entre los lados delanteros y traseros de una rueda.

Balance valve A regulating valve that controls a pressure of just the right value to balance other forces acting on the valve.

Válvula niveladora Una válvula de reglaje que controla a la presión del valor correcto para mantener el equilibrio contra las otras fuerzas que afectan a la válvula.

Ball bearing An antifriction bearing consisting of a hardened inner and outer race with hardened steel balls that roll between the two races, and supports the load of the shaft.

Rodamiento de bolas Un cojinete de antifricción que consiste de una pista endurecida interior e exterior y contiene bolas de acero endurecidos que ruedan entre las dos pistas, y sostiene la carga de la flecha.

Ball-type valve A valve that uses the movement of a ball to control fluid flow.

Válvula bola-tipo Una válvula que utiliza el movimiento de una bola para controlar el flujo de un líquido.

Band A steel band with an inner lining of frictional material. A device used to hold a clutch drum at certain times during transmission operation.

Banda Una banda de acero que tiene un forro interior de una materia de fricción. Un dispositivo que retiene al tambor del embrague en algunos momentos durante la operación de la transmisión.

Base A solution that has an excess of OH ions, also called an alkali. Also the center layer of a bipolar transistor.

Base Una solución que tiene un exceso de iones de OH, también denominada álcali. También la capa central de un transistor bipolar.

Bearing The supporting part that reduces friction between a stationary and rotating part or between two moving parts.

Cojinete La parte portadora que reduce la fricción entre una parte fija y una parte giratoria o entre dos partes que muevan.

Bearing cage A spacer that keeps the balls or rollers in a bearing in proper position between the inner and outer races.

Jaula del cojinete Un espaciador que mantiene a las bolas o a los rodillos del cojinete en la posición correcta entre las pistas interiores e exteriores.

Bearing caps In the differential, caps held in place by bolts or nuts which, in turn, hold bearings in place.

Tapones del cojinete En un diferencial, las tapas que se sujeten en su lugar por pernos o tuercas, los cuales en su turno, retienen y posicionan a los cojinetes.

Bearing cone The inner race, rollers, and cage assembly of a tapered roller bearing. Cones and cups must always be replaced in matched sets.

Cono del cojinete La asamblea de la pista interior, los rodillos, y el jaula de un cojinete de rodillos cónico. Se debe siempre reemplazar a ambos partes de un par de conos del cojinete y los anillos exteriores a la vez.

Bearing cup The outer race of a tapered roller bearing or ball bearing.

Anillo exterior La pista exterior de un cojinete cónico de rodillas o de bolas.

Bearing race The surface on which the rollers or balls of a bearing rotate. The outer race is the same thing as the cup, and the inner race is the one closest to the axle shaft.

Pista del cojinete La superficie sobre la cual rueden los rodillos o las bolas de un cojinete. La pista exterior es lo mismo que un anillo exterior, y la pista interior es la más cercana a la flecha del eje.

Belleville spring A tempered spring steel cone-shaped plate used to aid the mechanical force in a pressure plate assembly.

Resorte de tensión Belleville Un plato de resorte del acero revenido en forma cónica que aumenta a la fuerza mecánica de una asamblea del plato opresor.

Bell housing A housing that fits over the clutch components and connects the engine and the transmission.

Concha del embrague Un cárter que encaja a los componentes del embrague y conecta al motor con la transmisión.

Belt Alternator Starter (BAS) A combined assembly of a generator/motor that is driven by a belt and serves as the starter motor and an alternator.

Cinturón Alternador Motor de arranque (BAS) Un combinado asamblea de un generador/motor o sea unidad por una cinturón y servir como el motor de arranque y un alternador.

Bevel gear A form of spur gear that has its teeth cut at an angle.

Egranaje biselado Una forma de engranaje de estímulo que tiene sus dientes cortados en ángulo.

Bolt torque The turning effort required to offset resistance as a bolt is being tightened.

Torsión del perno El esfuerzo de torsión que se requiere para compensar la resistencia del perno mientras que esté siendo apretado.

Brake horsepower (bhp) Power delivered by the engine and available for driving the vehicle; bhp = torque × rpm/5252.

Caballo indicado al freno (bhp) Potencia que provee el motor y que es disponible para el uso del vehículo; bhp = de par motor × rpm/5252.

Bronze An alloy of copper and tin.

Bronce Una aleación de cobre y hojalata.

Burnish To smooth or polish by the use of a sliding tool under pressure.

Bruñir Pulir o suavizar por medio de una herramienta deslizando bajo presión.

Burr A feather edge of metal left on a part being cut with a file or other cutting tool.

Rebaba Una lima espada de metal que permanece en una parte que ha sido cortado con una lima u otro herramienta de cortar.

Bushing A cylindrical lining used as a bearing assembly; can be made of steel, brass, bronze, nylon, or plastic.

Buje Un forro cilíndrico que se usa como una asamblea de cojinete que puede ser hecho del acero, del latón, del bronce, del nylon, o del plástico.

Butt-end locking ring A locking ring whose ends are cut to butt up against or contact each other once in place. There is no gap between the ends of a butt-end ring.

Anillo de enclavamiento a tope Un anillo de enclavamiento cuyos extremidades son cortadas para toparse o ajustarse una contra la otra en lugar. No hay holgura entre las extremidades de un anillo de retén a tope.

C-clip A C-shaped clip used to retain the drive axles in some rear axle assemblies.

Grapa de C Una grapa en forma de C que retiene a las flechas motrices en algunas asambleas de ejes traseras.

Cage A spacer used to keep the balls or rollers in proper relation to one another. In a constant-velocity joint, the cage is an open metal framework that surrounds the balls to hold them in position.

Jaula Una espaciador que mantiene una relación correcta entre los rodillos o las bolas. En una junta de velocidad constante, la jaula es un armazón abierto de metal que rodea a las bolas para mantenerlas en posición.

CAN (Controller Area Network) Bus A commonly used multiplexing protocol for serial communication. The communication wire is a twisted-pair wire.

Bus CAN (Controller Area Network) Protocolo de multiplexión comúnmente usado para comunicaciones en serie. El cable de comunicación es un par trenzado.

Cantilever A projecting lever or beam that is supported on only one end.

Voladizo Una palanca o una viga de proyección que son supporte en solamente un extremo.

Cardan Universal Joint A nonconstant velocity universal joint consisting of two yokes with their forked ends joined by a cross. The driven yoke changes speed twice in 360 degrees of rotation.

Junta Universal Cardan Una junta universal de velocidad no constante que consiste de dos yugos cuyos extremidades ahorquilladas se unen en cruz. El yugo de arrastre cambia su velocidad dos veces en 360 grados de rotación.

Case harden To harden the surface of steel. The carburizing method used on low-carbon steel and other alloys to make the case or outer layer of the metal harder than its core.

Cementar Endurecer la superficie del acero. El método de carburación que se emplea en el acero de bajo carbono o en otros aleaciones para que el cárter o capa exterior queda más dura que lo que esta al interior.

Catalyst A compound or substance that can speed up or slow down the reaction of other substances without being consumed itself. In an automotive catalytic converter, special metals (e.g., platinum or palladium) are used to promote combustion of unburned hydrocarbons and a reduction of carbon monoxide while the gases are passing through the exhaust.

Catalizador Un compuesto o sustancia que puede acelerar o lentificar la reacción de otras sustancias sin consumirse. En un convertidor catalítico automotriz, se utilizan metales especiales (por ej., platino o paladio) para promover la combustión de los hidrocarburos que no se quemaron y una reducción de monóxido de carbono mientras los gases pasan por el escape.

Caustic Something that has the ability to eat away at something through chemical action.

Cáustico Una sustancia cáustica que tiene la capacidad de corroer, de quemarse o de disolver algo con la acción química.

CCD The common point or bus for Chrysler's multiplexing system; the acronym for "Chrysler Collision Detection" bus.

CCD Un conector común, o bus, del sistema de la multiplexación de Chrysler, llamado "Chrysler Collision Detection" bus (bus para detección de la colisión).

Centrifugal clutch A clutch that uses centrifugal force to apply a higher force against the friction disc as the clutch spins faster.

Embrague centrífugo Un embrague que emplea a la fuerza centrífuga para aplicar una fuerza mayor contra el disco de fricción mientras que el embrague gira más rapidamente.

Centrifugal force The force acting on a rotating body that tends to move it outward and away from the center of rotation. The force increases as rotational speed increases.

Fuerza centrífuga La fuerza que afecta a un cuerpo en rotación moviendolo hacia afuera y alejándolo del centro de rotación. La fuerza aumenta al aumentar la velocidad de rotación.

Chassis The vehicle frame, suspension, and running gear. On FWD cars, it includes the control arms, struts, springs, trailing arms, sway bars, shocks, steering knuckles, and frame. The driveshafts, constant-velocity joints, and transaxle are not part of the chassis or suspension.

Chasis El armazón de un vehículo, la suspensión, y el engranaje de marcha. En los coches de FWD, incluye los brazos de mando, los postes, los resortes (chapas), los brazos traseros, las estabilizadoras, las articulaciones de la dirección y el armazón. Los árboles de mando, las juntas de velocidad constante, y la flecha impulsora no son partes del chasis ni de la suspensión.

Check-ball valves A valve that uses a check ball to open and close a fluid circuit.

Valvulas con bola de control Las válvulas que utilizan una bola del cheque para abrirse y cierran un circuito flúido.

Circlip A split steel snap ring that fits into a groove to hold various parts in place. Circlips are often used on the ends of FWD driveshafts to retain the constant-velocity joints.

Grapa circular Un seguro partido circular de acero que se coloca en una ranura para posicionar a varias partes. Las grapas circulares se suelen usar en las extremidades de los árboles de mando en FWD para retener las juntas de velocidad constante.

Clearance The space allowed between two parts, such as between a journal and a bearing.

Holgura El espacio permitido entre dos partes, tal como entre un muñon y un cojinete.

Clutch A device for connecting and disconnecting the engine from the transmission or for a similar purpose in other units.

Embrague Un dispositivo para conectar y desconectar el motor de la transmisión o para tal propósito en otros conjuntos.

Clutch packs A series of clutch discs and plates installed alternately in a housing to act as a driving or driven unit.

Conjuntos de embrague Una seria de discos y platos de embrague que se han instalado alternativamente en un cárter para funcionar como una unedad de propulsión o arrastre.

Clutch slippage A situation in which engine speed increases but increased torque is not transferred through to the driving wheels.

Resbalado del embrague Una situación en el qual velocidad del motor aumenta pero la torsión aumentada del motor no se transfere a las ruedas de marcha por el resbalado del embrague.

Clutch volume index (CVI) This value represents the physical amount of fluid and time that is required to fill a transmission's clutch assembly and stroke the piston. The index is used to determine the amount of extra fluid that should be sent to a clutch to compensate for wear.

Indicador de volumen del embrague Valor que representa la cantidad física de fluido y tiempo que se necesita para llenar el ensamblaje del embrague de una transmisión y propulsar el pistón. El indicador se usa para determinar la cantidad de fluido extra que debería enviarse al embrague para compensar por el desgaste.

Coast A condition of deceleration when the engine is slowing down the vehicle.

Deceleración Una condición de la velocidad que disminuye cuando el motor está retrasando el vehículo.

Coefficient of friction The ratio of the force resisting motion between two surfaces in contact to the force holding the two surfaces in contact.

Coeficiente de la fricción La relación entre la fuerza que resiste al movimiento entre dos superficies que tocan y la fuerza que mantiene en contacto a éstas dos superficies.

Coil spring A heavy wire-like steel coil used to support the vehicle weight while allowing for suspension motions. On FWD cars, the front coil springs are mounted around the MacPherson struts. On the rear suspension, they may be mounted to the rear axle, to trailing arms, or around rear struts.

Muelles de embrague Un resorte espiral hecho de acero en forma de alambre grueso que soporte el peso del vehículo mientras que permite a los movimientos de la suspensión. En los coches de FWD, los muelles de embrague delanteros se montan alrededor de los postes Macpherson. En la suspensión trasera, pueden montarse en el eje trasero, en los brasos traseros, o alrededor de los postes traseros.

Combustion Rapid oxidation, with the release of energy in the form of heat and light.

Combustión Oxidación rápida, con la liberación de energía en forma de calor y luz.

Compound A mixture of two or more ingredients.

Compuesto Una combinación de dos ingredientes o más.

Computer An electronic device capable of following instructions to automatically alter the operation of other components.

Odenador o Computadora Un ordenador es un dispositivo electrónico que puede seguir instrucciones de alterar automáticamente la operación de otros componentes.

Concentric Two or more circles having a common center.

Concéntrico Dos círculos o más que comparten un centro común.

Conductor A material that allows electrical current to easily flow through it.

Conductor Un material que permite que la corriente eléctrica fluya fácilmente.

Constant-velocity joint A flexible coupling between two shafts that permits each shaft to maintain the same driving or driven speed regardless of operating angle, allowing for a smooth transfer of power. The constant-velocity joint (also called CV joint) consists of an inner and outer housing with balls in between, or a tripod and yoke assembly.

Junta de velocidad constante Un acoplador flexible entre dos flechas que permite que cada flecha mantenga la velocidad de propulsión o arrastre sin importar el ángulo de operación, efectuando una transferencia lisa del poder. La junta de velociadad constante (también llamado junta CV) consiste en un cárter interior e exterior entre los cuales se encuentran bolas, o de un conjunto de trípode y yugo.

Continuously variable transmission (CVT) Transmission that automatically changes torque and speed ranges without requiring a change in engine speed. A CVT is a transmission without fixed forward speeds.

Transmisión continua variable (CVT) Transmisión que cambia automáticamente el rango de la velocidad y el par de torsión sin necesidad de un cambio en las revoluciones del motor. Es una transmisión sin velocidad fija hacia delante.

Contraction A reduction in mass or dimension; the opposite of expansion.

Contración Una reducción en la masa o en la dimensión; el opuesto de expansión.

Controlled-load servo A type of servo that has two pistons and allows for the quick release of a band during shifting.

Servo de carga controlada Un tipo de servo que tiene dos pistones y permiten que desembraga rápidamente una banda al cambiar de velocidad.

Converter capacity An expression of a torque converter's ability to absorb and transmit engine torque with a limited amount of slippage.

Capacidad de convertidor Una expresión de la capacidad del convertidor de la torsión de absorber y de transmitir el esfuerzo de torsión del motor con una cantidad limitada de resbalamiento.

Converter clutch regulator valve A hydraulic valve that controls the operation of a torque converter clutch by regulating the amount of hydraulic pressure that is applied to the back (on) side of the torque converter clutch.

Válvula de regulación del embrague de convertidor par Válvula hidráulica que controla la operación del embrague de convertidor par al regular la cantidad de presión hidráulica que se aplica en el lado trasero (encendido) del embrague de convertidor par.

Converter clutch switch valve A hydraulic valve that controls the operation of a torque converter clutch by regulating the amount of hydraulic pressure that is applied to the front (off) side of the converter clutch.

Válvula de interrupción del embrague de convertidor par Válvula hidráulica que controla la operación del embrague de convertidor par al regular la cantidad de presión hidráulica que se aplica en el lado de adelante (apagado) del embrague de convertidor par.

Corrode To eat away gradually as if by gnawing, especially by chemical action.

Corroer Roído poco a poco, primariamente por acción químico.

Corrosion Chemical action, usually by an acid, that eats away (decomposes) a metal.

Corrosión Un acción químico, por lo regular un ácido, que corroe (descompone) un metal.

Corrosivity The characteristic of a material that enables it to dissolve metals and other materials or burn the skin.

La corrosividad La característica de un material que le permita disolver los metales y otros materiales o quemarse la piel.

Counterclockwise rotation Rotating in the opposite direction of the hands on a clock.

Rotación en sentido inverso Girando en el sentido opuesto de las agujas de un reloj.

Coupling A connecting means for transferring movement from one part to another; may be mechanical, hydraulic, or electrical.

Acoplador Un método de conección que transfere el movimiento de una parte a otra; puede ser mecánico, hidráulico, o eléctrico.

Coupling phase Point in torque converter operation in which the turbine speed is 90 percent of impeller speed and there is no longer any torque multiplication.

Fase del acoplador El punto de la operación del convertidor de la torsión en el cual la velocidad de la turbina es el 90 percent de la velocidad del impulsor y no queda ningún multiplicación de la torsión.

Crescent A half-moon–shaped part that isolates the inlet side of a fluid pump from the outlet side of the pump.

Creasciente Una parte formada medialuna que aísla la cara de la entrada de una bomba flúida de la cara de enchufe de la bomba.

Curb weight The weight of the vehicle when it is not loaded with passengers or cargo.

Peso en vacío El peso de un vehículo sin incluir pasajeros ni carga.

Current The flow of electrons through a conductor.

Corriente El flujo de electrones a través de un conductor.

Damper A device used to reduce or eliminate vibration.

Compuerta de tiro Un dispositivo que reduce o elimina vibraciones.

Deceleration The rate of decrease in speed.

Desaceleración Índice de disminución de la velocidad.

Deflection Bending or movement away from normal due to loading.

Desviación Curvación o movimiento fuera de lo normal debido a la carga.

Degree A unit of measurement equal to 1/360th of a circle.

Grado Una uneda de medida que iguala al 1/360 parte de un círculo.

Density Compactness; relative mass of matter in a given volume.

Densidad La firmeza; una cantidad relativa de la materia que ocupa a un volumen dado.

Detent A small depression in a shaft, rail, or rod into which a pawl or ball drops when the shaft, rail, or rod is moved. This provides a locking effect.

Detención Un pequeño hueco en una flecha, una barra o una varilla en el cual cae una bola o un linguete al moverse la flecha, la barra o la varilla. Esto provee un efecto de enclavamiento.

Detent mechanism A shifting control designed to hold the transmission in the gear range selected.

Aparato de detención Un control de desplazamiento diseñado a sujetar a la transmisión manual en la velocidad selecionada.

Diagnosis A systematic study of a machine or machine parts to determine the cause of improper performance or failure.

Diagnóstico Un estudio sistemático de una máquina o las partes de una máquina con el fín de determinar la causa de una falla o de un operación irregular.

Differential A mechanism between drive axles that permits one wheel to run at a different speed than the other while turning.

Diferencial Un mecanismo entre dos semiejes que permite que una rueda gira a una velocidad distincta que la otra en una curva.

Differential action An operational situation in which one driving wheel rotates at a slower speed than the opposite driving wheel.

Acción del diferencial Una situación durante la operación en la cual una rueda propulsora gira con una velocidad más lenta que la rueda propulsora opuesta.

Differential case The metal unit that encases the differential side gears and pinion gears, and to which the ring gear is attached.

Caja de satélites La unidad metálica que encaja a los engranajes planetarios (laterales) y a los satélites del diferencial, y a la cual se conecta la corona.

Differential drive gear A large circular helical gear that is driven by the transaxle pinion gear and shaft and drives the differential assembly.

Corona Un engranaje helicoidal grande circular que es arrastrado por el piñon de la flecha de transmisión y la flecha y propela al conjunto del diferencial.

Differential housing Cast iron assembly that houses the differential unit and the drive axles. Also called the rear axle housing.

Cárter del diferencial Una asamblea de acero vaciado que encaja a la unedad del diferencial y los semiejes. También se llama el cárter del eje trasero.

Differential pinion gears Small beveled gears located on the differential pinion shaft.

Satélites Engranajes pequeños biselados que se ubican en la flecha del piñon del diferencial.

Differential pinion shaft A short shaft locked to the differential case that supports the differential pinion gears.

Flecha del piñon del diferencial Una flecha corta clavada en la caja de satélites que sostiene a los satélites.

Differential ring gear A large circular hypoid-type gear enmeshed with the hypoid drive pinion gear.

Corona Un engranaje helicoidal grande circular endentado con el piñon de ataque hipoide.

Differential side gears The gears inside the differential case that are internally splined to the axle shafts and are driven by the differential pinion gears.

Planetarios (laterales) Los engranajes adentro de la caja de satélites que son acanalados a los semiejes desde el interior, y que se arrastran por los satélites.

Digital signal A voltage signal that is a series of on/off pulses.

Señal digital Una serie de señales del voltaje, en o apagado.

Diode A semiconductor that allows current to flow through it in one direction only.

Diodo Un semiconductor que permite que la corriente eléctrica fluya en una dirección.

Dipstick A metal rod used to measure the fluid in an engine or transmission.

Varilla de medida Una varilla de metal que se usa para medir el nivel de flúido en un motor o en una transmisión.

Direct drive System in which one turn of the input driving member corresponds to one complete turn of the driven member, such as when there is direct engagement between the engine and driveshaft in which the engine crankshaft and the driveshaft turn at the same rpm.

Mando directo Una vuelta del miembro de ataque o propulsión que se compara a una vuelta completa del miembro de arrastre, tal como cuando hay un enganchamiento directo entre el motor y el árbol de transmisión en el qual el cigueñal y el árbol de transmisión giran al mismo rpm.

Disengage A process in which a component is released or unattached from other components, such as the release of a clutch.

Desengranar Proceso en el cual un componente queda libre de otros, como al soltar el embrague.

Double-wrap band A transmission band that is split with overlapping ends.

Banda con dos envolvederos Una banda de la transmisión que está partida, con extremos que solapan.

Dowel A metal pin attached to one object that, when inserted into a hole in another object, ensures proper alignment.

Espiga Una clavija de metal que se fija a un objeto, que al insertarla en el hoyo de otro objeto, asegura una alineación correcta.

Downshift To shift a transmission into a lower gear.

Cambio descendente Cambiar la velocidad de una transmision a una velocidad más baja.

DRAC Digital Radio Adapter Controller, a device used to change digital signals to analog signals in computer control systems.

DRAC Regulador del adaptador de radio digital, un dispositivo que cambia señales digital a las señales analogicas en sistemas de control de ordenador.

Driveline The universal joints, driveshaft, and other parts connecting the transmission with the driving axles.

Flecha motríz Las juntas universales, el árbol de mando, u otras partes que conectan a la transmisión con los ejes impulsores.

Driveline torque Relates to rear-wheel driveline; the transfer of torque between the transmission and the driving axle assembly.

Potencia de la flecha motríz Se relaciona a la flecha motríz de las ruedas traseras y transfere la potencia de la torsión entre la transmisión y el conjunto del eje trasero.

Driven gear The gear meshed directly with the driving gear to provide torque multiplication, reduction, or a change of direction.

Engranaje de arrastre El engranaje endentado directamente al engranaje de ataque para proporcionar la multiplicación, la reducción, o los cambios de dirección de la potencia.

Drive pinion gear One of the two main driving gears located within the transaxle or rear driving axle housing. Together the two gears multiply engine torque.

Engranaje de piñon de ataque Uno de dos engranajes de ataque principales que se ubican adentro de la flecha de transmisión o en el cárter del eje de propulsión. Los dos engranajes trabajan juntos para multiplicar la potencia.

Driveshaft An assembly of one or two universal joints connected to a shaft or tube; used to transmit power from the transmission to the differential. Also called the propeller shaft.

Árbol de mando Una asamblea de uno o dos uniones universales que se conectan a un árbol o un tubo; se usa para transferir la potencia desde la transmisión al diferencial. También se le refiere como el árbol de propulsión.

Drum A cylinder-shaped device that is attached to or houses components. A friction unit is normally used to stop and hold a rotating drum.

Tambor Un dispositivo que tenga la forma del cilindro o tambor y se asocia a los componentes, o contiene los componentes. Un ensamblaje que utiliza la fricción esta utilizado para parar y sostener un tambor que gira.

Dry friction The friction between two dry solids.

Fricción seca Fricción entre dos sólidos secos.

DTC The acronym for diagnostic trouble code.

DTC La sigla en inglés por un código diagnóstico de averías.

Dual band Another name for a double-wrap band.

Banda dual Otro nombre para banda con dos envolvederos.

Dual-shaft transmission The name given to early helical gear, constant mesh automatic transmissions. These units have a main shaft and a countershaft.

Transmisión con eje-dual El nombre dado a la transmisión automática que tiene un engranaje helicoidal temprano, endentado o engranado constantemente. Estas transmisiones tienen un eje principal y un contraeje.

Dual-stage oil pump Transmission oil pump with two gears driven by the torque converter hub through a central drive gear. At low speeds, both gears supply fluid to the transmission and at higher speeds only the primary gear supplies fluid under pressure. This arrangement allows the pump to deliver the pressure and fluid flow of a large displacement pump at low engine speeds, and the economy of a small displacement pump at higher engine speeds.

Bomba de aceite de doble fase Bomba de aceite con dos engranajes impulsados por el cubo del convertidor par mediante el mecanismo central de transmisión. A velocidades bajas, ambos engranajes proporcionan fluido a la transmisión, y a velocidades más altas sólo el engranaje principal proporciona fluido bajo presión. Este sistema permite que la bomba proporcione la presión y el flujo de una bomba de desplazamiento grande a velocidades bajas, y el ahorro de una bomba de desplazamiento pequeña a velocidades más altas.

Dynamic In motion.

Dinámico En movimiento.

Dynamic balance The balance of an object when it is in motion; for example, the dynamic balance of a rotating driveshaft.

Balance dinámico El balance de un objeto mientras que esté en movimiento: por ejemplo el balance dinámico de un árbol de mando giratorio.

Dynamic pressure Pressure that changes and/or causes something to move when pressurized fluid flow is present.

Presión dinámica La presión que cambia o la presión que causa algo moverse cuando el flujo presurizado del líquido está presente.

Dynamic seal A seal used between two parts that move in relation to each other.

Sello dinámico Un sello que se utiliza entre dos partes que tiene el movimiento relativo el uno al otro.

Eccentric One circle within another circle wherein both circles do not have the same center; a circle mounted off center. On FWD cars, front-end camber adjustments are accomplished by turning an eccentric cam bolt that mounts the strut to the steering knuckle.

Excéntrico Se dice de dos círculos, el uno dentro del otro, que no comparten el mismo centro o de un círculo ubicado descentrado. En los coches FWD, los ajustes de la inclinación se efectuan por medio de un perno excéntrico que fija el poste sobre el articulación de dirección.

ECT Electronically controlled transmission.

ECT Una transmisión que se controle electrónicamente.

Efficiency The ratio between the power of an effect and the power expended to produce the effect; the ratio between an actual result and the theoretically possible result.

Eficiencia La relación entre la potencia de un efecto y la potencia que se gasta para producir el efecto; la relación entre un resultado actual y el resultado que es una posibilidad teórica.

Elastomer Any rubber-like plastic or synthetic material used to make bellows, bushings, and seals.

Elastómero Cualquiera materia plást parecida al hule o una materia sintética que se utiliza para fabricar a los fuelles, los bujes y las juntas.

Electricity A useable force caused by the directed movement of electrons.

Electricidad Una fuerza usable que es causada por el movimiento dirigido de electrones.

Electrolysis A chemical and electrical decomposition process that can damage metals such as brass, copper, and aluminum in the cooling system.

Electrólisis Un proceso de descomposición química y eléctrica que puede dañar los metales como el bronce, el cobre y el aluminio en el sistema de refrigeración.

Electrolyte A material whose atoms become ionized, or electrically charged, in solution. Automobile battery electrolyte is a mixture of sulfuric acid and water.

Electrólisis Un material cuyo átomos ponerse ionizar, o eléctricamente cargado, en solución. Automóvil batería electrólisis es un mezcla de ácido sulfúrico y agua.

Electromagnetism A principle of using electricity to align the electrons in some material to give it magnetic properties.

Electromagnetismo Un principio de usar electricidad para alinear los electrones en un material para darle características magnéticas.

Electronic controlled transaxle (ECT) An automatic transmission whose hydraulic activity is controlled by a computer or an electronic control unit.

Eje de la transmisión controlado electrónicamente (ECT) Una transmisión automática cuya actividad hidráulica es controlada por una computadora o por una unidad de control electrónica.

Element A substance containing only one type of atom.

Elemento Sustancia que contiene sólo un tipo de átomo.

EMCC Electronic Modulated Converter Clutch.

EMCC Un embrague del convertidor que se modula electrónicamente.

Engage A process in which a component is coupled or attached to other components, such as the locking of a clutch.

Engranar Proceso en el cual un componente se acopla o se une a otros componentes, como al accionar el embrague.

Engagement chatter A shaking, shuddering action that takes place as the driven disc makes contact with the driving members. Chatter is caused by a rapid grip and slip action.

Chasquido de enganchamiento Un movimiento de sacudo o temblor que resulta cuando el disco de ataque viene en contacto con los miembros de propulsión. El chasquido se causa por una acción rápida de agarrar y deslizar.

Engine The source of power for most vehicles. It converts burned fuel energy into mechanical force.

Motor El orígen de la potencia para la mayoría de los vehículos. Convierte la energía del combustible consumido a la fuerza mecánica.

Engine load The physical resistance placed on an engine's crankshaft.

Carga del motor La resistencia física puesta en el cigüeñal de un motor.

Engine torque A turning or twisting action developed by the engine, measured in foot-pounds or kilogram meters.

Torsión del motor Una acción de girar o torcer que crea el motor, ésta se mide en libras-pie o kilos-metros.

Engine vacuum The low pressure formed by an engine's pistons.

Vacío del motor La presión baja formada por los pistones de un motor.

EPC solenoid Electronic Pressure Control solenoid, used to control and maintain mainline pressure in some transmissions.

EPC Solenoide Un solenoide electrónico del control de presión se utiliza para controlar y para mantener la presión de la línea principal en algunas transmisiones.

Equilibrium Condition that exists when the applied forces on an object are balanced and there is no overall resultant force.

Equilibrio Estado en el que las fuerzas que se aplican sobre un objeto están balanceadas y no existe una fuerza resultante total.

Evaporate Atoms or molecules break free from the body of the liquid to become gas particles.

Evaporarse Átomos o moléculas se desprenden del cuerpo del líquido para volverse partículas de gas.

Expansion An increase in size. For example, when a metal rod is heated, it increases in length and perhaps also in diameter; the opposite of contraction.

Dilatación Aumento de tamaño. Por ejemplo, cuando se calienta una barra de metal, aumenta su longitud y quizá también su diámetro. Es lo opuesto de contracción.

Extension housing An aluminum or iron casting of various lengths that encloses the transmission output shaft and supporting bearings.

Cubierta de extensión Una pieza moldeada de aluminio o acero que puede ser de varias longitudes que encierre a la flecha de salida de la transmisión y a los cojinetes de soporte.

External gear A gear with teeth across the outside surface.

Engranaje exterior Un engranaje cuyos dientes estan en la superficie exterior.

Externally tabbed clutch plates Clutch plates that are designed with tabs around the outside periphery to fit into grooves in a housing or drum.

Placas de embrague de orejas externas Las placas de embrague que se diseñan de un modo para que las orejas periféricas de la superficie se acomoden en una ranura alrededor de un cárter o un tambor.

Face The front surface of an object.

Cara La superficie delantera de un objeto.

Fail-safe valve The valve found in some hydraulic circuits that allows limited transmission operation when a component or components have failed.

Válvula con capacidad de compensar errores automáticamente Una válvula que permite la operación limitada de la transmisión cuando un componente o componentes ha fallado.

Final drive ratio The ratio between the drive pinion and ring gear.

Relación del mando final La relación entre el piñon de ataque y la corona.

Fit The contact between two machined surfaces.

Ajuste El contacto entre dos superficies maquinadas.

Fixed displacement pump Fluid pumps that maintain a particular amount of fluid flow.

Bomba de desplazamiento fijo Una bomba que mantiene un nivel fijo del líquido.

Flammable Something that supports combustion. The flammability of a substance is a statement of how well the substance supports combustion.

Inflamable Inflamable describe una sustancia que soporta la combustión. La inflamabilidad de una sustancia es una declaración de cómo está bien la sustancia utiliza la combustión.

Flange A projecting rim or collar on an object that keeps it in place.

Reborde Una orilla o un collar sobresaliente de un objeto cuyo función es de mantenerlo en lugar.

Flexplate A lightweight flywheel used only on engines equipped with an automatic transmission. A flexplate is equipped with a starter ring gear around its outside diameter and also serves as the attachment point for the torque converter.

Placa articulada Un volante ligera que se usa solamente en los motores que se equipan con una transmisión automática. El diámetro exterior de la placa articulada viene equipado con un anillo de engranajes para arrancar y también sirve como punto de conección del convertidor de la torsión.

Flow-directing valves Valves that direct the pressurized fluid to the appropriate supply device to cause a change in gear ratios.

Válvulas que dirigen el líquido Valves que ordenan el líquido presurizado al dispositivo apropiado para causar un cambio en relaciones de transformación del engranaje.

Fluid coupling A device in the powertrain consisting of two rotating members; transmits power from the engine, through a fluid, to the transmission.

Acoplamiento de fluido Un dispositivo en el tren de potencia que consiste de dos miembros rotativos; transmite la potencia del motor, por medio de un fluido, a la transmisión.

Fluid drive A drive in which there is no mechanical connection between the input and output shafts, and power is transmitted by moving oil.

Dirección fluido Una dirección en la cual no hay conecciones mecánicas entre las flechas de entrada o salida, y la potencia se transmite por medio del aceite en movimiento.

Flux lines The invisible lines of attraction from the pole of a magnet.

Líneas de flujo magnético Líneas invisibles de la atracción del poste de un imán.

Flywheel A heavy metal wheel that is attached to the crankshaft and rotates with it; helps smooth out the power surges from the engine power strokes; also serves as part of the clutch and engine-cranking system.

Volante Una rueda pesada de metal que se fija al ciguñeal y gira con ésta; nivela a los sacudos que provienen de la carrera de fuerza del motor; también sirve como parte del embrague y del sistema de arranque.

Flywheel ring gear A gear fitted around the flywheel that is engaged by teeth on the starting-motor drive to crank the engine.

Engranaje anular del volante Un engranaje, colocado alrededor del volante que se acciona por los dientes en el propulsor del motor de arranque y arranca al motor.

Foot-pound (ft.-lb.) A measure of the amount of energy or work required to lift 1 pound a distance of 1 foot.

Pie libra Una medida de la cantidad de energía o fuerza que requiere mover una libra a una distancia de un pie.

Force Any push or pull exerted on an object; measured in pounds and ounces, or in newtons (N) in the metric system.

Fuerza Cualquier acción empujado o jalado que se efectua en un objeto; se mide en pies y onzas, o en newtones (N) en el sistema métrico.

Forward clutch A multiple-disc clutch that is normally applied during forward speed ranges.

Embrague de avance Un embrague de discos múltiples que habitualmente se aplica durante los rangos de velocidad de avance.

Four-wheel-drive (4WD) On a vehicle, driving axles at both front and rear, so that all four wheels can be driven.

Tracción a cuatro ruedas En un vehículo, se trata de los ejes de dirección fronteras y traseras, para que cada uno de las ruedas puede impulsar.

Frame The main understructure of the vehicle to which everything else is attached. Most FWD cars have only a subframe for the front suspension and drivetrain. The body serves as the frame for the rear suspension.

Armazón La estructura principal del vehículo al cual todo se conecta. La mayoría de los coches FWD sólo tiene un bastidor auxiliar para la suspensión delantera y el tren de propulsión. El carrocería del coche sirve de chassis par la suspensión trasera.

Freewheel To turn freely and not transmit power.

Volante libre Da vueltas libremente sin transferir la potencia.

Freewheeling clutch A mechanical device that will engage the driving member to impart motion to a driven member in one direction but not the other. Also known as an "overrunning clutch."

Embrague de volante libre Un dispositivo mecánico que acciona el miembro de tracción y da movimiento al miembro de tracción en una dirección pero no en la otra. También se conoce bajo el nombre de un "embrague de sobremarcha."

Frequency The number of complete cycles that occur within a given period of time.

Frecuencia El número de los ciclos completos que ocurren dentro de un período del tiempo dado.

Friction The resistance to motion between two bodies in contact with each other.

Fricción La resistencia al movimiento entre dos cuerpos en contacto el uno con el otro.

Friction bearing A bearing in which there is sliding contact between the moving surfaces. Sleeve bearings, such as those used in connecting rods, are friction bearings.

Rodamientos de fricción Un cojinete en el cual hay un contacto deslizante entre las superficies en movimiento. Los rodamientos de manguitos, como los que se usan en las bielas, son rodamientos de fricción.

Friction disc In the clutch, a flat disc, faced on both sides with frictional material and splined to the clutch shaft. It is positioned between the clutch pressure plate and the engine flywheel. Also called the clutch disc or driven disc.

Disco de fricción En el embrague, un disco plano al cual se ha cubierto ambos lados con una materia de fricción y que ha sido estriado a la flecha del embrague. Se posiciona entre el plato opresor del embrague y el volante del motor. También se llama el disco del embrague o el disco de arrastre.

Friction facings A hard-molded or woven asbestos or paper material that is riveted or bonded to the clutch driven disc.

Superficie de fricción Un recubrimiento remachado o aglomerado al disco de arrastre del embrague que puede ser hecho del amianto moldeado o tejido o de una materia de papel.

Front pump Pump located at the front of the transmission. It is driven by the engine through two dogs on the torque converter housing. It supplies fluid whenever the engine is running.

Bomba delantera Una bomba ubicado en la parte delantera de la transmisión. Se arrastre por el motor al través de dos álabes en el cárter del convertidor de la torsión. Provee el fluido mientras que funciona el motor.

Front-wheel-drive (FWD) Describes a vehicle with all drivetrain components located at the front.

Tracción de las ruedas delanteras (FWD) El vehículo tiene todos los componentes del tren de propulsión en la parte delantera.

Fulcrum The point at which a lever pivots.

Fulcro El punto en la cual una palanca gira.

FWD Abbreviation for front-wheel-drive.

FWD Abreviación de tracción de las ruedas delanteras.

Gasket A layer of material, usually made of cork, paper, plastic, composition, or metal, or a combination of these, placed between two parts to make a tight seal.

Empaque Una capa de una materia, normalmente hecho del corcho, del papel, del plástico, de la materia compuesta o del metal, o de cualquier combinación de éstos, que se coloca entre dos partes para formar un sello impermeable.

Gasket cement A liquid adhesive material or sealer used to install gaskets.

Mastique para empaques Una substancia líquida adhesiva, o una substancia impermeable, que se usa para instalar a los empaques.

Gear A wheel with external or internal teeth that serves to transmit or change motion.

Engranaje Una rueda que tiene dientes interiores o exteriores que sirve para transferir o cambiar el movimiento.

Gear lubricant A type of grease or oil blended especially to lubricate gears.

Lubricante para engranaje Un tipo de grasa o aceite que ha sido mezclado específicamente para la lubricación de los engranajes.

Gear ratio The number of revolutions of a driving gear required to turn a driven gear through one complete revolution. For a pair of gears, the ratio is found by dividing the number of teeth on the driven gear by the number of teeth on the driving gear.

Relación de los engranajes El número de las revoluciones requeridas del engranaje de propulsión para dar una vuelta completa al engranaje arrastrado. En una pareja de engranajes, la relación se calcula al dividir el número de los dientes en el engranaje de arrastre por el número de los dientes en el engranaje de propulsión.

Gear reduction When a small gear drives a large gear, there is an output speed reduction and a torque increase that result in a gear reduction.

Velocidad descendente Cuando un engranaje pequeño impulsa a un engranaje grande, hay una reducción en la velocidad de salida y un incremento en la torsión que resultan en una cambio descendente de los velocidades.

Gearshift A linkage-type mechanism by which the gears in an automobile transmission are engaged and disengaged.

Varillaje de cambios Un mecanismo tipo eslabón que acciona y desembraga a los engranajes de la transmisión.

Gear-type pump A fixed displacement pump that consists of two gears in mesh.

Bomba con engranaje Una bomba de desplazamiento fijo que consiste de dos engranajes en acoplamiento.

Gear whine A high-pitched sound developed by some types of meshing gears.

Ruido del engranaje Un sonido agudo que proviene de algunos tipos de engranajes endentados.

Governor pressure The transmission's hydraulic pressure; directly related to output shaft speed. It is used to control shift points.

Regulador de presión La presión hidráulica de una transmisión se relaciona directamente a la velocidad de la flecha de salida. Se usa para controlar los puntos de cambios de velocidad.

Governor valve A device used to sense vehicle speed. The governor valve is attached to the output shaft.

Válvula reguladora Un dispositivo que se usa para determinar la velocidad de un vehículo. La válvula reguladora se monta en la flecha de salida.

Gross weight The total weight of a vehicle plus its maximum rated payload.

Peso bruto El peso total de un vehículo más su carga útil nominal máxima.

Guide rings Rings built into a torque converter to direct the vortex flow and provide for smooth and turbulence-free fluid flow.

Anillos que dirigen Anillos que se construyen en un convertidor del esfuerzo de torsión para dirigir el flujo del vórtice y para prever el flujo flúido que es liso y sin turbulencia.

Heat A form of energy caused by the movement of atoms and molecules.

Calor Forma de energía causada por el movimiento de átomos y moléculas.

Heat treatment Heating, followed by fast cooling, to harden metal.

Tratamiento térmico Calentamiento, seguido por un enfriamiento rápido, para endurecer a un metal.

Helical gear A gear with teeth that are cut at an angle or are spiral to the gear's axis of rotation.

Engranaje helicoidal Un engranaje con los dientes cortados en ángulo o un ángulo que tuerce en espiral del eje de la rotación del engranaje

Herringbone gear A gear cut in a V-shaped pattern.

Engranaje espinapez Un engranaje cortado en forma de V.

Hertz (Hz) The unit that frequency is most often expressed as and is equal to one cycle per second.

Hercio La unidad en la que se suele expresar la frecuencia y que equivale a un ciclo por segundo.

High/reverse clutch A clutch that is engaged in low and reverse gears.

Embrague alto/de retroceso Un embrague que se acciona en marchas bajas y de retroceso.

Hook-end seals A metal sealing ring that has small hooks at each end.

Sellos con ganchos Un anillo de metal que está utilizado para sellar y que tiene ganchos pequeños en cada extremo.

Horsepower A measure of mechanical power, or the rate at which work is done. One horsepower equals 33,000 ft.-lbs. (foot-pounds) of work per minute. It is the power necessary to raise 33,000 pounds a distance of 1 foot in 1 minute.

Caballo de fuerza Una medida de fuerza mecánica, o el régimen en el cual se efectúa el trabajo. Un caballo de fuerza iguala a 33,000 lb.p. (libras pie) de trabajo por minuto. Es la fuerza requerida para transportar a 33,000 libras una distancia de 1 pie en 1 minuto.

Hub The center part of a wheel, to which the wheel is attached.

Cubo La parte central de una rueda, a la cual se monta la rueda.

Hybrid valve A spool valve that relies on spring tension and hydraulic force for movement.

Válvula híbrida Una válvula de carrete que necesita la tensión del resorte y la fuerza hidráulica para el movimiento.

Hybrid electric vehicle (HEV) A vehicle that is equipped with both an electric motor drive system and an internal combustion engine.

Vehículo eléctrico híbrido (HEV) Vehículo equipado a la vez con un sistema de propulsión de batería eléctrica y un motor de combustión interna.

Hydraulic pressure Pressure exerted through the medium of a liquid.

Presión hidráulica La presión esforzada por medio de un líquido.

Hypoid gear A type of spiral, beveled gearset that allows for the meshing to occur at a point other than the centerline of the gears.

Engranaje hipoide Un conjunto de engranajes biselados y en espiral, que permite para que el endentar ocurra en un punto con excepción de la línea central de los engranajes.

Hybrid valve A spool valve that relies on spring tension and hydraulic force for movement.

Valor híbrido Una válvula devanadora que cuenta con la tensión de un resorte y la fuerza hidráulica para lograr movimiento.

ID Inside diameter.

DI Diámetro interior.

Idle Engine speed when the accelerator pedal is fully released and there is no load on the engine.

Marcha lenta La velocidad del motor cuando el pedal accelerador esta completamente desembragada y no hay carga en el motor.

Ignitability The characteristic of a material that enables it to spontaneously ignite.

Capacidad de encender La característica de un material que lo permita encender espontáneamente.

Impeller The pump or driving member in a torque converter.

Impulsor La bomba o el miembro impulsor en un convertidor de torsión.

Impermeable Having the inability to adsorb fluids.

Impermeable Que es incapaz de absorber fluidos.

Increments Series of regular additions from small to large.

Incrementos Una serie de incrementos regulares que va de pequeño a grande.

Index To orient two parts by marking them. During reassembly the parts are arranged so the index marks are next to each other. Used to preserve the orientation between balanced parts.

Índice Orientar a dos partes marcándolas. Al montarlas, las partes se colocan para que las marcas de índice estén alinieadas. Se usan los índices para preservar la orientación de las partes balanceadas.

Inertia Constant moving force; applied to carry the crankshaft from one firing stroke to the next.

Inercia Fuerza de movimiento constante. Se aplica para llevar el cigüeñal de un tiempo de encendido al siguiente.

Input shaft The shaft carrying the driving gear by which the power is applied, as to the transmission.

Flecha de entrada La flecha que porta el engranaje propulsor por el cual se aplica la potencia, como a la transmisión.

Input speed sensor (ISS) A sensor that measures the speed of the turbine shaft as it enters the transmission.

Sensor de velocidad de entrada (ISS) Un sensor que mide la velocidad del árbol de la turbina a medida que entra en la transmisión.

Inspection cover A removable cover that permits entrance for inspection and service work.

Cubierta de inspección Una cubierta desmontable que permite a la entrada para inspeccionar y mantenimiento.

Insulator A material that does not allow current to flow through it easily.

Aislador Un material que no permite el flujo de la corriente eléctrica fácilmente.

Integral Built into, as part of the whole.

Integral Incorporado, una parte de la totalidad.

Integrated Motor Assist (IMA) The motor/generator assembly used in Honda hybrids; fits between the engine and the transmission.

Sistema de propulsión híbrida (IMA) Ensamblaje de motor y generador usado en los vehículos híbridos de Honda. Va instalado entre el motor y la transmisión.

Integrated starter alternator damper (ISAD) Similar to Honda's IMA, the unit replaces the flywheel, generator, and starter motor. It is placed between the engine and transmission in some hybrid vehicles.

Sistema integrado motor de arranque-alternador-reductor (ISAD) Similar al IMA de Honda, esta unidad reemplaza el volante, el gene-rador y el motor de arranque. Va instalado entre el motor y la transmisión en algunos vehículos híbridos.

Intermediate clutch A clutch assigned to the center planetary gearset.

Embrague intermedio Un embrague asignado al conjunto de marchas planetarias centrales.

Intermediate one-way clutch Normally a roller-type one-way clutch that is assigned to the center planetary gearset.

Embrague intermedio unidireccional Por lo general, un embrague unidireccional de tipo rodillo que se asigna al conjunto de marchas planetarias centrales.

Internal gear A gear with teeth pointing inward, toward the hollow center of the gear.

Engranaje internal Un engranaje cuyos dientes apuntan hacia el interior, al hueco central del engranaje.

Interaxle differential Another name for the center differential unit used in some 4WD vehicles.

Diferencial intereje Otro nombre para el ensamblaje diferencial central utilizado en algunos vehículos con tracción a cuatro ruedas.

Isolator springs Springs used in converter clutches to absorb the normal torsional vibrations of the engine.

Resortes del aislador Resortes utilizados en embragues del convertidor para absorber las vibraciones torsionales normales del motor.

Journal A bearing with a hole in it for a shaft.

Manga de flecha Un cojinete que tiene un hoyo para una flecha.

Key A small block inserted between the shaft and hub to prevent circumferential movement.

Chaveta Un tope pequeño que se meta entre la flecha y el cubo para prevenir un movimiento circunferencial.

Keyway A groove or slot cut to permit the insertion of a key.

Ranura de chaveta Un corte de ranura o mortaja que permite insertar una chaveta.

Kickdown Forced downshift.

Patada abajo Cambiar de velocidad a un engranaje más bajo sin usar el embrague.

Kinetic energy Energy in motion.

Energía cinética Energía en movimiento.

Knock A heavy metallic sound usually caused by a loose or worn bearing.

Golpe Un sonido metálico fuerte que suele ser causado por un cojinete suelto o gastado.

Knurl To indent or roughen a finished surface.

Moletear Indentar o desbastar a una superficie acabada.

Lands The raised area on a spool valve.

Tierras El área levantada en una válvula de carrete.

Latent heat The heat required to change a mass's state of matter.

Calor latente Calor necesario para cambiar el estado de la materia de una masa.

Lathe-cut seal Another name for a square-cut seal designed to withstand axial movement.

Sello cortado por torno Un sello cuadrado-cortó diseñado para soportar el movimiento axial.

LED The common acronym for a light-emitting diode.

LED La sigla común en inglés por un diódo emisor de luz.

Lepelletier gears A compound planetary gearset that is based on a simple planetary gearset connected to a Ravigneaux compound gearset.

Tren de engranaje Lepelletier Tren de engranaje planetario compuesto basado en un tren de engranaje planetario sencillo conectado a un tren de engranaje Ravigneaux compuesto.

Lever A device made up of a bar turning about a fixed pivot point, called the *fulcrum*, that uses a force applied at one point to move a mass on the other end of the bar.

Palanca de cambio Dispositivo que consta de una barra que se mueve sobre un pivote fijo, o punto de apoyo, y que usa una fuerza que se aplica sobre un punto para mover una masa en el otro extremo de la barra.

Limited-slip A differential unit that sends some torque to the axle that has resistance.

Diferencial de resbalón limitado Un diferencial que envía algo del esfuerzo de torsión al árbol que tiene resistencia.

Line pressure The hydraulic pressure that operates apply devices and is the source of all other pressures in an automatic transmission. It is developed by pump pressure and regulated by the pressure regulator.

Presión de línea La presión hidráulica que opera dispositivos de aplicación y que es la fuente de todas las demás presiones en una transmisión automática. Se desarrolla por presión de bomba y se regula mediante el regulador de presión.

Linkage Any series of rods, yokes, levers, and so on used to transmit motion from one unit to another.

Biela Cualquiera serie de barras, yugos, palancas, y todo lo demás, que se usa para transferir los movimientos de una unedad a otra.

Lip seals A type of seal used to seal parts that have axial movement.

Sello del labio Un tipo de sello utilizado para sellar las piezas que tienen movimiento axial.

Load The work an engine must do, under which it operates more slowly and less efficiently. The load could be that of driving up a hill or pulling extra weight.

Carga Trabajo que un motor debe hacer, con el cual funciona más lentamente y menos eficazmente. La carga podría ser la de subir una cuesta o tirar de un peso extra.

Load-sensing device A device that causes a change in operation of something in response to a change in engine load.

Dispositivo de Carga-detección Un dispositivo que causa un cambio en la operación algo en respuesta a un cambio en carga del motor.

Lock pin Used in some ball sockets (inner tie-rod end) to keep the connecting nuts from working loose. Also used on some lower ball joints to hold the tapered stud in the steering knuckle.

Clavija de cerrojo Se usan en algunas rótulas (las extremidades interiores de la barra de acoplamiento) para prevenir que se aflojan las tuercas de conexión. También se emplean en algunas juntas esféricas inferiores para retener al perno cónico en la articulación de dirección.

Locking ring A type of sealing ring that has ends that meet or lock together during installation. There is no gap between the ends when the ring is installed.

Anillo de enclavamiento Un tipo de anillo obturador que tiene las extremidades que se tocan o se enclavan durante la instalación. No hay holgura entre las extremidades cuando se ha instalado el anillo.

Lockplates Metal tabs bent around nuts or bolt heads.

Placa de cerrojo Chavetas de metal que se doblan alrededor de las tuercas o las cabezas de los pernos.

Lockwasher A type of washer which, when placed under the head of a bolt or nut, prevents the bolt or nut from working loose.

Arandela de freno Un tipo de arandela que, al colocarse bajo la cabeza de un perno, previene que el perno o la tuerca se aflojan.

Low speed The gearing that produces the highest torque and lowest speed of the wheels.

Velocidad baja La velocidad que produce la torsión más alta y la velocidad más baja a las ruedas.

Lubricant Any material, usually a petroleum product such as grease or oil, that is placed between two moving parts to reduce friction.

Lubricante Cualquier substancia, normalmente un producto de petróleo como la grasa o el aciete, que se coloca entre dos partes en movimiento para reducir la fricción.

Magnetism A force between two poles of opposite potential, caused by the alignment of electrons.

Magnetismo La fuerza entre dos postes de un imán, causados por la alineación de electrones.

Mainline pressure The hydraulic pressure that operates apply devices and is the source of all other pressures in an automatic transmission. It is developed by pump pressure and regulated by the pressure regulator.

Línea de presión La presión hidráulica que opera a los dispositivos de applicación y es el orígen de todas las presiones en la transmisión automática. Proviene de la bomba de presión y es regulada por el regulador de presión.

Main oil pressure regulator valve Regulates the line pressure in a transmission.

Válvula reguladora de la linea de presión Regula la presión en la linea de una transmisión.

Manual control valve A valve used to manually select the operating mode of the transmission. It is moved by the gearshift linkage.

Válvula de control manual Una válvula que se usa para escojer a una velocidad de la transmisión por mano. Se mueva por la biela de velocidades.

MAP sensor A manifold absolute pressure sensor; used to measure the pressure in an area and compare it to the pressure around it.

MAP sensor Un sensor de la presión en el colector del escape para medir la presión en un área y para compararla a la presión alrededor de ella.

Mass The amount of matter in an object.

Masa Cantidad de materia en un objeto.

Matter Anything that occupies space and has mass.

Materia Todo lo que ocupa un espacio.

MCCC Modulated converter clutch control.

MCCC Control Modulado Del Embrague Del Convertidor.

Mechanical advantage Based on the principles governing levers, force can be increased by increasing the length of the lever or by moving the pivot or fulcrum.

Ventaja Mecánica La ventaja mecánica se basa en el prinicple de palancas. La fuerza aumenta cuando la longitud de la palanca aumenta, o cuando se mueve el fulcum.

Meshing The mating or engaging of the teeth of two gears.

Engrane Embragar o endentar a los dientes de dos engranajes.

Micrometer A precision measuring device used to measure small bores, diameters, and thicknesses. Also called a mike.

Micrómetro Un dispositivo de medida precisa que se emplea a medir a los taladros pequeños y a los espesores. También se llama un mike (mayk).

MIL The malfunction indicator lamp for a computer control system. Prior to J1930, the MIL was commonly called a Check Engine or Service Engine Soon lamp.

MIL La lámpara de indicación de averías para un sistema de control de computadora. Antes del J1930, la MIL se llamaba la lámpara de Revise Motor o Servicia el Motor Pronto.

Misalignment A situation in which bearings are not on the same centerline.

Desalineamineto Una situación en el qual los cojinetes no comparten la misma linea central.

Modulator A vacuum diaphragm device connected to a source of engine vacuum. It provides an engine load signal to the transmission.

Modulador Un dispositivo de diafragma de vacío que se conecta a un orígen de vacío en el motor. Provee un señal de carga del motor a la transmisión.

Momentum A type of mechanical energy that is the product of an object's weight times its speed.

Momento Tipo de energía mecánica que es el producto del peso de un objeto por su velocidad.

Mounts Made of rubber to insulate vibrations and noise while they support a powertrain part, such as engine or transmission mounts.

Monturas Hecho de hule para insular a las vibraciones y a los ruidos mientras que sujetan una parte del tren de propulsión, tal como las monturas del motor o las monturas de la transmisión.

Multiple disc A clutch with a number of driving and driven discs as compared to a single plate clutch.

Discos múltiples Un embrague que tiene varios discos de propulsión o de arraste al contraste con un embrague de un sólo plato.

Multiplexing An electrical system in which voltage signals or information can be shared with two or more computers.

Multiplexación Un sistema eléctrico en el cual el voltaje señala, o la información, se puede compartir con dos o más ordenadores.

Needle bearing An antifriction bearing using a great number of long, small-diameter rollers. Also known as a quill bearing.

Rodamiento de agujas Un rodamiento (cojinete) antifricativo que emplea un gran cantidad de rodillos largos y de diámetro muy pequeños.

Needle check valve A valve that utilizes a straight rod to open and close a fluid passage.

Válvula de cheque de la aguja Una válvula que utiliza una barra recta para abrirse y cierra un paso para los líquidos.

Neoprene A synthetic rubber that is not affected by the various chemicals that are harmful to natural rubber.

Neoprene Un hule sintético que no se afecta por los varios productos químicos que pueden dañar al hule natural.

Neutral In a transmission, the setting in which all gears are disengaged and the output shaft is disconnected from the drive wheels.

Neutral En una transmisión, la velocidad en la cual todos los engranajes estan desembragados y el árbol de salida esta desconectada de las ruedas de propulsión.

Neutral-start switch A switch wired into the ignition switch to prevent engine cranking unless the transmission shift lever is in neutral or the clutch pedal is depressed.

Interruptor de arranque en neutral Un interruptor eléctrico instalado en el interruptor de encendido que previene el arranque del motor al menos de que la palanca de cambio de velocidad esté en una posición neutral o que se pisa en el embrague.

Newton-meter (Nm) Metric measurement of torque or twisting force.

Metro newton (Nm) Una medida métrica de la fuerza de torsión.

Nominal shim A shim with a designated thickness.

Laminilla fina Una cuña de un espesor especificado.

Nonhardening A gasket sealer that never hardens.

Sinfragua Un cemento de empaque que no endurece.

Nut A removable fastener used with a bolt to lock pieces together; made by threading a hole through the center of a piece of metal that has been shaped to a standard size.

Tuerca Un retén removable que se usa con un perno o tuerca para unir a dos piezas; se fabrica al filetear un hoyo taladrado en un pedazo de metal que se ha formado a un tamaño especificado.

OD Outside diameter.

DE Diámetro exterior.

Ohm's law The basic law of electricity that defines the behavior of voltage, current, and resistance in an electrical circuit.

La Ley de Ohm La ley del ohmio es una ley básica de la electricidad que define el comportamiento del voltaje, de la corriente, y de la resistencia en un circuito eléctrico.

Oil circuit A schematic drawing showing the paths of fluid flow and the valves necessary to perform a specific function.

Circuito del aceite Un gráfico esquemático que muestra los caminos del flujo flúido y las válvulas que son necesarios realizar una función específica.

Oil seal A seal placed around a rotating shaft or other moving part to prevent leakage of oil.

Empaque de aciete Un empaque que se coloca alrededor de una flecha giratoria para prevenir el goteo de aceite.

One-way clutch See Sprag clutch.

Embrague de una via Vea Sprag clutch.

On/off switch An electrical device that opens and closes an electrical circuit.

Interruptor conectar/desconectar Un dispositivo eléctrico que abre y cierra un circuito eléctrico.

Open An electrical problem that causes an incomplete circuit and zero current flow.

Abierto Un problema eléctrico que causa un circuito incompleto y reduce el flujo de la corriente eléctrica al zeroe.

Open-end seal A seal whose ends do not meet when it is installed.

Sello con los extremos abiertos Un sello, los extremos de el cual, no ensamblan cuando están instalados.

Orifice A small opening designed to restrict flow.

Orofice Una apertura pequeña diseñada para restringir flujo.

O-ring A type of sealing ring, usually made of rubber or a rubber-like material. In use, the O-ring is compressed into a groove to provide the sealing action.

Anillo en O Un tipo de sello anular, suele ser hecho de hule o de una materia parecida al hule. Al usarse, el anillo en O se comprime en una ranura para proveer un sello.

Oscillate To swing back and forth like a pendulum.

Oscilar Moverse alternativamente en dos sentidos contrarios como un péndulo.

OSS An output shaft speed sensor.

OSS Un sensor de velocidad para un eje de salida.

Outer bearing race The outer part of a bearing assembly on which the balls or rollers rotate.

Pista exterior de un cojinete La parte exterior de una asamblea de cojinetes en la cual ruedan las bolas o los rodillos.

Out-of-round Wear of a round hole or shaft which, when viewed from an end, will appear egg-shaped.

Defecto de circularidad Desgaste de un taladro o de una flecha circular, que al verse de una extremidad, tendrá una forma asimétrica, como la de un huevo.

Output shaft The shaft or gear that delivers the power from a device, such as a transmission.

Flecha de salida La flecha o la velocidad que transmite la potencia de un dispositivo, tal como una transmisión.

Output speed sensor (OSS) A sensor that measures the speed of the output shaft before it is transferred to the final drive unit.

Sensor de velocidad de salida (OSS) Un sensor que mide la velocidad del árbol de salida antes de que se transfiera a la unidad de transmisión final.

Overall ratio The product of the transmission gear ratio multiplied by the final drive or rear axle ratio.

Relación global El producto de multiplicar la relación de los engranajes de la transmisión por la relación del impulso final o por la relación del eje trasero.

Over-center spring Another name for a Belleville-type return spring.

Resorte sobre el centro Otro nombre por un resorte de vuelta del Belleville-tipo.

Overdrive Any arrangement of gearing that produces more revolutions of the driven shaft than of the driving shaft.

Sobremultiplicación Un arreglo de los engranajes que produce más revoluciones de la flecha de arrastre que los de la flecha de propulsión.

Overdrive band In some transmissions, this band is applied to provide an overdrive forward gear ratio.

Banda de sobremarcha En algunas transmisiones, esta banda se aplica para brindar un índice de sobremarcha de avance.

Overdrive one-way clutch In some transmissions, this one-way clutch freewheels only when overdrive gear is operating.

Embrague unidireccional de sobremarcha En algunas transmisiones, este embrague unidireccional lobera las ruedas sólo cuando está funcionando la sobremarcha.

Overdrive ratio Identified by the decimal point indicating less than one driving input revolution compared to one output revolution of a shaft.

Relación del sobremultiplicación Se identifica por el punto decimal que indica menos de una revolución del motor comparado a una revolución de una flecha de salida.

Overrun coupling A freewheeling device to permit rotation in one direction but not in the other.

Acoplamiento de sobremarcha Un dispositivo de marcha de rueda libre que permite las giraciones en una dirección, pero no en la otra dirección.

Overrunning clutch A device consisting of a shaft or housing linked together by rollers or sprags operating between movable and fixed races. As the shaft rotates, the rollers or sprags jam between the movable and fixed races. This jamming action locks together the shaft and housing. If the fixed race should be driven at a speed greater than the movable race, the rollers or sprags will disconnect the shaft.

Embrague de sobremarcha Un dispositivo que consiste de una flecha o un cárter eslabonados por medio de rodillos o palancas de detención que operan entre pistas fijas y movibles. Al girar la flecha, los rodillos o palancas de detención se aprietan entre las pistas fijas y movibles. Este acción de apretarse enclava el cárter con la flecha. Si la pista fija se arrastra en una velocidad más alta que la pista movible, los rodillos o palancas de detención desconectarán a la flecha.

Oxidation Burning or combustion; the combining of a material with oxygen. Rusting is slow oxidation, and combustion is rapid oxidation.

Oxidación Quemando o la combustión; la combinación de una materia con el oxígeno. El orín es una oxidación lenta, la combustión es la oxidación rápida.

Pascal's law The law of fluid motion.

Ley de pascal La ley del movimiento del fluido.

Parallel The quality of two items being the same distance from each other at all points; usually applied to lines and, in automotive work, to machined surfaces.

Paralelo La calidad de dos artículos que mantienen la misma distancia el uno del otro en cada punto; suele aplicarse a las líneas y, en el trabajo automotívo, a las superficies acabadas a máquina.

Parallel circuit An electrical circuit that allows current to flow in various branches without affecting the operation of the other circuits.

Circuito paralelo Un circuito elecritcal que permite que la corriente fluya en varias ramificaciones sin afectar la operación de los otros circuitos.

Pawl A lever that pivots on a shaft. When lifted, it swings freely and when lowered, it locates in a detent or notch to hold a mechanism stationary.

Trinquete Una palanca que gira en una flecha. Levantado, mueve sín restricción, bajado, se coloca en una endentación o una muesca para mantener sín movimiento a un mecanismo.

PCM The powertrain control module of a computer control system. Prior to J1930, the PCM was commonly called a ECA, ECM, or one of many acronyms used by the various manufacturers.

PCM El módulo de control del sistema de transmisión de fuerza de un sistema de control de una computadora. Antes d3l J1930, el PCM se llamaba un ECA, un ECM, o una de varias siglas usadas por los varios fabricantes.

PCS Pressure Control Solenoid.

PCS Un solenoide por el cual la presión es controlada.

Peen To stretch or clinch over by pounding with the rounded end of a hammer.

Martillazo Estirar o remachar con la extremidad redondeado de un martillo de bola.

Permeable Able to absorb fluids.

Permeable Que es capaz de absorber fluidos.

pH scale A scale used to measure how acidic or basic a solution is.

Escala de pH Una escala que se utiliza para medir la acidez o la alcalinidad de una solución.

Pinion gear The smaller of two meshing gears.

Piñón engranaje El más pequeño de dos engranajes que endientan.

Pitch The number of threads per inch on any threaded part.

Paso El número de filetes por pulgada de cualquier parte fileteada.

Pivot A pin or shaft on which another part rests or turns.

Pivote Una chaveta o una flecha que soporta a otra parte o sirve como un punto para girar.

Planetary gearset A system of gearing that is modeled after the solar system. A pinion is surrounded by an internal ring gear and planet gears are in mesh between the ring gear and pinion around which all revolve.

Conjunto de engranajes planetarios Un sistema de engranaje cuyo patrón es el sistema solar. Un engranaje propulsor (la corona interior) rodea al piñon de ataque y los engranajes satélites y planetas se endentan entre la corona y el piñon alrededor del cual todo gira.

Planet carrier The carrier or bracket in a planetary gear system that contains the shafts upon which the pinions or planet gears turn.

Perno de arrastre planetario El soporte o la abrazadera que contiene las flechas en las cuales giran los engranajes planetarios o los piñones.

Planet gears The gears in a planetary gearset that connect the sun gear to the ring gear.

Engranages planetarios Los engranajes en un conjunto de engranajes planetario que connectan al engranaje propulsor interior (el engranaje sol) con la corona.

Planet pinions In a planetary gear system, the gears that mesh with, and revolve about, the sun gear; they also mesh with the ring gear.

Piñones planetarios En un sistema de engranajes planetarios, los engranajes que se endentan con, y giran alrededor, el engranaje propulsor (sol); también se endentan con la corona.

Poppet valve A valve shaped like a "T" that is used to open and close a fluid circuit.

Válvula Poppet Una válvula que se forma como un "T que se utiliza" para abrirse y cierra un circuito flúido.

Porosity A statement of how porous or permeable to liquids a material is.

Porosidad Una expresión de lo poroso o permeable a los líquidos es una materia.

Potential energy Stored energy.

Energía potencial Energía almacenada.

Potentiometer A variable resistor that is used to change voltage without affecting main circuit current.

Potenciómetro Un resistor variable que se utiliza para cambiar voltaje sin afectar la corriente del circuito principal.

Pounds-foot (lbs.-ft.) The correct expression for the amount of torque present at a point is pounds per foot, but some literature lists torque in units of foot-pounds (ft.-lbs.). One pound-foot is the torque obtained by a force of 1 pound applied to a wrench handle 12 inches long.

Libra/pie (lb/pie) La expresión correcta para la cantidad de par de torsión presente en un punto es libras por pie, pero también se encuentra como libra/pie (lb/pie). Una libra/pie es el par de torsión que se obtiene con una fuerza de una libra aplicada en el mango de una llave de 12 pulgadas de largo.

Power A measure of work being done.

Potencia Medida de un trabajo que se está realizando.

Power split device The basic name for the CVT transmission used in Ford and Toyota hybrid vehicles. This device is based on planetary gears and divides the output of the engine and the electric motors to drive the wheels or the generator.

Dispositivo divisor de potencia Nombre básico que se usa para la transmisión CVT de los vehículos híbridos de Ford y Toyota. Este dispositivo se basa en trenes de engranajes planetarios y divide las potencias del motor y la batería eléctrica para impulsar las ruedas o el generador.

Powertrain The mechanisms that carry the power from the engine crankshaft to the drive wheels; these include the clutch, transmission, driveline, differential, and axles.

Tren impulsor Los mecanismos que transferen la potencia desde el cigueñal del motor a las ruedas de propulsión; éstos incluyen el embrague, la transmisión, la flecha motríz, el diferencial y los semiejes.

Preload A load applied to a part during assembly so as to maintain critical tolerances when the operating load is applied later.

Carga previa Una carga aplicada a una parte durante la asamblea para asegurar sus tolerancias críticas antes de que se le aplica la carga de la operación.

Press-fit Forcing a part into an opening that is slightly smaller than the part itself to make a solid fit.

Ajustamiento a presión Forzar a una parte en una apertura que es de un tamaño más pequeño de la parte para asegurar un ajustamiento sólido.

Pressure Force per unit area, or force divided by area. Usually measured in pounds per square inch (psi) or in kilopascals (kPa) in the metric system.

Presión La fuerza por unidad de una area, o la fuerza divida por la area. Suele medirse en libras por pulgada cuadrada (lb/pulg2) o en kilopascales (kPa) en el sistema métrico.

Pressure plate That part of the clutch that exerts force against the friction disc; it is mounted on and rotates with the flywheel.

Plato opresor Una parte del embraque que aplica la fuerza en el disco de fricción; se monta sobre el volante, y gira con éste.

Pressure regulator valve A valve used to maintain fluid pressure by not allowing the pressure to build beyond a specified amount.

Válvula que regula la presión Una válvula utilizada para mantener la presión del líquido, no permitiendo la presión de aumentar más allá de una medida especificada.

Pressure sensors Sensors used to monitor fluid pressure.

Sensores de la presión Sensores que vigilan la presión del líquido.

Programmable Controller Interface Data Bus (PCI Bus) A multiplex system that uses a single wire; the serial data comprises bits and frames.

Bus PCI (controlador de interfaz programable) Sistema de multiplexión que usa un solo cable. Los datos en serie constan de bits y grupos de bits.

Propeller shaft See Driveshaft.

Flecha de propulsion Vea Flecha motríz.

psi Abbreviation for *pounds per square inch*, a measurement of pressure.

Lb/pulg2 Una abreviación de libras por pulgada cuadrada, una medida de la presión.

Pulley A drive or driven component that serves the same purpose as a gear except it is driven by a belt and because pulleys don't mesh, the driven pulley rotates in the same direction as the drive pulley.

Polea Un mecanismo impulsor, o el componente impulsor, que responde al mismo propósito que un engranaje a menos que sea conducido por una correa y porque las poleas no endientan, la polea impulsora rota en la misma dirección que la polea impulsora.

Pulsation To move or beat with rhythmic impulses.

Pulsación Moverse o batir con impulsos rítmicos.

Pulse width The length of time in milliseconds that a component is energized.

Anchura del pulso La longitud del tiempo, en los milisegundos, que el componente se energiza.

PWM Pulse width modulated, a term used to define the operation of a solenoid that is turned on and off by a computer to control its output.

PWM La anchura del pulso modulada es una frase que define la operación de un solenoide, la salida de el cual es controlada por un ordenador que pueda encender el solenoide y apagarlo.

Race A channel in the inner or outer ring of an antifriction bearing in which the balls or rollers roll.

Pista Un canal en el anillo interior o exterior de un cojinete antifricción en el cual ruedan las bolas o los rodillos.

Radial The direction moving straight out from the center of a circle. Perpendicular to the shaft or bearing bore.

Radial La dirección al moverse directamente del centro de un círculo. Perpendicular a la flecha o al taladro del cojinete.

Radial clearance Clearance within the bearing and between balls and races perpendicular to the shaft. Also called radial displacement.

Holgura radial La holgura en un cojinete entre las bolas y las pistas que son perpendiculares a la flecha. También se llama un desplazamiento radial.

Radial load A force perpendicular to the axis of rotation.

Carga radial Una fuerza perpendicular al centro de rotación.

Ratio The relation or proportion that one number bears to another.

Relación La correlación o proporción de un número con respeto a otro.

Ravigneaux geartrain A compound gearset that combines two planetary units with a common ring gear.

Ravigneaux tren de engranaje Un conjunto compuesto de engranajes combina dos conjuntos planetarias con una corona común.

Reaction area The area around a spool valve that allows pressurized fluid to move the valve.

Area de la reacción El área alrededor de una válvula de carrete que permite líquido presurizado para mover la válvula.

Reaction members The members of a gearset that are held stationary so the other members can react to them.

Reacción miembros Los miembros del conjunto de los engranajes que son inmóviles de modo que los otros miembros puedan reaccionar a ellos.

Reactivity The characteristic of a material that enables it to react violently with water and other materials.

Reactividad La característica de un material que lo permita reaccionar violentamente con agua y otros materiales.

Reactor Another name for the stator in a torque converter.

Un reactor Otro nombre para el estator en un convertidor del esfuerzo de torsión.

Rear-wheel-drive A term associated with a vehicle in which the engine is mounted at the front and the driving axle and driving wheels are at the rear of the vehicle.

Tracción trasera Un término que se asocia con un vehículo en el cual el motor se ubica en la parte delantera y el eje propulsor y las ruedas propulsores se encuentran en la parte trasera del vehículo.

Redox A term used to describe the process of something being reduced while another object is being oxidized.

Redox Un término que se utiliza para describir el proceso de algo que se reduce mientras que otro objeto se oxida.

Reduction The removal of oxygen from exhaust gases.

Reducción La extracción de oxígeno de una molécula.

Reference voltage sensor An electrical device that responds to changes by altering the voltage (reference voltage) it receives.

Sensore de voltaje referencia Un dispositivo eléctrico que responde a los cambios por alterando el voltaje que recibe.

Relay valve A spool valve used to control the direction of fluid flow without affecting the pressure of the fluid.

Relé valvula Una válvula del carrete que se utiliza para controlar la dirección del flujo flúido sin alterar la presión del líquido.

Relief valve A valve used to protect against excessive pressure in the case of a malfunctioning pressure regulator.

Válvula de seguridad Una válvula que se usa para guardar contra una presión excesiva en caso de que malfulciona el regulador de presión.

Resistance An electrical term for something that opposes current flow.

Resistencia Una palabra para algo que opone el flujo de la corriente eléctrica.

Retaining ring A removable fastener used as a shoulder to retain and position a round bearing in a hole.

Anillo de retén Un seguro removible que sirve de collarín para sujetar y posicionar a un cojinete en un agujero.

RFI Radio frequency interference. This acronym is used to describe a type of electrical interference that may affect voltage signals in a computerized system.

RFI Interferencia de frecuencias de radio. Esta sigla se usa en describir un tipo de interferencia eléctrica que puede afectar los señales de voltaje en un sistema computerizado.

Ring gear The outside internal gear of a planetary gearset.

La corona El engranaje interior más externo de un conjunto de engranajes planetarios.

Roller bearing An inner and outer race upon which hardened steel rollers operate.

Cojinete de rodillos Una pista interior y exterior en la cual operan los rodillos hecho de acero endurecido.

Roller clutch A type of overrunning clutch that uses rollers to engage or disengage.

Embrague del rodillo Un tipo de sobrecorriendo embrague que utiliza los rodillos para engranar y para desengranar.

Rollers Round steel bearings that can be used as the locking element in an overrunning clutch or as the rolling element in an antifriction bearing.

Rodillos Articulaciones redondos de acero que pueden servir como un elemento de enclavamiento en un embrague de sobremarcha o como el elemento que rueda en un cojinete antifricción.

Rotary flow A fluid force generated in the torque converter that is related to vortex flow. The vortex flow leaving the impeller is not only flowing out of the impeller at high speed but is also rotating faster than the turbine. The rotating fluid striking the slower turning turbine exerts a force against the turbine that is defined as rotary flow.

Flujo rotativo Una fuerza fluida producida en el convertidor de torsión que se relaciona al flujo torbellino. El flujo torbellino saliendo del rotor no sólo viaja en una alta velocidad sino también gira más rápidamente que el turbino. El fluido rotativo chocando contra el turbino que gira más lentamente, impone una fuerza contra el turbino que se define como flujo rotativo.

Rotor-type pump A fluid pump that uses an inner and outer rotor inside a housing to move the fluid.

Bomba de rotor-tipo Una bomba que utiliza un rotor interno y un rotor externo para mover los líquidos.

rpm Abbreviation for revolutions per minute, a measure of rotational speed.

rpm Abreviación de revoluciones por minuto, una medida de la velocidad rotativa.

RTV Room Temperature Vulcanizing. A type of aerobic sealer used in place of a gasket.

RTV Vulcanización en la temperatura ambiente, Un tipo de sellador aerobio utilizado en lugar de un empaque.

RWD Abbreviation for rear-wheel-drive.

RWD Abreviación de tracción trasera.

SAE Society of Automotive Engineers.

SAE La Sociedad de Ingenieros Automotrices.

SBEC Single Board Engine Controller; used in Chrysler vehicles to monitor and control the engine and driveline components.

SBEC Sola tarjeta regulador del motor utilizó en vehículos para vigilar y para controlar el motor y los componentes del tren de propulsión.

Scarf-cut rings A name for Teflon locking-end seals.

Annillos empalme cortado Un nombre para los sellos de Teflon con extremos que cierran.

Seal A material shaped around a shaft, used to close off the operating compartment of the shaft, preventing oil leakage.

Sello Una materia, formado alrededor de una flecha, que sella el compartimiento operativo de la flecha, previniendo el goteo de aceite.

Semiconductor A material that is neither a good conductor nor a good insulator.

Semiconductor Un material que es ni un buen conductor ni un buen aislador.

Separator plate A plate found in valve body assemblies that is used to seal off some fluid passages as well as to control and direct fluid flow through other passages.

Plancha separador Un plancha encontró en ensamblajes del cuerpo de válvula que se utiliza para aislar algunos pasos de líquido, y también controlar y dirigir el líquido en otros canales.

Series circuit An electrical circuit that has the loads arranged so that circuit current must flow through each one whenever the circuit is activated.

Serie circuito Un circuito de Serie es un circuito eléctrico que tiene las cargas dispuestas de modo que la corriente eléctrica deba atravesar cada vez que se active el circuito.

Servo A device that converts hydraulic pressure into mechanical movement, often multiplying it. Used to apply the bands of a transmission.

Servo Un dispositivo que convierte la presión hidráulica al movimiento mecánico, frequentemente multiplicándola. Se usa en la aplicación de las bandas de una transmisión.

Shank The part of a shoe that protects the ball of the foot.

Enfranque La pieza de un zapato que proteja la bola del pie.

Sheaves The sides of a pulley.

Roldanas Las caras de una polea.

Shift feel The quality of a shift into another forward gear.

Sensación del cambio de velocidad La calidad de un cambio a otro engranaje delantero.

Shift l.ever The lever used to change gears in a transmission. Also the lever on the starting motor that moves the drive pinion into or out of mesh with the flywheel teeth.

Palanca del cambiador La palanca que sirve para cambiar a las velocidades de una transmisión. También es la palanca del motor de arranque que mueva al piñon de ataque para engranarse o desegranarse con los dientes del volante.

Shift schedule The prescribed time and conditions that must be met before a transmission upshifts or downshifts.

Horario de cambios entre las velocidades El tiempo y las condiciones prescritos que deben ser satisfechas antes de que una transmisión cambia ascendente o cambia descendente.

Shift valve A valve that controls the shifting of the gears in an automatic transmission.

Válvula de cambios Una válvula que controla a los cambios de las velocidades en una transmisión automática.

Shim Thin sheets used as spacers between two parts, such as the two halves of a journal bearing.

Laminilla de relleno Hojas delgadas que sirven de espaciadores entre dos partes, tal como las dos partes de un muñón.

Short An electrical problem that results from an unwanted path to ground or to power.

Cortocircuito Un problema eléctrico que resulta de un camino indeseado a la tierra o a la potencia.

Side clearance The clearance between the sides of moving parts when the sides do not serve as load-carrying surfaces.

Holgura lateral La holgura entre los lados de las partes en movimiento mientras que los lados no funcionan como las superficies de carga.

Simpson geartrain A compound gearset in which two planetary units work together through a common sun gear.

Simpson tren de engranajes Un conjunto compuesto de los engranajes en los cuales dos ensamblajes planetarios funcionan juntos a través de un engranaje común.

Single-wrap band A solid, one-piece brake band.

Banda de envolvedero solo Una banda del freno compuesta de una sola pieza sólida.

Sliding-fit Where sufficient clearance has been allowed between the shaft and journal to allow free running without overheating.

Ajuste corredera Donde se ha dejado una holgura suficiente entre la flecha y el muñón para permitir una marcha libre sin sobrecalentamiento.

Slip yoke The transmission end of a driveshaft that fits into the extension housing and onto the splines of the output shaft.

Deslice yugo El extremo (hacia la transmisión) de un eje impulsor que encaja en el cubierta de extensión tambien sobre las ranuras de la flecha del salida.

Snap ring Split spring-type ring located in an internal or external groove to retain a part.

Anillo de seguridad Un anillo partido tipo resorte que se coloca en una muesca interior o exterior para retener a una parte.

Solenoid An electrical device that works on principles of magnetism and has a moveable center core that causes some action.

Solenoide U dispositivo eléctrico que utiliza los principios del magnetismo, y tiene un cono movible en el centro que causa una cierta acción cuando se mueve.

Solution Formed when a solid dissolves into a liquid and its particles break away from this structure and mix evenly in the liquid.

Solución Se forma cuando un sólido se disuelve en un líquido y sus partículas abandonan su estructura y se mezclan en forma uniforme en el líquido.

Solvent The liquid in a solution.

Solvente El líquido en una solución.

Specific gravity The weight of a volume of any liquid divided by the weight of an equal volume of water at the same temperature and pressure. The ratio of the weight of any liquid to the weight of water, which has a specific gravity of 1.000.

Gravedad específica El peso de un volumen de cualquier líquido dividido por el peso de un volumen igual de agua a igual temperatura y presión. El índice del peso de cualquier líquido respecto del peso del agua, que tiene una gravedad específica de 1,000.

Speed The distance an object travels in a set amount of time.

Rapidez Distancia que recorre un objeto en un período determinado.

Speed reduction The result of meshing two dissimilar sized gears. When a small gear drives a larger gear, speed reduction takes place.

Velocidad reducción Una reducción en velocidad es el resultado de endentar dos engranajes de tamaño distinto. Cuando un engranaje pequeño conduce un engranaje más grande, la reducción de velocidad sucede.

Spindle The shaft on which the wheels and wheel bearings mount.

Husillo La flecha en la cual se montan las ruedas y el conjunto del cojinete de las ruedas.

Spline Slot or groove cut in a shaft or bore; a splined shaft onto which a hub, wheel, gear, and so on, with matching splines in its bore is assembled so that the two must turn together.

Acanaladura (espárrago) Una muesca o ranura cortada en una flecha o en un taladro; una flecha acanalada en la cual se asamblea un cubo, una rueda, un engranaje, y todo lo demás que tiene un acanaladura pareja en el taladro de manera de que las dos deben girar juntos.

Split-band Another name for a double-wrap band.

Banda separada Otro nombre de banda con dos envolvederos.

Split lip seal Typically, a rope seal sometimes used to denote any two-part oil seal.

Sello hendido Típicamente, un sello de cuerda que se usa a veces para demarcar cualquier sello de aceite de dos partes

Split pin A round split spring steel tubular pin used for locking purposes; for example, locking a gear to a shaft.

Chaveta hendida Una chaveta partida redonda y tubular hecho de acero para resorte que sirve para el enclavamiento; por ejemplo, para enclavar un engranaje a una flecha.

Spool valve A cylindrically shaped valve with two or more valleys between the lands. Spool valves are used to direct fluid flow.

Válvula de carrete Una válvula de forma cilíndrica que tiene dos acanaladuras de cañon o más entre las partes planas. Las válvulas de carrete sirven para dirigir el flujo del fluido.

Sprag clutch A member of the overrunning clutch family using a sprag to jam between the inner and outer races used for holding or driving action.

Embrague de puntal Un miembro de la familia de embragues de sobremarcha que usa a una palanca de detención trabada entre las pistas interiores e exteriores para realizar una acción de asir o marchar.

Spring A device that changes shape when it is stretched or compressed, but returns to its original shape when the force is removed; the component of the automotive suspension system that absorbs road shocks by flexing and twisting.

Resorte Un dispositivo que cambia de forma al ser estirado o comprimido, pero que recupera su forma original al levantarse la fuerza; es un componente del sistema de suspensión automotívo que absorba los choques del camino al doblarse y torcerse.

Spring retainer A steel plate designed to hold a coil or several coil springs in place.

Retén de resorte Una chapa de acero diseñado a sostener en su posición a un resorte helicoidal o más.

Spur gear A gear with teeth that are cut straight across the gear.

Estímulo Engranaje Un engranaje con dientes que fueron cortados derecho a través del engranaje.

Square-cut seal A fixed seal for a caliper piston that has a square cross section.

Junta cuadrada Junta fija para un pistón de la mordaza que tiene una sección transversal cuadrada.

Squeak A high-pitched noise of short duration.

Chillido Un ruido agudo de poca duración.

Squeal A continuous high-pitched noise.

Alarido Un ruido agudo continuo.

Stall A condition in which the engine is operating and the transmission is in gear, but the drive wheels are not turning because the turbine of the torque converter is not moving.

Paro Una condición en la cual opera el motor y la transmisión esta embragada pero las ruedas de impulso no giran porque no mueva el turbino del convertidor de la torsión.

Stall speed The maximum engine speed allowed by a torque converter when the engine is operating at full throttle with the gear selector in D range and the vehicle stationary.

Velocidad de la parada La velocidad del motor máxima permitida por un convertidor del esfuerzo de torsión cuando el motor está funcionando en la válvula reguladora llena con el selector del engranaje en rango de D y el vehículo inmóvil.

Stall test A test of the one-way clutch in a torque converter.

Prueba de paro Una prueba del embrague de una vía en un convertidor de la torsión.

Static pressure Pressure that is present and does not change.

Presión estática La presión que es presente y no cambia.

Static seal A seal that prevents fluid from passing between two or more parts that are always in the same relationship to each other.

Junta estática Un junta que evita que el líquido pase entre dos o más porciones que estén siempre en el mismo lazo con uno a la otra.

Stator The reaction member of a torque converter. It redirects the fluid back toward the impeller.

Estator El miembro reacciónario de un convertidor del esfuerzo de torsión. Hace el líquido fluir detrás hacia el impulsor.

Steel discs The discs in a multiple-disc assembly that do not have a frictional coating.

Discos de acero Los discos de un conjunto de discos múltiples que no tienen un revestimiento de rozamiento.

Stress The force to which a material, mechanism, or component is subjected.

Esfuerzo La fuerza a la cual se somete a una materia, un mecanísmo o un componente.

Sun gear The central gear in a planetary gear system around which the rest of the gears rotate. The innermost gear of the planetary gearset.

Engranaje principal (sol) El engranaje central en un sistema de engranajes planetarios alrededor del cual giran los otros engranajes. El engranaje más interno del conjunto de los engranajes planetarios.

TCM Transmission control module.

TCM Módulo de control de la transmisión.

Temper To change the physical characteristics of a metal by applying heat.

Templar Cambiar las características físicas de un metal mediante la aplicación de calor.

Tensile strength The amount of pressure per square inch the bolt can withstand just before breaking when being pulled apart.

Resistencia a la tracción La cantidad de presión por pulgada cuadrada que puede soportar un objeto antes de romperse cuando se lo estira.

Tension Effort that elongates or "stretches" a material.

Tensión Un esfuerzo que alarga o "estira" a una materia.

Thermistors Electrical resistors that change in resistive value according to their temperature.

Termistores Los termistores son los resistores eléctricos que cambian en valor resistente según su temperatura.

Throttle valve A valve that responds to throttle position and/or engine load.

Valvula de admisión Una válvula que responde a la posición de la válvula reguladora y/o a la carga del motor.

Thrust bearing A bearing designed to resist or contain side or end motion as well as reduce friction.

Cojinete de empuje Un cojinete diseñado a detener o reprimir a los movimientos laterales o de las extremidades y también reducir la fricción.

Thrust load A load that pushes or reacts through the bearing in a direction parallel to the shaft.

Carga de empuje Una carga que empuja o reacciona por el cojinete en una dirección paralelo a la flecha.

Thrust washer A washer designed to take up end thrust and prevent excessive endplay.

Arandela de empuje Una arandela diseñada para rellenar a la holgura de la extremidad y prevenir demasiado juego en la extremidad.

Tolerance A permissible variation between the two extremes of a specification or dimension.

Tolerancia Una variación permisible entre dos extremos de una especificación o de un dimensión.

Toroidal A word used to note that something is doughnut shaped.

Toroidal Una palabra usada para observar que es algo tiene la dimensión de una variable de un buñuelo.

Torque A twisting motion, usually measured in ft.-lb. (Nm).

Torsión Un movimiento giratorio, suele medirse en pies/libra (Nm).

Torque converter A turbine device utilizing a rotary pump, one or more reactors (stators), and a driven circular turbine or vane, whereby power is transmitted from a driving to a driven member by hydraulic action. It provides varying drive ratios; with a speed reduction, it increases torque.

Convertidor de la torsión Un dispositivo de turbino que utilisa a una bomba rotativa, a un reactor o más, y un molinete o turbino circular impulsado, por cual se transmite la energía de un miembro de impulso a otro arrastrado mediante la acción hidráulica. Provee varias relaciones de impulso; al descender la velocidad, aumenta la torsión.

Torque converter clutch Located inside the torque converter, this device provides a mechanical link between the crankshaft and the input shaft of the transmission.

Embrague del convertidor par Ubicado dentro del convertidor par, este dispositivo proporciona una conexión mecánica entre el cigüeñal y el árbol de entrada de la transmisión.

Torque converter limit valve Controls the hydraulic pressure in the torque converter and keeps it below a specified value.

Válvula limitadora del convertidor par Controla la presión hidráulica en el convertidor par y lo mantiene por debajo de un valor especificado.

Torque curve A line plotted on a chart to illustrate the torque personality of an engine. When the engine operates on its torque curve, it is producing the most torque for the quantity of fuel being burned.

Curva de la torsión Una linea delineada en una carta para ilustrar las características de la torsión del motor. Al operar un motor en su curva de la torsión, produce la torsión óptima para la cantidad del combustible que se consuma.

Torque multiplication The result of meshing a small driving gear and a large driven gear to reduce speed and increase output torque.

Multiplicación de la torsión El resultado de engranar a un engranaje pequeño de ataque con un engranaje más grande arrastrado para reducir la velocidad y incrementar la torsión de salida.

Torque steer An action felt in the steering wheel as the result of increased torque.

Dirección la torsión Una acción que se nota en el volante de dirección como resultado de un aumento de la torsión.

TOT Transmission oil temperature sensor.

TOT Un sensor que mide la temperatura del aceite de la transmisión.

Toxicity The term used to describe how poisonous a substance is.

Toxicidad Toxicidad describe cómo venenoso una sustancia es.

Traction The gripping action between the tire tread and the road's surface.

Tracción La acción de agarrar entre la cara de la rueda y la superficie del camino.

Tractive force The amount of power available from the drivetrain. It is calculated by multiplying the engine's torque by the overall gear ratio.

Fuerza de tracción Cantidad de potencia disponible de la transmisión. Se calcula multiplicando el par de torsión del motor por la relación general de transmisión.

Tractive resistance The forces that work against the movement of a vehicle, such as rolling resistance and aerodynamic drag.

Resistencia de tracción Fuerza que trabaja contra el movimiento del vehículo, como la resistencia al rodamiento y la resistencia aerodinámica.

Transaxle Type of construction in which the transmission and differential are combined in one unit.

Flecha de transmisión Un tipo de construcción en el cual la transmisión y el diferencial se combinan en una unidad.

Transfer assembly An arrangement used in some transaxles to connect the output to the final drive gears.

Ensamblaje transferencia Un arreglo utilizado en algunas flechas impulsoras para conectar la salida con los engranajes impulsores finales.

Transfer case An auxiliary transmission mounted behind the main transmission. Used to divide engine power and transfer it to both front and rear differentials, either full time or part time.

Cárter de la transferencia Una transmisión auxiliar montada detrás de la transmisión principal. Sirve para dividir la potencia del motor y transferirla a ambos diferenciales delanteras y traseras todo el tiempo o la mitad del tiempo.

Transfer plate Another name for the separator plate found in some valve body assemblies.

Plancha transferencia Otro nombre de plancha separador encontró en algunas ensamblajes del cuerpo de válvula.

Transmission The device in the powertrain that provides different gear ratios between the engine and drive wheels as well as reverse.

Transmisión El dispositivo en el trén de potencia que provee las relaciones diferentes de engranaje entre el motor y las ruedas de impulso y también la marcha de reversa.

Transmission oil temperature (TOT) sensor A sensor that monitors the temperature of the ATF in a transmission, it is normally a thermistor.

Sensor de temperatura del aceite de la transmisión (TOT) Un sensor que controla la temperatura del ATF en una transmisión, habitualmente es un termistor.

Transverse Powertrain layout in a front-wheel-drive automobile extending from side to side.

Transversal Una esquema del tren de potencia en un automóvil de tracción delantera que se extiende de un lado a otro.

Turbine A finned wheel-like device that receives fluid from the impeller inside a torque converter and forces it back to the stator.

Turbina Una turbina es un dispositivo con las aletas en forma de una rueda que reciba el líquido del impellor y lo fuerce de nuevo al estator.

Two-mode hybrid system A hybrid system that fits into a standard transmission housing and is basically two planetary gearsets coupled to two electric motors, which are electronically controlled. This combination results in a continuously variable transmission and motor/generators for hybrid operation. This arrangement has two distinct modes of hybrid drive operation: one mode for low-speed and low-load conditions, and the other for use while cruising at highway speeds.

Sistema híbrido de dos modos Sistema híbrido que encaja en el montaje de una transmisión estándar y es básicamente un par de trenes de engranajes planetarios acoplados con un par de baterías eléctricas, controladas electrónicamente. Esta combinación resulta en una transmisión continuamente variable y motores/generadores para operación híbrida. Este sistema tiene dos modos diferentes de impulso híbrido: un modo para baja velocidad y baja carga, y el otro para velocidad de autopista.

Two-way check valve A ball-type check valve without a return spring. This type valve is used where pressure from two different sources is sent to the same outlet port.

Válvula de doble paso de cheque Una válvula bola-tipo que falta un resorte de vuelta. Este tipo de válvula se utiliza donde la presión a partir de dos diversas fuentes se envía al mismo orificio de salida.

U-joint A four-point cross connected to two U-shaped yokes that serves as a flexible coupling between shafts.

Junta de U Una cruceta de cuatro puntos que se conecta a dos yugos en forma de U que sirven de acoplamientos flexibles entre las flechas.

Universal joint A mechanical device that transmits rotary motion from one shaft to another shaft at varying angles.

Junta Universal Un dispositivo mecánico que transmite el movimiento giratorio desde una flecha a otra flecha en varios ángulos.

Upshift To shift a transmission into a higher gear.

Cambio ascendente Cambiar a la velocidad de una transmisión a una más alta.

Vacuum Any pressure lower than atmospheric pressure.

Vacío Un vacío es cualquier presión que sea más baja que la presión de la atmósfera.

Vacuum modulator A device that causes fluid pressure in an automatic transmission to increase when engine vacuum is low.

El modulador del vacío Un dispositivo que causa la presión del líquido en una transmisión automática al aumento cuando el vacío del motor es bajo.

Vacuum transducer An electronic device that changes vacuum signals into electrical signals.

Transductor del vacío Un dispositivo electrónico que cambia señales del vacío en señales eléctricas.

Valley The area on a spool valve that is between the valve's lands.

Valle El área de una válvula de carrete el área que está entre las tierras de la válvula de carrete.

Valve body Main hydraulic control assembly of a transmission containing the components necessary to control the distribution of pressurized transmission fluid throughout the transmission.

Cuerpo de la válvula Asamblea principal del control hidráulico de una transmisión que contiene los componentes necessarios para controlar a la distribución del fluido de la transmisión bajo presión por toda la transmisión.

Vane-type pump A type of fluid pump that uses several sliding vanes that seal against a slide in the housing to push fluid out.

Bomba del paleta-tipo Un tipo de bomba flúida que utilice varias paletas que resbalan contra una superficie rezbaladiza en la cubierta para empujar el líquido hacia fuera.

Variable capacity pump Type of fluid pump whose output is controllable.

Bombas con capacidad variable Estas bombas están en una clasificación de la bomba, la salida de la cual es controlable.

Variable Force Motor (VFM) An electro-hydraulic actuator made of a variable force solenoid and a regulating valve. It controls main line pressure by moving a pressure regulator valve against spring pressure.

Motor de fuerza variable (VFM) Un actuador electrohidráulico compuesto por un solenoide de fuerza variable y una válvula reguladora. Controla la presión de la línea principal al mover una válvula reguladora de presión contra la presión del resorte.

Variable force solenoid (VFS) Another name for a VFM.

Solenoide de fuerza variable (VFS) Otro nombre para un VFM.

VCC Viscous converter clutch.

VCC Embrague del convertidor viscoso.

Vehicle identification number (VIN) The number assigned to each vehicle by its manufacturer, primarily for registration and identification purposes.

Número de identificacíon del vehículo El número asignado a cada vehículo por su fabricante, primariamente con el propósito de la registración y la identificación.

Velocity The speed of an object in a particular direction.

Velocidad Rapidez de un objeto en una dirección en particular.

VFM Variable force motor, used to change and maintain line pressure.

VFM Una sigla de un motor con la fuerza variable que se utiliza para cambiar y para mantener la presión de la línea.

VFS Variable force solenoid, used to maintain line pressure.

VFS Una sigla de un solenoide que se utiliza para mantener la presión de la línea.

Vibration A quivering, trembling motion; felt in a vehicle at different speed ranges.

Vibración Un movimiento de estremecer o temblar que se siente en el vehículo en varios intervalos de velocidad.

Viscosity The resistance to flow exhibited by a liquid. A thick oil has greater viscosity than a thin oil.

Viscosidad La resistencia al flujo que manifiesta un líquido. Un aceite espeso tiene una viscosidad mayor que un aceite ligero.

Viscous clutch A clutch assembly that engages and disengages in response to the viscosity of the fluid in the clutch assembly.

Embrague viscoso Un ensamblaje de embrague que engancha y desune en respuesta a la viscosidad del líquido en el ensamblaje de embrague.

Volatile The characteristic of a material that allows it to vaporize quickly.

Volátil La característica de un material que permita que se vaporice rápidamente.

Voltage Electrical force or pressure that causes and allows for current flow.

Voltaje La fuerza o la presión eléctrica que causa y permite el flujo de la corriente eléctrica.

Voltage drop The amount of voltage or energy consumed when pushing current through a resistance.

Caída de voltaje La cuantía de voltaje o energía consumida de cuando empuja la corriente eléctrica con una resistencia.

Voltage generator A device that generates electricity. Commonly used as a speed sensor.

Generador del voltaje un dispositivo que genera electricidad. Utilizado comúnmente como sensor de velocidad.

Vortex Path of fluid flow in a torque converter. The vortex may be high, low, or zero depending on the relative speed between the pump and turbine.

Vórtice La vía del flujo de los fluidos en un convertido de torsión. El vórtice puede ser alto, bajo, o cero depende de la velocidad relativa entre la bomba y la turbina.

Vortex flow Recirculating flow between the converter impeller and turbine that causes torque multiplication.

Flujo del vórtice El fluyo recirculante entre el impulsor del convertidor y la turbina que causa la multiplicación de la torsión.

Wave Any single swing of an object back and forth between the extremes of its travel through matter or space.

Onda Cualquier oscilación de un objeto hacia adelante y hacia atrás entre los extremos de su desplazamiento a través de la materia o el espacio.

Wavelength The distance between each compression of a sound or an electrical wave.

Longitud de onda La distancia entre cada compresión de una onda de sonido o eléctrica.

Weight A force exerted on a mass by gravity.

Peso Fuerza que ejerce la gravedad sobre una masa.

Wet-disc clutch A clutch in which the friction disc (or discs) is operated in a bath of oil.

Embrague de disco flotante Un embrague en el cual el disco (o los discos) de fricción opera en un baño de aceite.

Wheel A disc or spokes with a hub at the center that revolves around an axle, and a rim around the outside on which the tire is mounted.

Rueda Un disco o rayo que tiene en su centro un cubo que gira alrededor de un eje, y tiene un rim alrededor de su exterior en la cual se monta el neumático.

Work The product of a force and the distance through which it acts.

Trabajo El producto de una fuerza y de la distancia a través de la cual actúe.

Yoke In a universal joint, the drivable torque-and-motion input and output member attached to a shaft or tube.

Yugo En una junta universal, el miembro de la entrada y la salida que transfere a la torsión y al movimiento, que se conecta a una flecha o a un tubo.

INDEX

Note: Figures and tables are denoted by f and t, respectively.